A CITIZEN'S DISCLOSURE ON UFOS AND ETI

VOLUME FOUR

IN SEARCH OF EXTRATERRESTRIAL INTELLIGENCE

TERENCE M. TIBANDO

Copyright Page

In writing this book, I sought out the best possible evidence available on this subject whether that was from numerous UFO and ETI related books, networking with other UFO authors, researchers and first-hand witnesses to UFO sightings, from films and TV documentaries, from internet searches, or just from my personal sightings and contact experiences.

When material is quoted in this book full acknowledgement is given to the author or source of that material as indicated by the extensive bibliography, webliography and videography at the back of the book.

When photographic images are used in this book that are obtained from the internet, usually from Google Images, a full search was made to determine copyright information, the author's name, or address or email address or phone number or copyright mark in order to asked permission to use their photographs. In almost all cases where such images are posted to the internet, there was no satisfactory way to identify the owner of the image even through Google because they left no identification of themselves to be found. When an author's name does appear on an image, written permission was sought or it was not used at all; more often than not, there usually was no reply or response back from the owner.

This lack of due diligence to place a copyright mark or the owner's name is all it would take for that person to claim ownership of a picture, yet the lack of it creates major problems for many people, especially for other authors who lawfully seek their permission to use their photo images.

Because this book is one of six volumes in a series created as public educational material and is of a transitional nature, and I have quoted or referenced the websites from where I obtained the photo images and therefore, I am invoking the Fair Use Doctrine also known as Fair Usage Clause to publish these images in my book.

I will of course give full acknowledgement and credit to the author's and owners of such images in all my future book publications in recognition of their work if they come forward to be identified.

A CITIZEN'S DISCLOSURE ON UFOS AND ETI

VOLUME FOUR

IN SEARCH OF EXTRATERRESTRIAL INTELLIGENCE

A CITIZEN'S DISCLOSURE ON UFOS AND ETI

VOLUME FOUR

IN SEARCH OF EXTRATERRESTRIAL INTELLIGENCE

TERENCE M. TIBANDO

"Hggna"

A Cosmic Cousin
Publication

Other Publications by the Author Page

Although, this book is the author's first publication it forms a part of six smaller books or volumes that was originally written as one massive tome of UFO and ETI information entitled: **"A Citizen's Disclosure On UFOs And ETI "** which began in March 2009 and was completed in August 2016.

Other books/volumes by the author in this series:

1. Book One (Volume One): **"Global Evidence of the UFO and ETI Presence"**

2. Book Two (Volume Two): **"UFO Disclosure and Covert Programs of Deception"**

3. Book Three (Volume Three): **"Military Intelligence Industrial Complex, USAPs and Covert Black Projects"**

4. Book Four (Volume four): **"In Search of Extraterrestrial Intelligence"**

5. Book Five (Volume Five): **"Evidence of a Type Two ET Civilization in Our Solar System"**

6. Book Six (Volume Six): **"The Rosetta Stone of ETI Contact and Communications"**

Dedication

I wish to dedicate this book and the other remaining books in this series to my loving wife Annice and our daughter Annika Tibando.

As anyone who has ever written a book well knows, it is rarely just a one person that writes a book to full completion without seeking the assistance or advice of someone else or a few helpful friend or colleagues, or in my case, it's a family member to ensure the successful completion of my literary work.

My wife, Annice keeps my feet on the ground while my mind soars into the limitless space and beyond.

My daughter, Annika is truly an amazingly creative and talented daughter, who took time away from her busy schedule in creating a new fashion line of clothing to assistance me with a more polished professional appearance to the book cover images. She accomplished in a month what had taken me a year to do because I thought I could do everything myself, obviously I was wrong.

I thank God every day that you are both in my life as my wife and daughter, you are always in my heart!

I Love you both always to infinity and beyond!

Introduction

This book is the fourth volume of "A Citizen's Disclosure on UFOs and ETI" which originally formed part of a 3500 page encyclopedic tome. It provides the latest and best evidence on this subject matter that comes from my 65years of personal experience dating back to 1953 when I and my mother witnessed flying saucer type craft hover over a Canadian Air Force Base in St. Jeans, Quebec. This was followed by an ET visitation by three ghost-like beings that came into my bedroom a couple of days later.

This became a lifetime of UFO sightings and ETI encounters and interactions witnessed from Quebec to British in Canada and in parts of western US, either as an individual or in group settings as a part of CSETI field expeditions.

Regardless of what you may have read or been told by the mainstream media or the science community, this phenomenon is real "as the airplanes that fly above." We expect our elected officials to be transparent in their public disclosure but instead, there has been deception, lies and cover-ups. Our trust in officialdom has been betrayed and increasingly, this distrust toward officialdom has been undermined relentlessly leading to a public awaking and self-investigation of the truth for themselves.

This fourth volume in a series of six volumes will add to the public's resource for knowledge, truth and honesty on this subject. The unfettered self-investigation of truth is a God-given right of everyone to understand the world in which they live in.

As you read through this book, the seasoned UFO researcher will find much that is new as well as familiar, especially with some alternative and controversial perspectives that re-evaluate some long, widely-held traditional beliefs held within the UFO Community.

In the series, "A Citizen's Disclosure on UFOs and ETI" we connect the dots and assemble the pieces of the UFO and ETI puzzle to view the larger picture of this phenomenon. An unfettered knowledge of this extraterrestrial phenomenon has profound implications for our future, it is pregnant with potentialities to revolutionize human civilization for millennia for to come. The current task at hand is for our governments, its officials, the science, religious, military and intelligence communities to be honest and transparent in its disclosure on this subject with the general public.

The UFO database is unfortunately corrupted and has been so for decades, and no matter how careful the research and entry of evidence, inevitably errors will creep in because new evidence comes to light to prove an alternative reality. It is my hope that any further errors by myself in content are minimal or non-existent. Corruption of the UFO database has been perpetuated due to poor research or the deliberate ignoring or marginalizing of real facts. With this in mind, corrections have been made where necessary while still maintaining the integrity of the evidence as it was originally reported.

The reader should view this book as an interactive reference manual perhaps, the first of its kind, to permit the reader to do further research of their own with its many colour coded web links, video links and bibliography. In essence, it extends a 3500-page textbook into a 10,000-page tome!

The reader will be able to follow the web link references to see where I have collected the information for this book. Unfortunately, at the time of publication of this book, some of these web links may no longer be active, it is the nature of the internet and it ephemeral web links!

My apologies for any inconvenience this may cause, however, try "googling" a specific word or phrase; this may reveal the same or a similar website!

Up and until this volume, our focus has been on what has occurred on Earth through eye-witness accounts military cover up and secrecy and the restructure of humanity civilization through constant wars, corrupt politics, massive environmental damage and an economic imbalance globally. Now, while there is yet still time, we explore our Solar System for other signs of life other than our own.

We find that life is everywhere far beyond the "Habitable Zone" that NASA and astronomers have postulated in our star system. Life is adaptable to each particular planetary environment and water does not have to be the main ingredient for life, although, it helps. We explore nearby space, the Moon the inner planets closest to the Sun, including the Sun, and the outer gaseous planets, its satellites and beyond.

We find that ETI are present almost everywhere and appear to be surveilling, tracking and monitoring our space explorations as we are becoming increasingly aware of their presence.

As we begin to study other planets especially Mars, the Asteroid Belt, the systems of Jupiter and Saturn, UFO craft have been discovered with some of an incredible planetary size. Life forms of various kinds have been discovered on Mars making it a true planetary cousin to the Earth.

NASA exploration of Mars with its orbiters and land rovers continue to search for life, however, insider knowledge suggests that a search for alien technology is also a part of NASA's agenda, yet, the public still remains in the dark as to everything that has been discovered in space!

Only an awaking humanity can demand a different outcome from their elected leaders and government service branches.

This book can help in changing the outcome of that 'chess game' in favour of the common people and the laying of the foundation of a good hopeful future! **Enjoy the ride!**

Terry Tibando - September 2015

BOOK FOUR (VOL UME FOUR)
IN SEARCH OF EXTRATERRESTRIAL INTELLIGENCE

CHAPTER 68
ARE ASTRONAUTS DEBRIEFED AFTER EACH MISSION USING MIND CONTROL METHODS? 165

CHAPTER 69
THAT'S ONE ITTY BITTY, SMALL STEP FOR A MAN, ONE HUGE, GIGANTIC LEAP FOR WHOM... NASA? 174

CHAPTER 70
ONE OF THESE DAYS ALICE, JUST ONE OF THESE DAYS --- POW! RIGHT TO THE MOON!!! 181

IN SEARCH OF EXTRATERRESTRIAL INTELLIGENCE

CHAPTER 59

"WE ARE NOT ALONE; WE NEVER HAVE BEEN"!

Stargazing has become for most people a rare pastime experience unless you happen to live out in the countryside away from the commotion of the big cities with their pervasive polluting nightlights that brighten the skyline. At such times, for those who are fortunate to take the time to look heavenward, one cannot help but be awe-struck by the majestic vista of stars, planets, nebulae and of course, our **Milky Way galaxy**. The universe is a big place and it gives a person time to reflect upon our place in it and upon the meaning of life or whether we are alone in the universe.

Space is filled with more than just stars, planets, and galaxies; it has become in five decades crowded with thousands of manmade objects orbiting our planet. It is not unusual today for someone to see the **International Space Station (ISS)** orbiting overhead every 90 minutes or to see many satellites both telescopic exploratory devices as well as military spy satellites monitoring activity down here, on Earth. If a person stays out stargazing long enough, once and awhile in puzzlement, he will catch sight of something unusual that does not readily obey the laws of physics or conventionality. He may by chance spot an object that zigzags or makes an abrupt right or left hand turn or simply come to a complete dead stop in space. In the rational mind of the observer, this mysterious object moving through space is like no previously known manmade device seen before. Even so, there are some manmade space vehicles that do operate unconventionally!

It is at these times, we ask ourselves, what is it? What could it be? Is there a natural explanation for it being there? Is it a secret manmade spacecraft or something from another world? These simple questions beget more complex questions like, are we alone in the universe? Is there other life in this immense vastness? How could an all-divine being, God have created just us? *"It would seem to be a lot of wasted space if it was just us in the universe!"* - from the movie "Contact"

In our ruminations, we search for the rational as well as the divine to answer these age-old questions. The late **Carl Sagan** in his television series "Cosmos" and from his book of the same name, he would often say that *"there are billions and billions of stars in the universe. and we are the stuff of stars!"* A truly thought provoking perception and perhaps, it is the rational answer we are searching for, however, it is not emotionally satisfying or comforting to us. The truth is at times, a hard pill to swallow and may go down tasting like bitter medicine, but it is often what we need. We are, therefore, one with the universe and with all life within it!

The manifestation of God for this day and age, **Baha'u'llah** has said in His holy writings: *"Dost thou reckon thyself only a puny form when within thee the universe is folded?"* Here, we find that there is more to our reality than just our physical being. This is more of a comforting response that is both spiritually and physically satisfying; in truth, we are one with the universe and more. Science and religion, therefore, are in agreement!

We are not only, one people on one planet but, we are also, one people in one universe! We share the universe with all intelligent life in a common bond of sentience, spirituality, and physicality, we are all one!

In uncovering the truth surrounding the mystery of the UFO/ETI phenomenon then, we must leave no stone or rock left unturned, no door must bar our way or remain locked toward the truth and no vail must obscure our gaze, our sight must be free, unhindered and penetrating! Then and only then, can we answer with absolute conviction, that age-old question: ***"Are we alone in the universe?"*** In our virtual journey through the Solar System and beyond we will discover that the answer to that question will reaffirm and support the hard-core evidence we have shown from the first section of this mighty tome to be a resounding: ***"We have never been alone"!!***

Throughout this textbook, the best available proofs and evidence will establish once and for all, that not only is there life in the universe but, it is highly intelligent and is currently visiting our planet in spacecraft referred to as UFOs. ***In the minds of the general public, UFOs simply refers to ET spacecraft (ETS) therefore, the terms are used interchangeably. A better term would be ETVs (Extraterrestrial Vehicles).***

There have been countless observations of ETI visitations to Earth, whether seen in our skies or out in space or on the ground and there have been numerous encounters with them going back many millennia. On the other hand, the cover-up and suppression of the ETI existence by the **Military Industrial Complex** has been less than a hundred years yet, it has quickly developed into multiple levels of secrecy affecting many branches of science. Chiefly, the sciences of astronomy, physics, exobiology and medicine in association with aviation and space exploration, as well as genetic research have also become an integral part of the cover-up and this list continues to grow. https://www.youtube.com/watch?v=lkswXVmG4xM

However, no matter how strong the dam walls of secrecy appear to be in containing the truth, insider leaks inevitably appear and slowly cracks develop in that wall of secrecy, first as a steady trickle and then as an unstoppable mighty torrent. Truth in all things cannot be contained, it always seeks its own level. ***"The truth shall set you free, but first, it will piss you off!"***

To understand the universe we live in, we must first understand the reality of that universe. Is it a physical construction or is it interdimensional or something else? Is it a living entity or is it a reflection of something greater and more mysterious? Is the universe limited by boundaries and borders or is it truly infinite in size beyond the comprehension of mankind? Does the answer lie with quantum physics or spirituality or both?

In the next chapter, we explore the known and theoretical states of the universe, often referred to as the Holographic Universe.

CHAPTER 60

THE HOLOGRAPHIC UNIVERSE AND
THE HOLOGRAPHIC PARADIGM

The Holographic Universe

Back in 1982, a physicist by the name of **Alain Aspect** and his research team performed an experiment at the University of Paris that would have a redounding effect on the world of science. An experiment of such monumental importance that for the most part went unnoticed by the public, except in the academia of science.

Aspect's experiment is related to the **EPR Experiment (or Einstein-Podolsky-Rosen Experiment)** also, known as the **EPR Paradox**, a consciousness or thought experiment devised in 1935 by **Albert Einstein**, and his colleagues, **Podolsky and Rosen,** in order to disprove **Quantum Mechanics** is not a complete physical theory.

A dichotomy arose in the EPR experiment that either: The result of a measurement performed on one part **A** of a quantum system has a non-local effect on the physical reality of another distant part **B**, in the sense that quantum mechanics can predict outcomes of some measurements carried out at **B**; or …

Quantum mechanics is incomplete in the sense that some element of physical realty corresponding to **B** cannot be accounted for by quantum mechanics (that is, some extra variable is needed to account for it.)

Aspect and his team discovered that under certain circumstances subatomic particles such as electrons are able to instantaneously communicate with each other regardless of the distance separating them. It doesn't matter whether they are 10 feet or 10 billion miles apart.

Somehow, each particle always seems to know what the other is doing. The problem with this feat is that it violates Einstein's long-held tenet that no communication can travel faster than the speed of light ($E = mc^2$). Since traveling faster than the speed of light is tantamount to breaking the time barrier.

Aspect's findings imply that *objective reality does not exist,* **that despite its apparent solidity,** *the universe is at heart a phantasm,* a gigantic and splendidly detailed **Hologram**.

To understand this assertion, one must first understand a little about holograms. A hologram is a three- dimensional photograph made with the aid of a laser. To make a hologram, the object to be photographed is first bathed in the light of a laser beam. Then a second laser beam is bounced off the reflected light of the first and the resulting interference pattern (the area where the two laser beams commingle) is captured on film. When the film is developed, it looks like a meaningless swirl of light and dark lines. But as soon as the developed film is illuminated by another laser beam, a three-dimensional image of the original object appears. The three-dimensionality of such images is not the only remarkable characteristic of holograms. If a hologram of a rose for example is cut in half and then illuminated by a laser, each half will still be found to contain the entire image of the rose.

Indeed, even if the halves are divided again, each snippet of film will always be found to contain a smaller but intact version of the original image. Unlike normal photographs, every part of a hologram contains all the information possessed by the whole.

The "whole in every part" nature of a hologram provides us with an entirely new way of understanding organization and order. For most of its history, Western science has labored under the bias that the best way to understand a physical phenomenon, whether a frog or an atom, is to dissect it and study its respective parts.

A hologram teaches us that some things in the universe may not lend themselves to this approach. If we try to take apart something constructed holographically, we will not get the pieces of which it is made; we will only get smaller wholes.

This insight suggested to University of London physicist, **David Bohm** another way of understanding Aspect's discovery. Bohm believes the reason subatomic particles are able to remain in contact with one another regardless of the distance separating them is not because they are sending some sort of mysterious signal back and forth, but because their separateness is an illusion. He argues that at some deeper level of reality such particles are not individual entities, but are actually extensions of the same fundamental something. This fundamental connectedness would correlate with **The Fifth Element**, and its mathematical proof of all aspects of the universe being energetically connected - **Hal Puthoff's** assertion in his work on **Zero-Point Energy** of all charges in the universe being connected and that further mass is in all likelihood an illusion as well -- and both of these modern day theories of physics being in accordance with ancient traditions and philosophies, which claim the same connectedness of the diverse parts of the universe.

Bohm suggests, imagining an aquarium containing a fish and that you are unable to see the aquarium directly but your only source of knowledge about it and what it contains comes from two television cameras, one directed at the aquarium's front, and the other directed at its side.

Looking at the two TV monitors, you might assume that the fish on each of the screens are separate entities. After all, because the cameras are set at different angles, each of the images will be slightly different. But, as you continue to watch the two fish, you will eventually become aware that there is a certain relationship between them.

When one turns, the other also makes a slightly different but corresponding turn; when one faces the front, the other always faces toward the side. If you remain unaware of the full scope of the situation, you might even conclude that the fish must be instantaneously communicating with one another, but this is clearly not the case.

This, says Bohm, is precisely what is going on between the subatomic particles in Aspect's experiment. According to Bohm, the apparent faster-than-light connection between subatomic particles is really telling us that there is a deeper level of reality we are not privy to, a more complex dimension beyond our own that is analogous to the aquarium. And, he adds, we view objects such as subatomic particles as separate from one another because we are seeing only a portion of their reality.

Such particles are not separate "parts", but facets of a deeper and more underlying unity that is ultimately as holographic and indivisible as the previously mentioned rose. And since everything in physical reality is comprised of these **"eidolons"** (phantoms), the universe is itself a projection, a hologram. http://www.crystalinks.com/holographic.html

In addition to its phantom-like nature, such a universe would possess other rather startling features. If the apparent separateness of subatomic particles is illusory, it means that at a deeper level of reality all things in the universe are infinitely interconnected.

The electrons in a carbon atom in the human brain are connected to the subatomic particles that comprise every salmon that swims, every heart that beats, and every star that shimmers in the sky. Everything interpenetrates everything, and although human nature may seek to categorize and pigeonhole and subdivide, the various phenomena of the universe, all apportionments are of necessity artificial and all of nature is ultimately a seamless web.

In a holographic universe, even time and space could no longer be viewed as fundamentals. Because concepts such as location break down in a universe in which nothing is truly separate from anything else, time and three-dimensional space, like the images of the fish on the TV monitors, would also have to be viewed as projections of this deeper order.

The Super Hologram

At its deeper level reality is a sort of **super hologram** in which the past, present, and future all exist simultaneously. This suggests that given the proper tools it might even be possible to someday reach into the super holographic level of reality and pluck out scenes from the long-forgotten past.

What else the super hologram contains is an open-ended question. Allowing, for the sake of argument, that the super hologram is the matrix that has given birth to everything in our universe, at the very least, it contains every subatomic particle that has been or will be -- every configuration of matter and energy that is possible, from snowflakes to quasars, from blue whales to gamma rays. It must be seen as a sort of cosmic storehouse of "All That Is." To this should be added a philosophical motto: **"All this is that!" The beginning and the end are one and the same, the Alpha and Omega!**

Although **Bohm** concedes that we have no way of knowing what else might lie hidden in the super hologram, he does venture to say that we have no reason to assume it does not contain more. Or as he puts it, perhaps the super holographic level of reality is a "mere stage" beyond which lies "an infinity of further development" http://www.crystalinks.com/holographic.html

The Holographic Brain

The **Holographic Brain** is a model of reality that also works well in describing brain function and memory as Standford neurophysiologist **Karl Pribram** discovered. Pribram wanted to know how and where memories are stored in the brain. For it had been shown for decades that memories are dispersed throughout the brain rather than being confined to a specific location.

!n the 1920s, brain scientist **Karl Lashley** found that fractionating portions of a rat's brain he was unable to eradicate its memory of how to perform complex tasks it had learned prior to surgery. If only someone could originate a mechanism that would explain this curious "whole in every part" nature of memory storage.

Enter Pribram, who in the 1960s encounters the concept of holography and finds the explanation that brain scientists had been searching for. Pribram explanation of memories are encoded not in neurons, or small groupings of neurons, but in patterns of nerve impulses that crisscross the entire brain in the same way that patterns of laser light interference crisscross the entire area of a piece of film containing a holographic image. In other words, Pribram believes the brain is itself a hologram.

Pribram's theory also explains how the human brain can store so many memories in so little space. It has been estimated that the human brain has the capacity to memorize something on the order of 10 billion bits of information during the average human lifetime (or roughly the same amount of information contained in five sets of the Encyclopedia Britannica). Holograms he realizes also, possess an astounding capacity for information storage--simply by changing the angle at which the two lasers strike a piece of photographic film, it is possible to record many different images on the same surface. One cubic centimeter of film is capable of holding as many as 10 billion bits of information.

Indeed, one of the most amazing things about the human thinking process is that every piece of information seems instantly cross-correlated with every other piece of information--another feature intrinsic to the hologram. Because every portion of a hologram is infinitely interconnected with ever other portion, it is perhaps nature's supreme example of a cross-correlated system.

The brain is able to translate the avalanche of frequencies it receives via the senses (light frequencies, sound frequencies, and so on) into the concrete world of our perceptions. Encoding and decoding frequencies is precisely what a hologram does best. Just as a hologram functions as a sort of lens, a translating device able to convert an apparently meaningless blur of frequencies into a coherent image, Pribram believes the brain also comprises a lens and uses holographic principles to mathematically convert the frequencies it receives through he senses into the inner world of our perceptions. http://www.crystalinks.com/holographic.html

Holophonic Sound

Argentinean-Italian researcher **Hugo Zucarelli** recently extended the holographic model into the world of acoustic phenomena. Puzzled by the fact that humans can locate the source of sounds without moving their heads, even if they only possess hearing in one ear, Zucarelli discovered that holographic principles can explain this ability.

Zucarelli has also developed the technology of **Holophonic Sound**, a recording technique able to reproduce acoustic situations with an almost uncanny realism.

Pribram's belief that our brains mathematically construct "hard" reality by relying on input from a frequency domain. It has been found that each of our senses is sensitive to a much broader range of frequencies than was previously suspected.

Researchers have discovered, for instance, that our visual systems are sensitive to sound frequencies, that our sense of smell is in part dependent on what are now called "cosmic frequencies", and that even the cells in our bodies are sensitive to a broad range of frequencies. Such findings suggest that it is only in the holographic domain of consciousness that such frequencies are sorted out and divided up into conventional perceptions.

But the most mind-boggling aspect of Pribram's holographic model of the brain is what happens when it is put together with **Bohm's theory**. For if the concreteness of the world is but a secondary reality and what is "there" is actually a holographic blur of frequencies, and if the brain is also a hologram and only selects some of the frequencies out of this blur and mathematically transforms them into sensory perceptions, what becomes of objective reality?

Put quite simply, it ceases to exist. As the religions of the East have long upheld, the material world is **Maya**, an illusion, and although we may think we are physical beings moving through a physical world, this too is an illusion. We are really "receivers" floating through a kaleidoscopic sea of frequency, and what we extract from this sea and transmogrify into physical reality is but one channel from many extracted out of the **super hologram**.
http://www.crystalinks.com/holographic.html

The Holographic Paradigm

This striking new picture of reality, the synthesis of **Bohm** and Pribram's views, has come to be called the **Holographic Paradigm**, and although many scientists have greeted it with skepticism, it has galvanized others. A small but growing group of researchers believe it may be the most accurate model of reality science has arrived at thus far.

More than that some believe it may solve some mysteries that have never before been explainable by science and even establish the paranormal as a part of nature. Numerous researchers have concluded that many para-psychological phenomena become much more understandable in terms of the holographic paradigm.

In a universe in which individual brains are actually indivisible portions of the greater hologram and everything is infinitely interconnected, telepathy may merely be the accessing of the holographic level. It is obviously much easier to understand how information can travel from the mind of individual 'A' to that of individual 'B' at a far distance point and helps to understand numerous unsolved puzzles in psychology. In particular, understanding many of the baffling phenomena experienced by individuals during altered states of consciousness.

The Holographic State of Being and the Rational Concept of Re-incarnation

In the 1950s, during the course of his research, **Stansilov Grof** encountered examples of patients regressing and identifying with virtually every species on the evolutionary tree from dinosaurs to apes (research findings which helped influence the man-into-ape scene in the movie, "Altered States").

Regressions into the animal kingdom were not the only puzzling psychological phenomena Grof encountered. He also had patients who appeared to tap into some sort of collective or racial unconscious. Individuals with little or no education suddenly gave detailed descriptions of Zoroastrian funerary practices and scenes from Hindu mythology, (sometimes referred to as a false perception of the *"re-incarnated state of being"*). In other categories of experience, individuals gave persuasive accounts of out-of-body journeys, of precognitive glimpses of the future, of regressions into apparent past-life incarnations (re-incarnation).

It should be stated here, with no deliberate denigration of individual religious belief systems that such acceptance of re-incarnation is a delusional state of mind and a misunderstanding of the relationship of deceased souls who have moved on beyond this physical realm of existence. People are free to choose whatever they wish to believe in, whether that be life after death, re-incarnation, the sacredness of cows, dragons, giants, the tooth fairy, Santa Claus or "little green men," they should however, know the facts and truths that separates the subjective and the irrational from their belief.

There are absolute truths that are inviolable, eternal and non-negotiable in their meaning. There are also, relative truths and transactional truths that are relevant to the exigencies of this current age as both are transitory in nature. The difference is in knowing which truth you are adhering to.

The **Ten Commandments**, life and death, and planetary gravitational forces, electro-motive force are examples of absolute truths; planting seeds will produce a mature plant, unless the soil is infertile or if an animal or bird eats the seed, this is relative truth. The creation and abrogation of political laws is a transactional truth.

When we examine the concept of re-incarnation that of having lived past lives from former ages, this is a relative transitional truth, where the belief is based upon subjective information usually indentified in western societies through the procedure of regressive hypnosis. There can only be one Cleopatra, one Caesar and not multitudes of people claiming to be those same re-incarnated personages!

The **"Life Spectrum"** is a lineal continuum and not a repetitive close-looped bio-system of physical incarnations. There are three states of existence and all are multi-leveled. The first is a **Pre-existence** which is without time or space and usually, very few of us can remember anything of that existence, though it is possible through meditation and reflection to re-call some aspects of it. The Second is the **Temporal existence** of which we are all part of as we dwell within this phenomenal physical reality and this plane of existence is for learning in preparation for the next existence. The final plane of existence is the **After-existence** and it too is without time or space. In fact, the temporal existence has no time and space, as this too is an illusion.

Most of these concepts are a part of the holy teachings of most of the world's religions which are described as the spiritual existences of mankind. Even science has accepted that there are multiple levels of existence, often termed as higher dimensions of reality. However, as it applies to mankind, there are three planes of existence with an infinite number of levels in each existence from which we may partake. Our true reality is spiritual or inter-dimensional or multi-dimensional by nature and our journey through these existences is infinite, linear, and not repetitively close-looped.

On a scientific perspective, a form of physical re-incarnation is possible only in the following manner, i.e. a person dies, his body decomposes into its basic molecular constituents which then mingle with other minerals of the ground and soil, the plants in the soil absorb these molecular particles into the plant structure by means of osmosis through its roots and leaves. The plant grows, matures and the animals eat the plant as sustenance and we humans eat the animal as food. The atoms and molecules of the human have now transitioned through the different kingdoms and have returned, again to the human form. The original molecular structure of one human through a cycle of integration, disintegration and re-integration has returned to a human form. This is physical re-incarnation and not spiritual re-incarnation. The qualities and the intellect and the personality of the individual which made that person unique have not returned, merely some of the physical constituents of the individual have returned. This then, is the true meaning of re-incarnation which is a relative truth.

The spiritual, rational soul, the experiences and intellectual nature of the human is indestructible, indivisible, and eternal and does not return to human form after the physical decomposition of the body. It is a one way trip to the after-life, to the next plane of existence, not an endless looping bus ride around to the point of "bodily origin" or "home" again. This is an absolute truth, a fact! The proof is self-evident.

So, how is it that people claim that they are the re-incarnated personage of "Queen so and so" or a mighty Spartan warrior, a run-away slave, a prince or a president, a war pilot, an artist, etc., etc?

It is because they do not understand their non-local spiritual natures. We are spirit beings who evince wondrous psyche powers and abilities which for the most part, many of us are ignorant about our true natures. We are capable of soul or astral travel particularly during our sleep and dreamtime and in this state; the space-time continuum breaks down and becomes meaningless. We are in this state in the realm of infinite consciousness! We are able to travel vast distances in a blink of an eye and travel through time both into the past and into the future, because reality is a non-linear and non-local construct. People who existed in the past still survive in a spiritual state, and are accessible in this non-local state, where it is possible to associate with these departed souls to the extent that their very life experiences can imprint profoundly upon us, both subconsciously and sometimes consciously when we are in this astral soul travelling form. Upon our conscious awakened state, we may recall that union of souls but, confuse its reality believing it to be our own past life experience when in reality, it is someone else's experience. This is due to the fact, that astral travel is for the most part experienced by most people on a subconscious level. For this reason, we tend not to remember it as this unique experience but, rather as bits and pieces of a dream that may surface to conscious recollection or by means of hypnotic regression, where the therapist may be no wiser in understanding its spiritual implications or reality than the one who has been hypnotized.

The Holographic Conscious Universe

To what degree does this holographic reality extend, to what parameters? Such words imply a limited perception and understanding of the nature of the universe that we humans typically confine ourselves in out thinking and behaviour to just three dimensions.

What we have found so far is that the universe is far more complex than we can imagine and that it is more than just three or four dimensions if we consider time as a fourth dimension. Everything scientists have discovered to date points to a universe that is infinitely multi-dimensional and that time and space really are non-existent and are simply devices to measure the third dimension in which we all currently, but temporarily exist.

As **Nikola Tesla** has stated over a century ago, everything in existence is explainable by *frequency, vibration, and resonance.*

Everything that we know is a fractal set of some larger fractal-set of energy or consciousness within an infinite universe. If the brain (mind), sound, the universe, and the super universe are all holographic, i.e., all fractals of something more infinite, then as religions are correct when they have stated for many millennia, that there is a supreme, divine consciousness at work from which all things derive their very existence from!

"Be thou not surprised at this. Reflect upon the inner realities of the universe, the secret wisdoms involved, the enigmas, the interrelationships, the rules that govern all. For every part of the universe is connected with every other part by ties that are very powerful and admit of no imbalance, nor any slackening whatever. In the physical realm of creation, all things are eaters and eaten: the plant drinketh in the mineral, the animal doth crop and swallow down the plant, man doth feed upon the animal, and the mineral devoureth the body of man. Physical bodies are transferred past one barrier after another, from one life to another, and all things are subject to transformation and change, save only the essence of existence itself— since it is constant and immutable, and upon it is founded the life of every species and kind, of every contingent reality throughout the whole of creation." Selections from the Writings of Abdu'l-Baha; Compiled by the Research Department of the Universal House of Justice; © 1978 The Universal House of Justice; ISBN 85398 081 0; Printed in Great Britain by W & J MacKay Ltd, Chatham
https://www.bahai.org/library/authoritative-texts/search#q=life%20in%20the%20universe

This infinite consciousness is termed as all embracing, all encompassing, omniscient, omnipotent, omnipresent and omnibenevolent and are the theological attributes we associate with the supreme being or intelligence of the universe, God!

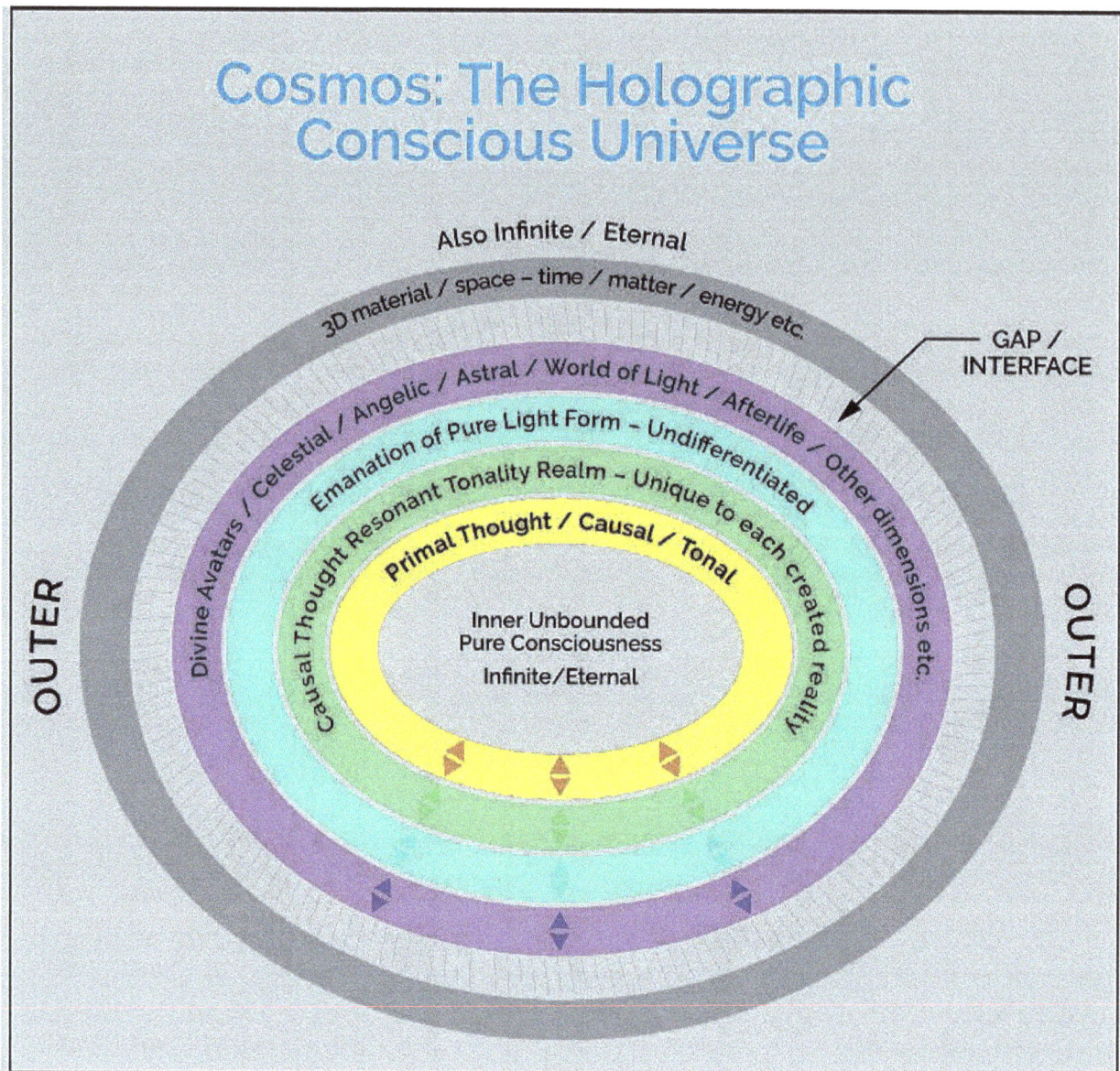

**The Cosmos as the Holographic Conscious Universe with its relationship
to all other realms of existence that are infinite and eternal**
(copyrighted Sirius Technology Advanced research LLC – 2018)
https://siriusdisclosure.com/wp-content/uploads/2018/07/Petaluma-Slides.pdf

Dr. Steven Greer explores this holographic cosmic consciousness further as it relates to human life and extraterrestrial intelligence and the connection to each other through the common consciousness of the universe.

All conscious sentient beings whether human or ETI or other dimensional entities are connected through various states of consciousness such as waking, sleeping, dreaming be it lucid and precognition, and through meditation of unbounded pure consciousness.

There is also, as Dr. Greer states higher states of consciousness: **Cosmic Consciousness** (the pure consciousness while waking, dreaming and sleeping); **God Consciousness or Celestial Perception** (which also include waking, dreaming and sleeping); **Unity Consciousness** (in that all that exists directly perceived as pure consciousness) and finally, **Brahmin or Universal Consciousness (**where experience of the entire Cosmos as One with self-operatively).
https://siriusdisclosure.com/wp-content/uploads/2018/07/Petaluma-Slides.pdf

Knowing these states of consciousness, we realize that all life in the universe that is sentient, aware and intelligent is connected via the realms of higher consciousness and I turn connected to the Supreme Divine Creator of the universe. That thought is unbounded, beyond all space and time and that *a thought that is sent can be received before it is even sent,* because it is non-linear and unbounded!

It is no stretch of the imagination to realize that as the Moody Blues song states in their album "In Search of the Lost Chord", *"thinking is the best way to travel!"* and it's also, *the best way to communicate* with other intelligent beings across vast distances!!

A Universe is Folded Within Us!

To this attests **Baha'u'llah** in His book, **"The Seven valleys and the Four Valleys,"**

"Indeed, O Brother, if we ponder each created thing, we shall witness a myriad perfect wisdoms and learn a myriad new and wondrous truths. One of the created phenomena is the dream. Behold how many secrets are deposited therein, how many wisdoms treasured up, how many worlds concealed. Observe, how thou art asleep in a dwelling, and its doors are barred: on a sudden thou findest thyself in a far-off city, which thou enterest without moving thy feet or wearying thy body; without using thine eyes, thou seest: without taxing thine ears, thou hearest: without a tongue, thou speakest, and perchance when ten years are gone, thou wilt witness in theouter world the very things thou hast dreamed tonight.

Now there are many wisdoms to ponder in the dream, which none but the people of this Valley can comprehend in their true elements. First, what is this world, where without eye and ear and hand and tongue, a man puts all of these to use? Second, how is it that in the outer world thou seest today the effect of a dream, when thou didst vision it in the world of sleep some ten years past? Consider the difference between these two worlds and the mysteries which they conceal, that thou mayest attain to divine confirmations and heavenly discoveries and enter the regions of holiness.

God, the Exalted, hath placed these signs in men, to the end that philosophers may not deny the mysteries of the life beyond nor belittle that which hath been promised them. For some hold to reason and deny whatever the reason comprehendeth not, and yet weak minds can never grasp the matters which we have related, but only the supreme, Divine Intelligence can comprehend them:"

"Likewise, reflect upon the perfection of man's creation and that all these planes and states are folded up and hidden away within him.

"Dost thou reckon thyself only a puny form when within thee the universe is folded?"
(The Seven Valleys and the Four Valleys) by Baha'u'llah

In later research, **Grof** found the same range of phenomena manifested in therapy sessions which did not involve the use of drugs. Because the common element in such experiences appeared to be the transcending of an individual's consciousness beyond the usual boundaries of ego and/or limitations of space and time, Grof called such manifestations transpersonal experiences, and in the late '60s he helped found a branch of psychology called **Transpersonal Psychology** devoted entirely to their study. http://en.wikipedia.org/wiki/Transpersonal_psychology

Although Grof's newly founded **Association of Transpersonal Psychology** garnered a rapidly growing group of like-minded professionals and has become a respected branch of psychology, for years neither, Grof nor any of his colleagues were able to offer a mechanism for explaining the bizarre psychological phenomena they were witnessing. But that has changed with the advent of the holographic paradigm.

The Holographic Universe- a universe is folded witin each of us
http://www.wakingtimes.com/2014/11/11/unlock-door-holographic-reality/ and http://blogs.discovermagazine.com/d-brief/tag/exoplanets/#.WKzFQvIuyLE

As Grof noted, if the mind is actually part of a continuum, a labyrinth that is connected not only to every other mind that exists or has existed, but to every atom, organism, and region in the vastness of space and time itself, the fact that it is able to occasionally make forays into the labyrinth and have transpersonal experiences no longer seems so strange. Perhaps, in creating reality, we have already become, as in **Star Trek - The Next Generation** like the *"Q of the Continuum"* or we are part of a consciousness virtual reality experiment.

The holographic paradigm also has implications for so-called hard sciences like biology. Keith Floyd, a psychologist at Virginia Intermont College, has pointed out that if the concreteness of reality is but a holographic illusion, it would no longer be true to say the brain produces consciousness. Rather, it is Consciousness that creates the appearance of the brain as well as the body and everything else around us we interpret as physical.

Such a turnabout in the way we view biological structures has caused researchers to point out that medicine and our understanding of the healing process could be transformed by the holographic paradigm. If the apparent physical structure of the body is but a holographic projection of consciousness, it becomes clear that each of us is much more responsible for our health than current medical wisdom allows. What we now view as miraculous remissions of disease may actually be due to changes in consciousness, which in turn effect changes in the hologram of the body. Similarly, controversial new healing techniques such as visualization may work so well because in the holographic domain of thought images are ultimately as real as reality. Even visions and experiences involving non-ordinary reality become explainable under the holographic paradigm.

Through conscious awareness of the dynamics of a holographic reality and the power of the rational mind, it may be possible to "turn off physical reality as we know it and to be able to turn it back on again" for the universe is a non-local construct! This is reminiscent of Carl Jung's understanding of the proto-archetype of reality. **Be careful of what you believe in, for what you conceive, so will you perceive!**

Perhaps, we agree on what is "there" or "not there" because what we call **consensus reality** is formulated and ratified at the level of the human unconscious at which all minds are infinitely interconnected. If this is true, it is the most profound implication of the holographic paradigm of all for it means that such experiences as "turning off and on again" reality are not commonplace, only because we have not programmed our minds with the beliefs that would make them so.

In a holographic universe, there are no limits to the extent to which we can alter the fabric of reality. What we perceive as reality is only a canvas waiting for us to draw upon it any picture we want. Anything is possible, for Magic is our birthright, no more or less miraculous than our ability to compute the reality we want when we are in our dreams.

Indeed, even our most fundamental notions about reality become suspect, for in a holographic universe, as **Pribram** has pointed out, even random events would have to be seen as based on holographic principles and therefore determined.

Synchronicities or meaningful coincidences suddenly make sense, and everything in reality would have to be seen as a metaphor. Even the most haphazard events would express some underlying symmetry. http://www.crystalinks.com/holographic.html

We should now have a pretty good understanding into what life is all about in the holographic universe, its inter-connectedness, its multi-dimensional or spiritual nature and the oneness we share with all created sentient life forms throughout this universe. So, as we stated at the beginning of this section, what are the differences and similarities that exist between extraterrestrial experiences and paranormal experiences?

It now appears that we and all Extraterrestrial Intelligences throughout the universe share one super holographic universe, one holographic state of being, one holographic paradigm, and one holographic mind! The universe is alive and sentient; we live in a holographic conscious universe and dare it be stated...it is the one **Holographic Soul**!

We and They are One or as **Dr. Steven Greer** has uniquely coined: **"We are One People, in One Universe!"**

If you do not understand the differences and similarities then when an observation of an unusual event occurs right in front of you and perhaps other witnesses, your interpretation of what is happening may be confused with events of some other reality. This is no different than when a doctor is trying to diagnose a full-blown heart attack from an over- strained chest muscle pain caused from hard exercise or work. Tests need to be run on the patient to see if the symptoms are heart attack related or chest muscle strain because there are similarities between the two but each is a different problem and the proper diagnosis will lead to the correct treatment and cure. Understanding the difference and discovering the similarities comes from a sound knowledge base but mostly from experience.

There are overlaps between the paranormal and the extraterrestrial experiences that create confusion in the minds of the observers because they either lack the basic knowledge or previous experience in this strange paradigm of reality. For example, apparitions of the departed from this mortal life may appear as a ball of light either floating around a cemetery or in a house which materializes as a ghostly apparitional form of a person. The ability for a departed soul to appear in this manner is a state or condition of consciousness or as a condition of that level of existence in which it is as familiar to the soul as when it walked the earth and breathed the free air of physical existence.

Extraterrestrial beings and their spacecraft may also appear as a ball of light flying around an open field or any other location as well as in a home. They can materialize into a visible, physical being or be visible but, in a holographic or apparitional form like a ghost or astral entity. Their ability to appear in this manner may be due to higher consciousness and/or mental development and learning or from advance technology that interfaces with the mind or consciousness. There are even reports of ET beings assuming an animal form or as a bird.

Astral beings may also, appear to be apparitional and have the ability to move through walls like some ETI or float or fly like ET spacecraft and even travel vast distances and enter other realms of existence in the twinkling of an eye.

Spirit beings may also be departed animal spirits, which Native American people referred to as "Spirit Guides" particularly when someone is on a "vision quest" or as a rite of passage from adolescence to adulthood or maturity.

The reader may want to research further the similarities and differences between UFOs, ETI and the paranormal in an excellent paper on this subject that can found on the website http://cseti.org/ or at http://siriusdisclosure.com/ entitled: "ETs and the New Cosmology" by Dr. Steven Greer.

Also, the reader should view Dr. Greer's recent 2018 YouTube lecture on *"Cosmos: The Holographic Conscious Universe"* and as well as read his paper by the same title. In his video lecture, Dr. Greer shows that the holographic universe, the holographic brain, the holographic sound, the holographic paradigm and the holographic state of being are all really one and the same thing of a much larger **Holographic Conscious Universe.** Within this living conscious universe and its infinite levels of awareness and consciousness, our physical outer 3D reality is only one aspect of that infinite universe which includes all life forms both visible and invisible, extraterrestrial, interdimensional and spiritual.

CHAPTER 61

THE PRESENCE OF EXTRATERRESTRIAL INTELLIGENCE IN OUR SOLAR SYSTEM: A RESOLUTION TO THE FERMI PARADOX

Next time there is a clear starry night, go out into your backyard or to your favourite park site (or make-out spot) and look up. Like most people, you will be awe struck by the beauty and majesty of the sky above and you will also feel dumbfounded by the sheer immensity of the universe. It is enough to drive you into a small sphere of insignificance, however, that is not the intention of this exercise. You are not insignificant but rather, a great and noble creation that is able to comprehend the universe to some extent.

Baha'u'llah, the Manifestation of God for this day and age says: ***"Noble I made you, wherewith dost thou abase thyself?"*** The Hidden Words of Baha'u'llah; Baha'u'llah; copyright 1954; by National spiritual assembly of the Baha'is of the United States; printed in the USA; Library of Congress Card No. # 54-7328

And once again, **Baha'u'llah** affirms, ***"Dost thou reckon thyself only a puny form when within thee the universe is folded?*** The Seven Valleys and the four Valleys; Baha'u'llah; copyright 1945 and 1952; by National spiritual assembly of the Baha'is of the United States; printed in the USA; Library of Congress Card No. # 53-12275

We are an amazing emulation in miniature of that greater universe above us and which surrounds and interfaces with us! With further observation and reflection that age old question," are we alone in the universe?" and "Are there other intelligent beings in the universe?"

No doubt, **Physicist Enrico Fermi** must have felt something similar when as the story goes that, one day back on the 1940's, a group of atomic scientists, including the famous Enrico Fermi, were sitting around talking, when the subject turned to extraterrestrial life. Fermi is supposed to have then asked, *"So? Where is everybody?"* What he meant was: If there are all these billions of planets in the universe that are capable of supporting life, and millions of intelligent species out there, then how come none has visited earth? This has come to be known as **The Fermi Paradox.** http://waitbutwhy.com/2014/05/fermi-paradox.html and http://abyss.uoregon.edu/~js/cosmo/lectures/lec28.html

Nowadays, Ufologists reviewing what Fermi had originally stated back then would have considered Fermi to be cut out of the loop of insider information held by a small handful of select individuals, who really knew the big picture about UFOs and the presence of **Extraterrestrial Intelligence (ETI)** already visiting the planet! Fermi's peers and contemporaries would have promoted his theory while those in the know have utilized his hypothesis and maybe used it to keep the general public sidetracked away from figuring out the reality of the UFO/ETI phenomenon with the scientific rationale of Enrico Fermi's paradox.

Currently for those who subscribe to **Space.com** will often see its writers promoting the Fermi Paradox as proof that ETI do not exist in our corner of the universe or at least in our solar system. Nothing could be further from the truth as this chapter will reveal!

Fermi's Paradox goes something like this, that any civilization with a sufficiently advanced technology like rocket technology, (nuclear, ion or ramjet) and an immodest amount of opportunistic imperial imperative could rapidly colonize the entire Galaxy. Because interstellar distances are vast and living creatures have finite lifetimes, perhaps, the best way for an advanced civilization to colonize the Milky Galaxy is construct self-replicating, autonomous robots. The idea of self-reproducing automaton was proposed by mathematician **John von Neumann** in the 1950's. The idea is that a device/space probe with **artificial intelligence (AI)** could, 1) perform tasks in the real world and 2) make copies of itself (like bacteria).
http://www.daviddarling.info/encyclopedia/V/vonNeumannprobe.html

Growth of the number of probes would occur exponentially and the Galaxy could be explored in 3.5 to 4 million years. While this time span seems long compared to the age of human civilization, remember the Galaxy is over 10 billion years old and any past extraterrestrial civilization could have explored the Galaxy 250 times over. Within a few million years, every star system could be brought under the newly developed galactic empire of humanity or whatever other species who had similar imperialistic agendas of colonization. A few million years may sound long, but in fact, it's quite short compared with the age of the Galaxy, which is roughly ten thousand million years. Colonization of the Milky Way should be a quick exercise, a blink of an eye in time! http://abyss.uoregon.edu/~js/cosmo/lectures/lec28.html

So what Fermi immediately realized was that the aliens have had more than enough time to disseminate the Galaxy with their presence. But looking around, he didn't see any clear indication that they're out and about. This prompted Fermi to ask (what was to him) an obvious question: *"where is everybody?"* http://abyss.uoregon.edu/~js/cosmo/lectures/lec28.html

Now, considering that the Galaxy has been around for over 10 billion plus years and the speed of technological advancement in our own culture, (we have gone from horse-drawn wagon to landing men on the Moon to exploring the surface of another planet, all within a 100 years), then a more relevant point is where are all the super-advanced alien civilizations? Russian astrophysicist **Nikolai Kardashev** proposed a useful scheme to classify advanced civilizations; he argues that ET would posses one of three levels of technology. A **Type I Civilization** is similar to our own, one that uses the energy resources of a planet. A **Type II Civilization** would use the energy resources of a star, such as a **Dyson Sphere**. A **Type III Civilization** would employ the energy resources of an entire galaxy. A Type III civilization would be easy to detect, even at vast distances. http://abyss.uoregon.edu/~js/cosmo/lectures/lec28.html\

Given our current exponential growth in human technology particularly in space exploration and flight, is it very conceivable that the first attempts in space exploration and the quest for colonization of the galaxy would be overtaken in a thousand years and superceded with more highly advanced spacecraft, most likely capable of travelling at luminal or super luminal speeds while also carrying human beings to other star systems. They would literally be bypassing the original **John von Neumann** self-reproducing AI probes making them obsolete before they could even complete 5% of their mission!

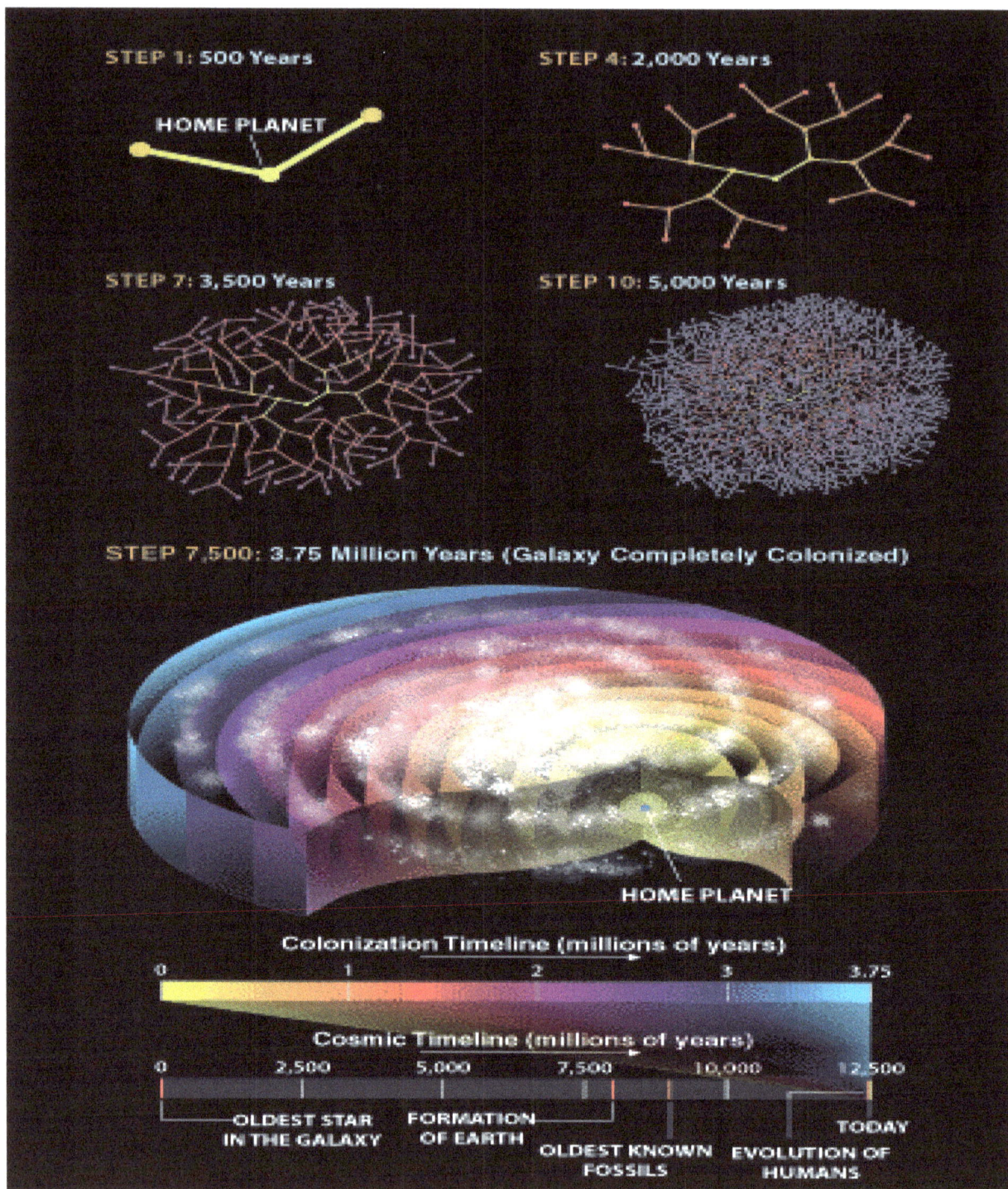

The exponential growth of imperialistic colonization of the Milky Way Galaxy
http://abyss.uoregon.edu/~js/cosmo/lectures/lec28.html

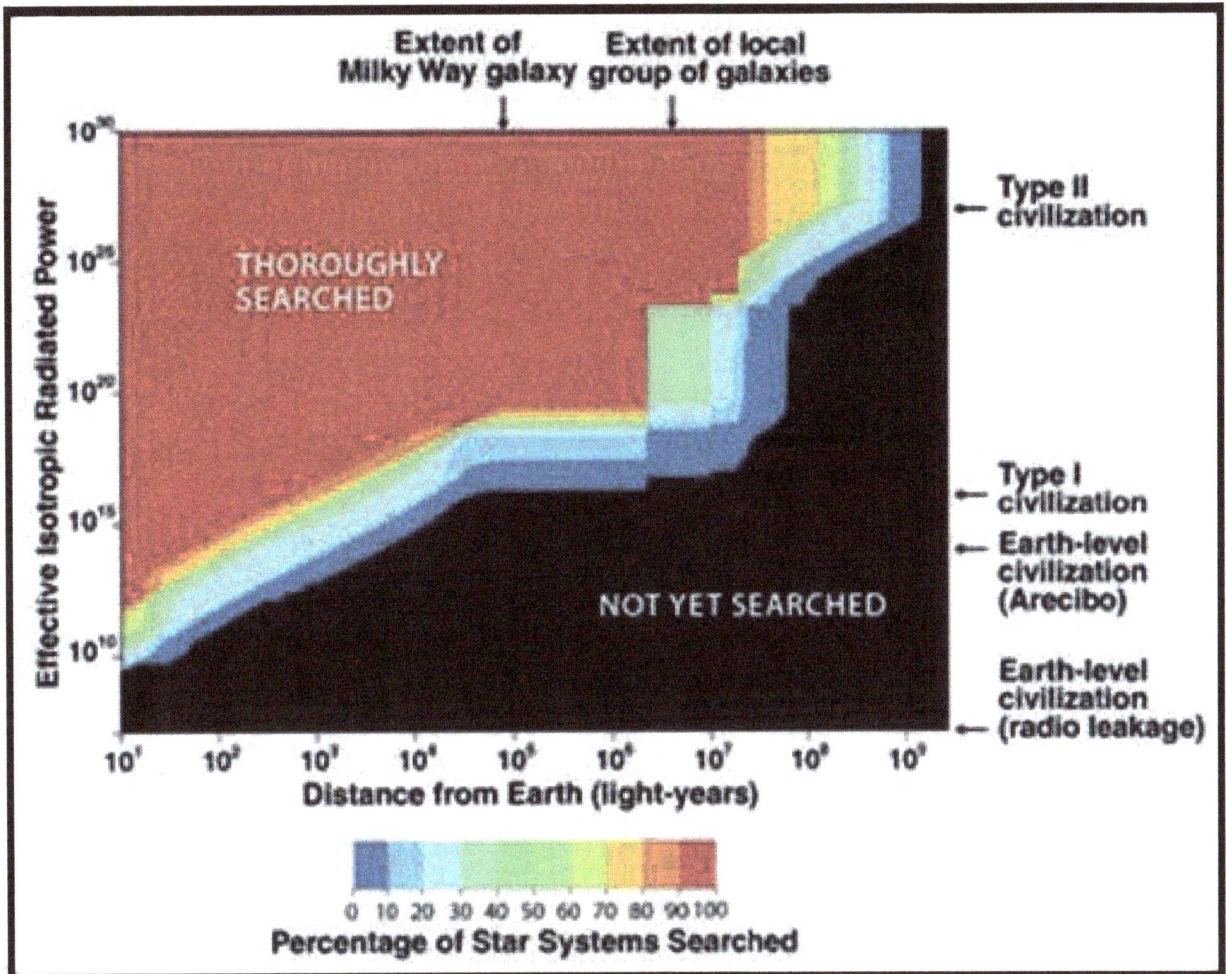

The growth of space exploration and possible galactic colonization based upon Earth-level and Type I and II Civilizations
http://abyss.uoregon.edu/~js/cosmo/lectures/lec28.html

Oddly, this has lead to the misperceived fact that because aliens don't seem to be walking about on our planet, this apparently implies rather illogically that there are no extraterrestrials anywhere among the vast tracts of the Galaxy. This seems to mirror the US Air Force's own position in its assertion that UFOs are misidentified everyday objects seen by the untrained public and because there is no evidence of an Extraterrestrial Intelligence presence therefore, UFOs and ETs don't exist! ***((Of course, this hasn't been a reason or a deterrent to the US Military from stopping its tracking, targeting and shooting down of UFOs or ETSs (Extraterrestrial spacecraft), thereby, recovering them and their interstellar pilots!))*** Many researchers consider this to be a radical conclusion to draw from such a simple observation. Surely, there is a straightforward explanation for what has become known as the **Fermi Paradox**. There must be some way to account for our apparent loneliness in a galaxy that we assume is filled with other clever beings. (Bold italics added by author for emphasis)
http://abyss.uoregon.edu/~js/cosmo/lectures/lec28.html

The assertion that **Extraterrestrial Intelligences (ETI)** do not exist, based on the apparent contradictions inherent in the **Fermi Paradox**, rests upon an unproven and untenable presumption: That ETI are not now present in the Solar System. The current observational status of the Solar System is insufficient to support the assumption that ETI are not here. Most advanced civilisations also would be either invisible or unrecognizable using current human observational methods, so millions of advanced societies may exist and still not be directly detectable by us. Thus, the Fermi Paradox cannot logically be raised as an objection to the existence of ETI until these major observational deficiencies have been corrected.

The **Fermi Paradox**, attributed to a question from Enrico Fermi after a discussion of the possibility of extraterrestrial life during the 1940's, is traditionally formulated as follows: If there are intelligent beings elsewhere, then in time they must achieve the technology of nuclear power and space flight and would explore and colonize the Galaxy, as humanity has explored and colonized the Earth. Thus, they should have been able to travel to Earth, but we see no evidence of such visitations, ergo they cannot exist. Fermi's question "Where are they?" implicitly construes the absence of extraterrestrials on Earth as positive evidence of their nonexistence elsewhere in the Universe. Dormant for many decades, recently the Fermi Paradox has re-emerged as a major modern challenge to the existence of Extraterrestrial Intelligences (ETI). http://www.rfreitas.com/Astro/ResolvingFermi1983.htm

One could say that the *"undead" Fermi paradox has risen as a "rotting corpse" of its former self, like a "walking stinking zombie" spreading its plague of faulty thinking and its pestilent, folderol hypothesis among the living, viable evidence of UFO/ETI visitations to this planet.*

Robert Freitas, scientist and a Senior Research Fellow at the **Institute for Molecular Manufacturing (IMM)** in California from which most of this topic originates, argues that such scientists as **Dyson, Hart,** and **Tipler** support the Fermi paradox that respectively, the Galaxy shows no evidence of ETI technological developments of civilizations, it's still a "wild" place; that Hart's "Fact A", there no ET intelligence from outer space existing on Earth currently; space travel is infeasible, ETI choose not to explore or colonize, ETI are coming but have not arrived as yet, and Earth may have been visited in the past but not currently; advanced ETI would employ mining, organize and colonized the entire galaxy with self replicating machines in the less than 300 million years and because there is no sign of them in our Solar system, hence they do not exist.

Past and recent discussions of the Fermi Paradox make one critical assumption challenged in the present work: That the absence of extraterrestrials or their artifacts on Earth or in the Solar System is an undisputed fact, and, more generally, that advanced technology invariably leads to observable alteration of the large-scale physical environment. Cox implicitly recognizes this when he notes that Hart's "Fact A" is not an empirical fact but rather a theory and Schwartzman rejects the assumption to best reconcile Hart's arguments with those of the advocates of large **N (the number of communicative civilizations in the Galaxy)**. But the extent of our ignorance of potential evidence of ETI in the Solar System is not generally appreciated. We have not yet looked very hard for extraterrestrials in the Solar System, or elsewhere. Until these observational deficiencies are corrected, the Fermi Paradox cannot logically be raised as a valid objection to the existence of ETI.

Freitas in order to establish his resolution arguments against the Fermi Paradox concedes that continuing reports of UFOs in our skies, however controversial, cannot be taken as convincing evidence of extraterrestrial visitation or technology. Hence his discussion is restricted to a brief review of the evidence that ETI are absent from the human observational Universe, excluding Earth. http://www.rfreitas.com/Astro/ResolvingFermi1983.htm

By conceding, what **Freitas** says is the unconvincing evidence of the UFO phenomenon and thus, evidence of any extraterrestrial visitation or technology, he is playing it safe. His decision to not argue in support of this important UFO aspect, merely follows in the same footsteps by those in the government, the military and intelligence communities who marginalize or simply dismiss outright, the whole UFO/ETI subject matter, because it is too highly controversial; filled with too many recriminating counter measures against all those in the professional and scientific communities who would investigate the phenomenon in an unbiased scientific manner. Such has been the intimidations, threats, the loss of tenure, and loss of reputation to those who would dare challenge the status quo and secret agenda of the military industrial complex. http://www.rfreitas.com/Astro/ResolvingFermi1983.htm

Author's Rant: This book however, represents in its entirety, the overwhelming hardcore evidence that should the reader follow from cover to cover and from page to page, he will find that it not only addresses all the arguments brought up in the Fermi paradox but, proves conclusively beyond any doubt, that UFOs are real, that ETI are currently visiting and engaged in contact and communications with random as well as selected individuals and groups of people on this planet, and that there has been past historical contact and ancient involvement in human societal development. But also, there are outposts and possibly large scale colonies and settlements on the Moon and on Mars. In addition, there is even evidence of a Type II Civilization presence within our Solar System. Their massive planet-size spacecraft have already been photographed repeatedly around Saturn and some of the Outer Planets of the Solar System by the Cassini space probe, as well as also being routinely imaged by the SOHO satellite, indicating that these massive objects are in very close orbits around the Sun and appear to be interacting with the Sun in some form of energy transfer!!!

Failure to take into serious consideration the UFO/ETI existence and presence in our Solar System is to set the investigation by real science backwards by decades and delay the technological and social advancement of civilization on this planet. However, with that being stated, let's see where Feitas' arguments in resolving the **Fermi's Paradox** leads us and inject with comments that not only support his position but actually delivers the proof of ETI existence and their presence in our Solar System.

Non-Observability of Advanced Civilizations - Advanced Civilizations May Not be Visible

Why might ETI move out into the Galaxy, and not just stay at home? Possible motives might include physical resource shortages, the need for new habitats for expanding populations *(Lebensraum),* the drive for sociopolitical power or prestige, the thirst for adventure by certain segments of the population, mischief or criminal activity, genuine curiosity about the cosmos and the search for knowledge, religious proselytization of other sentient races, propagation of biological genetic material, and so forth. All of these mechanisms, in one guise or another have proved very effective in driving mankind to colonize his entire planet. Similar drives are likely to motivate a highly successful, intelligent extraterrestrial species, but most may not produce astronomically visible effects. http://www.rfreitas.com/Astro/ResolvingFermi1983.htm

Author's Rant: This argument implies that aliens are afflicted with the same anthropocentric thinking as humans and for humans to even entertain the notion that ETI may even think like us is to illustrate how unprepared we are to engage in contact and communications with them. They are alien after all and their thinking processes will probably seem strangely odd to us. However, we may also be pleasantly surprised to find that they do share with us some of the same higher spiritual qualities for life and a worship of a supreme deity or an all-powerful cosmic force or power. The best position for us is to remain open, receptive and positive, since they have already arrived at our doorstep. We should consider them, the lead dancer in the dance of contact and communications with us as the new dance partner willing to learn the new dance steps, since we have not had the opportunity to engage in this type of contact before.

There are also many reasons why advanced ETI might not engage in colonization, missionary, or exploitative activities. Unless interstellar cargo transport is feasible, the most important resource is knowledge, implying non-invasive exploration rather than physical exploitation. Self-replicating systems would then be used only to build probes and libraries. Even if the material needs of a planetary system full of sentient beings requires the rapacious strip-mining of an entire galaxy, our Solar System would probably be spared if stellar systems with life are exceedingly rare. http://www.rfreitas.com/Astro/ResolvingFermi1983.htm

Author's Rant: Consider the axiom that "Life is where you find it" meaning that life is prevalent everywhere and why not? Why would this seem so strange and unscientific a proposition when most astronomers and scientists have accepted Stephen Hawking's "Big Bang Theory" of the formation of the universe? That from out of nothingness, a single point of quantum explosive matter emerged into the physical matrix of being to become the ever expanding macro cosmic universe of physical reality. Wherein all atomic particles of matter of the microverse bonded to become the known universe and therefore, all matter has its equals and opposites, evenly distributed throughout the universe. What we find and know of on this planet and on some of the other planets in our Solar System, are also replicated in similar forms elsewhere in this Galaxy and throughout the macro universe. So, since life exists on this planet and because all similar life-sustaining matter also exists throughout the universe, because it derives its origin from one single of reality as a result of the Big Bang, therefore, by de facto, life must exist everywhere throughout the universe!

Genes are environment-specific survival instructions, so transmission of pure genetic information seems pointless unless the target planet is terraformed. Interstellar ovum arks, generation ships, and automated bio-regeneration only generate more independent competitors for exactly the same limited galactic resources. Like the bottle-babies of **Huxley's** *Brave New World,* there might be little "parental" attachment or sense of community with cloned societies whose cultural interactions have information feedback loops with delays of 10^4-10^5 years, comparable to the timescale of speciation. In these cases each society would remain isolated from the others and never forge a pan-galactic union, or might choose not to spread their clones throughout the Galaxy.

These kinds of arguments suggest that the mere possession of advanced technology and civilization need not imply astronomically observable effects. **Planetary (Type 1), Stellar (Type II),** and benign or diminutive **Galactic (Type III) Civilisations** are not precluded by existing data(see Section 2.3). Our Galaxy may have millions of such societies. Both ETI and very advanced technologies may exist - indeed, Type II cultures should possess any technical skill we can now imagine. But large-scale observational effects of the astroengineering activities of these organizations lie, for the most part, below the threshold of detectability of instruments employedby present-day mankind. http://www.rfreitas.com/Astro/ResolvingFermi1983.htm

Is There a Cosmic Censorship of Rapacious Galactic Civilizations?

Such explanations for the absence of obvious astroengineering activity in the Galaxy, if ETI exist, cannot conclusively resolve the **Fermi Paradox** because they cannot apply to all ETI at all times after they achieve interstellar flight as required by **Hart**. In theory, some civilizations might be unable to resist their rapacious sociological urges to colonize and to overpopulate, and might proceed to do both at a phenomenal rate. Others may be driven by motives too alien to contemplate, resulting in a "cancer of purposeless technological exploitation". Any one such species in theory would press to the physical limits of the Galaxy, creating an observably artificially-ordered, "tamed" Galaxy which, because of our failure to observe it, supposedly demonstrates the nonexistence of ETI.

This reasoning is unconvincing because it assumes that "one bad apple spoils the lot." Perhaps there can be no bad apples, because they self-destruct before damaging any others. The interstellar abyss may prove an excellent quarantine mechanism. We may see no spoilative **Type III Civilizations** either because some unknown selective process rules out their existence, or because they have never survived long enough to complete their galactiforming programme and leave any major observable effects.

Apparently Type III civilizations that make extensive and exploitative use of highly visible, very advanced technology are very rare or nonexistent. Why might this be? Extinction as a natural phenomenon is quite common on Earth, where biological life has a 99.9% species extinction rate. Social pessimists might argue that the almost instinctual voracity of humans, coupled with our technological capability for self-destruction, may eventually lead to the downfall of our planetary civilization. It is logically acceptable to infer the existence of selective mechanism(s) which result in the "cosmic censorship" of civilizations who exploit glaringly obvious, rapacious technologies to the utmost.

The ultimate outcome of a mastery of fusion, starflight, machine replication and other advanced technologies is not necessarily an exploitative, colonial galactic civilization, nor is the outcome necessarily astronomically observable. An information-intensive culture, for instance, would be very difficult to see. The present observational record can only support the much more restricted conclusion that no rapacious galactic civilizations are currently loose in the Galaxy. http://www.rfreitas.com/Astro/ResolvingFermi1983.htm

Author's Rant: We must also accept the probability of a *"cosmic pecking order"* in the universe. On Earth, much of the population accepts and acknowledges that there is a Supreme Being in the universe that made all that we know, see and understand, including the universe and everything in it and all that is yet to be discovered by humans. We call this Supreme Being by many names, but it is best known as the Divine Creator, GOD! To look at it in another way, since we are talking about the existence of Extraterrestrial Intelligence and life in the universe and the ability to travel and colonize the known universe or at least this Milky Way Galaxy then, think of GOD as the Supreme Extraterrestrial Intelligence (SETI) in the universe!!! By the very existence of a SETI, there must be varying level of intelligences of extraterrestrial life elsewhere all conforming or obeying and carrying out the will of GOD, the Supreme Extraterrestrial Intelligence. There is thus, a pecking order in the universe to carry out the will of the Supreme Being and we, humans are on or near the bottom of the "universal pecking order" (UPO). What this all means, is that most if not all interstellar travelling civilizations must have a moral or spiritual compass that conforms to the universal will of GOD or SETI. Malevolent or hostile imperialistic aliens or ETI are not permitted to interact with less developed intelligent beings or interfere in their development. It also means that the universe simply cannot have thousands of imperialistic agenda-driven ETI running around colonizing every star system in sight, as soon or later, such agendas will bring more than one or two or many civilizations into hostile militaristic confrontation with each other. The universe simply doesn't operate that way and it is contradictory to the Will of GOD and his creation of the universe.

Limits on Observability of Advanced Civilizations

Visual observations of the Galaxy along the plane are limited to a few kiloparsecs; because of interstellar gas and dust obscuration. Even radio VLBI cannot resolve one AU **Dyson Sphere-like structures** beyond a range of a few kpc, and few celestial objects have been mapped to this extreme resolution. High technology activities by ETI such as terraforming and astrophagy might not be immediately obvious at interstellar distances, even to radio astronomers, and certainly a major fraction of the Galaxy could have been colonized without us being able to observe it. Advanced races could have made many starts on transforming the Galaxy but failed for various reasons, yet the observational consequences to humanity would be minimal.

Perhaps sheer scale in constructing observable artifacts is restricted by the Square-Cube law, gravitational tidal forces, available energy sources, and other fundamental physical limitations to astro-engineering projects which we may not yet fully appreciate. For example, large-scale artificial habitats may be more likely to consist of less-observable swarms of small **O'Neill colonies** rather than gigantic monolithic architectures, as this maximizes the ratio of living area to mass. Also, potential but unknown linkage between the existence of ETI and such unexplained astronomical phenomena as the galactic "missing mass," quasars, and exploding galaxies like Cygnus A cannot be overlooked.

A well-ordered galaxy could imply an intelligence at work, but the absence of such order is insufficient evidence to rule out the existence of galactic ETI. The incomplete observational record at best can exclude only a certain limited class of extraterrestrial civilization - the kind that employs rapacious, cancer-like, exploitative, highly-observable technology. Most other galactic-type civilizations are either invisible to or unrecognizable by current human observational methods, as are most if not all of expansionist interstellar cultures and **Type** I or **Type II societies**. Thus millions of extraterrestrial civilizations may exist and still not be directly observable by us. http://www.rfreitas.com/Astro/ResolvingFermi1983.htm

Author's Rant: The absence of an astro-engineered architectural universe by Type II or Type III Civilizations on a scale of gigantic monolithic proportions is another indication that humans simply do not understand the concepts of advance civilizations and their motives for being. We humans still live in a three dimensional paradigm of reality where everything we do and live for is based upon that 3D reality, even though our great Divine Teachers from every religion we follow on Earth have been telling us that are true nature is a spiritual (multi-dimensional) reality. Our primary purpose for life then, is to know God, to know and worship Him, to accept and follow the teachings of His Divine Manifestations for the day and age in which we live in and to carry forward an ever-advancing civilization! It's really that simple!!!

Baha'u'llah in His holy Writings, verse 40 of The Hidden Words of Baha'u'llah proclaims:

"O Son of Man! *Wert thou to speed through the immensity of space and traverse the expanse of heaven, yet thou wouldst find no rest save submission to Our command and humbleness before Our Face."* The Hidden Words of Baha'u'llah by Baha'u'llah; copyright 1954; The National spiritual Assembly of The United States; Printed in the USA; Library of Congress Catalog Card No.54-7328

They May Be in the Solar System

Two general categories of evidence are commonly used to argue that ETI are not present in the Solar System. These are (a) lack of purposeful communications, and (b) lack of physical observables. However, we shall argue that null data in each of these two evidentiary classes *may be due more to anthropocentric assumptions and the poor observational record* than to the absence of ETI in the Solar System. *Finally, a correct assumption by Freitas!! Well done!* (Bold italics added by author for emphasis). http://www.rfreitas.com/Astro/ResolvingFermi1983.htm

Lack of Purposeful Communications

The first evidentiary category is lack of purposeful communications: If ETI were here, they would have used their superior technology to contact us, or to get us to notice them. If they are here, where are they?

Author's Rant: This is a rather presumptuous position by humans towards ETI. We are expecting them to reveal themselves to us which they have done, by their presence in our skies with their spacecraft (UFOs). Yet, we deny their existence and argue among ourselves that we have mistaken their presence for some other everyday common event or object because, the people who expect ETI to reveal their presence would do so before those who represent officialdom!

In other words, if it didn't happen before official circles or wasn't announced officially by those in power, then it simply didn't happen! They are here, you just have to pay attention and maybe they might just communicate to you!!

Ball proposes that we may be part of an interstellar zoo or wilderness preserve, carefully isolated for our own good. **Stephenson** suggests that if intelligence is rare in the Galaxy, it is more likely to be handled with greater circumspection and care. **Kuiper** and **Morris** reiterate the old science fiction notions of culture shock and the idea that we might not be contacted until we reach some intellectual threshold to avoid "extinguishing the only resource on this planet that could be of any value to ETI." **Papagiannis** suggests "confusion and indecision... they might be debating on whether to crush us or help us, postponing their decisions, waiting to see what we are going to do with ourselves." http://www.rfreitas.com/Astro/ResolvingFermi1983.htm

Author's Rant: Again, Freitas is moving in the right direction, although, we are no one's zoo experiment or part of a interstellar wilderness preserve, but we are however, unique as a species, not because we are the only humans in the universe but, because of our uniqueness to being native to this planet and our diversity of spiritual beliefs, our cultures and societies, our many interesting languages, levels of development and technological achievements. The fact that ETI are not chatting up a storm with us is due largely in part to our spiritual development. We are like adolescence still immature, bickering and fighting among ourselves over the resources of this planet. The good news is that handfuls of people globally are maturing a little faster than the majority of humanity and some of them are communicating with ETI. The more we mature and get our act together collectively as a world commonwealth, the more open the contact and communications from ETI will happen on a planetary level. Be Prepared!! The best is yet to come!!!

The author has suggested another more likely possibility - that there is no reason why. They should *not* be silent. It is anthropocentric to assume, for example, that alien spacecraft entering the Solar System on a mission of reconnaissance or self-replication will feel the obligation to announce their presence to us or to request permission to proceed. Probes will probably just ignore us and go on about their business.

It is entirely conceivable that some ETI may not particularly care whether we find them or not, or may actually be interested in communicating with us yet be unwilling to initiate contact unconditionally. For example, they may refuse to speak to beings who occupy only one planet, fearing culture shock to a "One Earth" mentality unaccustomed to dealing with the environmental and cultural relativities of many worlds. The exercise of finding them would serve as an initiation, a kind of "entrance examination" to pass before opening contact. **Platt** suggests that advanced civilizations may be "like the parents who do not talk to the baby until the baby wakes up."

Or, if life is not especially rare in the Universe, then, rather than carefully shepherding our development with extreme circumspection like an endangered species, ETI may adopt a much more casual approach to contact. They may have tremendous confidence in their ability to manage potential contact events to successful resolution, based on numerous previous similar encounters. In this case, humankind might rate neither ultraconservative wildlife management nor heedless astrophagic exploitation. The most likely response would be careful and unobtrusive observation, with no special effort to conceal the alien presence. Base site would be chosen for reasons of efficiency, maintainability, and low environmental risk. http://www.rfreitas.com/Astro/ResolvingFermi1983.htm

Author's Rant: This too, may be closer to the truth of the situation, but it is more likely for the already stated reasons given by this author. We are a maturing species that is about to enter into adulthood in a big way based upon spiritual moral values and this is an exciting time for them to see any emerging intelligent species transition out of imperialistic, resource-base driven nationhood into becoming a global community, a one world order based upon spiritual values in proper balance with a material world.

Of course, these arguments may be multiplied endlessly, illustrating the probative weakness of this particular category of null evidence. While we have no convincing evidence of purposeful communication between humanity and advanced ETI, neither do the null data constitute compelling evidence that ETI are not here because this conclusion is contingent upon unknown choices that may be available to such beings and our assumption that they would wish to communicate.

Lack of Physical Observables

The only true compelling category of evidence is physical observables. For instance, we might observe their spacecraft, their self-replicating machinery, their habitats or strip mines, or a host of other physical manifestations which are the hallmark of technological activity. This evidence also depends on alien choice - ETI might conceal their technology for the same reasons they choose not to communicate. Some activities might be too obvious to disguise, such as planetary xenoforming or astrophagy by voracious swarms of replicating machines, but given a sufficiently advanced technology ETI should be able to conceal most of their mechanisms and activities to high perfection.

The present analysis thus concerns only those ETI who, for whatever reason, choose not to use perfect means to hide all physical evidence of their technological activities. This is the most conservative assumption possible from the standpoint of the **Fermi Paradox**, as the more restrictive assumption that some or all activities are perfectly disguised leads to a trivial explanation for the apparent absence of ETI in the Solar System.

Physical Evidence in the Solar System

What sort of physical evidence might be available for us to find? Material artifacts of a commonplace nature including small tools, debris, garbage dumps, radioactive hot spots, and passive monoliths located on planetary surfaces would be less than 1-10 metres in size. Nobody in the Solar System has yet been comprehensively mapped to this resolution except Earth, which is excluded from the present analysis; hence none of these commonplace items is observable by our present instrumentation. http://www.rfreitas.com/Astro/ResolvingFermi1983.htm

Author's Rant: This is not true! Contrary to Freitas thinking and hypothesizing at this point, Freitas may be either, totally unaware of, or is deliberately not taking into consideration (for reasons known only to himself), the current massive exploration efforts by many nations from Earth of the planet Mars with their high resolution orbiting satellites and the Mars Rovers that are currently exploring the Martian terrain. Next to the Moon, Mars is now the third most mapped planetary body in the Solar System! Venus is fourth. Photographic imaging and mobile ground-based instrumentation are also revealing numerous diverse life forms, both fossilized and currently extant on the Red Planet, along with discoveries of artificial artifacts, both small and monolithic in scale all over the planet surface indicating the possible presence of one or more intelligent species on Mars! (Bold text added by author for emphasis and response purposes).

If an exploitative Type II civilization exists or had ever existed in our vicinity, then the Solar System would have been wholly converted to replicating machines mass, our planets sorted into their constituent elements for transshipment or industrial use and the Sun-stripped of its fuel - and of course we would not be here to discuss the outcome of these activities. Since humanity exists, rapacious nearby stellar civilizations are ruled out, much as exploitative galactic civilizations are provisionally excluded by the observational evidence as discussed earlier. http://www.rfreitas.com/Astro/ResolvingFermi1983.htm

Author's Rant: Here again, there is far too much anthropocentric thinking going on, this is not to deny the possibility of such engineering missions or concepts should be ruled out. In fact, NASA's Hubble, Cassini and SOHO satellites and other space probes have repeatedly shown through raw photographic data that massive planetary-size ETI space vehicles or potentially robotic spacecraft are indeed in orbit or very near to the Sun and to some of the Outer Planets for reasons as yet, unknown to scientists but, their observability and existence is without question, REAL! (Refer to the many pictures in the section titled "In Search of Extraterrestrial Life" within this book which illustrates this reality).

On the other hand, **Kuiper** and **Morris** and **Stephenson** argue that the only plausible interstellar mission which would be launched by a benign civilization would be one of pure exploration, the pursuit of knowledge as a source of wealth. If this is true, then the most noticeable extraterrestrial artifact we might expect to find in the Solar System would be either a self-reproducing machine system which is building and launching interstellar probes bound for other star systems (or the aftermath of such activity) or just the probes themselves parked here in some convenient locale or orbit. http://www.rfreitas.com/Astro/ResolvingFermi1983.htm

Author's Rant: This theory is probably closer to the truth than Freitas realizes but, the reality of benign civilizations extends far beyond this explanation.

Observables of Interstellar Messenger Probes

Detection of probes would be especially challenging, as these could in theory be located almost anywhere. A typical alien probe might be 1-10 metres in size - this is large enough to house a microwave antenna to report back to the senders, and to survive micrometeorite impacts for millions of years, but light enough to fly across the interstellar gulf without consuming unreasonable amounts of energy. http://www.rfreitas.com/Astro/ResolvingFermi1983.htm

Author's Rant: Currently, there is in orbit around the Earth an alien craft called the _"Black Knight"_ (a least the size of the Space Shuttle and possibly larger in size) by Ufologists which has been photographed repeatedly by astronauts from various Space Shuttles and no doubt by the Russian Space Agency and their Cosmonauts. Its reality has been verified but not publicly acknowledged! Other much smaller alien monitoring devices have also been seen and photographed and one was alleged to have been captured by NASA, according to the Michael D. Swords, PhD, a retired professor of Natural Science at Western Michigan University, a Ufologist and a board member of the J. Allen Hynek Center for UFO Studies.

A spherical Solar System boundary enclosing the orbit of Pluto consists of $260,000 \text{ AU}^3$ of mostly empty interplanetary space and 10^{11} km^2 of planetary and asteroidal surface area. To be able to say with any certainty that there is no alien presence in the Solar System, you have to have carefully combed most of this space for artifacts.

Currently the sky has been exhaustively surveyed to perhaps magnitude +14, the **Palomar Schmidt Sky Survey** extends to +21, and the best available magnitude limit for any telescope on Earth is about $m_v = +24$. This means that at best, current surveys from Earth might have detected an unmoving, mirror-shiny, optimally-oriented 10-metre object orbiting 0.01, 0.25, and 1 AU from Earth, respectively. If the artifact is smaller, moving, black, or canted at a different angle then it will be even harder to see.

So we can only scan the nearest 4 AU^3 of space for probes, but we have at least $260,000 \text{ AU}^3$ to search. Even if the Palomar 200-inch telescope was employed exclusively to search for alien artifacts it could reach at most one-millionth of the necessary volume. Orbital space, in other words, is _at least_ 99.999% unexplored for 1-10 metre objects.

A more realistic assessment suggests that the visual detection threshold for alien probes for present-day humanity includes only 10^{-5}-10^{-11} of the potential probe residence volume. This estimate assumes a random search pattern typical of past serendipitous observations which might have discovered alien artifacts if they were present. Radar and infrared measurements cannot substantially improve this current limit.

How about probes parked on planetary surfaces? Of the 0.1 trillion square kilometres of Solar System territory other than Earth, less than 50 million has been examined to 1-10 metre resolution. So 99.95% is still virgin territory as far as a search for extraterrestrial artifacts is concerned. If objects are buried somewhere or floating in a Jovian atmosphere, there is almost zero chance we could have found them up to now. Even huge 1-10 kilometre artificial alien habitats occupying the **Asteroid Belt** would appear visually indistinguishable from asteroids to terrestrial observers, and the Belt population itself is poorly catalogued. So it is exceedingly unlikely that we would have spotted an extraterrestrial artifact anywhere in the Solar System unless it was desperately trying to get our attention. And why should it bother to do that? http://www.rfreitas.com/Astro/ResolvingFermi1983.htm

Author's Rant: See the last stated response above. When UFOs or more correctly, ETVs (Extraterrestrial Vehicles) piloted by visiting ETI as reported by millions of people world-wide are factored into this discussion, We see that not only are they present here and now, but they have been trying to get our attention and there is now a growing number of accounts where ETI are communicating with humans1 This author can testify to this fact having had contact and communications with ETI!

Observables of Self-Replicating Systems

Observation of an operating replicating machine system would be only marginally easier. Likely sites are the **Asteroid Belt** and the outer Jovian and Saturnian moons. Recent technical studies suggest individual replicating systems may be 100 metres in diameter or less, so a factory system for building probes should not exceed 0.1-1 km in size, again well beyond our ability to see it except on the Moon and portions of Mars. Ignition of fusion rockets to propel daughter probes out of the Solar System is detectable using amateur telescopes, but the observation window is very small and of very short duration. Self-reproducing probes should be able to replicate a whole generation in 1000 years or less, and be quickly on their way, so only mining pits and small debris may remain at this late date.
http://www.rfreitas.com/Astro/ResolvingFermi1983.htm

Author's Rant: Lunar photographs taken by the US, Russia, China that show what only can be described to be buildings, colonies settlements and mining complexes and open mining pits on the Moon as well as massive equipment of some particular function as yet unknown but, their proportions is on scale a few kilometers in size! We are talking about gigantic equipment, here!

The total mass of probes needed to explore even the entire Galaxy is astonishingly small. If each self-replicating probe, mass fully-fueled about 10^{10} kg makes 10 replicas during each of 11 generations, enough to span the entire Galaxy, that is 10^{11} x 10^{10} kg = 10^{21} kg or about the mass of Ceres, the largest known asteroid. If the Solar System carried the burden of manufacturing all 10^{11} probes to explore the entire Galaxy, how could we know if one Ceres-size asteroid had ever been removed from the Asteroid Belt?
http://www.rfreitas.com/Astro/ResolvingFermi1983.htm

Author's Rant: It is hard to image self-replicating space probes operating with conventional rocket powered propulsion as a viable method to explore the galaxy in order to colonize. The technology needed would have to be far more advance, capable of operating with a minimum light speed propulsion system.

What has not been considered is that the universe is of such an incomprehensible size with many bodies like galaxies that are also massively huge which means that numerous stars and numerous planets are far larger than our own Sun or our Earth. In fact gigantism is the norm; therefore, giant beings may also be the norm, if indeed there is a universe filled with Extraterrestrial sentient life! When travelling between stars or across the galaxy or even between galaxies, why think small? Thinking in scales of gigantic proportions is more likely the reality of the universe. That means that spacecraft may also, be mammoth in size,

like the size of a small or large moon or even planet-size. An advanced civilization of giant beings may actually take uninhabitable and internally magnetically-dead moons and convert them into travelling spaceship worlds filled with their kind. Far fewer such worlds would be required to populate the galaxy than would be required with the type spacecraft as has already been hypothesized in this discussion. One or two worlds could be sent into a star system with its represented inhabitants from the old home world on board have been living on or inside the lunar body. These moon-size worlds would not need to be self-replicating as the inhabitants could mine the natural resources of the moon to build smaller spacecraft and for other life sustaining needs.

As they entered the star system they could park or orbit their lunar spaceship worlds around any planet and then they could begin exploring the planets in that solar system with smaller spacecraft that may be just a few miles or kilometers in size. Mission accomplished! One or two of these new worlds may be suitable for habitation, even with some minor to major re-adjustments through terraforming.

Having proposed this other possibly as perhaps, more realistic, the fact is that a Type I or Type II may have reasons as theorized above for self-preservation, imperialistic expansion needs or the acquisition for natural resources, biological or minerals, etc. However, it is more likely that civilizations between Type II and Type III may actually have less need for the physicality of the universe and may through technological means or higher development of consciousness, spirituality or intellect have transition into to an altered reality beyond the physical universe. They may have evolved to the point of taking constructs of ideas and bring them into physical reality thus, bypassing the need to mine raw material for construction purposes. They would seemingly manufacture anything they want out of nothingness or from the pure thought energy matrix.

We are talking about highly advanced technological and spiritual beings that have evolved to living and operating from the spiritual or interdimensional realm of existence that has little need of a physical universe except as explorers and possibly to ensure that a developing and promising species advances forward in civilization and to fulfill their destination!

And take the argument one step further. Assume that one million extraterrestrial civilizations each pillage the Solar System for materials to build and launch their own million independent probe networks, each covering every star in the Galaxy. The total requirement is still only

10^6 x 10^{21} kg = 10^{27} kg, about the mass of Jupiter. It is doubtful we could say for certain if even this much matter had been stolen away sometime in the remote past.

More likely, starfarers won't be greedy and will require each target star system to supply no more than one new generation of replicants. This is only 10^{11} kg, enough to fill one 1-kilometre crater 40 metres deep or to make one 400-metre-wide asteroid. We'd never miss the mass.

Even more likely, ETI will erect their self-replicating probe factories in uninhabitable star systems to avoid disturbing us and just send nonreproducing exploratory probes here. In this

case, no local -mass would be missing and there'd be no surface debris either.
http://www.rfreitas.com/Astro/ResolvingFermi1983.htm

Freitas' Summary and Conclusions

In this paper the critical assumption that ETI are not present in the Solar System, essential to the logic of the **Fermi Paradox** argument for the nonexistence of extraterrestrial intelligence, has been challenged. Observations suggest that the Galaxy may be devoid only of rapacious, exploitative civilizations, not of all ETI, at the present time, and that current knowledge of the Solar System is insufficient to support the presumption that ETI are not here. Hence the conclusion that we are alone, based on the Fermi Paradox, **is without foundation**.

Author's Rant: We have proven that ETI of the non-rapacious kind are now currently visiting the Earth as recounted in the voluminous UFOs cases reported annually by every strata of society and by every ethnic culture, globally.

The observational record is sufficiently incomplete that major galactic technologies may yet have gone unnoticed. The strongest conclusion justified by the data is: Due to some special selection effect, galactic civilizations that make extensive and exploitative use of large-scale, astronomically observable, very advanced technology are very rare or nonexistent. This does not rule out planetary- or stellar-scale technical civilizations, nor does it preclude benign, appropriate-technology" galactic civilizations. A well-ordered galaxy would imply the existence of ETI, but the absence of such order cannot prove the nonexistence of ETI.

Author's Rant: We have also proven that ETI civilizations do exist on some planetary bodies within our Solar System with evidence of their presence in the form of their spacecraft in and around planets and the Sun also, artifacts on the planet along with many large monuments as indication of their ancient and current existence.

Two categories of evidence might show that ETI are not present in the Solar System: The admitted lack of purposeful communications (uncompelling because it depends upon unknown choices available to ETI and our assumption that they would wish to communicate) and physical observables.

Author's Rant: We have also demonstrated that ETI are already in contact with some people on Earth and not through the conventional ways of radio telescopes or through radio or television, etc. which also demonstrate their willingness to communicate through physical presence of their spacecraft and in person, through telepathy and by crop circle pictographs.

Astronomical searches to date are extremely unlikely to have observed physical evidence of ETI if it exists. The most easily observable extraterrestrial artifacts would be self-reproducing machine systems, the aftermath of their activity, or interstellar messenger probes. Orbiting probes could be located almost anywhere within a search volume of which only 10^{-5}-10^{-11} has been reliably scanned by terrestrial telescopes, and few observations to date would have been capable of detecting operating replicating machine systems. Even building and launching probes to every star in the Galaxy generates an unobservably small local mass deficit, but most likely the probe factory and its mining pits would be located elsewhere so probes would remain the only observables.

Author's Rant: The astronomical search for ETI life and their artifacts within the Solar System, in the form of Galactic self-replicating space probes while a laudable and meritorious effort of science is nevertheless misdirected and thus far, a failed mission, as the UFO/ETI phenomenon has evidentially been determined to be much closer to home. It seems they have already found us and not we, who have discovered them out in the far reaches of space! While mainstream science will always be a part of an advancing civilization, it has in this particular instance been trumped by the amateur investigative efforts of the common people!

Evidence from astronomy that could confirm or deny the presence of ETI in the Solar System is scanty or nonexistent. Cosmological data arguably also are incomplete, consequently cannot prove the nonexistence of all galactic-scale ETI. The Fermi Paradox cannot logically be raised as an objection to the existence of ETI until these major observational deficiencies have been corrected.
http://www.rfreitas.com/Astro/ResolvingFermi1983.htm

So, if Extraterrestrial Intelligences do exist based upon Freitas' argument in rebuttal to the Fermi Paradox and given the recent conclusion reached by NASA that there are trillions upon trillions of star systems each with a system of planets too numerous to calculate, then by reasonable logic there are billions or trillions upon trillions of life forms in the universe. With this assessment many of these planets are populated with intelligent life forms and some of those Extraterrestrial intelligence must have by now reached and landed on Earth.

The obvious question is why haven't they landed say on the White House Lawn or in Empire Stadium or near Parliament Hill or any other political area or large public venue? The fact is they have landed upon our Earth as testified by millions of witness accounts and reports from around the globe. This book is all about those ETI who have landed and engaged humanity; however, the best answer may actually come from Dr. Steven Greer in the YouTube video link below: https://www.youtube.com/watch?v=aYk_nPoIu8E

CHAPTER 62

I SPY AN ASTRONOMICAL COVER-UP

In the nineteenth century, there was a new spirit of the age which fill people the world over with a new hope for a brighter future pregnant with the possibilities that a new day had dawned upon mankind. It was a time of fulfillment and promise. People were exploring the world around them and searching the heavens above like never before, in search of answers to questions since that first time, when we came down from out of the trees and put on fur-skins and discovered fire.

In that century of new beginnings, there were many books and articles written by many world famous authors like **Jules Verne**, **Herbert George Wells** and **Edgar Rice Burrows** who prophetically wrote about life in the near and distant future and about the possible inhabitants on other planets. Many people caught up in the excitement of the age of possibilities came to believe that intelligent beings may live on the Moon, Mars, and Venus. Since manned space travel was still nothing more than entertaining science fiction and actual spaceflight to other planets was more than a century away from becoming fulfilled prophetic reality, some people suggested some possible ways to signal the extraterrestrials that may be on the Moon or Mars, long before the invention of radio.

Johann Carl Friedrich Gauss

One of these individuals was a gentleman named **Johann Carl Friedrich Gauss,** an 18th century German mathematician and physical scientist who contributed significantly to many fields, including number theory, algebra, statistics, analysis, differential geometry, geodesy, geophysics, electrostatics, as well as astronomy, and optics. He became known as "the Prince of Mathematicians" or "the foremost of mathematicians" and considered as the "greatest mathematician since antiquity". Gauss had a remarkable influence in many fields of mathematics and science and is ranked as one of history's most influential mathematicians.

Gauss proposed a possible method to communicate with extraterrestrial life in the solar neighbourhood in constructing a giant triangle and three squares in the form of the **Pythagoras** could be drawn on the Siberian tundra. The outlines of the shapes would have been ten-mile wide strips of pine forest; the interiors could be rye or wheat.
http://en.wikipedia.org/wiki/Carl_Friedrich_Gauss

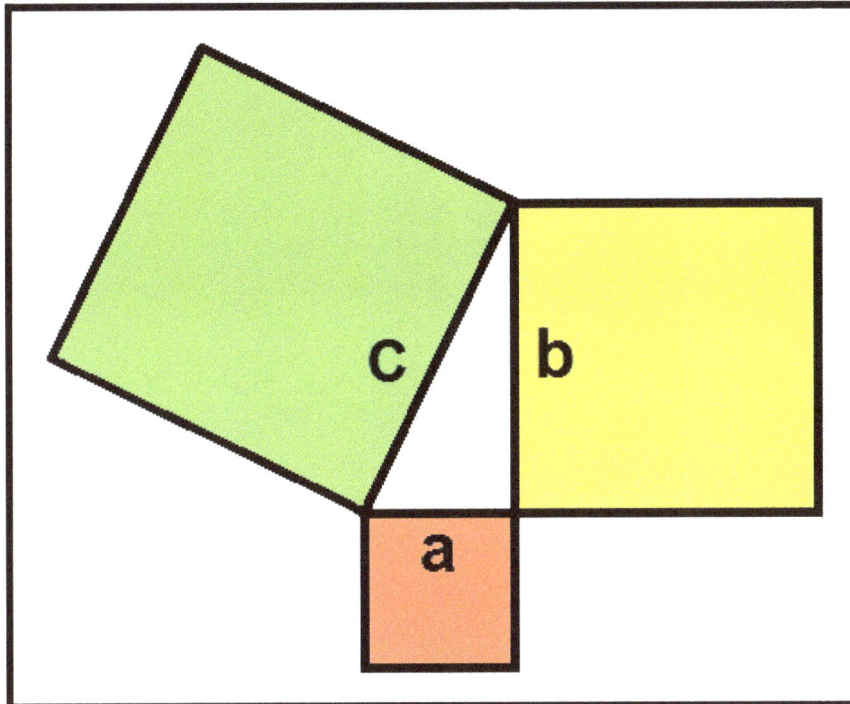

The theorem of Pythagoras was considered as one way to communicate to Extraterrestrials that there are signs of intelligent life on Earth.
(Google Image)

Joseph Johann Littrow

Joseph Johann Littrow proposed using the Sahara as a blackboard. Giant trenches several hundred yards wide could delineate twenty-mile wide shapes. Then the trenches would be filled with water, and then enough kerosene could be poured on top of the water to burn for six hours. Using this method, a different signal could be sent every night.

Franz von Gruithuisen

Meanwhile, other astronomers were looking for signs of life on other planets. In 1822, **Franz von Gruithuisen** thought he saw a giant city and evidence of agriculture on the moon, but astronomers using more powerful instruments refuted his claims. Gruithuisen also believed he saw evidence of life on Venus. Ashen light had been observed on Venus, and he postulated that it was caused by a great fire festival put on by the inhabitants to celebrate their new emperor. Later he revised his position, stating that the Venusians could be burning their rainforest to make more farmland.

Needless, to say, people's imaginations including astronomers and mathematicians were filled with all kinds of outlandish possibilities but, some would be proven much later to have a basis of truth behind these earlier observations. Astronauts landing on the Moon would not only confirm the presence o Extraterrestrial spacecraft following them to the Moon but landing nearby or hovering in close proximity to Apollo astronauts but, also their discoveries of remarkable lunar structures and buildings on the Moon's surface. But, let's continue with what history records from what astronomers thought they knew in their time.

Giovanni Schiaparelli

By the late 1800s, the possibility of life on the moon was put to rest. Astronomers at that time believed in the **Kant-Laplace hypothesis**, which stated that the farthest planets from the sun are the oldest—therefore Mars was more likely to have advanced civilizations than Venus. It was evident that Venus was perpetually shrouded in clouds, so the Venusians probably would not be very good astronomers. Subsequent investigations focused on contacting Martians. In 1877 **Giovanni Schiaparelli** announced he had discovered **"canali"** ("channels" in Italian, which occur naturally, and mistranslated as "canals", which are artificial) on Mars—this was followed by thirty years of Mars enthusiasm. (See earlier work in the previous section on Schiaparelli).

As it turns out, as seen in the previous section of this book, that from photographs taken by various Mars reconnaissance orbiters, that there are tube-like structures crisscrossing much of the Martian landscape. There appears to be glass-like tunnels perhaps transporting water which in fact may be a type of "canali" or canal. Schiaparelli, it seems may have been right after all!

Charles Cros

The inventor **Charles Cros** was convinced that pinpoints of light observed on Mars and Venus were the lights of large cities. He spent years of his life trying to get funding for a giant mirror with which to signal the Martians. The mirror would be focused on the Martian desert, where the intense reflected sunlight could be used to burn figures into the Martian sand.

Nikola Tesla

Perhaps, there is no other inventor-scientist before or since the time of **Nikola Tesla** has embodied his attributes of scientific brilliance and showmanship, as a pioneer making so many inroads into the frontiers of new science. A one man "tour de force" who dazzled the world with his incredible genius and exceptional insights giving us some of the most amazing discoveries and inventions developed well over a hundred years ago that are still benefitting mankind today! His inventions were so far ahead of current thinking by his contemporaries that he was considered a man light years ahead of his time, to the point that many of his astounding inventions have either been suppressed by the military industrial complex and U.S. patent office under the "catch-all blanket" of **National security** or have been quietly utilized in secret covert military armourment and weaponry!

Nikola Tesla was a Serbian-American inventor, electrical engineer, mechanical engineer, physicist, and futurist best known for his contributions to the design of the modern **alternating current (AC)** electricity supply system.

Nikola Tesla, A genius light years ahead of his time

Tesla gained experience in telephony and electrical engineering before immigrating to the United States in 1884 to work for **Thomas Edison**. He soon struck out on his own with financial backers, setting up laboratories and companies to develop a range of electrical devices. His patented AC induction motor and transformer were licensed by George Westinghouse, who also hired Tesla as a consultant to help develop a power system using alternating current. Tesla is also known for his high-voltage, high-frequency power experiments in New York and Colorado Springs which included patented devices and theoretical work used in the invention of radio communication, for his X-ray experiments, and for his ill-fated attempt at intercontinental wireless transmission in his unfinished **Wardenclyffe Tower project**.

Nikola Tesla mentioned many times during his career that he thought his inventions such as his **Tesla coil**, used in the role of a **"resonant receiver**, could communicate with other planets and even observed repetitive signals of what he believed were Extraterrestrial radio communications coming from Venus or Mars in 1899.

Tesla sitting in his Colorado Springs laboratory with his "Magnifying transmitter" generating millions of volts and producing 7-metre (23 ft) long arcs.

At his lab, Tesla observed unusual signals from his receiver (which he interpreted as 1—2—3—4), which he later believed were extraterrestrial radio wave communications coming from Mars. The signals were substantially different from the signals those that he had noted from the noise of storms and the earth. Specifically, he later recalled that the signals appeared in groups of one, two, three, and four clicks together. Tesla was highly criticized upon revealing his finding. Tesla had mentioned that he thought his inventions could be used to talk with other planets. It is debatable what type of signals Tesla received or whether he picked up anything at all. Research has suggested that Tesla may have had a misunderstanding of the new technology he was working with, *(This is not likely as the man was a genius and knew about electricity and frequencies, signals, etc. more than any other man at the time)* or that the signals Tesla observed may have been from a non-terrestrial natural radio source such as the Jovian plasma torus signals. *(This is a reasonable possibility).* Other sources hypothesize that he may have intercepted **Marconi's** European experiments—in December 1901, Marconi successfully transmitted the letter S (dot/dot/dot, the same three impulses that Tesla claimed to have received from outer space while at Colorado in 1899) from Poldhu, England to Signal Hill,

Newfoundland, Canada—or signals from another experimenter in wireless transmission.
http://en.wikipedia.org/wiki/Nikola_Tesla

Whatever the real reason for these "signals", they were eventually considered to be terrestrial radiation, at least that what the public was told at the time.

Around 1900, the **Guzman Prize** was created; the first person to establish interplanetary communication would be awarded 100,000 francs under one stipulation: Mars was excluded because **Madame Guzman** thought communicating with Mars would be too easy to deserve a prize.

When the Martian canals proved illusory, it seemed that humans were alone in the solar system.
http://en.wikipedia.org/wiki/Communication_with_Extraterrestrial_Intelligence

Astronomers posit that there must be life in the universe but, there has been no evidence for its existence as yet. Why is that? Because astronomers haven't personally made the discovery of life outside of this planet, regardless of the fact that there are voluminous accounts of UFOs piloted by Extraterrestrial Intelligences have already discovered us, here on Earth. If no astronomer or scientist has made the discovery in their chosen field of investigation then, it simply doesn't exist! This is hardly what you would call the scientific method of investigation!

We are told by such world-renowned astronomers like **Seth Shostak**, leading astronomer of the **SETI (Search for Extraterrestrial Life)** program that even, with the use of a worldwide array of radio telescopes listening for any telltale interstellar communication from ETs using a similar earth-based technology, no reply or response has yet been received. No photographs from any orbiting space telescope like Hubble have captured a single grinning alien face, no NASA Mars Rover has discovered any signs of Martian life and no interstellar travelling spacecraft like **Voyager I** and **II** have encountered any alien spacecraft or alien civilizations during their long space odysseys. Even, if there were aliens trying to communicate to Earth, Shostak figures it would be him who would be the first to decode their signals and communicate back to them!

Oooh! The height arrogance!!

The best that astronomers will say or indicate today, is that they do expect perhaps, at some point in the near future, to discover Extraterrestrial life at least at the microbial level. Making contact and greeting an intelligent and sentient ET life form may still be hundreds or even, thousands of years away. Sounds pretty much small-minded and bleak for future prospects!

Here, once again, we perceive the squeaky cogs and gears of the machinery of big deception and cover-up are in motion dispensing more lies, denials, disinformation and misinformation to the general public even when the public's gaze is cast towards the far reaches of outer space for answers.

Some astronomers who happen to be in the employment of NASA or working for the Departments of Defense or with national intelligence agencies, like the NRO, have already discovered life on other planets in our Solar System. At this point, we should not be surprised to

learn that radio signals have been received from alien signals using large arrays of radio telescopes; that alien spacecraft currently orbiting our planet have also been photographed by various space telescopes and satellites orbiting the Moon, Mars, Saturn and even the Sun! Even, NASA's **Mars Exploration Rover (MER) Mission**, an ongoing robotic space mission involving three rovers, *Spirit* and *Opportunity*, and more recently, *Curiosity* that are currently exploring the planet Mars have found evidence not only of former ancient life on Mars but, that it still exists there to this day. This information, however, at this time is being covered up and suppressed from public awareness by NASA and its scientists and is only coming to light by private UFO researchers and amateur astronomers.
http://en.wikipedia.org/wiki/Mars_Exploration_Rover

Once again, **Dr. Steven Greer** in an interview with **Art Bell** on his late night "Coast to Coast" radio program tells how one of his witnesses revealed that ETI signals have been routinely received by radio telescopes, in particular, the **BETA Radio Observatory** at Harvard University. It was reported to Greer that the editor of **Sky and Telescope** magazine happen to walk into the control room at the observatory one night at the instant an unusual signal was received that was determined not to be from any known earth-based or satellite sources nor from any planetary or natural astronomical origin. The editor was informed that this observatory had received artificial signals on several occasions that were of Extraterrestrial in origin yet, they had never made any public announcement that this was routinely occurring fully realizing of course that these telemetry events were of monumental importance to science.

The fact is that many planetary bodies and their orbiting satellites or moons within our solar system may all have life on them in some shape or form. This is based upon the fact that our Sun or **Sol** is a fixed or stable star. The statement made by **Baha'u'llah:**

"Know thou that every fixed star hath its own planets, and every planet its own creatures, whose number no man can compute." (Gleanings, No. LXXXII)

is scientifically correct yet, science has still to prove this assertion through discovery! And, if they have made this discovery then, they have not disclosed anything about its existence to the press or the general public. Gleanings from the Writings of Baha'u'llah; 1939; Baha'i Publishing Trust; Wilmette, Illinois, USA; Library of Congress Card No. 52-14896

Brookings Institute Report to NASA

NASA sponsored a study called the **Proposed Studies on the Implications of Peaceful Space Activities for Human Affairs**, also known as the **Brookings Institute Report to NASA** in the late 1950s by the **Brookings Institution** to identify long-range goals of the United States space program and their impact on American society. The final resulting report was submitted to NASA in 1960 included a discussion of the implications of the discovery of extraterrestrial life and intelligence. The **"think tank"** at Brookings pointed out that the reactions of both individuals and governments would probably depend on their social, cultural, and religious backgrounds, as well as on the nature of the discovery. The finding of lower life forms or "subhuman intelligence," it was thought, could quickly be assimilated. However, more profound effects might follow from the discovery of intelligence that was superior to our own.

It was pointed out that in anthropological files, there are many examples of societies, sure of their place in the universe, which have disintegrated when they had to associate with previously unfamiliar societies espousing different ideas and different life ways; others that survived such an experience usually did so by paying the price of changes in values and attitudes and behavior. Since intelligent life might be discovered at any time via the radio telescope research presently underway, and since the consequences of such a discovery are presently unpredictable because of our limited knowledge of behavior under even an approximation of such dramatic circumstances, two research areas can be recommended:

1. Continuing studies to determine emotional and intellectual understanding and attitudes – and successive alterations of them if any – regarding the possibility and consequences of discovering intelligent extraterrestrial life.

2. Historical and empirical studies of the behavior of peoples and their leaders when confronted with dramatic and unfamiliar events or social pressures ...

Such studies, the report recommended, should take account of public reactions to past hoaxes (see **"Moon Hoax),** waves of unidentified flying objects (see saucer flap of 1947; **"Washington Invasion**), and events such as the **"1938 War of the Worlds" radio play**. They should also consider how best to inform the public of contact with an extraterrestrial intelligence, or whether such knowledge should be withheld. International relations, the report concluded, might be permanently altered because of *"a greater unity of men on earth, based on the 'oneness' of man or on the age-old assumption that any stranger is threatening."*

The report warned that America should prepare to meet the psychological impact of such a revelation; it could bring about profound changes, or even the collapse of our civilization. Of particular interest is that the report outlines the need to investigate the possible social consequences of an Extraterrestrial discovery as already mentioned but also, ***that such a discovery should be kept from the public*** in order to avoid political change and the possible "devastating" effect on Scientists themselves due to the discovery that many of their own most cherished theories could be at risk.

In reality, the public is far more adaptive to new potentially life-altering changes than the institutions of officialdom that act on their behalf. **The institutions that would be most greatly affected are the governments, the military departments, the science community and the religious institutions of every nation.** It would appear that from such a world-shaking revelation that any announcement or disclosure of the existence of alien life found in the universe, particularly one visiting our planet, people it seems would fair far better than our hallowed institutions.

It would mean that **those who are in the powerful seats of officialdom would lose those long cherished seats of power and authority**, as the common folk turn to new leaders for vision and direction. ***This would be the real panic and fear that would grip our institutions*** and one of the reasons for the excessive cover-up and secrecy behind so much of the existence of the UFO/ETI reality!

This is nothing new, however, in the history of mankind as such world-shaking events have occurred repeatedly, particularly when a new religious revelation comes along to displace or makes obsolete a former religious belief system. The people in power and authority, primarily the leaders of nations, the monarchies and ecclesiastics are threatened by the potential reformation of social structure that they resist such changes vehemently, in order to maintain and buttress their status of power and control over the people. Prophets of such revelations and revolutionary social change have ultimately paid the price with their death or their imprisonment or their exile.

Consider for a moment, what is currently taking place globally by the powerful, wealthy corporate elite and the **Military Industrial Complex** in securing and buttressing their seats of power over the masses, by maintaining an ongoing secrecy and suppression of knowledge related to unidentified flying objects and Extraterrestrial Intelligences visiting the planet. This is no mere coincidence but, an orchestrated agenda!

The **Brookings Report** has done more to shaping the direction of NASA and space exploration that any other scientific document. Ufologists feel strongly that when the Military Industrial Complex started to exert pressure upon the public space agency to implement the recommendations of the Brookings Institute then, censorship and honest public disclosure of all of NASA's discoveries were covered up and suppressed. NASA was no longer under civilian control but rather, by the military and *in reality, it has always been that way from its inception.* NASA was never a civilian program but, a military program from the start and it has become more obvious over the years when you recognize that almost all astronauts have a military position as opposed to those few who have a professional or doctorate designation. NASA is an adjunct, a division of the **Department of Defense (DOD)** and thus, beholding to the Department of Defense and not to the taxpaying public!
http://www.bibliotecapleyades.net/brooking/brookings_report.htm and
http://www.daviddarling.info/encyclopedia/B/BrookingsStudy.htmlnd

The Brooking Report argument simply doesn't fly and that document is nothing more than their political and legal cover for what they are doing because there is more going on here, than is being revealed.

When you partner the **Brookings Report** with **Executive Order 10501**which is a document for safeguarding official information in the interests of national defense of the United States, you end up with two very powerful legal documents that support the arguments, protocols and agendas of **National Defense** (which is the same as saying **National Security**). When considering National Defense perspectives it also includes not only the Military but, geo-economic and geopolitical interests as well. These two documents gives them legal justification to cover up the greatest story in the history of mankind.

CHAPTER 63

DO ASTRONOMERS SEE UFOS?

For astronomers who wanted to work in the space program, they had to abide by NASA's party line protocols which meant not disclosing publicly any discoveries or information related to Extraterrestrial Intelligence whether encountered in space, on a planetary body or elsewhere. When astronomers are formerly asked to give their position on Extraterrestrial life visiting the Earth they will deny the probability of it happening yet, when they are by themselves and the subject of **UFOs** and **ETs** comes up in private discussions, they do indeed believe that ETs are visiting us but, will not state so publicly.

The late astronaut, **Brian O'Leary** tells an interesting story of when several hundred astronomers met in the summer of 1968 at an evening reception at the **University of Victoria** in British Columbia to discuss matters of astronomy, astrophysics, space exploration and the eventual discovery of life in the universe. At some point during the night session, a young astronomer decided to take a break outdoors for some fresh air when he spotted a UFO/ ET craft hovering and maneuvering above the science building where the meeting was being held. Excitedly, he immediately ran back into the conference room exclaiming that a UFO was hovering over the building and that everyone should come out to witness it.

Here was a large contingent of scientists and astronomers who wanted to know if other life existed in the universe and when the proof presented itself, merely sat there laughing with incredulity, thinking it was some kind of prank and went back to their somber lectures and discussions. Perhaps, they were too embarrassed to admit to the possibility that Extraterrestrial intelligent life had arrived to give them the answer they were searching for.

It seems that when proof is offered to men of science, no matter how incredible the information or evidence may be, it would appear that investigation and the scientific method falls to the wayside and is ignored. **Erwin Schrodinger**, pioneer in quantum mechanics and a philosopher of science, wrote:

"The first requirement of a scientist is that he be curious. He should be capable of being astonished and eager to find out." **J. Allen Hynek; The UFO Experience; published 1972.**

Here was an opportunity for scientists who require physical proof of almost anything which can be observed, measured, weighed, and reproduced and yet, just sat there without any interest or curiosity to investigate a claim. For a group of scientists, physical proof would allow them to get their hands dirty, to be able to kick the tires on their experiments and stand around congratulating themselves from the practical applications of their theories. This would have been the end of a good day's work or lecture session, even, a good week or month or year of research for any scientist but, unfortunately, it never became reality for these astronomers. When it comes to UFOs and **ETI**, they don't fit into this nice, neat equation of reality at least, the reality of science that we know!

Can we paint all astronomers with the same paint brush of apathy, timidity, and complacency when it comes to investigating claims of UFOs? Perhaps, only those astronomers whose academic tenure or employment in the military or with NASA hangs precariously by a thin thread or while for believing in such things as UFOs and aliens. It may be that astronomers fear professional or public ridicule.

Lord Martin Rees

A **Huffington Post** article about some astronomers' professional beliefs, that only kooks and cranks see UFOs and not astronomers is indicative of professional arrogance but nevertheless, quite revealing of the basic attitude of scientists toward the subject. (Posted: 09/19/2012 9:28 am Updated: 09/19/2012 9:28 am):

"LORD MARTIN REES: ALIENS FASCINATE EVERYONE, BUT ONLY KOOKS SEE UFOS"

Lord Martin Rees
http://www.ast.cam.ac.uk/~mjr/

Astronomers -- the men and women who study the stars, galaxies and beyond -- have an almost universal agreement that there's nothing of scientific interest when it comes to the subject of UFOs.

In fact, many professionals who gaze into the heavens actually speak about UFO sightings with disdain -- even though countless military officials, heads of state, newsmakers and commercial pilots have said they've seen unidentified flying objects.

"No serious astronomer gives any credence to any of these stories," said **Lord Martin Rees,** the official U.K. astronomer royal.

Rees, author of the new book, "From Here To Infinity: A Vision For The Future Of Science," told **The Huffington Post** that, while "everyone's fascinated by aliens," he's in favor of the ongoing SETI Institute program -- the search for extraterrestrial intelligence.

"We should look at all possible techniques," Rees said. "We've no idea what's out there, and so we should look for anything that might seem to be some sort of artifact rather than something natural."

While Rees said he hopes real extraterrestrials will be detected within the next 40 years, he's completely and "utterly unconvinced" that any ETs have been visiting Earth.

"I think most astronomers would dismiss these," Rees said. "I dismiss them because if aliens had made the great effort to traverse interstellar distances to come here, they wouldn't just meet a few well-known cranks, make a few circles in corn fields and go away again."

Seth Shostak

Nevertheless, in a recent **HuffPost** blog, SETI Institute senior astronomer **Seth Shostak** wrote about the daily emails he receives from people describing "alien sightings, extraterrestrial plans for Earth, and agitated screeds about the reluctance of scientists to take the whole subject seriously."

Shostak said very few of those emails "are penned by hoaxers. The correspondents are sincere, and many simply wish to help us in our search for evidence of extraterrestrial intelligence."

While Shostak may at first appear sympathetic to folks claiming that if aliens are already here, he said he sees no real evidence.

"It's hard to believe that the aliens have cleverly arranged things so that only governments can find convincing evidence of their presence......The fact is if you're certain that our planet is hosting alien visitors, the way to gain acceptance for your point of view is to prove it."

To which may be added that *"absence of proof is not proof of absence"!*

The science of the unknown requires the scientific investigation of the scientist to prove the reality of the unknown that has entered and impacted its presence upon our reality. It is not up to the common man to do the work of the scientist but, he may be the recipient of the scientist's discoveries.

Seth Shostak

James McGaha

Another astronomer, **James McGaha,** said he agrees with Rees' contention that UFO reports should be dismissed.

"I totally agree with that. Rees is making what I call an elegant argument," McGaha, a retired Air Force pilot and director of the Grasslands Observatory in Tucson, Ariz., told HuffPost.

McGaha held a top secret security clearance during his career and was directly involved with classified projects at the now-legendary Area 51 military facility in Nevada. He's a fellow of the Royal Astronomical Society and has discovered numerous asteroids and comets.

McGaha is also a **UFO debunker**.

"I've got over 40,000 hours looking at the night sky, and I've never seen -- not once in 50 years -- something that I didn't know what it was. I've seen many strange things, but never anything I couldn't identify," McGaha said.

"Most professional astronomers don't look at the sky. Most of them sit in a room with computers and image the sky, so they don't ever go outside and look, and many of them don't know how to actually identify the constellations in the sky -- it's unfortunate, but that's the way it is."

McGaha gives no credence to any of the tens of thousands of UFO sightings or encounters that reliable people have reported.

James McGaha
http://www.centerforinquiry.net/speakers/mcgaha_james

"I don't think there's a single observation or report that I'm aware of that indicates an alien spacecraft -- not one," he said. "And I've looked at all of the important cases.

"Scientists are very sensitive to evidence and data. What physical evidence is there that an alien spacecraft has actually ever visited Earth? There is none, zero. And until there is -- an artifact, a piece, an alien -- scientists will not even think about investigating this," he said.

McGaha concedes -- just a little -- that UFOs, or alien ships, "are not impossible. If there's an intelligent race out there, they could build a spacecraft and get here. It would be very difficult, would require enormous resources and motivation to do it. It's not impossible, but highly unlikely."

Nick Pope

Others disagree about the rift between scientists and the UFO "community." *"I'm dismayed by the lack of understanding between the two groups,"* said journalist **Nick Pope**, who used to investigate UFOs for the **U.K. Ministry of Defense**.

"Unless they've studied the phenomenon, astronomers are no more qualified to talk about UFOs than Ufologists are to talk about, say, titanium oxide production in K-type stars. It's like marine biologists and oil company executives saying they understand each other's fields because they're both looking for something in the sea. The sea is about the only thing they have in common. So it is with astronomers, Ufologists, and the sky," Pope wrote in an email to HuffPost.

Initially, a skeptic, Pope's years at the Ministry of Defense led him to believe that at the very least, the UFO subject was *"worthy of proper scientific research. Especially, where one has for example photos, videos or radar data because such things can be studied in a proper scientific manner. It's unscientific to ignore data simply because it doesn't fit your worldview."*
November 19, 2012; Huff Post – weird News; Canada and
http://www.huffingtonpost.com/2012/09/19/lord-martin-rees-aliens-ufos_n_1892005.html?utm_hp_ref=mostpopular

Nick Pope
http://www.nickpope.net/

The arguments presented above by some of today's leading astronomers are based more on a limited worldview as Nick Pope has stated and fortunately, it is not the prevailing opinion of all scientists and astronomers. There is actually as much disparity in opinion on the subject matter as there is between Ufologists and professional debunkers. The fact is that numerous astronomers down through the centuries have recorded sightings that were not conventionally astronomical by nature, in other words, what they saw and reported was considered to be unidentified flying objects.

Such sweeping statements from well-regarded scientists are endlessly frustrating to the UFO researcher. Particularly given that interest in UFOs actually drives some people to study astronomy! Unfortunately, the idea that only kooks see UFOs is prevalent. Not only do astronomers see UFOs in America, but many astronomers see UFOs at a dramatically greater rate than the general population. *(I am one of those people who had many UFO sightings early in my life that when I went to university my major in science was Astronomy).* [Author's bold italics added for emphasis]. http://www.huffingtonpost.com/dan-mack/astronomers-ufo_b_1901480.html

J. Allen Hynek

On August 6, 1952, Astronomer **J. Allen Hynek** offered the USAF's **Project Blue Book** a **"Special Report on Conferences with Astronomers on Unidentified Aerial Objects."**

**J. Allen Hynek was the astronomer and chief UFO Investigator
to the US Air Force's Project Blue Book**
http://www.ufoaliendata.com/project-blue-book/

Hynek interviewed some 45 astronomers on their experiences and opinions about UFOs during and following the meeting of the American Astronomical Society that June. Hynek provides some notes on each individual astronomer and their opinions. Here's what some astronomers thought in 1952:

Astronomer Y (no sightings) said, "If I saw one, I wouldn't say anything about it."

Astronomer II (two sightings) "is willing to cooperate but does not wish to have notoriety," Hynek reports. http://www.huffingtonpost.com/dan-mack/astronomers-ufo_b_1901480.html

Astronomer OO: (one sighting) was a new observer at the Harvard Meteor Station in New Mexico. He saw two lights moving in parallel that were too fast for a plane and too slow for a meteor. He had not reported his observation.

Hynek concluded: *"Over 40 astronomers were interviewed of which five had made sightings of one sort or another. This is a higher percentage than among the populace at large. Perhaps this is to be expected, since astronomers do, after all, watch the skies."*

Dr. Peter Sturrock

The next data point came in 1973, when Peter A. Sturrock conducted a surveys among members of the San Francisco chapter of the **American Institute of Aeronautics and Astronautics (AIAA)** (1175 questionnaires mailed, 423 returned) and found no consensus concerning the nature and scientific importance of the UFO phenomenon, with views equally ranging from "impossible" to "certain" in reply to the question, "Do UFOs represent a scientifically significant phenomenon?"

In a later larger survey conducted among the members of the **American Astronomical Society** (2611 questionnaires mailed, 1356 replies), Sturrock found out that opinions were equally diverse, with 23% replying "certainly", 30% "probably", 27% "possibly", 17% "probably not", and 3% "certainly not" to the question if the UFO problem deserves scientific study. Sturrock also asked in the same survey if the persons surveyed had witnessed any event which they could not have identified and which could have been related to the UFO phenomenon, with around 5% replying affirmatively. Sixty-two astronomers responded that they had observed something they could not explain which could be relevant to the UFO phenomenon. Eighteen of those witnesses said they had previously reported their sightings, and Sturrock notes that a 30% reporting rate is greater than what is assumed for the average population.
http://en.wikipedia.org/wiki/Ufology#Surveys_of_scientists_and_amateur_astronomers_concer

In 1977 when **Dr. Peter Sturrock** made a questionnaire about UFO attitudes and experiences. Again the target was the members of the American Astronomical Society. The paper was eventually printed in 1994 in the ***Journal of Scientific Exploration***, a peer-reviewed but decidedly non-mainstream publication.

In 1980, a survey of 1800 members of various amateur astronomer associations by Gert Herb and J. Allen Hynek of the **Center for UFO Studies (CUFOS)** found that 24% responded "yes" to the question "Have you ever observed an object which resisted your most exhaustive efforts at identification?"
http://en.wikipedia.org/wiki/Ufology#Surveys_of_scientists_and_amateur_astronomers_concer

Section 3.2 of the paper titled "Comparison of Witnesses and Non-Witnesses" contains a table showing that UFO witnesses were actually more likely to be night sky observers (professional or amateur) while non-witnesses are more likely to not even be observing the skies at all!

**British astronomer Peter Sturrock conducted many surveys
to find out if astronomers do see UFOs**

http://www.aka-shakespeare.com/peter-sturrock.php

TABLE 3.2

Percentages of each group by category. "Night observers" may also observe sun. "Others" observe sun but not night sky.

	Non-Witnesses	Witnesses
Not Observers	35	16
Professional Night Observers	50	63
Amateur Night Observers	8	16
Other Observers	7	5

Sturrock also includes commentary from the astronomers, and again a sample is illuminating:

C1. *"I object to being quizzed about this obvious nonsense. Unidentified = unobserved or factually unrecorded: modern mythology. Too much respectability given to it."*

C10. *"I find it tough to make a living as an astronomer these days. It would be professionally suicidal to devote significant time to UFOs. However, I am quite interested in your survey."*

C16. *"**Menzel** and **Condon** have made further investigation unnecessary unless some really new phenomena are reported ... There is no pattern to UFO reports except that they predominantly come from unreliable observers."*http://www.huffingtonpost.com/dan-mack/astronomers-ufo_b_1901480.html

At the 1969 **AAAS UFO Debate** organized by **Carl Sagan, Dr. Franklin Roach** delivered a paper on "Astronomers' Views on UFOs." He focuses on the lack of publicized UFO reports from major astronomical research programs that constantly monitor vast swaths of the sky. He offered a quote from the famous astronomer **Gerard Kuiper**:

"I should correct a statement that has been made that scientists have shied away from UFO reports for fear of ridicule... A scientist chooses his field of inquiry because he believes it holds real promise. If later his choice proves wrong, he will feel very badly and try to sharpen his criteria before he sets out again. Thus, if society finds that most scientists have not been attracted to the UFO problem, the explanation must be that they have not been impressed with the UFO reports."

As the comments from the above surveys show, Kuiper was idealizing the behavior of younger scientists. Lacking his prestige and tenure, they were less willing to suffer mockery from their peers.

The final data point comes from the Soviet Union. "Observations of Anomalous Atmospheric Phenomena in the USSR: Statistical Analysis" is a report by **L.M. Gindilis, D.A. Men'Kov, and I.G. Petrovkaya.** It was published by the **Soviet Academy of Sciences** in 1979 and translated into English by NASA as Technical Memorandum no. 75665 in 1980 and later distributed by CUFOS. It is a statistical analysis of over 200 raw UFO reports in the Soviet Union. Three-quarters of the reports come from their massive wave of UFO reports in 1967.

Section 3, "Observers and Witnesses of Observations," contains some very interesting data. They note that "contrary to the widespread fallacy, there is a highly significant percentage of astronomers among the observers." By comparing the number of UFO observers from a given occupation with census data, the authors arrive at an **"Activity Coefficient."** A higher coefficient indicates a group is reporting more UFOs than expected by population.

At the time, approximately .002% of Soviets over the age of nine were astronomers. Yet they accounted for 10 reports in the Soviet dataset. This yields an activity coefficient of 7500 [Note: NASA's translation reads 7000]. Undergraduates had a coefficient of 3, maintenance workers .9 and Students .02. The Soviet numbers are clear: astronomers report UFOs at astronomical rates. http://www.huffingtonpost.com/dan-mack/astronomers-ufo_b_1901480.html

Astronomers see UFOs. Unless we think they are kooks *simply because they saw a UFO*, the data shows that Lord Rees is incorrect. In the United States, astronomers who observe UFOs on their instruments fear ridicule from other scientists and the press. Despite the aura of illegitimacy around UFOs, the data indicates that astronomers even report UFOs at noticeably greater rates than laypeople.

Better and more recently survey data is clearly desirable. Hynek's survey was informal, Sturrock's is 35 years old, and the Soviet analysis is done on unvetted reports. Only Sturrock's paper was subject to peer-review. But as we have seen, it takes considerable courage for a scientist to brave "career suicide" and study UFOs despite proclamations that the subject is off-limits.

There is a lot of hard work that needs to be done if science hopes to understand the UFO phenomenon. It would be particularly useful to adopt the Soviet activity coefficient and apply it to other databases. Hopefully, Lord Rees hasn't scared too many people away from applying the scientific method to UFO reports. http://www.huffingtonpost.com/dan-mack/astronomers-ufo_b_1901480.html

There is a list of UFO sightings by astronomers that was developed from one that was published in **U.F.O.I.C. Newsletter**, Australia, and reprinted in 1973, the Fall issue of "***Flying Saucers***". It is an extensive list *48 pages long* that summarizes sightings (about 372 as of September 2nd, 2004) by many famous and not so famous astronomers from November 17, 1623, to October 1, 2000. It is well worth reading and certainly dispels the myth by debunkers that astronomers don't see UFOs. Some of the more well-known astronomers who have seen UFOs down through the ages are **Johannes Kepler, Edmund Halley, Tiberius Cavallo, Francis Arago, Pastorff, Schmidt, Dr. Lincoln La Paz, Clyde Tombaugh, Dr. Seymour I. Hess**, and amateur astronomer and UFO contactee: **George Adamski.** http://www.scribd.com/doc/16805639/A-List-of-UFO-Sightings-by-Astronomers.

The list comes from many sources as indicated below:

Richard L. Thompson 'Alien Identities' (orig. source 'The UFO Evidence');
'Flying Saucers Over America'(collection of newspaper cuttings);
Robert Loftin 'Identified Flying Saucers';
'Flying Saucer Review's World Roundup of UFO Sightings and Events';
Desmond Leslie's section of 'Flying Saucers Have Landed';
 Richard Hall 'From Airships To Arnold';
Harold T. Wilkins 'Flying Saucers on The Moon' and 'Flying Saucers Uncensored';
Jimmy Guieu 'Flying Saucers Come From Another World';
Kevin Randle 'Project Blue Book Exposed';
 '165 Little Known UFO Sightings';
Loren E. Gross' series 'UFOs: A History';
Lt. Col. Wendelle C. Stevens (Ret.) & Paul (Moon Wai) Dong 'UFOs Over Modern China';
M. K. Jessup 'The UFO Annual';
Paris Flammonde 'UFO Exist!';
G. McWane & D. Graham 'The New UFO Sightings';
Michael Hervey 'UFOs Over the Southern Hemisphere' & 'UFOs: The American Scene';
Wendelle C. Stevens & August C. Roberts 'UFO Photographs Around The World';
Michael David Hall 'UFOs: A Century of Sightings';
Jan L. Aldrich 'Project1947';
Peter Paget 'UFO-UK' & 'The Welsh Triangle';
Ion Hobana & Julien Weverbergh 'UFO's From Behind The Iron Curtain';
Charles Fort 'New Lands' & 'Book Of The Damned';
'The Dictionary of Scientific Biography';
Irene Granchi 'UFOs and Abductions in Brazil';
Coral & Jim Lorenzen 'UFOs: The Whole Story';
Jacques &Janine Vallee 'Challenge To Science';
Nicholas Redfern 'A Covert Agenda';

Kenneth Arnold & Raymond Palmer 'The Coming of the Saucers';
Michael Hesemann 'UFOs The Secret History';
Paul Stonehill 'The Soviet UFO Files';
Roger H. Stanway & Anthony R. Pace 'Flying Saucer Report - UFOs Unidentified Undeniable';
George Leonard 'Somebody Else is on the Moon';
Richard M. Dolan 'UFOs and the National Security State';
Randles 'The Pennine UFO Mystery'
Arthur Shuttlewood 'Warnings From Flying Friends' & 'The Flying Saucers' & 'UFO Magic In Motion' & 'UFOs: Key To The New Age';
Michael Hesemann 'UFOs: The Secret History';
Frank Edwards 'Flying Saucers - Serious Business';
Fred Steckling' We Discovered Alien Bases On The Moon';
William R. Corliss 'Mysterious Universe';
L'Astronomie Bulletin de la Societe Astronomique De France;
Flying Saucer News;
Mundo Monitor;
The Emergency Press;
Spacelink;
UFO Chronicle;
Uranus;
UFO Universe;
Awareness;
Bufora Journal;
Understanding Yearbooks 1965 /68;
Flying Saucer Review
Also: National UFO Reporting Center, edited by Peter Davenport (NUFORC)
UFO Roundup, ed. Joseph Trainor (UFORup)
Filer's Files, ed. George Filer (FF)

It should be obvious to anyone that most astronomers are professional observers and are able to distinguish between astronomical phenomena and manmade flying objects from those that are unidentified flying objects. Astronomers watch the skies more often than the general public because it is their profession to do so.

They photograph stellar objects constantly, take astrophysical measurements of distant stars and galaxies, analyze the data, develop theories and mathematical formulae and draw conclusions about the universe we live in. They search for life in the universe by listening for radio signals and sorting out the potential WOW signals thought to be Extraterrestrial communications (using radio communication technology no less!) from all the background noise of the universe. They work with other scientists, physicists and engineers to develop and build satellites and spacecraft to send to the Moon and other planets in our Solar System. These spacecraft orbit other worlds or land upon planetary surfaces, travel over the land testing for mineral composition or for the existence of water; even searching for any signs of life, microbial or more complex higher life forms.

CHAPTER 64

RADIO TELESCOPES (SETI), SATELLITES, SPACECRAFT, RADIO AND TELEVISION, ETC.

If You Send a Signal Into Space Will Anyone Answer It?

Imagine that you've dropped a round stone into a placid lake and you observe the ever-expanding ripples generated by the stone's initial point of entry into the lake. What you would see is many ripples spreading out into ever larger concentric rings and you will also notice is that the size and strength of the ripples are diminishing in inverse proportion to the distance from the epicentre. In like manner when a radio signal leaves the Earth out into space, it too expands as a sphere or bubble as it travels further away from the Earth. This expanding sphere of radio signals, however, also diminishes in frequency strength the further it travels out into interstellar space. This science fact is a science fiction staple often portrayed in TV shows and Hollywood movies like a time machine or as time travel. The further you get from earth, the further back you go in the history of radio and TV broadcasts. http://zidbits.com/2011/07/how-far-have-radio-signals-traveled-from-earth/

This concept is depicted in the beginning of the movie "Contact", as an ever expanding "bubble" of broadcast radio signals from the Earth travels outward at the speed of light. The first of these early radio transmissions were short range experiments that used simple clicks and interrupts to show transmission of information in the 1890s. In 1900, **Reginald Fessenden** made the first — though incredibly weak — voice transmission over the airwaves. The next year saw a step up in power as **Guglielmo Marconi** made the first ever transatlantic radio broadcast.

This means that at 110 light-years away from earth — the edge of a radio "sphere" which contains many star systems — our very first radio broadcasts are beginning to arrive. At 74 light-years away, television signals are being introduced. Star systems at a distance of 50 light-years are now entering the *"Twilight Zone"*.

The question that needs to be asked at this point is *"Will any Extraterrestrial life within that radio sphere detect us?"*

While it's interesting to imagine how far our radio signals have traveled into space, it's extremely unlikely that an alien civilization will be able to catch the latest episode of *"I Love Lucy"*. This is thanks to the **inverse square law**. In layman's term, it's a form of signal degradation.

As radio signals leave earth, they propagate out in a waveform, much like our example of dropping a stone in a lake, the waves diffuse or "spread out" over distance thanks to the exponentially larger area they must encompass. The area can be calculated by multiplying length times width which is why we measure it in square units – square centimeters, square miles, etc. This means that the farther away from the source, the more square units of area a signal has to "illuminate". http://zidbits.com/2011/07/how-far-have-radio-signals-traveled-from-earth/

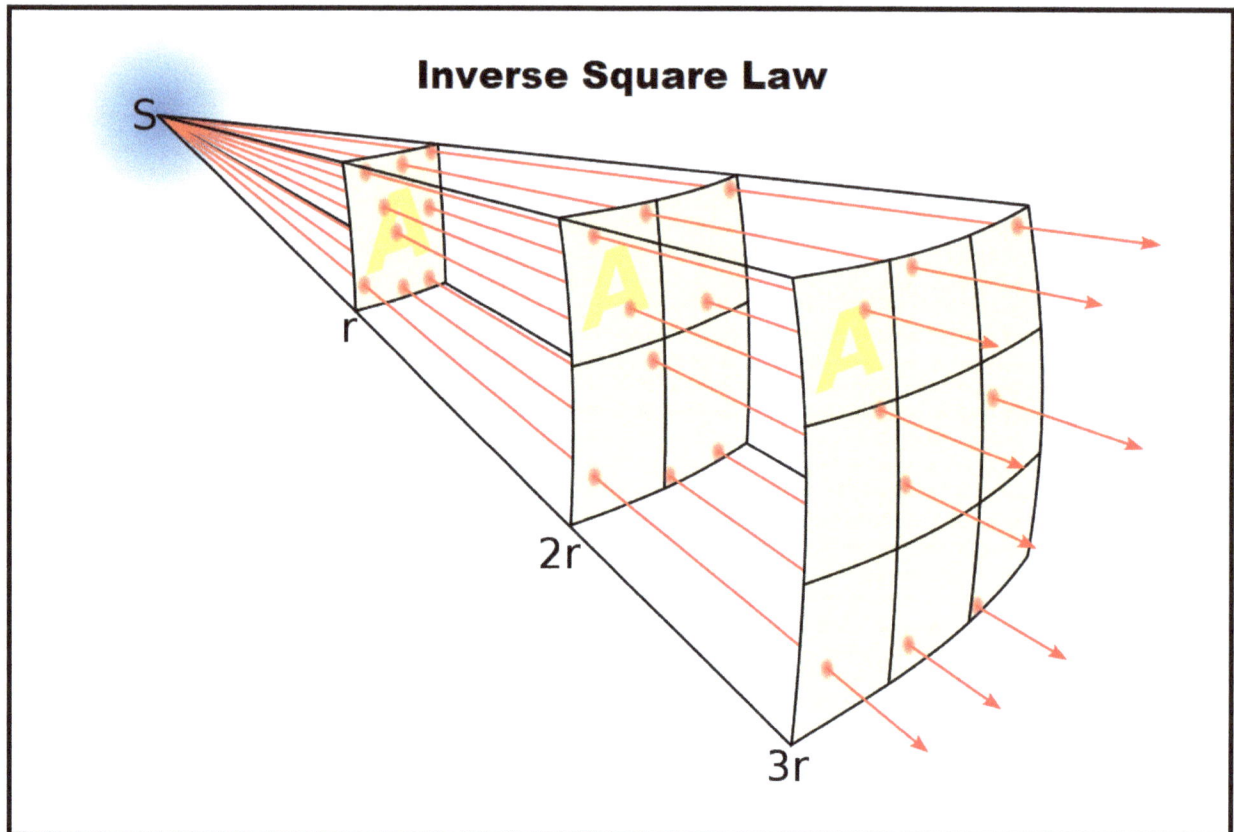

Inverse Square Law

A radio signal's strength is inversely proportional to the square of the distance it travels in which case it diminishes over ever greater distances as indicated in the above sketch
(Google Image)

Because of this inverse square law, all our terrestrial radio signals become indistinguishable from background noise at around a few light-years from earth. For a civilization, only a couple hundred light-years away, trying to listen to our broadcasts would be like trying to detect the small ripple from a pebble dropped in the Pacific Ocean off the coast of California – from Japan.

The simple mathematics, physics, and obvious truth that every first year university student knows is that any radio or microwave or light-wave transmissions, whether laser amplified or not, cannot exceed the speed of light and therefore, is not the best method to communicate between stars with any interstellar civilization!

Seth Shostak, NASA, and every astronomer with a degree also knows this basic law of physics! Surely, this has got to be the height of audacity and arrogance from a poorly conceived scientific program that was doomed to failure from the outset!

So, the really *Big Question* that jumps to mind is *"Why does SETI bother to listen to radio signals in space?"* http://zidbits.com/2011/07/how-far-have-radio-signals-traveled-from-earth/

"And why does SETI still plead and expect another financial handout from the US Government when it knows that it will never ever receive any signals back from an alien civilization that may intercept it, even at 100 light years away?

"And if ETI have answered Earth's transmissions, have they replied with a personal visit in spacecraft that we call UFOs to our planet and if so, has SETI or the government acknowledged their visit?"

While no alien civilization is likely to pick up our television or radio broadcasts unless they're within a few light-years, radio signals can be focused and amplified. *Most of our broadcasts were not intended for "detection in space."* Radio signals can be aimed, focused and amplified to mitigate signal degradation for interstellar communication. *These signals would also eventually degrade* but, are able to travel much, much further before degradation occurs. Hundreds of light-years or more depending on how much power is used.

It's now becoming possible to detect the atmospheric composition of extrasolar planets. This breakthrough has allowed researchers to narrow down our hunt for earth-like worlds. It's quite possible that an advanced alien culture can also do this, and detected an abundance of water in *our* atmosphere. If they have, they may have sent a focused radio message in our direction. If we're not listening, though, we may just miss it. http://zidbits.com/2011/07/how-far-have-radio-signals-traveled-from-earth/

The Drake /Green Bank Equation

Astronomers are essentially explorers in search of life on other planets! The search for extraterrestrial life has become more scientifically fashionable these days among astronomers, biologists, and the general public due in part to two physicists, **Giuseppe Cocconi** and **Philip Morrison** who published in September 1959, an intriguing article titled, "Searching for Interstellar Communications" in the British weekly journal Nature.

Radio telescopes argued Cocconi and Morrison had become sensitive enough that it might be possible to pick up interstellar transmissions broadcasted into space by an Extraterrestrial civilization. They suggested that such messages may be might be transmitted at a wavelength of 21 centimeters (1,420.4 megahertz), which is the wavelength of radio emission from neutral hydrogen, the most common element in the universe. Other intelligences, who may think like us, might see this as a logical landmark in the radio spectrum where Earth scientists would think to look.

Some months later, radio astronomer **Frank Drake** began the first systematic search for intelligent signals emanating elsewhere in the universe. Drake used the 25-meter dish of the National Radio Astronomy Observatory in Green Bank, West Virginia, to slowly scan frequencies in the 21-cm wavelength range. In July 1960 after a few months of listening to two nearby Sun-like stars: Epsilon Eridani and Tau Ceti, his program known as **Project Ozma** ended unsuccessfully.

Not being discouraged, **Drake** organized a meeting with a select group of scientists to discuss the prospects and pitfalls of the "search for extraterrestrial intelligence" — nowadays abbreviated SETI. In November 1961, ten radio technicians, astronomers, and biologists convened for two days at Green Bank.

In attendance was a young **Carl Sagan** and Nobel Prize winner in chemistry from Berkeley, **Melvin Calvin**. http://www.ufoevidence.org/documents/doc270.htm

It was in preparing for this meeting that Drake came up with his famous equation known as the **Drake Equation** or the **Green Bank Equation** named after the place in which the meeting was held:

$$N = R \times fp \times ne \times fl \times fi \times fc \times L$$

The simple but fascinating multiplication equation breaks down a great unknown into a series of smaller, more approachable unknowns expressed thusly:

N   is the **number** of "observable civilizations" that currently exist in our Milky Way Galaxy.
R   is the **rate** at which stars have been born in the Milky Way per year,
fp   is the **fraction of** these **stars that have** solar systems of **planets**,
ne is the average **number of "Earth-like" planets** (potentially suitable for life) in the typical solar system,
fl   is the **fraction** of those planets on which **life** actually **forms**,
fi   is the **fraction** of life-bearing planets where **intelligence evolves**,
fc   is the **fraction** of intelligent species that produce interstellar **radio communications**, and
L   is the average **lifetime** of a communicating civilization in years.

There are some obvious problems with this equation that are not readily agreed upon by astronomers or biologists like the number of communicating intelligences cannot be so easily discerned. http://www.ufoevidence.org/documents/doc270.htm

The rate of star formation in our galaxy is approximately one per year, **R = 1** this is with certainty. In fact, astronomers have recently determined that stars formed at a higher rate several billion years ago, when the stars that might now bear intelligent life were being born. So, a value of **R = 3** is more realistic.

The next factor, **fp**, is probably smaller than one: not every star can have planets. On the other hand, if a star has a planetary system, it seems plausible that two or three of its planets and moons will have liquid water and be potentially suitable for the origin of life, so maybe the product of **fp** and **ne** is close to **1**. *(There is a problem here. A star that is not stable may go nova before the planets have had a chance to form completely or to develop life. A stable star may have developed and formed life but, water may not be the constituent substance of life on the planets. Astronomers and biologists need to start to think outside the box!)* [Bold italics added by author].

Baha'u'llah, the current Manifestation of God (prophet) for this day and age and founder of the **Baha'i Faith** makes an intriguing pronouncement regarding life in the universe:

"Know thou that for every fixed star it hath its planets and every planet, its creature whose number no man can compute"!

By this divine insight, Baha'u'llah makes it clear that every fixed star is a stable star and as such, a star that is stable will have planets and "*those planets will have life*" to such a degree that calculating that numerical reality would be an impossibility!

Now, before everyone goes off into the deep end of contentious debate, realize that divine knowledge which is spiritual knowledge is the same as scientific knowledge which comes from one source! In other words, spiritual knowledge is knowledge through revelation and inspiration and scientific knowledge is knowledge through discovery and invention, yet these twin pillars of knowledge come from that divine source of all knowledge, God!

Therefore, science and religion are one. They agree with each other and are in harmony. They are in truth, one and the same and this is another divine law and teaching from the Baha'i Faith.

From this understanding, science now has a stepping stone in which to prove the reality that life exists everywhere in the universe and is not unique to our planet!

Recent discoveries that many or most young stars are surrounded by planet-forming disks, and detections of scores of actual planets orbiting nearby Sun-like stars since 1995, confirm what astronomers had already suspected: planets are common.

So-called **"proto-planetary disks"** are routinely detected by infrared observations and are seen directly in, for instance, **Hubble Space Telescope** photographs of the **Orion Nebula**, one of the most prolific star-forming regions in our part of the Milky Way. Sub-millimeter wave observations have shown much more tenuous dust disks around many older stars, including Drake's first target, Epsilon Eridani. Many of these disks are doughnut shaped. According to many theorists, the central holes can only be swept clear by planets accreting gas and dust from the disk's inner portion. In addition, some of the disks (including Epsilon Eridani's) show distortions that may directly indicate a planet circling in their outer regions.
http://www.ufoevidence.org/documents/doc270.htm

Optimistically, Darwinian process of natural selection eventually favors the evolution of intelligence, life will form wherever it can **(fl = 1),** and that no intelligent civilization would exist for long without discovering electricity and radio and feeling the urge to communicate **(fc = 1).** *(Again, this is an assumption that aliens with an ounce of intelligence will develop along the same lines as humans, this is anthropocentric thinking. ETs may have developed some other form or type of communication technology that may not be purely electrical, it could be a telepathic/consciousness interface with electronics!)*

In this most optimistic case, the Drake equation boils down to the simple observation that **N = L** (the average lifetime of technological civilizations, in years). If **L** is, say, 100,000 years, there would be about 100,000 chatty civilizations in our galaxy. And that's assuming that only one arises during a planet's entire multi-billion-year lifetime. *(The chances that 100,000 alien civilizations use radio communications technology is too incredible to even contemplate, it is* *more astronomical in its concept than the Drake equation suggests! I would suggest that another interstellar civilization using this type of technology is thousands of light years away from us, that they would have switched to some other communication technology early in their development by the time we discovered that they had used it at one time).*

That figure of 100,000 would mean there is one radio-emitting civilization right now per 4 million stars — reason enough to tune in on the heavens and start hunting for them. If they were scattered at random throughout the Milky Way, the nearest one would probably be about 500 light-years from us. A two-way conversation would require a time equal to a good fraction of recorded human history, but a one-way broadcast might be audible. *(Here, we come to the crux of the problem in using radio communication; it's just too damn slow for interstellar communication! Radio frequencies or electrical energy can only travel at the speed of light – 186,000 miles per second or 300,000 kph! Saying, "Hello, How are you?" would mean that we and possibly another generation or two or more will have ceased to exist before we received a reply assuming they speak English or one of the traditional Earth languages or a mathematical language that's recognizable!)*

However, in the 50+ years of SETI's lifetime, it has failed to find anything, even though radio telescopes, receiver techniques, and computational abilities have improved enormously since the early 1960s, *including SETI @home.com utilizing the power of the home computer to aid in the search for ETs*. (Italics added for emphasis by author). Granted, the "parameter space" of possible radio signals (the possible frequencies, locations on the sky, signal strengths, frequency drift rates, on-off duty cycles, etc.) is vastly larger than the tiny bit that has yet been searched. But we have discovered, at least, that our galaxy is not teeming with very powerful alien transmitters continuously broadcasting near the 21-centimeter hydrogen frequency. No one could say this in 1961. http://www.ufoevidence.org/documents/doc270.htm

Author's Rant: I would suggest that the possibility of using similar communication technology is not out of the question between diverse interstellar or intergalactic civilizations but, the probability is unlikely and most advanced civilizations would recognize the limitations of light speed technology and would have developed a civilization based on transluminal and consciousness interfaced technology. It would be easy to overestimate the values of one or more of the Drake parameters. Star stability is a factor, as is the possibility that life may not need to originate or grow on a planet. Recently, astronomers have discovered water molecules and even, the building blocks of life in meteorites, comets and the gaseous clouds of nebulae which has lead to the Hypothesis of Panspermia. Life it seems is where you find it or where you look for it! Life it seems is everywhere in the universe!

Consider this concept: if **Stephen Hawking** is correct about the **"Big Bang Theory",** as to how the universe began, then by default, if everything began from one point of origin and spread out into the known cosmos then, logically it follows that whatever, we have here on Earth, must also exist in the rest of the universe in some similar shape and form, allowing also, for a multitude of diversification. Ergo, life is everywhere and we are one with the universe!! Now, it is up to the astronomers, the physicists, the biologists and the NASA officials to prove it! **It is this author's contention that scientists (or some rogue scientists) have already proven that life exists**

throughout the universe and not just by theory alone but with actual proof, however, they are not disclosing the evidence as yet!

Type I , Type II and Type III Civilizations

What about the spiritual aspects of a space-faring civilization? A technologically advanced civilization without the pre-requisite spiritual development to provide a counter balance is to unwittingly unleash a runaway civilization that becomes an uncontrollable Frankenstein monster of material excess set loose upon an innocent and unsuspecting global populace. This, unfortunately, is where we are in our own evolutionary development on this planet.

The **Kardashev Scale** is a method of measuring a civilization's level of technological advancement, based on the amount of energy a civilization is able to utilize. The scale has three designated categories called *Type I, II, and III*. A **Type I Civilization** uses all available resources impinging on its home planet, **Type II Civilization** harnesses all the energy of its sun and a **Type III Civilization** of its galaxy. The scale is only hypothetical and in terms of an actual civilization, highly speculative; however, it puts energy consumption of an entire civilization in a cosmic perspective. It was first proposed in 1964 by the Soviet astronomer **Nikolai Kardashev.** Others have extended the scale to even more hypothetical **Type IV** beings who can control or use the entire universe, or **Type V** that control collections of universes. Metrics other than pure power usage have also been proposed, such as 'mastery' of a planet, a system or a galaxy rather than considering energy alone, or considering the amount of information controlled by a civilization rather than the amount of energy. http://en.wikipedia.org/wiki/Kardashev_scale

In 1964, Kardashev defined three levels of civilizations, based on the order of magnitude of the amount of power available to them:

- **Type I**: *"Technological level close to the level presently* (here referring to 1964) *attained on earth, with energy consumption at $\approx 4 \times 10^{19}$ erg/sec* (4×10^{12} watts.) Guillermo A. Lemarchand stated this as "*A level near contemporary terrestrial civilization with an energy capability equivalent to the solar insolation on Earth, between 10^{16} and 10^{17} watts.*"

- **Type II**: "*A civilization capable of harnessing the energy radiated by its own star* (for example, the stage of successful construction of a Dyson sphere), *with energy consumption at $\approx 4 \times 10^{33}$ erg/sec.* Lemarchand stated this as "*A civilization capable of utilizing and channeling the entire radiation output of its star. The energy utilization would then be comparable to the luminosity of our Sun, about 4×10^{26} watts.*"

Type III: "*A civilization in possession of energy on the scale of its own galaxy, with energy consumption at $\approx 4 \times 10^{44}$ erg/sec.*" Lemarchand stated this as "*A civilization with access to the power comparable to the luminosity of the entire Milky Way galaxy, about 4×10^{37} Watts.*" http://en.wikipedia.org/wiki/Kardashev_scale and http://www.youtube.com/watch?feature=player_detailpage&v=4Imd_0iCucg

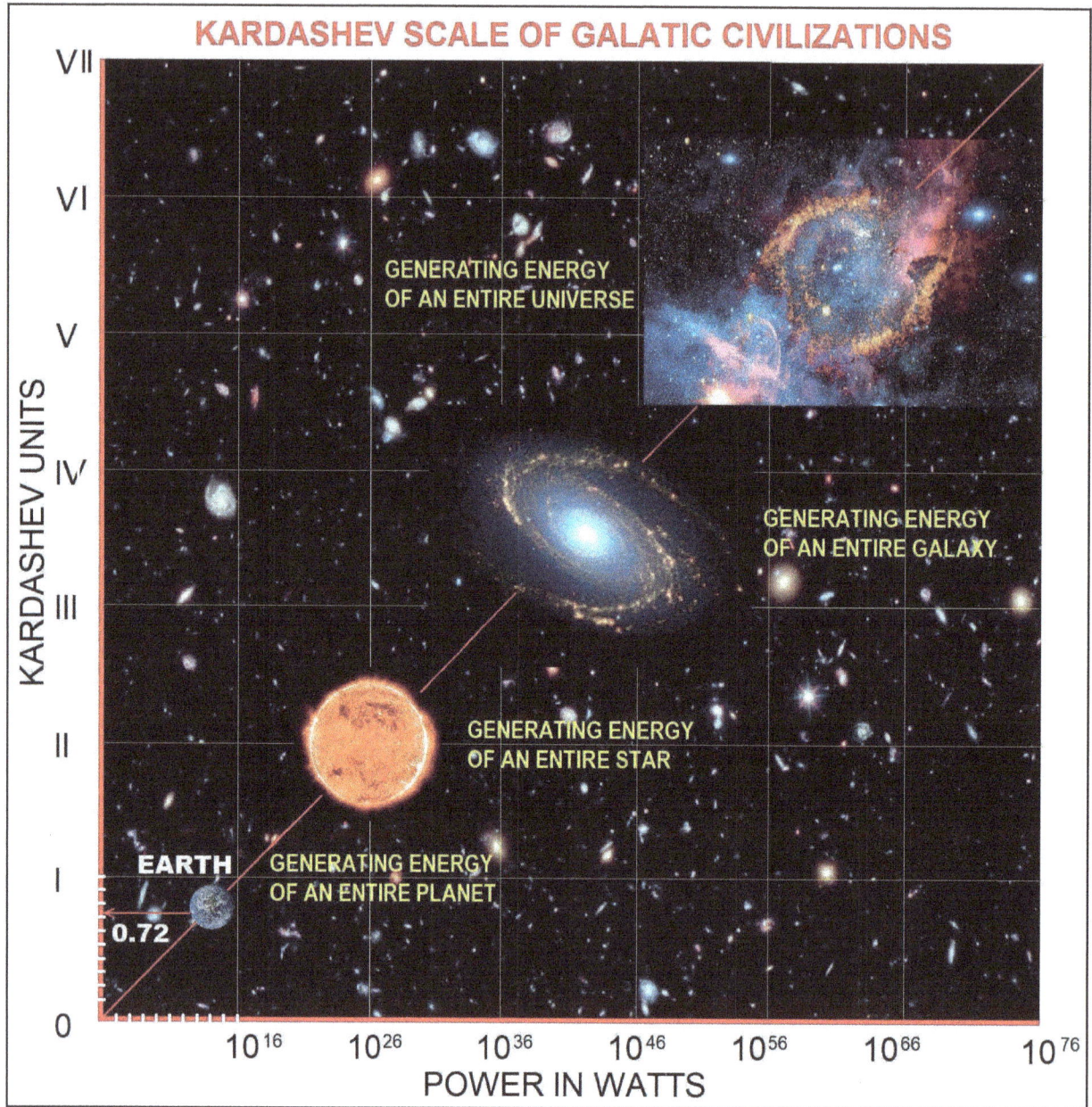

The Kardashev Scale indicating Type I, II, and III Civilizations

**Michio Kaku, American theoretical physicist, and promoter
of the concept of Type III, and III Civilizations**
http://www.intellectualtakeout.org/blog/world-famous-scientist-god-created-universe

To communication with other stellar civilizations or to travel to another star system requires more than the technical ability or development of what appears on the surface to be an advanced Type I civilization. It requires that a high degree of planetary unity and spiritual morality be an established cornerstone of that civilization. Space-faring travellers as representatives of their planet should also possess a reverence for other life forms different than their own. American theoretical physicist, **Michio Kaku**, the most recognized supporter of the Kardashev Scale has so often stated, in order to join the **"Galactic Club"** of starfaring civilizations, we must first become a **Type 1 Civilization** that is technologically advanced enough, without first blowing ourselves up in the process. But, before that can happen, we must first become a **Type 1 Spiritual Civilization.** The necessary prerequisite required by any intelligent species before they are able to travel to another neighbouring stellar civilization is to be spiritually civilized!

For an intelligent species such as humans to be acknowledged by other stellar civilizations as a Type 1 Civilization, there must first be the establishment of a world commonwealth. There must be a sustained way of life that demonstrates the ability of an intelligent species to live in peace and harmony among themselves, free from prejudices of all kinds. In our current state of planetary societal values with our war-mongering traditions to fight among ourselves or with anything different than ourselves, we must not go travelling off-planet to other worlds in this condition. We could easily be perceived as an infectious, spiritually backward species moving through the universe from one planet to another like a cancer.

The people on this planet must first need, to establish a common vision, a common unified direction, and purpose, before we can travel among the stars. Our spiritual values must keep pace with our technological achievements. We are not as yet, a Type I civilization! We need to grow up as a species and put aside our childish and adolescence behaviours and become the adults that we are destined to be! Until we do, we will not be permitted to travel beyond our own solar system. (We will come back to this aspect of global maturity, shortly).

https://www.youtube.com/watch?v=6GooNhOIMY0 and
https://www.youtube.com/watch?v=1w6u3ZVhSvY

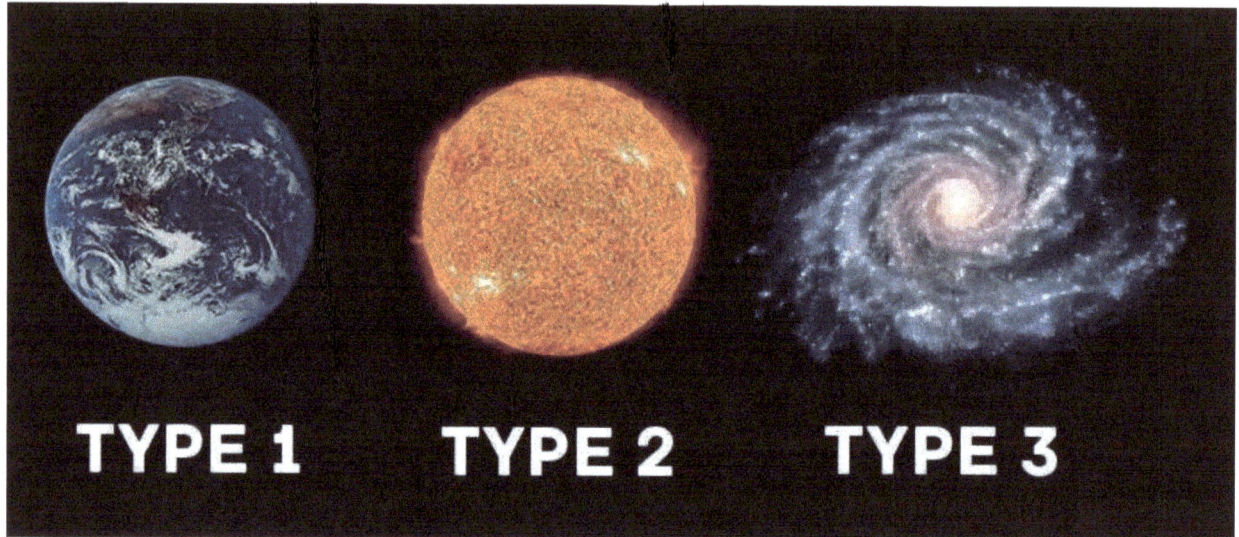

Civilization Types based upon the Kardashev Scale. The Earth is not as the picture portrays even a Type One Civilization but a Type .72 or Type .75 Civilization
https://briankoberlein.com/2014/08/30/alien-glow/

CHAPTER 65

SETI (SEARCH FOR EXTRATERRESTRIAL INTELLIGENCE) OR IS IT THE SILLY EFFORT TO INVESTIGATE?

The **Search for Extraterrestrial Intelligence (SETI)** is the collective name for a number of activities people undertake to search for intelligent extraterrestrial life. Some of the most well-known projects are run by Harvard University, the University of California, Berkeley, and the **SETI Institute.** SETI projects use scientific methods to search for intelligent life on other planets. For example, electromagnetic radiation is monitored for signs of transmissions from civilizations on other worlds. The United States government contributed to early SETI projects, but recent work has been primarily funded by private sources.

There are great challenges in searching across the cosmos for a first transmission that could be characterized as intelligent, since its direction, spectrum, and method of communication are all unknown beforehand. SETI projects necessarily make assumptions to narrow the search, the foremost being that electromagnetic radiation would be a medium of communication for advanced extraterrestrial life.
http://en.wikipedia.org/wiki/Search_for_extraterrestrial_intelligence

The **SETI Institute** is a not-for-profit organization whose mission is to "explore, understand and explain the origin, nature and prevalence of life in the universe". One program is the use of both radio and optical telescopes to search for deliberate signals from extraterrestrial intelligence, this is the **SETI** program that the public is most familiar with. Other research pursued within the **Carl Sagan Center for the Study of Life in the Universe**, includes the discovery of extrasolar planets, potentials for life on Mars and other bodies within the Solar System, and the habitability of the galaxy (including the study of extremophiles). The SETI Institute's public outreach efforts include working with teachers and students in promoting science education and the teaching of evolution, working with NASA on exploration missions such as **Kepler** and **SOFIA**, and producing a weekly science program: *Are We Alone?*

Within the SETI Institute, **Jill Tarter** heads the SETI effort; **Dr. David Morrison** is the Director of the Carl Sagan Center, and **Edna DeVore** is the Director of Education and Public Outreach. **Dr. Seth Shostak** is Senior Astronomer and host of *Are We Alone?* The SETI Institute is headquartered in Mountain View, California. http://en.wikipedia.org/wiki/SETI_Institute
It is Seth Shostak who is perhaps, the best publically known SETI astronomer and science educator having appeared on many TV news programs with people like Larry King, science/astronomy television documentaries and radio talk programs such as Coast to Coast with interviewers like Art Bell and George Noory. Shostak is also a well-known debunker of UFOs and hosts the monthly "Skeptic Check" show focused on debunking pseudo-science, UFOs and practices such as astrology and dowsing. http://en.wikipedia.org/wiki/Seth_Shostak

Seth Shostak is SETI's chief spokesman and promoter of the SETI Institute's program and strongly believes that no physical evidence exists for ET visitation or landing upon the Earth. Although he does feel confident that Extraterrestrial life does exist in the universe, it's just not visiting our planet at this time. He feels that the best way to make contact with ETs is via communication with electromagnetic radiation by listening or monitoring for signs of transmissions from civilizations on other worlds.

As early as 1896, **Nikola Tesla** suggested that radio could be used to contact extraterrestrial life. In 1899 while investigating atmospheric electricity using a Tesla coil receiver in his Knob Hill lab, Tesla observed repetitive signals, substantially different from the signals noted from storms and Earth noise, that he interpreted as being of extraterrestrial origin. He later recalled the signals appeared in groups of one, two, three, and four clicks together. Tesla thought the signals were coming from Mars. Analysis of Tesla's research has ranged from suggestions that Tesla detected nothing, he simply was misunderstanding the new technology he was working with, to claims that Tesla may have been observing naturally occurring Jovian plasma torus signals. In the early 1900s, **Guglielmo Marconi, Lord Kelvin,** and **David Peck Todd** also stated their belief that radio could be used to contact Martians, with Marconi stating that his stations had also picked up potential **Martian radio signals**.
http://en.wikipedia.org/wiki/Rio_Scale#Post_detection_disclosure_protocol

The US government and the military quickly realized that these inventors were onto something significant which could have not only profound scientific and social implications for humanity but, for the US military and thus, became actively involved in the listening program.

On August 21–23, 1924, Mars entered an opposition closer to Earth than any time in a century before or since. In the United States, a **"National Radio Silence Day"** was promoted during a 36-hour period from the 21–23, with all radios quiet for five minutes on the hour, every hour. At the **United States Naval Observatory,** a radio receiver was lifted 3 kilometers above the ground in a dirigible tuned to a wavelength between 8 and 9 kilometers, using a "radio-camera" developed by Amherst College and **Charles Francis Jenkins**. The program was led by **David Peck Todd** with the military assistance of **Admiral Edward W. Eberle** (Chief of Naval Operations), with **William F. Friedman** (chief cryptographer of the US Army), assigned to translate any potential Martian messages.
http://en.wikipedia.org/wiki/Rio_Scale#Post_detection_disclosure_protocol

In the March 1955 issue of Scientific American, **John D. Kraus** described a concept to scan the cosmos for natural radio signals using a flat-plane radio telescope equipped with a parabolic reflector. With $71,000 total in grants from the **National Science Foundation**, the first Kraus-style radio telescope was powered up in 1963. This funding would be considered a bargain by today's standards in science research grants as later requests for radio astronomy exploration would total in the billions of dollars.

In 1971, *NASA funded a SETI study* that involved **Drake, Bernard Oliver** of **Hewlett-Packard Corporation**, and others. The resulting report proposed an Earth-based radio telescope array with 1,500 dishes known as **"Project Cyclops"**. The price tag for the Cyclops array was $10 billion USD. Cyclops was not built, but the report formed the basis of much SETI work that followed.

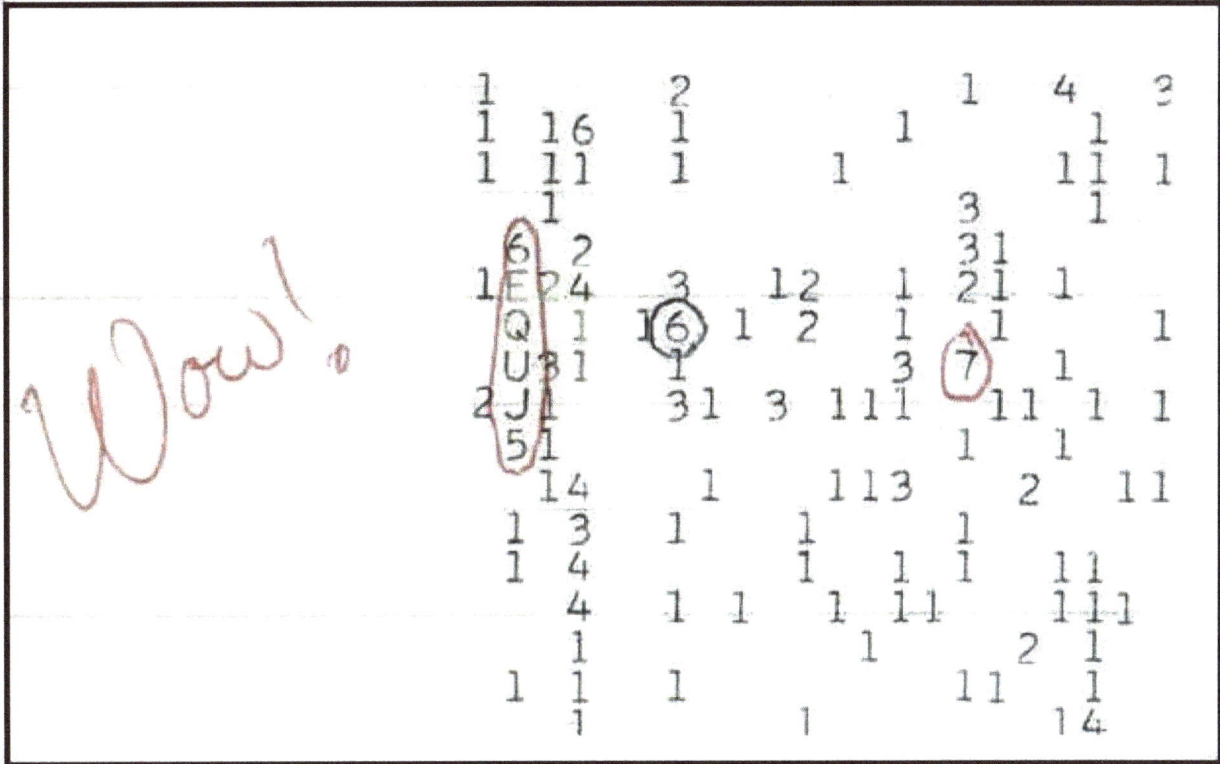

The WOW! Signal
https://en.wikipedia.org/wiki/Wow!_signal

This Ohio State University SETI program gained fame on August 15, 1977, when **Jerry Ehman**, a project volunteer, witnessed a startlingly strong signal received by the telescope. He quickly circled the indication on a printout and scribbled the phrase "Wow!" in the margin. Dubbed the **Wow! signal,** it is considered by some to be the best candidate for a radio signal from an artificial, extraterrestrial source ever discovered, but it has not been detected again in several additional searches. http://en.wikipedia.org/wiki/Search_for_extraterrestrial_intelligence

Interstellar scintillation of a weaker continuous signal—similar, in effect, to atmospheric twinkling—could be a possible explanation, although this still would not exclude the possibility of the signal being artificial in its nature. However, even by using the significantly more sensitive **Very Large Array**, such a signal could not be detected, and the probability that a signal below the Very Large Array level could be detected by the **Big Ear radio telescope** due to interstellar scintillation is low. Other speculations include a rotating lighthouse-like source, a signal sweeping in frequency, or a one-time burst.

Ehman has stated his doubts that the signal is of intelligent extraterrestrial origin: *"We should have seen it again when we looked for it 50 times. Something suggests it was an Earth-sourced signal that simply got reflected off a piece of space debris."*

He later recanted his skepticism somewhat, after further research showed an Earth-borne signal to be very unlikely, due to the requirements of a space-borne reflector being bound to certain unrealistic requirements to sufficiently explain the nature of the signal. Also, the 1420 MHz signal is problematic in itself in that it is "protected spectrum": it is bandwidth in which terrestrial transmitters are forbidden to transmit due to it being reserved for astronomical purposes. In his most recent writings, Ehman resists "drawing vast conclusions from half-vast data"—acknowledging the possibility that the source may have been military in nature or otherwise, may have been a production of Earth-bound humans.

In 1979 the University of California, Berkeley launched a SETI project named **"Search for Extraterrestrial Radio Emissions from Nearby Developed Intelligent Populations (SERENDIP)"**. In 1986, UC Berkeley initiated their second SETI effort, SERENDIP II, and has continued with four more SERENDIP efforts to the present day. The latest incarnation of the SERENDIP project is SERENDIP V.v, a commensal all-sky survey using the Arecibo radio telescope began in June 2009.
http://en.wikipedia.org/wiki/Search_for_extraterrestrial_intelligence

In 1980, **Carl Sagan**, **Bruce Murray**, and **Louis Friedman** founded the **U.S. Planetary Society**, partly as a vehicle for SETI studies.

In the early 1980s, Harvard University physicist Paul Horowitz took the next step and proposed the design of a spectrum analyzer specifically intended to search for SETI transmissions. This work led in 1981 to a portable spectrum analyzer named **"Suitcase SETI"** that had a capacity of 131,000 narrow band channels. Suitcase SETI was followed in 1985 by **Project "META" (Megachannel Extra-Terrestrial Assay).** The META spectrum analyzer had a capacity of 8.4 million channels and a channel resolution of 0.05 hertz. An important feature of META was its use of frequency doppler shift to distinguish between signals of terrestrial and extraterrestrial origin. The project was led by Horowitz with the help of the Planetary Society and was partly funded by movie maker **Steven Spielberg**. *Spielberg's involvement in the Planetary Society and with radio astronomy may explain his growing interest in UFO research which has become a touchstone in many of his sci-fi movies.* (Bold italics added by author for emphasis).The follow-on to META was named **"BETA (Billion-channel Extraterrestrial Assay),** capable of receiving 250 million simultaneous channels with a resolution of 0.5 hertz per channel. http://en.wikipedia.org/wiki/Search_for_extraterrestrial_intelligence

In 1978, the NASA SETI program was heavily criticized by **Senator William Proxmire**, and funding for SETI research was removed from the NASA budget by Congress in 1981, however, funding was restored in 1982, after **Carl Sagan** talked with Proxmire and convinced him of the program's value. In 1992, the U.S. government funded an operational SETI program, in the form of the NASA **Microwave Observing Program (MOP).**

MOP drew the attention of the U.S. Congress, where the program was ridiculed and canceled a year after its start. SETI advocates continued without government funding, and in 1995 the non-profit SETI Institute of Mountain View, California resurrected the MOP program under the name of **Project Phoenix,** backed by private sources of funding. Project Phoenix, under the direction of **Jill Tarter**, is a continuation of the targeted search program from MOP and studies roughly 1,000 nearby Sun-like stars.

Founded in 1994 in response to the US Congress cancellation of the NASA SETI program, the **SETI League, Inc.** is a membership-supported nonprofit organization with 1,500 members in 62 countries. This is a grass-roots alliance of amateur and professional radio astronomers. **SETI@home** was conceived by **David Gedye** along with **Craig Kasnoff** and is a popular volunteer distributed computing project that was launched by the University of California, Berkeley in May 1999. It was originally funded by **The Planetary Society** and **Paramount Pictures**, and later by the state of California. Any individual can become involved with SETI research by downloading the **Berkeley Open Infrastructure for Network Computing (BOINC)** software program, attaching to the **SETI@home project**, and allowing the program to run as a background process that uses idle computer power.
http://en.wikipedia.org/wiki/Search_for_extraterrestrial_intelligence

The **SETI Institute** has been collaborating with the Radio Astronomy Laboratory at UC Berkeley to develop a specialized radio telescope array for SETI studies, something like a mini-cyclops array. The array concept is named the **Allen Telescope Array (ATA)** (formerly, **One Hectare Telescope (1HT)** after the project's benefactor **Paul Allen**. In April 2011, the ATA was forced to enter "hibernation" due to funding shortfalls. Regular operation of the ATA was resumed on December 5, 2011.

Arecibo Observatory in Puerto Rico with its 300 m dish- the world's largest.
A small fraction of its observation time is devoted to SETI searches.
https://en.wikipedia.org/wiki/Arecibo_Observatory

In 1974, a largely symbolic attempt was made at the **Arecibo Observatory** to send a message to other worlds. It was sent towards the globular star cluster M13, which is 25,000 light years from Earth. The first **Interstellar Radio Message (IRM), the "Arecibo Message",** was transmitted in Nov 1974 from **Arecibo Radar Telescope**. Further, IRMs Cosmic Call, Teen Age Message, Cosmic Call 2, and A Message From Earth were transmitted in 1999, 2001, 2003 and 2008 from **Yevpatoria** or **Eupatoria Planetary Radar**.
http://en.wikipedia.org/wiki/Search_for_extraterrestrial_intelligence

Additional information on messages sent outward from Earth at Communication with Extraterrestrial Intelligence, Active SETI, List of interstellar radio messages. These other Earth-based messages to the stars will be dealt with in the section of how best to communicate with ETs. https://www.youtube.com/watch?v=SFFi3sWq5jE

Arecibo Message

The **Arecibo Message** was broadcast into space a single time via frequency modulated radio waves at a ceremony to mark the remodeling of the Arecibo radio telescope on 16 November 1974. It was aimed at the globular star cluster M13 some 25,000 light years away because M13 was a large and close collection of stars that was available in the sky at the time and place of the ceremony. The message consisted of 1679 binary digits, approximately 210 bytes, transmitted at a frequency of 2380 MHz and modulated by shifting the frequency by 10 Hz, with a power of 1000 kW. The "ones" and "zeros" were transmitted by frequency shifting at the rate of 10 bits per second. The total broadcast was less than three minutes.

The cardinality of 1679 was chosen because it is a semiprime (the product of two prime numbers), to be arranged rectangularly as 73 rows by 23 columns. The alternative arrangement, 23 rows by 73 columns, produces jumbled nonsense. The message forms the image shown below, or its inverse when translated into graphics characters and spaces.
http://en.wikipedia.org/wiki/Arecibo_message

, then at Cornell University and creator of the Drake equation, wrote the message, with help from **Carl Sagan**, among others. The message consists of seven parts thatencode the following (from the top down):

1. The numbers one (1) to ten (10)
2. The atomic numbers of the elements hydrogen, carbon, nitrogen, oxygen, and phosphorus, which make up deoxyribonucleic acid (DNA)
3. The formulas for the sugars and bases in the nucleotides of DNA
4. The number of nucleotides in DNA, and a graphic of the double helix structure of DNA
5. A graphic figure of a human, the dimension (physical height) of an average man, and the human population of Earth
6. A graphic of the Solar System
7. A graphic of the Arecibo radio telescope and the dimension (the physical diameter) of the transmitting antenna dish

Because it will take 25,000 years for the message to reach its intended destination of stars (and an additional 25,000 years for any reply), the Arecibo message was more a demonstration of human technological achievement than a real attempt to enter into a conversation with extraterrestrials. In fact, the stars of M13, that the message was aimed at, will no longer be in that location when the message arrives. According to the Cornell News press release of November 12, 1999, the real purpose of the message was not to make contact, but to demonstrate the capabilities of the newly installed equipment. http://en.wikipedia.org/wiki/Arecibo_message

"Arecibo Answer"

The "**Arecibo Answer**" is a crop circle (well, crop rectangle, to be accurate) that is purported to be a response to the "Arecibo message", a piece of coded information about Earth and humanity which was first beamed into space in 1974. It appeared in 2001 near the **Chilbolton Radio Telescope** in Hampshire, UK. Despite the fact crop circles has divided researchers, scientists and the public into two opposing camps, i.e. those who think all crop circles are hoaxes by two old town drunks by the name of Doug and Dave or those who say that crop circles are the communicative products of Extraterrestrials visiting our planet, it certainly has provoked an investigative response to determine if this crop circle is evidence of an Extraterrestrial answer to our message.

The crop circle is a near replica of the Arecibo message, which contained various pieces of information such as the numbers of chemical elements, the composition of DNA, the position of Earth in the Solar System, and a depiction of a human being. The "answer" itself doesn't expand much upon this and still forms the same 23 x 73 grid (because these numbers are primes) and most of the chemical data remains the same. The changes to the message to create the response are straight from existing alien folklore and science fiction. In the section detailing important chemical elements, the main focus is altered from carbon to silicon, and the diagram of DNA is re-scribbled slightly. At the bottom, the pictogram of a human is replaced with a shorter figure with a large, bulbous head. This is in clear reference to the "grey" type of alien, and as a depiction can only be something that a human would come up with.

The likelihood of the Arecibo message ever being picked up is very, very low. Although, the message was targeted for the globular star cluster M13, 25,000 light years away. The message does pass close enough to a few nearby stars to have been potentially "received" in their vicinity. Perhaps, some debunkers would wonder why the recipients wouldn't "simply send back a message via radio, instead of coming here and altering some unsuspecting farmer's crops at night and then streaking away without being detected?"

Is it reasonable for ET Intelligence to respond in this way to a human-initiated signal that was sent into space back in 1974? **Dr. Frank Drake** in a recorded interview in June 2009 was asked about the detection of signals from Extraterrestrial Intelligence and the feasibility of a reply to the Arecibo message in the form of crop circles.

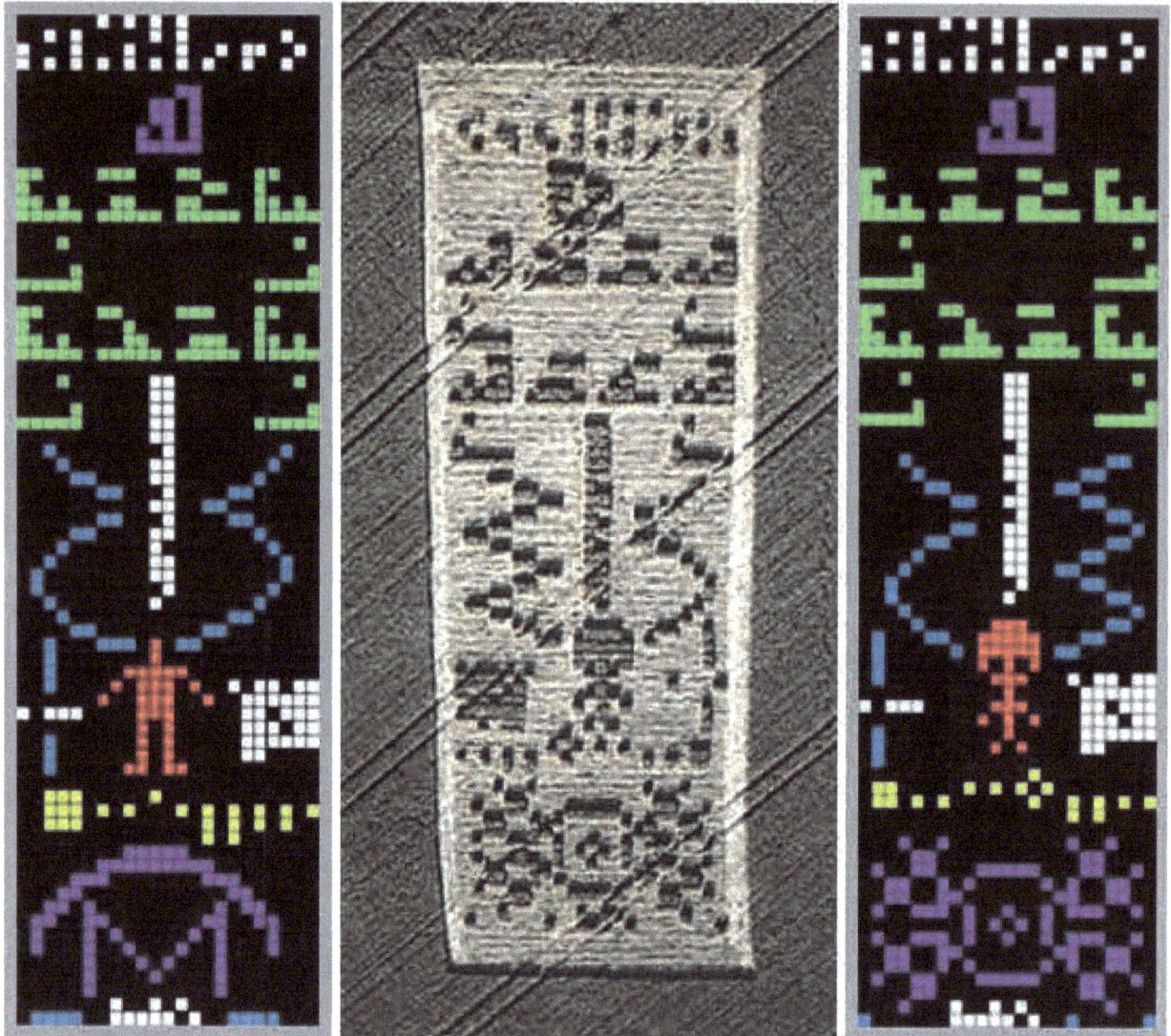

The Arecibo message (left) in colourized mosaics and the "answer" (right). The original message was transmitted as a series of 1s and 0s, and the answer was found as a crop circle (middle) in 2001 near the Chilbolton Radio Telescope in Hampshire, UK.
(Google Images)

He stated his view that radio signals are the optimal way in which extraterrestrials would communicate with humanity. He gave NASA communications with satellites and space missions as an example of the appropriateness of radio signals. He said that the **Search for Extraterrestrials Intelligence (SETI)** should focus on those parts of the electromagnetic spectrum that are the most appropriate for radio signals from extraterrestrials. In answer to the question of whether some crop circles were a possible extraterrestrial response to the 1974 Arecibo message, Drake said: *"aliens would know better than to communicate with humans in this ridiculous way"*.

Dr. Drake said that the *"universe has made it so that radio is the preferred method for interstellar communication."* Radio signals, however, would form an extremely cumbersome and impractical means of interstellar communication. The distance between the Earth and the closest star, Proxima Centauri, would require approximately 4 years for a radio signal to travel the distance. This would be very impractical and so another form of communications would be necessary. http://www.examiner.com/article/seti-pioneer-fails-to-debunk-crop-circles-as-et-messages

With statements like these from such a famous scientist, nuclear physicist and long time Ufologist, **Stanton T. Freidman** felt compelled to challenge this and other foolish perspectives with some scientific fact. He asserts that the best we humans are capable of at our current state of technological development is limited to the confines of the speed of light when it comes to communications or anything else for that matter. We cannot exceed this physical speed limitation at this time, even though, scientists say that it may be possible theoretically to exceed by multiples the speed of light in the near future with transluminal technology.

Communication by electromagnetic radiation, if developed by another interstellar civilization would recognize in very short order of their technological development that this was not a viable way to talk to worlds in other star systems and therefore, would scrap it very quickly for something more advanced. Even, the fact that communicating with the orbiting satellites around Mars or the Mars Rovers takes about 11 minutes to send instructions to them and 11 minutes in finding out if it responded correctly. A total of 22 minutes for a round trip in communications is a long time, particularly if your Rover is travelling near the rim of a crater. Committing technological suicide suddenly and unexpectedly is not only professionally embarrassing but a terrible way to end a 2.4 billion dollar mission by inadvertently not stopping in time thus, driving an expensive land rover over the precipice of a crater, all because of a time delay in communications. This situation becomes increasingly more axiomatic, the further satellites travel to the outer planets and the far reaches of our Solar System. Simply put, radio communications is not a very effective means to communicate with over long interstellar distances! It is no wonder that Freidman has referred to SETI as the **"Silly Effort To Investigate"**!

In response to Drake's comments veteran crop circle researcher, **Colin Andrews** pointed to various features of crop circles that reflect an extraterrestrial origin. *"I don't believe there will be a signal of any kind because ET is already here, right under SETI's noses and are interacting and influencing earth events, by means of some crop circles."*
http://www.examiner.com/article/seti-pioneer-fails-to-debunk-crop-circles-as-et-messages

Michael Salla in his article *"SETI Pioneer fails to debunk crop circles as ET messages"* found in the *Examiner.com* says that **crop circles** could be created by Extraterrestrials using torsion fields for remote communications and other purposes is a plausible means in which Extraterrestrials could communicate with humanity. The swirling clockwise or counterclockwise patterns of the stalks used to make crop circles could be the principle of rotation in making crop circles and how extraterrestrial communications occur over interstellar distances?

Professor Nikolai Kozyrev, a Russian pioneer in the study of torsion fields, show that torsion fields impact on gravity and travel faster than electromagnetic radiation or radio signals. Kozyrev experimentally confirmed that torsion fields travel at least 109 times the speed of light and was observed to be even greater when it came to detecting. The work of Kozyrev and other scientists point to torsion fields as the optimal means for Extraterrestrials to communicate over interstellar and even intergalactic distances. If torsion field communication devices are the standard means of Extraterrestrial communications then, Salla says, there is an answer to how crop circles are formed. Using some kind of torsion field generator and monitor, a crop circle could be created at great distances, even interstellar, by advanced extraterrestrial civilizations. Salla states this may explain videos of what appear to be the formation a crop circle by spinning plasma balls of light that appear to be under remote intelligent control.

The idea that advanced extraterrestrials use radio signals for communications, as Frank Drake suggests, is a 20th century anachronism. *It is anthropocentric thinking that many scientists fall prey to when trying to predict the outcome to events before they happen by projecting those same historical values onto other cultures or in this instance, upon Extraterrestrial Intelligence.* (Bold italics added by author for emphasis).

Russian astronomers have clearly demonstrated that torsion fields move at superluminal speeds and thus form the optimal way for interstellar communications to occur. Insofar as crop circles are created by some rotational generating principle or torsion field, they may well be, as Colin Andrews suggests, a form of communication by extraterrestrials. Remotely generated torsion fields may also explain some of the paranormal phenomena associated with crop circles that Andrews claims have drawn much official interest. http://www.examiner.com/article/seti-pioneer-fails-to-debunk-crop-circles-as-et-messages

ETI simply won't use a slow radio communication system of electromagnetic signals but rather, a more highly effective means that would communicate a message on many levels. By manipulating live crops into recognizable patterns without killing the plants (something that humans are not capable of doing without first destroying the plants), this would be indicative of a presence of ET intelligence visiting the Earth!

Curiously, prior to the discovery of the **Arecibo "Answer" crop glyph** originally dubbed the **"Persian Carpet"**, about five days earlier, another crop glyph or "photogram" was found in the same field a few hundred feet away, it appeared to be a "human face" done in pixel relief (similar to newsprint). It was found by the Observatory staff but, not reported right away because it was not obvious until viewed from the air by aircraft. Close examination of this crop glyph "Face" bore a striking resemblance to the July 1976 **Viking image** of the **"Face on Mars"** monument in the **Cydonia region of Mars**.

**The Chilbolton Man crop glyph found next to the Chilbolton Radio Observatory
in the same field as the Arecibo "Answer" crop glyph**
http://www.enterprisemission.com/glyph.htm

This latest crop circle discovery caused an immediate sensation when announced on the Internet in England and worldwide, due in part to the fact that it "struck many observers as a deliberate effort to remind everyone of the infamous *Face on Mars*. For one thing, the Face crop effort seemed to be designed to replicate the light and shading of the original Viking Cydonia image (below)." http://www.bibliotecapleyades.net/circulos_cultivos/esp_circuloscultivos12.htm

We are, however, getting ahead of ourselves so, we will come back to this Martian enigma when we do a virtual trip to each planet in our Solar System.

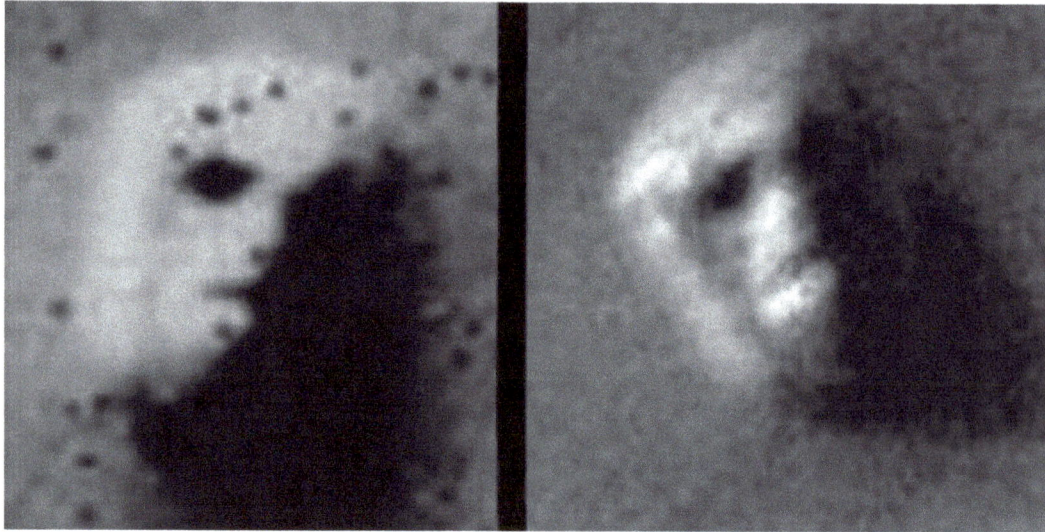

The July 1976 Viking image of the "Face on Mars" in the Cydonia region of Mars.
http://www.drraytalks.com/content/face-mars

The fact is that many efforts were made to send messages out into the universe from our planet to any and all Extraterrestrials that may be listening or if travelling through space and by chance happen to encounter one of our Voyager spacecraft destined to eventually leave our star system.

Arecibo Radio Telescope Now Lays in Ruins

On 10 August 2020, an 8-centimeter-thick steel cable, one of 18 suspending a 900-ton instrument platform high above the **Arecibo radio telescope** dish, had pulled out of its socket at one end and fallen, slicing into the dish. A second support cable snapped 3 months later, on 6 November, and the **National Science Foundation (NSF)**, which owns the observatory, said attempting repairs was too dangerous: Arecibo would be dismantled. On 1 December, fate took control as more cables snapped and the platform, as heavy as 2000 grand pianos, came crashing down into the dish.

Although 57 years old, Arecibo was a scientific trailblazer. Its powerful radar could bounce radio waves off other planets and asteroids, revealing the contours of their surfaces. And for most of Arecibo's life, it was the biggest radio dish in the world, (now, China's **FAST radio telescope** hold that distinction), able to sense the faintest emissions, from the metronomic beats of distant stellar beacons called pulsars to the whisper of rarefied gases between galaxies.

Decades earlier, staff noted cable wires snapping and suspected that corrosion from water was to blame. In 1976, managers tackled the problem by painting the cables to seal them off from the elements and installing fans to blow dry air through the length of the cables. Natural disasters hastened the end, Lugo says. Hurricane Maria battered Puerto Rico in 2017. Phoenix says it was "an opportunity for trouble," because the storm's winds could have picked up seawater, whose salt makes it especially corrosive, and dumped some on the telescope. The observatory was also shaken by a series of earthquakes in December 2019 and January 2020.

On 1 December 2020, the 900-ton instrument platform of the Arecibo Observatory lays in ruins at the bottom of its 1000 foot due to corrosion from the elements.

Credit: Ricardo Arduengo/AFP VIA Getty Images

https://www.sciencemag.org/news/2021/01/how-famed-arecibo-telescope-fell-and-how-it-might-rise-again

There is talk about cleaning up the debris site and rebuilding the telescope with an 8 million dollar infusion from Puerto Rico Governor Wanda Vázquez Garced, but more is needed to see it rebuilt to completion but this may be some time off in the future. https://www.sciencemag.org/news/2021/01/how-famed-arecibo-telescope-fell-and-how-it-might-rise-again and https://www.youtube.com/watch?v=u34uwFOCM-4

The world's largest single-dish radio observatory, the **Five-hundred-meter Aperture Spherical Radio Telescope (FAST)** in southern China is preparing to open to astronomers around the world, ushering in an era of exquisitely sensitive observations that could help in the hunt for gravitational waves and probe the mysterious fleeting blasts of radiation known as **Fast Radio Bursts (FRB).**

The 1.2-billion-yuan (US $171 million) telescope, also known as **Tianyan** or **'Eye of Heaven'**, took half a decade to build (2011 to 2016) in the remote Dawodang depression in the Guizhou province of southwest China. Although, it is open to astronomers from around the world, it will be seen whether the Chinese government which controls everything in that country will impose censorship on any new findings or discoveries. China is a nation not known to have an open-door policy of sharing anything with the rest of the world.

The complex project has a radical design and the pay-off for science will be immense. FAST will collect radio waves from an area twice the size of the next-largest single-dish telescope, the

Arecibo Observatory in Puerto Rico, which began to collapse in late August and had total collapse in November 2020..

The Chinese observatory's massive size means that it can detect extremely faint radio-wave whispers from an array of sources across the Universe, such as the spinning cores of dead stars, known as pulsars, and hydrogen in distant galaxies. It will also explore a frontier in radioastronomy — using radio waves to locate exoplanets, which may harbour extraterrestrial life. **https://www.nature.com/articles/d41586-019-02790-3** and **http://www.spaceopedia.com/astronomy/telescopes/tianyan/**

Tianyan or 'Eye of Heaven' aka. FAST will enable highly sensitive measurements of astronomical phenomenon. Credit: Ou Dongqu/Xinhua/ZUMA https://www.nature.com/articles/d41586-019-02790-3

Communication with Extraterrestrial Intelligence (CETI)

Communication with extraterrestrial intelligence (CETI) is a branch of the search for extraterrestrial intelligence that focuses on composing and deciphering messages that could theoretically be understood by another technological civilization. The best-known CETI experiment was the 1973 Arecibo message composed by Frank Drake and Carl Sagan. There are multiple independent organizations and individuals engaged in CETI research; the abbreviations CETI and SETI alone should not be taken as referring to any particular organization (such as the SETI Institute).

CETI research has focused on four broad areas: mathematical languages, pictorial systems such as the Arecibo message, **algorithmic communication systems (ACETI)** and computational approaches to detecting and deciphering "natural" language communication. There remain many undeciphered writing systems in human communication, such as **Linear A**, discovered by archeologists. Much of the research effort is directed at how to overcome similar problems of decipherment which arise in many scenarios of interplanetary communication.
http://en.wikipedia.org/wiki/Communication_with_Extraterrestrial_Intelligence

If astronomers were to communicate to any ETs in the local galactic neighbourhood, how would they accomplish such a task? It has long been considered that the language of mathematics and science would be the perfect and preferred way to communicate as there is no ambiguity in the language of math that is based on concepts of physics as there would be with traditional verbal and written languages.

Initial attempts were made using short and long pulses as described in the 1953 publication of **Astraglossa** by **Lancelot Hogbon**. Short pulse represented numbers and a series of long pulses denoted symbols for addition and subtraction, etc.

This concept was expanded upon in *Lincos: Design of a Language for Cosmic Discourse* by **Hans Freudenthal** in 1960 to create a general-purpose language derived from basic mathematics and logic symbols.

Carl Sagan explored the concept of **prime numbers (2, 3, 5, 7, 11, 13, 17, 19, 23, 29 etc.)** as a starting point, followed by various universal principles and facts of mathematics and science. He used such a concept in his science fiction novel "Contact" which later became a popular movie by the same name. http://en.wikipedia.org/wiki/Prime_number and
http://en.wikipedia.org/wiki/Communication_with_Extraterrestrial_Intelligence

Perhaps, the best-known communication system has been the pictorial messages such as the **Arecibo Message** as discussed earlier which utilized simple diagrams or pictograms sent via bitmaps to describe fundamental mathematical and physical concepts. Assuming that another intelligent civilization has a similar visual and technological capability (both very weak assumptions based upon anthropocentric thinking). Technologies on two separate and far-flung planets may have similar evolutionary developments but, when it comes to interstellar distances, radio or electromagnetic radiation as previously explained is too slow for communication purposes.

The spacecraft Pioneer 10 and Pioneer 11 launched in 1972 and 1973 carried gold plaques depicting the location of the Earth in the galaxy and the solar system, and the form of the human body.

The Voyager 1 and Voyager 2 probes launched in 1977 carried two golden plaques inscribed with diagrams depicting the human form, our solar system, and its location. Also included were recordings of pictures and sounds from Earth.

The *Cosmic Call* messages consisted of a few digital sections - **"Rosetta Stone",** copy of **Arecibo Message**, **Bilingual Image Glossary**, the **Braastad message**, as well as text, audio, video and other image files submitted for transmission by everyday people around the world. The "Rosetta Stone" was composed by **Stephane Dumas** and **Yvan Dutil** and represents a multi-page bitmap that builds a vocabulary of symbols representing numbers and mathematical operations. The message proceeds from basic mathematics to progressively more complex concepts, including physical processes and objects (such as a hydrogen atom). The message is designed with noise resistant format and characters, which make it resistant to alteration by noise. These messages were transmitted in 1999 and 2003 from **Evpatoria Planetary Radar** under scientific guidance of **Alexander L. Zaitsev**. **Richard Braastad** coordinated the overall project. http://en.wikipedia.org/wiki/Communication_with_Extraterrestrial_Intelligence

There are many other forms of communication as Earth beings search out the cosmos in ways both mathematical and musical. The **Teen-Age Message** consists of three sections:

Section 1 represents a coherent-sounding radio signal with slow Doppler wavelength tuning to imitate transmission from the Sun's center to help extraterrestrials detect the **TAM** and diagnose the radio propagation effect of the interstellar medium.

Section 2 is analog information and represents musical melodies, performed on the Theremin that produces a quasi-monochromatic signal, which is easily detectable across interstellar distances. Analog on-line transmission Theremin Concert for Aliens is shorter to broadcast than the digital method.

Section 3 represents a well known Arecibo-like binary digital information: the logotype of the TAM, bilingual Russian and English Greeting to Aliens, and Image Glossary.

There is another **Cosmic Call-2 (2003) message** similar to the first **Cosmic Call message** and **Algorithmic communication systems**. These systems build upon the early work of mathematical languages, whereby the sender describes a small set of mathematics and logic symbols that form the basis for a rudimentary programming language that the recipient can run on a **virtual machine.** Algorithmic communication has a number of advantages over static pictorial and mathematical messages, including localized communication, forward error correction and the ability to embed proxy agents within the message. In principle, a sophisticated program when run on a fast enough computing substrate may exhibit complex behavior and perhaps intelligence.

Cosmic OS describes a virtual machine that is derived from lambda calculus and *Logic Gate Matrices* **(LGM)** describes a universal virtual machine that is constructed by connecting coordinates in an n-dimensional space via mathematics and logic operations. Using this method, one can describe an arbitrarily complex computing substrate as well as the instructions to be executed on it.

Finally, in the family of electronic signals is the research into the possibility of receiving a signal /message not directed at us (where we are eavesdropping) or one that is in its natural communicative form. This event has already occurred as previously mentioned with the

Harvard Beta Radio Observatory. This method determines if the signal detected has intelligent-like structure, categorize the type of structure detected and then decipher its content: from its physical level encoding and patterns to the parts-of-speech, which encodes internal and external ontologies.

Primarily, this structure modeling focuses on the search for generic human and inter-species language universals to devise computational methods by which language can be discriminated from non-language and core structural syntactic elements of unknown languages can be detected. Aims of this research include: contributing to the understanding of language structure and the detection of intelligent language-like features in signals, to aid the search for extraterrestrial intelligence.
http://en.wikipedia.org/wiki/Communication_with_Extraterrestrial_Intelligence

Some of the star systems or constellations in which signals have been sent are 16 Cyg A Cygnus, 15 Sge Sagitta, GI 777 Cygnus, Cassiopeia, Orion, 55Cnc Cancer, Andromeda, 47 UMa Ursa Major, Delphinus, 47UMa Ursa Major, 37 Gem Gemini, Virgo, Hydra, and Draco.

Stephen Hawking, the Big Bang Theory and Everything Goes Bust!

The imperative for scientific exploration is so strong among astronomers and scientists that it has drawn its cautionary detractors from the same field of scientific endeavour. Enter Stephen Hawking, a British theoretical physicist, cosmologist, and author. He is best known for his collaboration work with **Roger Penrose** on theorems on gravitational singularities in the framework of general relativity, and the theoretical prediction that black holes should emit radiation, often called **Hawking radiation**.

He is an Honorary Fellow of the Royal Society of Arts, a lifetime member of the Pontifical Academy of Sciences, and a recipient of the Presidential Medal of Freedom, the highest civilian award in the United States. Hawking was the Lucasian Professor of Mathematics at the University of Cambridge between 1979 and 2009. Subsequently, he became research director at the university's Centre for Theoretical Cosmology.

Hawking has achieved success with works of popular science in which he discusses his own theories and cosmology in general; his *A Brief History of Time* stayed on the British *Sunday Times* best-sellers list for a record-breaking 237 weeks. Hawking has a motor neuron disease related to amyotrophic lateral sclerosis (ALS), a condition that has progressed over the years. He is now almost entirely paralyzed and communicates through a speech generating device. He married twice and has three children.

Most people know of Hawking because of his **Big Bang Theory** of the origin of the universe. A theory that asserts that time and the physical universe began from a single point and exploded or expanded outward to its current state as we know it. It is a theory that suggests that the universe expands from the original big bang and then contracts in on itself due to gravitational forces; when the universe stopped expanding and eventually collapsed, time would also run backwards. This theory was not shared with his fellow mathematicians and other physicists and later, Hawking developed a model in which the universe had no boundary in space-time, replacing the initial singularity of the classical Big Bang models with a region akin to the North Pole. One cannot travel north of the North Pole, but there is no boundary there – it is simply the point where all north-running lines meet and end. The no-boundary proposal predicted a **closed universe** but, it led to the realization that it is also compatible with an **open universe**.

Author's Rant: Personally, I think that the physical universe did have a beginning point of singularly, of one point in space and time that came into being by forces from other realms of existence or dimensionality which caused the universe to expand to its current state. In its present form, it has reached a steady state of harmonic resonance, neither expanding nor contracting but acting more like the human body, a living being! Is the universe a living being? I do not know but, it does have many similarities to a living conscious organism!

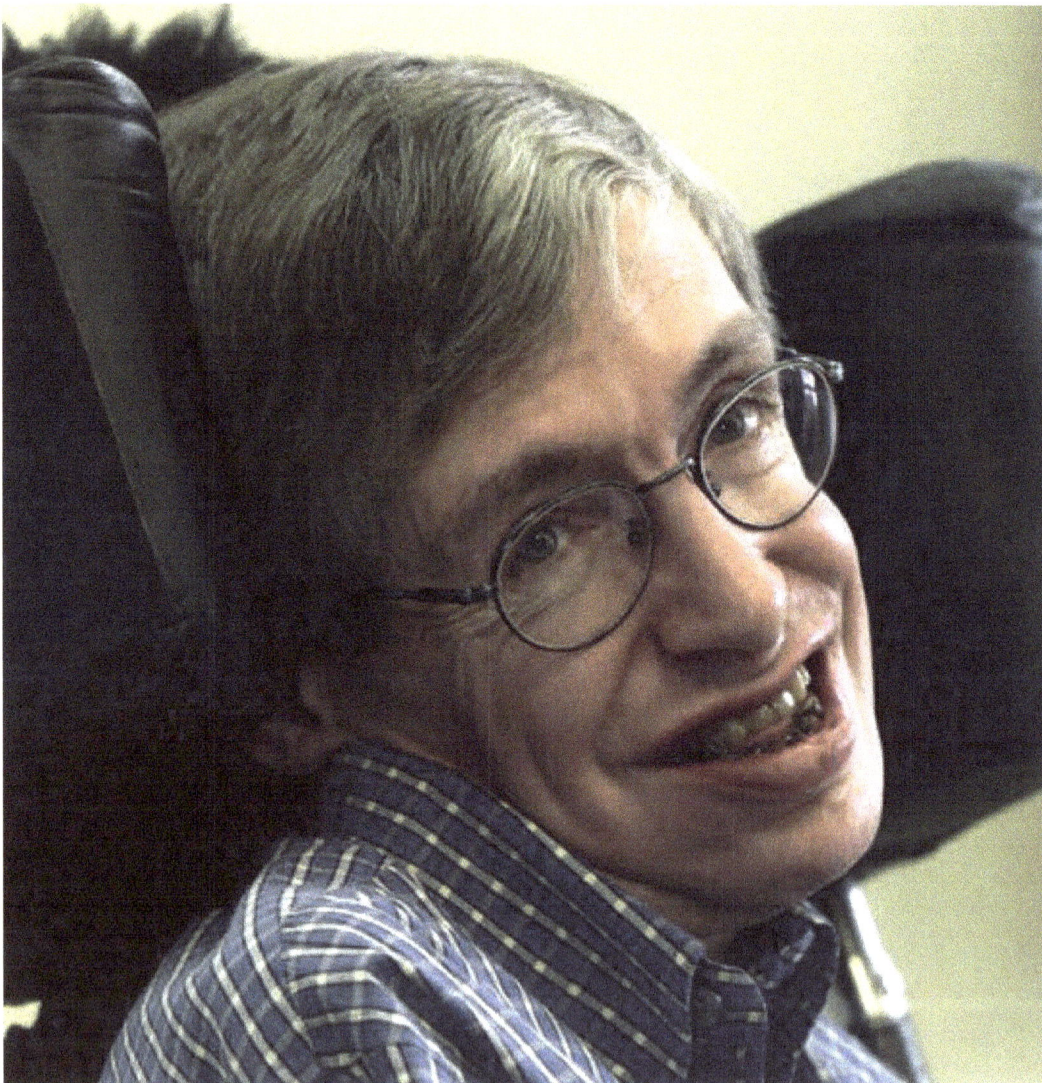

Stephen Hawking, British theoretical physicist, cosmologist, and author

To say that Stephen Hawking is a brilliant man has to be an understatement as he has been recognized worldwide for his theoretical work receiving many awards, honours and buildings named after him.

On 19 December 2007, a statue of Hawking by artist Ian Walters was unveiled at the Centre for Theoretical Cosmology, University of Cambridge. Buildings named after Hawking include the Stephen W. Hawking Science Museum in San Salvador, El Salvador, the Stephen Hawking Building in Cambridge, and the Stephen Hawking Centre at Perimeter Institute in Canada. In 2002, following a UK-wide vote, the BBC included him in their list of the 100 Greatest Britons. He has received the following award and honours:

- 1975 Eddington Medal
- 1976 Hughes Medal of the Royal Society
- 1979 Albert Einstein Medal
- 1981 Franklin Medal
- 1982 Commander of the Order of the British Empire
- 1985 Gold Medal of the Royal Astronomical Society
- 1986 Member of the Pontifical Academy of Sciences
- 1988 Wolf Prize in Physics
- 1989 Companion of Honour
- 1999 Julius Edgar Lilienfeld Prize of the American Physical Society
- 2003 Michelson Morley Award of Case Western Reserve University
- 2006 Copley Medal of the Royal Society
- 2008 Fonseca Prize of the University of Santiago de Compostela
- 2009 Presidential Medal of Freedom, the highest civilian honour in the United States

Hawking's first popular science book, *A Brief History of Time*, was published on 1 April 1988. It stayed on the British *Sunday Times* best-sellers list for a record-breaking 237 weeks. *A Brief History of Time* was followed by *The Universe in a Nutshell* (2001). A collection of essays titled *Black Holes and Baby Universes* (1993) was also popular. His book, *A Briefer History of Time* (2005), co-written by Leonard Mlodinow, updated his earlier works to make them accessible to a wider audience. http://en.wikipedia.org/wiki/Stephen_Hawking

Much of what Stephen Hawking has accomplished has set the tone for other astronomers and physicists as well as government organizations like NASA. Hawking made a statement back in 2010 that caught many people off guard including Ufologists. He stated that "*Of course it is possible that UFO's really do contain aliens as many people believe, and the Government is hushing it up.*" This was not, however, the statement that causes people in the UFO community and the public to get their hackles up. He stated that we needed to be careful even, fearful when transmitting radio signals or any other kinds of signals out into space, in case they are picked up by some neighbouring Extraterrestrials beings.

Hawking drew parallels to exploration and colonization on Earth, noting that the arrival of more technologically advanced Europeans in the Americas did not bode well for the natives. Hawking also mentioned that any aliens landing their UFOs on our planet will be looking to deplete and exploit our resources.

The news media snapped up these comments from the world-renowned theoretical physicist and cosmologist, like a tech-hungry crowd lining up at the nearest Best Buy to be the first to acquire the latest iPhone 5 and 6 or the Ipad, not caring if it has a few technical clichés in its overall design or operation.

One would expect a more sober and rational perspective from so talented a mind as Hawking's yet, there are serious flaws in his logic with regard to these comments, unless, of course, they are intentional.

Author's Rant: I have stated this repeatedly but, I will say it again: any comparison in contact with Extraterrestrial Intelligence visiting the Earth to the first medieval European explorers and the Native American Indians (North, Central, and South) is purely anthropocentric in thinking! There is no comparison! Humans are humans in thinking and behaviour and Aliens are alien in thinking and behaviour; they are not human! We humans assume too much, base upon our own history and in projecting our notions of morality onto all would-be visiting ET civilizations. This is nothing more than fear mongering! A negative mindset established by the military industrial complex, fostered by the Big Media and Hollywood moguls, irrationally given credibility by scientists from many disciplines notably by historians, archaeologists, and sociologists, based upon religious overtones of Christian superiority and under unrelenting pressure from the M.I.C. It is a close loop of negativity and anthropocentric thinking!

The reality has been that throughout our long recorded history, we have had constant visitations and interactions with Extraterrestrial Intelligences for thousands to tens of thousands of years and perhaps, even longer. Therefore, it would be safe to state that ETI have not shown any hostility or malevolence towards us in all that time. Any hostility by ETs may have been our misinterpretation of their attempts to halt our violent and aggressive actions in times of war. There have been many instances and case studies of their concern for our predilection for war and they have been monitoring our wartime activities wherever they occur on this planet. Pilots and ground infantry have routinely reported their appearances over the battlefields where we have fought. They are certainly not here to engage us in an all-out global conflict, one in which they could have done millennia ago and won with ease. It would make no sense at this time in our technological development to engage us in war, now that we are brimming to the teeth with nuclear and scalar weapon arsenals ready to launch from anywhere over the world with the ability to unleash **Armageddon** a thousand times over! No intelligent species or interstellar civilization would want that to happen anywhere in the universe!

This would also pre-suppose that ETs have solved the limitations of the light speed barrier and been able to traverse the vast distances of interstellar or galactic space utilizing the capability of non-linear transluminal technology and consciousness. Getting here from there would be no problem at all and if you had that type of technology then, you would not need the mineral resources from another planet that is tens to hundreds to thousands of light years away to solve your geo-socio-political planetary needs. You either find it closer to home or you employ the non-linear, non-local technology of consciousness and ideation to create what you require for your civilization. It is simply a matter of extracting the physical material from the realm of causation and ideation into physical reality; it is how their spacecraft are created among otherthings!

Visiting Extraterrestrial Intelligences have nothing that they need from us; they are not even hereto eat us for lunch. We would have tasted better about two hundred years ago than today, as our bodies are tainted with pollutants, toxins and radioactive Strontium 90. Eating us would either give them food poison or kill them. However, they are very curious about us as a species, our spiritual condition and the way we live upon the Earth, as well as an interest in our vast diverse range of multi-complex life forms on this planet, something that is not too common in the universe.

We must stop buying into the scary alien concept and genre as propagated in the Hollywood movies like *Alien (et al), Independence Day, The Fourth Kind, SkyLine, Fire in the Sky, Thing, Predator (et al), Signs, The Blob, Invasion of the Body Snatchers, Starship Troopers, Wars of the Worlds, Species* and all the rest of their ilk. These movies are purely for entertainment value, where we don't mind being scared for an hour or two but, not as somethingwhereby we are conditioned to being frightened of the day when real ET contact does eventuallyoccur. In particular is the overuse and exposure to the **Grey alien** type of being that is seen so often in films, TV shows, magazines, cartoons because of its supposed association with human- alien abduction scenarios. It is a deliberate psychological conditioning of **xenophobic** fear-base negativity by covert intelligence towards Extraterrestrial intelligences and in particular towards the **"Greys"** and the **"Reptilians"** with a future unfoldment of a *"false flag" alien invasion* … ala **"Independence Day"** involving these alien beings!

Stephen Hawking's comments though rash and irrational does force us to put the whole eventual human–alien contact scenario into a more rational perspective that is less self-righteousabout our position in the universe and away from the sentiment that everything different from ourselves namely, aliens are evil and intend in doing us and the Earth harm. *If there are "Klingons" in the universe then, we need to look no further than at ourselves for we are the ones with a history of malevolence, of evil intent, of malignant ideologies and outworn shibboleths along with a rapacious hunger for planetary resources.*

Professor Hawking's has merely held up the mirror and shown us who we really are, at thecurrent time the picture is neither pretty nor does it reflect our nobility as species.

When the day arrives that Earth has its first official contact with Extraterrestrial beings from another planet, we must ensure that we possess a positive mindset that is commensurate with our maturity, unity of purpose and one that reflects our spiritual nobility as a civilization that is ready to engage other intelligent civilizations on a path that is mutually peaceful, prosperous and sustainable. When that first contact and communications with ETs does occur, chances are, it will be a whole new learning curve in our understanding about the universe; it will be they who will be the lead dancers in a remarkable new relationship; it will be they who will teach us how to value ourselves and the life on this planet. We need to prepare for that day!

CHAPTER 66

THE SPACE RACE AND THE QUEST FOR ALL THINGS EXTRATERRESTRIAL

As we begin to venture further and further out into the Solar System and beyond the outermost most planets into the vast void of interstellar space, many nations are joining in the **"Space Race"** of planetary and stellar exploration. There is much to be gained for every nation and hopefully for all of humanity as long as there are no cover-ups of discoveries kept from the general public. Transparency and full disclosure should be the order of the day from every nation's space agency. That means the discovery of extraterrestrial life or of an extinct or extantplanetary civilization or its technology should be placed into the public domain right away and not sequestered away for the military or the private industrial to research for their own personalgain. Such discoveries belong to all mankind so that all may benefit from the new knowledge uncovered allowing for global civilization to advance, thereby.

When we think of the space race, we think of two nations competing, the US and Russia. In thelast couple of decades, we can add the combined efforts of most European nation under the **European Space Agency(ESA),** the **Chinese Space Agency(CSA)** and the **Indian Space Agency(ISA)** who are all major players. There are, however, many more nations, some with spaceports, launch sites, and/or space centres and to this growing list, we must add increasing number of privately own space agencies and programs owned by wealthy individual like **Elon Musk**.

Let's look first at the locations of the US and Russian spaceports and space centres and then seewhat other nations are beginning their leap off the planet into space.

Spaceports and Cosmodromes

A spaceport or cosmodrome (Russian) is a site for launching (or receiving) spacecraft, by analogy with seaport for ships or airport for aircraft. The word *spaceport*, and even more so*cosmodrome*, has traditionally been used for sites capable of launching spacecraft into orbit around Earth or on interplanetary trajectories. However, rocket launch sites for purely sub-orbital flights are sometimes called spaceports, as in recent years new and proposed sites for suborbital human flights have been frequently referred to or named 'spaceports'. Space stations and proposed future bases on the moon are sometimes called spaceports, in particular, if intended as a base for further journeys. http://en.wikipedia.org/wiki/Spaceport

Spaceports operate around the world, offering different capabilities and scales of operation - ranging from a basic control center, transportation infrastructure and launch platform to highly sophisticated facilities such as **Kennedy Space Center** in Florida. KSC and the **Russian Bikaner Commodore** are the oldest and most advanced. But, they are not alone. Following is a list of spaceports in the U.S.

US Orbital Spaceports

Launch sites for manned and unmanned missions, able to put payloads into Earth orbit and beyond; available to civilian and military government customers, commercial companies, and non-profit organizations. This also includes autonomous spaceport drone ship (an ocean-going, barge-derived, floating landing platform):

Cape Canaveral, Florida, USA
Cecil Airport
Clinton-Sherman Industrial Airpark
Corn Ranch
Edwards Air Force Base
Greater Green River Intergalactic Spaceport
Kennedy Space Center, Florida, USA
Kodiak Launch Complex, Alaska, USA
List of Cape Canaveral
Mid-Atlantic Regional Spaceport
Mid-Atlantic Regional Spaceport
Launch Pad 0
Mojave Air and Space Port
Ronald Reagan Ballistic Missile Defense Test Site, Kwajalein Atoll, Marshall Islands (USA)
Spaceport America
Spaceport Sheboygan
SpaceX private launch site
Vandenberg Air Force Base, California, USA
Wallops Flight Facility/Mid-Atlantic Regional Spaceport, Virginia, USA
White Sands
http://en.wikipedia.org/wiki/Category:Spaceports_in_the_United_States

US Suborbital Spaceports

Launch facilities that can place objects in space but not at orbital velocities; generally lack complex launch pads and ground equipment, or have geographic restrictions such as an inadequate buffer area between their facilities and populated areas; lower costs and fewer restrictions make these facilities attractive to private investors in new space programs focused on suborbital spaceflights, such as personal spaceflight.

California Spaceport, California, USA
Spaceport America, New Mexico, USA
Mojave Spaceport, California, USA
http://www.spacefoundation.org/media/space-watch/spaceports-can-be-found-throughout-world

A launch at Vandenberg Air Force Base, in California
https://militarybases.com/vandenberg-afb-air-force-base-in-lompoc-ca/

In particular, **Vandenberg Air Force Base** in California is considered a secret launch facility because it is military, while **Edwards Air Force Base** and **Mojave Air and Space Port** are Runway facilities for returning spacecraft, all in California. Mojave Space Port can launch manned sub-orbital vehicles but not heavy moon rockets. Area 51 is also a return Runway facility, though it is considered to be so secret as to be non-existent! (Yeah, right! It's the best kept secret military base known to the public!!)

Spaceports Around the World

Alcântara Launch Center
Andøya Space Center
Anhueng
Baikonur Cosmodrome
Barreira do Inferno Launch Center
Biak Spaceport

British commercial spaceport
Broglio Space Centre
Caisson Nemo
Dombarovsky (air base)
Esrange
Fort Churchill (rocket launch site)

Guiana Space Centre
Hammaguir
Jiuquan Satellite Launch Center
Kapustin Yar
Naro Space Center
Odyssey (launch platform)
Palmachim Airbase
Plesetsk Cosmodrome
Ras Al Khaimah spaceport
Salto di Quirra
Satish Dhawan Space Centre
Singapore spaceport
Sonmiani (space facility)
Svobodny Cosmodrome
Spaceport Sweden
Taiyuan Satellite Launch Center
Tanegashima Space Center
Thumba Equatorial Rocket Launching Station
Tilla Satellite Launch Centre
Uchinoura Space Center
Vostochny Cosmodrome
Wenchang Satellite Launch Center
Woomera, South Australia
Xichang Satellite Launch Center
Yoshinobu

http://en.wikipedia.org/wiki/Category:Spaceports

Space Centres

A space center is a place dedicated to space-related activity. It may be in public or private ownership.

These activities may concern:

- Research
- Manufacturing of major parts of space vehicles
- Launch of space vehicles
- In orbit control of space vehicles
- Government programs of a space agency
- Public education at a specialist science museum

Below is a list of International Space Centres by country, alphabetically:

Belgium
- Liege Space Center
- Euro Space Center

Canada
- John H. Chapman Space Centre
- Telus World of Science, Edmonton

Germany
- European Space Operations Centre
- Hubble European Space Agency
- European Astronaut Centre

France
- Toulouse Space Center
- Guiana Space Centre
- Cannes Mandelieu Space Center

India
- Vikram Sarabhai Space Centre
- Space Applications Centre
- Satish Dhawan Space Centre

Italy
- Broglio Space Centre

Japan
- Uchinoura Space Centre
- Tanegashima Space Centre

South Korea
- Naro Space Center

Netherlands
- European Space Research and Technology Centre

Norway
- Norwegian Space Centre

Philippines
- Mabuhaysat Subic Space Center
- Mabuhaysat Zamboanga Space Center[1]

Pakistan
- Sonmiani (space facility)
- Tilla Satellite Launch Center

Russia
- Babakin Space Centre
- Baikonur Cosmodrome
- Titov Main Test and Space Systems Control Centre

United Kingdom
- Leicester Space Centre
- Surrey Space Centre
- Harwell Science and Innovation Campus
- Jodrell Bank Observatory

United States
- Goddard Space Flight Center
- John C. Stennis Space Center
- Kennedy Space Center
- Lyndon B. Johnson Space Center and Space Center Houston
- Marshall Space Flight Center

http://en.wikipedia.org/wiki/Space_center

NASA (NACA)

The **National Aeronautics and Space Administration (NASA)** (or is that *"Never A Straight Answer"* or *"Not Another Stupid Announcement"* as NASA is sometimes referred to in the **UFO Community?** is the agency of the United States government that is responsible for the nation's *"civilian space program"* and for aeronautics and aerospace research. Since February 2006, NASA's mission statement has been to "pioneer the future in space exploration, scientific

discovery and aeronautics research." (Bold italic added for emphasis by author).
http://en.wikipedia.org/wiki/NASA

President Eisenhower established the **National Aeronautics and Space Administration (NASA)** in 1958 with a *"distinctly civilian (rather than military) orientation"* encouraging peaceful applications in space science. The **National Aeronautics and Space Act** was passed on July 29, 1958, replacing its predecessor, the **National Advisory Committee for Aeronautics (NACA)**. The agency became operational on October 1, 1958. (Bold italic added for emphasis by author). Interestingly, NACA's main focus was the development of the spaceplane like the X15 which was capable of attaining attitudes of low Earth orbit.

Since 1946, the National Advisory Committee for Aeronautics (NACA) had been experimenting with rocket planes such as the supersonic Bell X-1. In the early 1950s, there was a challenge to launch an artificial satellite for the **International Geophysical Year** (1957–58) under the American **Project Vanguard** program. However, the US alarmed by the launch of the world's first artificial satellite (*Sputnik*) on October 4, 1957, by the Soviet Union, perceived this as a threat to national security and technological leadership (known as the **"Sputnik crisis"**) and urged the U.S. Congress to immediate and swift action. Counseled by his advisers, **President Eisenhower** signed an agreement that a new federal agency based mainly on NACA was needed to conduct all non-military activity in space. The **Defense Advanced Research Projects Agency (DARPA)** was created in February 1958 *"to develop space technology for military application"*. (Bold italic added for emphasis). http://en.wikipedia.org/wiki/NASA and http://www.eisenhowermemorial.org/onepage/IKE%20&%20Science.Oct08.EN.FINAL%20%28v2%29.pdf

On July 29, 1958, Eisenhower signed the National Aeronautics and Space Act, establishing **NASA**. When it began operations on October 1, 1958, NASA absorbed the 46-year-old **NACA** intact; its 8,000 employees, an annual budget of US$100 million, three major research laboratories **(Langley Aeronautical Laboratory, Ames Aeronautical Laboratory,** and **Lewis Flight Propulsion Laboratory)** and two small test facilities. A NASA seal was approved by President Eisenhower in 1959. Elements of the **Army Ballistic Missile Agency** and the **United States Naval Research Laboratory** were incorporated into NASA. A significant contributor to NASA's entry into the **Space Race** with the Soviet Union was the technology from the German rocket program (led by **Wernher von Braun**, who was now working for ABMA) which in turn incorporated the technology of American scientist **Robert Goddard**'s earlier works. Earlier research efforts within the U.S. Air Force and many of DARPA's early space programs were also transferred to NASA. In December 1958, NASA gained control of the **Jet Propulsion Laboratory**, a contractor facility operated by the **California Institute of Technology**.
http://en.wikipedia.org/wiki/NASA

**The logos of the National Aeronautics and Space Administration (NASA)
and the former National Advisory Committee for Aeronautics (NACA)**
http://www.adelantesciences.com/ and https://commons.wikimedia.org/wiki/File:US-NACA-Logo.svg

All of these military acquisitions and transfers of technology was the result of the mid-twentieth century **Space Race** between the U.S.A. and the Soviet Union of Socialist Republics (U.S.S.R.), a competition for supremacy in space exploration. Between 1957 and 1975, the **Cold War** rivalry between the two nations focused on attaining firsts in space exploration, which were seen as necessary for national security and symbolic of technological and ideological superiority. The **Space Race** involved pioneering efforts to launch artificial satellites, sub-orbital and orbital human spaceflight around the Earth, and piloted voyages to the Moon.
http://en.wikipedia.org/wiki/Space_Race

The Space Race had its origins in Nazi Germany, beginning in the 1930s and continuing during World War II when Germany researched and built operational ballistic missiles. The Soviet Union and the United States became involved in the space race just after the end of the World War II when both captured advanced German rocket technology and personnel.

During the Second World War, General Dornberger was the military head of the army's rocket program, Zanssen became the commandant of the Peenemünde army rocket centre, and von Braun was the technical director of the ballistic missile program. They would lead the team that built the **Aggregate-4 (A-4) rocket**, which became the first vehicle to reach outer space during its test flight program in 1942 and 1943. By 1943, Germany began mass producing the A-4 as the **Vergeltungswaffe 2 ("Vengeance Weapon" 2)**, or more commonly, **V2),** a ballistic missile. Its supersonic speed meant there was no defense against it, and radar detection provided little warning. Germany used the weapon to bombard southern England and parts of Allied-liberated Western Europe from 1944 until 1945. After the war, the V-2 became the basis of early American and Soviet rocket designs. http://en.wikipedia.org/wiki/Space_Race

At war's end, American, British, and Soviet scientific intelligence teams competed to capture Germany's rocket engineers along with the German rockets themselves and the designs on which they were based. Each of the Allies captured a share of the available members of the German rocket team, but the United States benefited the most with **Operation Paperclip**, recruiting von Braun and most of his engineering team, who later helped develop the American missile and space exploration programs. The United States also acquired a large number of complete V2 rockets.

108

The German rocket center at **Peenemünde** was located in the eastern part of Germany, which became the Soviet zone of occupation. On Stalin's orders, the Soviet Union sent its best rocket engineers to this region to see what they could salvage for future weapons systems. The Soviet rocket engineers were led by **Sergei Korolev**, a rocket designer since the 1930s, who after the war became the USSR's chief rocket and spacecraft engineer, essentially the Soviets' counterpart to von Braun. His identity was kept a state secret throughout the Cold War, and he was identified publicly only as "**the Chief Designer.**" In the West, his name was only officially revealed when he died in 1966. http://en.wikipedia.org/wiki/Space_Race

In America, Von Braun and his team were sent to the United States Army's **White Sands Proving Ground**, located in New Mexico, in 1945. They set about assembling the captured V2s and began a program of launching them and instructing American engineers in their operation. These tests led to the first rocket to take photos from outer space, and the first two-stage rocket, the WAC Corporal-V2 combination, in 1949. The German rocket team was moved from Fort Bliss to the Army's new Redstone Arsenal, located in Huntsville, Alabama, in 1950. From here, Von Braun and his team would develop the Army's first operational medium-range ballistic missile, the Redstone rocket that would, in slightly modified versions, launch both America's first satellite and the first piloted Mercury space missions. It became the basis for both the Jupiter and Saturn family of rockets. http://en.wikipedia.org/wiki/Space_Race

In reality, NASA has never truly been just a civilian space program, it has always been a military space program! When you consider initially, all the military assets and the technical know-how that have gone into and have been developed chiefly by the military for NASA and the technical personnel that come from military backgrounds, not to mention that almost all the first astronauts from the Mercury Program up to the current time are all military personnel with the few exceptions of civilians tossed into the mix for good measure, you realize immediately, that the civilian space program that is so often touted, is mere publicity and propaganda that has been spin-doctored for the public's benefit. The only thing civilian about the space program, is that it is funded by civilian tax dollars and they have no say as to the direction of the program, nor do they reap the benefits of disclosure of any new knowledge that has come out of the space program, other than those things that do not reveal any possible discoveries of alien artifacts, technology or alien contact.

The Soviet space program included rocketry and space exploration programs by the former **Union of Soviet Socialist Republics (the Soviet Union or U.S.S.R.)** from the 1930s until its dissolution in 1991. Like the American space program, the Soviets operated primarily a classified military space program which helped the Soviet Union to lead in the Space Race with many successful firsts, causing no amount of public controversy and military consternation in the United States.

The USSR launched Sputnik 1, the first artificial satellite to orbit the Earth
http://animationstudiotoronto.com/portfolio/sputnik/

Two days after the United States announced its intention to launch an artificial satellite, on July31, 1956, the Soviet Union announced its intention to do the same. Sputnik was launched on October 4, 1957, beating the United States and stunning people all over the world.

The Soviets appeared technologically invincible when in 1961 they launched the first human intospace aboard the space capsule, **Vostok,** a young, handsome cosmonaut by the name of **Yuri Gagarin,** who appeared to symbolize the epitome of Russian manliness and handsomeness.
Although, it was stated that he was officially recognized as the first human being in space nevertheless, rumours flew around that the soviets had lost 3 or 4 cosmonauts in space disastersbefore they succeeded with the first successful safe launch and return made by Gagarin. Read further on for details of Russian space disasters.

Yuri Gagarin, the first human in space

The United States called their space travelers **astronauts** ("star sailors" from the Greek), and it was 3 weeks later, on 5 May 1961, when **Alan Shepard** became the first one in space, launched on a suborbital mission **Mercury**, in a spacecraft named *Freedom 7*.

Almost a year after the Soviets put a human into orbit, astronaut **John Glenn** became the first American to orbit the Earth, on 20 February 1962. His **Mercury** mission completed three orbits in the *Friendship 7* spacecraft, and splashed down safely in the Atlantic Ocean, after a tense reentry, due to what falsely appeared from the telemetry data to be a loose heat shield.
http://en.wikipedia.org/wiki/Space_Race

It was now possible to look up into the starry night and view for the first time in history, manmade spacecraft routinely orbiting the Earth approximately every 90 minutes.

Author's Rant: It was a wonderful time to live in particularly for a young child like myself at that time. I was nearly seven years old living in England and I recalled when we heard the news that people would be able to see the Russian satellite flying overhead, my family and I ran outside to view Sputnik 1. It was an amazing sight that filled my young mind with

thoughts of space travel. When I heard later, that a Russian cosmonaut had been launched into space aboard a spaceship my mind went into overdrive with excitement and a passion to grow up one day to become a cosmonaut or astronaut and orbit the Earth as well.

The Soviet space program successfully pioneered many accomplishments in space exploration:

- 1957: First intercontinental ballistic missile, the **R-7 Semyorka**
- 1957: First satellite, **Sputnik**
- 1957: First animal in Earth orbit, the dog **Laika** on **Sputnik**
- 1959: First rocket ignition in Earth orbit, first man-made object to escape Earth's gravity, **Luna**
- 1959: First data communications, or telemetry, to and from outer space, **Luna**.
- 1959: First man-made object to pass near the Moon, first man-made object in Heliocentric orbit, **Luna**
- 1959: First probe to impact the Moon, **Luna**
- 1959: First images of the moon's far side, **Luna**
- 1960: First animals to safely return from Earth orbit, the dogs **Belka** and **Strelka** on **Sputnik**.
- 1961: First probe launched to Venus, **Venera 1**
- 1961: First person in space (International definition) and in Earth orbit, **Yuri Gagarin** on **Vostok**, Vostok program.
- 1961: First person to spend over 24 hours in space **Gherman Titov**, **Vostok** (also first person to sleep in space).
- 1962: First dual manned spaceflight, **Vostok and Vostok 4**
- 1962: First probe launched to Mars, **Mars 1**
- 1963: First woman in space, **Valentina Tereshkova, Vostok**
- 1964: First multi-person crew (3), **Voskhod 1**
- 1965: First extra-vehicular activity (EVA), by **Aleksei Leonov, Voskhod 2**
- 1965: First probe to hit another planet of the Solar system (Venus), **Venera 3**
- 1966: First probe to make a soft landing on and transmit from the surface of the moon, **Luna**
- 1966: First probe in lunar orbit, **Luna**
- 1967: First unmanned rendezvous and docking, **Cosmos 186/Cosmos 188**.
- 1968: First living beings to reach the Moon (circumlunar flights) and return unharmed to Earth, Russian tortoises on **Zond 5**
- 1969: First docking between two manned craft in Earth orbit and exchange of crews, **Soyuz 4 and Soyuz 5**
- 1970: First soil samples automatically extracted and returned to Earth from another celestial body, **Luna**
- 1970: First robotic space rover, **Lunokhod** on the Moon.
- 1970: First data received from the surface of another planet of the Solar system (Venus), **Venera 7**
- 1971: First space station, **Salyut 1**
- 1971: First probe to impact the surface of Mars, **Mars 2**
- 1975: First probe to orbit Venus, to make soft landing on Venus, first photos from surface of Venus, **Venera 9**

112

- 1980: First Hispanic and Black person in space, **Arnaldo Tamayo Méndez on Soyuz 38**
- 1984: First woman to walk in space, **Svetlana Savitskaya (Salyut 7 space station)**
- 1986: First crew to visit two separate space stations **(Mir and Salyut 7)**
- 1986: First probes to deploy robotic balloons into Venus atmosphere and to return pictures of a comet during close flyby **Vega 1, Vega 2**
- 1986: First permanently manned space station, **Mir**, 1986–2001, with permanent presence on board (1989–1999)
- 1987: First crew to spend over one year in space, **Vladimir Titov** and **Musa Manarov** on board of **Soyuz TM-4 - Mir** http://en.wikipedia.org/wiki/Soviet_space_program

America may have been slow at the start of the Space Race but, they were more careful in their approach to entering a new environment with better-engineered spacecraft, better-trained astronauts and far better ethics toward safety and precautions, not letting anything jeopardize the complete success of their space missions. They may not have had the prestige of all the firsts that the Soviets had claimed but, in the end, the dogged yet, careful determination of the American space program which was inspired by **President John F. Kennedy's "We Choose to go to the Moon"** speech on 12 September 1962, eventually witnessed the Americans landing on the Moon before the Soviets.

"We choose to go to the moon in this decade and do the other things, not because they are easy, but because they are hard, because that goal will serve to organize and measure the best of our energies and skills, because that challenge is one that we are willing to accept, one we are unwilling to postpone, and one which we intend to win, and the others, too.

It is for these reasons that I regard the decision last year to shift our efforts in space from low to high gear as among the most important decisions that will be made during my incumbency in the office of the Presidency". **President John F. Kennedy** Speech at Rice University, Houston, 12 September 1962.

Neil Armstrong (left) working on the Moon's surface by the Lunar Module (LM) and Edwin "Buzz" Aldrin (right) during the first test walk on the moon in 1969.
https://www.nasa.gov/mission_pages/apollo/apollo11.html

Apollo 11 was the spaceflight that landed the first humans, Americans **Neil Armstrong** and **Buzz Aldrin**, on the Moon on July 20, 1969. Armstrong became the first to step onto the lunar surface 6 hours later on July 21 followed shortly afterwards by Aldrin and together they collected 47.5 pounds (21.5 kg) of lunar material for return to Earth. A third member of the mission, **Michael Collins**, piloted the command spacecraft alone in lunar orbit until Armstrong and Aldrin returned to it for the trip back to Earth.

The America had won the race to the Moon and had now leapt ahead of the Soviets in space exploration!

Most U.S. space exploration efforts have been lead by NASA, including the Apollo moon-landing missions, the **Skylab space station**, and later the **Space Shuttle**. Currently, NASA is supporting the **International Space Station** and is overseeing the development of the **Orion and Commercial Crew vehicles**. The agency is also responsible for the **Launch Services Program (LSP)** which provides oversight of launch operations and countdown management for unmanned NASA launches. Most recently, NASA announced a new **Space Launch System** that it said would take the agency's astronauts farther into space than ever before and provide the cornerstone for future human space exploration efforts by the U.S

Lost Cosmonauts or Phantom Cosmonauts

Apart from many Soviet firsts in space exploration, they were fraught with serious problems, disasters and outright catastrophes which went unheard of by the general public and unacknowledged by the Soviet Union until 1991, when the Communist regime fell and Russia in a new air openness or détente began to let the world in on some these well-kept secrets and programs.

Not long after men started to go into space, rumours and reports began to surface out of the Soviet union of catastrophic launch failures and cosmonauts being lost in space. A conspiracy theory called the **Lost Cosmonauts** or **Phantom Cosmonauts** alleged that Soviet cosmonauts entered outer space but, met with disaster with crash landings or simply vanishing into space, without their existence having been acknowledged by either the Soviet or Russian space authorities.

Proponents of the **Lost Cosmonauts Theory** concede that Yuri Gagarin was the first man to *survive* space travel, but claim that the Soviet Union attempted to launch two or more manned space flights prior to Gagarin's and that at least two cosmonauts died in the attempts. Another cosmonaut, **Vladimir Ilyushin** is believed to be **the first successful launch and return of a human being into space** but, his off-course landing ended in China where he was held by the Chinese government. The Government of the Soviet Union supposedly suppressed this information, to prevent bad publicity during the height of the Cold War.

The evidence cited to support Lost Cosmonaut theories is generally not regarded as conclusive, and several cases have been confirmed as hoaxes. In the 1980s, American journalist **James Oberg** researched space-related disasters in the Soviet Union but found no evidence of these Lost Cosmonauts. Since the collapse of the Soviet Union in the early 1990s, much previously restricted information is now available, including on **Valentin Bondarenko**, an early cosmonaut whose death on Earth, the Soviet government covered up. Even with the availability of published Soviet archival material and memoirs of Russian space pioneers, no hard evidence has emerged to support the Lost Cosmonaut stories. ***Bear in mind that not finding evidence particularly between Cold War adversaries is not proof that no evidence exists.*** (Bold italics added by author for emphasis). http://en.wikipedia.org/wiki/Lost_Cosmonauts

On the website *Cracked.com*, an article by **Evan V. Symon** titled: "5 Soviet Space Programs That Prove Russia Was Insane" tells of five tragedies that the Soviet space program experienced in the Space Race to get a man into space first, there were no doubt others that we may never know. The Soviets had the ability to cover up every single failure and destroy all evidence of incompetence for propaganda purposes, which is why no one can actually prove conclusively what really happened.

Also in 1959, pioneering space theoretician **Hermann Oberth** claimed that a pilot had been killed on a sub-orbital ballistic flight from **Kapustin Yar** in early 1958. He claimed at least four Soviet cosmonauts had already died between the years 1957 and 1959. According to **Hermann Oberth**, the Soviets were converting rockets to manned spaceships, and he would know because he was working with NASA and had seen the intelligence to prove it. http://www.cracked.com/article_19142_5-soviet-space-programs-that-prove-russia-was-insane.html

In December 1959, an alleged high-ranking Czech Communist official leaked information about many purported unofficial space shots. **Aleksei Ledovsky** was mentioned as being launched inside a converted **R-5A rocket**. Three more names of alleged cosmonauts claimed to have perished under similar circumstances were **Andrei Mitkov, Sergei Shiborin**, and **Maria Gromova**.

In December 1959, the Italian news agency **Continentale** repeated the claims that a series of cosmonaut deaths on suborbital flights had been revealed by a high-ranking Czech communist. No other evidence of Soviet sub-orbital manned flights ever came to light. http://en.wikipedia.org/wiki/Lost_Cosmonauts

But it wasn't until two Italian brothers with a knack for radio got in the act that the story really gained some ground. After cobbling together an improvised listening system comprised of scavenged equipment and sheer audacity, Achilles and Giovanni Battista Judica-Cordiglia started picking up some telemetry. Specifically, an SOS signal in Morse Code and the dying gasps and fading heartbeat of a cosmonaut whose signal was getting farther and farther away from Earth. And a Russian woman who said:

Transmission begins now. Forty-one. Yes, I feel hot. I feel hot, it's all... it's all hot. I can see a flame! I can see a flame! I can see a flame! Thirty-two... thirty-two. Am I going to crash? Yes, yes I feel hot... I am listening, I feel hot, I will re-enter. I'm hot!

Just as signs of screaming began, the transmission was cut off.
http://www.cracked.com/article_19142_5-soviet-space-programs-that-prove-russia-was-insane.html

On May 19, 1961, the **Torre Bert** listening station in northern Italy purportedly picked up a transmission of a woman's voice, sounding confused and frightened as her craft began to break up upon re-entry. One argument against this claim is that a transmission could not be heard of the re-entry stage of a flight, as there is a communications blackout when a vehicle enters Earth's atmosphere. Additionally, the voice on the recording does not adhere to Soviet cosmonauts' standard communications protocol. According to the official records, there were no launches from any Soviet launch sites that could have corresponded to this event. The two closest events were suborbital test launches of the **R-16 ICBM** on the 16th and the 24th.

Another recording from Torre Bert purports to be the sounds of labored breathing and a failing heartbeat. This combined with reports in the French and Italian press, claiming that **Sputnik** was a manned mission, gave rise to claims that a cosmonaut named **Gennady Mikhailov** was the first man in orbit and died there due to heart failure. According to the **TASS news agency** it was a failed Venus probe. Opponents of this theory claim these recordings are of highly doubtful veracity, as data on heart rate and breathing patterns were not transmitted via audio on Vostok spacecraft, but via telemetric data. However, breathing into a mic and an audible heartbeat monitor would still allow this theory.

The third Torre Bert recording claims to have heard a couple launched on February 17, 1961, aboard a **Lunik spacecraft** orbiting the earth, reporting *"Everything is satisfactory, we are orbiting the earth"* at regular intervals. On February 24, 1961, there were some garbled verbal transmissions about something the couple could see outside their ship that they urgently had to communicate to Earth. What happened is unclear, but communication was lost. Around the same time, the listening station at Torre Bert reportedly picked up an SOS signal from a craft in space. As the signal got weaker, it was assumed whatever craft it was *disappeared into deep space.* **Alexey Belokonev** is reportedly one of three (two men and a woman) cosmonauts aboard a November 1962 flight. The Torre Bert tower in Italy allegedly picked up a frantic set of messages relayed by the three occupants*. 'Conditions growing worse why don't you answer? . . . we are going slower . . . the world will never know about us........* (Italics and Bold added by author for emphasis) http://en.wikipedia.org/wiki/Lost_Cosmonauts and https://www.youtube.com/watch?v=2PYae8Pyzcc

Vladimir Ilyushin, son of Soviet airplane designer Sergey Ilyushin, was a Soviet pilot and is purported to have been a cosmonaut, alleged by some to have actually been the **first successful man in space** on April 7, 1961—an honor generally attributed to **Yuri Gagarin** on April 12, 1961.The theories surrounding this alleged orbital flight are that a failure aboard the spacecraft caused controllers to bring the descending capsule down several orbits earlier than intended, resulting in its landing in the **People's Republic of China**. The pilot was then held by Chinese authorities for a year before being returned to the Soviet Union. The international embarrassment that would have resulted from such an incident is cited as the Soviets' reason for not publicizing this flight—they reportedly focused their publicizing efforts on the subsequent successful flight of Yuri Gagarin instead. The theory gained some credibility in 1999 due to a documentary on the subject titled *The Cosmonaut Cover-up*. Interviewed in English, **Sergei Khrushchev**, son of former Soviet leader **Nikita Khrushchev**, said that it was true and that **Vladimir Ilyushin** was actually held in China for over a year as a "guest" of the People's Republic of China. He was later returned to the Soviet Union, but by then the Gagarin legend was in place and the bizarre incident was covered up. The main reason for concealment was to not let the West see the schismbetween China and the USSR.

http://en.wikipedia.org/wiki/Lost_Cosmonauts

Vladimir Ilyushin was Russia's 1st successful, but unacknowledged cosmonaut in space April 7, 1961

https://en.wikipedia.org/wiki/Vladimir_Ilyushin

These knuckle-biting and gut-wrenching tragedies forced the space programs of both countries into major setbacks, sometimes scraping unreliable rocket and capsule designs or whole space missions. Space flight was and still is a remarkable new technology fraught with many unforeseen variables, any of which could turn initial triumph into disastrous misfortune. The Soviet space agency begrudgingly admitted to a few these failures in their space program yearslater, as it was hard to hide the telemetry that was being monitored by both the British and Americans, who kept track of every launch activity behind the Iron curtain, both manned or unmanned.

117

Even, the Americans suffered their own catastrophes with the **Apollo 1 Command Module fire**. A cabin fire during a launch pad test on January 27 at Launch Pad 34 at **Cape Canaveral** killed all three crew members—**Command Pilot Virgil "Gus" Grissom**, **Senior Pilot Edward H. White** and **Pilot Roger B. Chaffee**—and destroyed the Command Module. http://en.wikipedia.org/wiki/List_of_spaceflight-related_accidents_and_incidents

There was also, the problem-plagued **Apollo 13** mission which barely made it back to Earth. Each leg of its journey toward the Moon and it return flight back to Earth was fraught with one or more technical complications, any of which could have ended the life of its crew, all the whilcflying in a severely damaged space command module. The world literally held its breath while prays were offered up for its safe return home. Apollo 13 eventually splashed down safely in thePacific Ocean and the mission was deemed a "successful failure" in that more was learned aboutengineering design, safety, and emergency protocols, all to avoid similar problems in future Moon missions. It was one for the history books!

Other US space tragedies were the **Space Shuttle Challenger** which disintegrated during launchin January 28,1986 because of faulty **O-ring seal** allowed hot gases from the shuttle **solid rocket booster (SRB)** to impinge on the external propellant tank and booster strut. The strut andaft end of the tank failed, allowing the top of the SRB to rotate into the top of the tank.

The Space Shuttle Columbia disintegrated upon re-entry on February 1, 2003, after a two-weekmission, **STS-107**. Damage to the shuttle's **thermal protection system (TPS)** led to structural failure of the shuttle's left wing and the spacecraft ultimately broke apart. Investigation revealeddamage to the reinforced carbon-carbon leading edge wing panel resulted from the impact of a piece of foam insulation that broke away from the external tank during the launch. http://en.wikipedia.org/wiki/List_of_spaceflight-related_accidents_and_incidents

These historical triumphs and tragedies in space by both countries during the Space Race and the Moon Race underscore one hidden fact that has never been disclosed by either of these two super powers. The real reason for the Space Race which on the surface appeared to be based on national security and military technical superiority over the other nation was, in fact, another **BIG LIE!** Again, recall the testimony given by **Dr. Carol Rosin** (stated earlier in this book) astold to her repeatedly by **Dr. Wernher von Braun** that the Russians or the Soviet Union who would be painted as the **"Bad Guys"**!

Both countries had been experiencing numerous UFO incursions and overflights of their militarybases as well as many large cities, large construction sites and the thousands of reports that camein from the general public to local police and military bases officials, plus the historic accounts made by pilots and ground troops during the Second World War of unusual flying objects seen inthe sky around aircraft. The Roswell Saucer Crash and similar crashes of other unearthly flying objects were well-established realities certainly by the end of WWII in the higher echelons of military officialdom and by the end of 1947 it was a cold hard fact.

The US and the Russians knew that these were not simply manmade craft but something Extraterrestrial in nature and getting into space for a closer look at these strange spacecraft or to travel to their home world or moon of origin was of paramount importance. Astronomers had reported strange objects or lights or other anomalies on or near the Moon for centuries since the first telescopes were invented. NASA scientists and the military took all these reports seriously, particularly whenever astronauts and cosmonauts were routinely going up into space to advance their space exploration efforts. It was no coincidence that every spaceflight into Earth orbit or towards the Moon was being monitored by an ET presence. ETs it seems were interested in everything that humans were doing, the more so when we started leaving the Earth. Our capabilities increased, our spacecraft development advanced and our missions grew more complicated; from the perspective of an interstellar civilization visiting the Earth, humans were on the move and they had agenda.

CHAPTER 67

NASA, ASTRONAUTS, COSMONAUTS, AND UFONAUTS

When UFOs were sighted by civilians, by military personnel or when they are pursued by air force pilots or tracked by radar, either on the ground or in the air then are seen to zip off at greatspeed into space, it should not then, come as any surprise that when astronauts and cosmonauts are circling the Earth in their spacecraft, that they would not be alone in space. In fact, almost every space mission, astronauts, and cosmonauts have reported sightings of UFOs or alien spacecraft in orbit with them or off in the distance.

When astronauts and cosmonauts see UFOs in space, it is not a delusion or the psychological effects of a microgravity environment, space sickness, radiation from space or long durations in cramped quarters. Astronauts are trained to adapt and work in this type of environment and theyare closely monitored by flight command and ground radar. Nothing can go unnoticed as cameras are usually focused on the astronauts most of the time in their spacecraft or the **International Space Station (ISS)** and on all **extravehicular activity (EVA)** even, the Earth isroutinely photographed or video recorded.

So, when they say that they see an unidentified flying object outside their space shuttle window that is exactly what they are seeing. This is observation is not open for dispute or re- interpretation of the facts. It is what they say it is! It is also being picked up on ground radar at flight control so, they know that when an astronaut reports a UFO near the **Space Shuttle** or thespace station, ground control is tracking it and the shuttle or the ISS. Sometimes, when astronauts see alien spacecraft either from the window of the shuttle or during a spacewalk working session and report their observation, command centre will simply order them to continue with their schedule task and to ignore the UFO that they are aware of its presence.

There have been occasions when the astronaut is excited and insistent about the UFO(s) presence and they lose focus of their task; command centre has in an irritated fashion ordered them to forget it and to get back to work.

American Astronauts

What do astronauts and cosmonauts see when they are up in space to orbit the Earth? Accordingto American astronaut **Scott Carpenter**, they see UFOs monitoring their space activities every time they go up!

Scott Carpenter probably said it best: *"At no time, when the astronauts were in space were they alone: there was a constant surveillance by UFOs."*

Initially, it was thought that astronauts were seeing small particles of ice that would break off the space capsule every time the capsule emerged around the Earth from the night side and enter intothe day side. Ice would form while on the dark side and when entering the day side every 90 minutes, the heat from the Sun's rays would heat up the skin surface of the capsule causing the ice to flake off. The rays of the Sun would reflect brilliantly off the ice particles creating a dazzling display of "fireflies" that would appear to move in and around the spacecraft. **John Glenn** in *Friendship 7* reported such an encounter on one of his three orbits describing as beingquite animated in their move around his craft. Waste dumps of urine were also considered as likely causes of these *"fireflies"* as reported by other astronauts on later missions. Astronauts were new to space so, it is possible to mistakenly identify normal things during unusual circumstances.

Mercury astronaut Scott Carpenter
https://astronautscholarship.org/Astronauts/scott-carpenter/

However, in the section on actual ET contact, we shall discover that some ET intelligences do appear to show themselves as bright sparkling light similar to fireflies! It may still be possible that Glenn did see live ET beings surrounding his spacecraft and not the mistaken ice particles orthe waste dumps of high flying streams of piss in orbit with him.

Of all the astronauts and cosmonauts that have gone up into space and returned safely, none have been more outspoken about the existence of UFOs and ETI than have **Major Gordon Cooper, Dr. Edgar Mitchell, Scott Carpenter, Eugene Cernan, Dr. Brian O'Leary, Edwin "Buzz" Aldrin, Story Musgrave** as well as Russian Cosmonauts **Marina Popovich, Major General**

Pavel Popovich and Vladimir Azhazha. There are many others from their corps that have also come forward with their testimony; each is recognized for the high caliber of their character andprofessionalism, and all are considered as heroes of their countries.

A closer look at their testimony is both startling and very revealing. If there were no other witnesses or evidence to the UFO/ETI phenomenon, their accounts would be more than enoughto prove the validity that Extraterrestrial Intelligences are visiting the Earth at this time.

Scott Carpenter

"At no time, when the astronauts were in space were they alone: there was a constantsurveillance by UFOs." http://www.syti.net/UFOSightings.html

Carpenter photographed a UFO while in orbit on May 24, 1962. NASA still has not released thephotograph.

Neil Armstrong

"We have no proof, But if we extrapolate, based on the best information we have available to us,we have to come to the conclusion that ... other life probably exists out there and perhaps in many places ..." From a statement in October 1999. http://www.syti.net/UFOSightings.html

Apollo 11 astronaut Neil Armstrong, first man to walk upon the gives a talkat the White House about *"removing one of truth protective layers"*
https://www.youtube.com/watch?v=PUx1SURbb3g

It should be noted here that Neil Armstrong was very tight lipped for most of his career and throughout his retired life remaining patriotically loyal to his country and to his military oaths of secrecy which cause him a lot of stress and consternation. He knew that Extraterrestrial life existed as he had witnessed their spacecraft (UFOs) up close on the Moon. In one of his last rare public appearances on the 25th anniversary of the first **Apollo Moon Landing,** he gave a brief tearful talk at the **White House** to some taxpayers in attendance in which he revealed in a somewhat carefully worded, cryptic language that the UFO subject matter was real and that *"… we leave you with much that is undone. There are great ideas undiscovered. Breakthroughs available to those who can remove one of truth's protective layers".*

Eugene Cernan

"…I've been asked about UFOs and I've said publicly I thought they were somebody else, some other civilization." **Cernan commanded the Apollo 17 Mission --- The quote is from a 1973 article in the Los Angeles Times.** http://www.syti.net/UFOSightings.html

Apollo 17 astronaut Eugene Cernan
https://www.pinterest.com/indigo1123/gone-but-not-forgotten/?lp=true

Colonel L. Gordon Cooper

Colonel L. Gordon Cooper is one of the original Mercury Astronauts and the last American to fly in space alone. On May 15, 1963, he shot into space in a Mercury capsule for a 22 orbit journey around the world. During the final orbit, Major Gordon Cooper told the tracking station at Muchea (near Perth Australia) that he could see a glowing, greenish object ahead of him quickly approaching his capsule. The UFO was real and solid because it was picked up by Muchea's tracking radar. Cooper's sighting was reported by the National Broadcast Company, which was covering the flight step by step; but when Cooper landed, reporters were told that they would not be allowed to question him about the UFO sighting.

Colonel L. Gordon Cooper
https://astronautscholarship.org/Astronauts/l-gordon-cooper-jr/

Major Cooper was a firm believer in UFOs. Ten years earlier, in 1951 he had sighted a UFO while piloting an F-86 Sabre jet over Western Germany. They were metallic, saucer-shaped discs at considerable altitude and could out-maneuver all American fighter planes. Major Cooper also testified before the **United Nations**: *"I believe that these extra-terrestrial vehicles and their crews are visiting this planet from other planets... Most astronauts were reluctant to discuss UFOs." "I did have occasion in 1951 to have two days of observation of many flights of them, of different sizes, flying in fighter formation, generally from east to west over Europe."*

And according to, a taped interview by J. L. Ferrando, **Major Cooper** said:

"For many years I have lived with a secret, in a secrecy imposed on all specialists in astronautics. I can now reveal that every day, in the USA, our radar instruments capture objects of form and composition unknown to us. And there are thousands of witness reports and a quantity of documents to prove this, but nobody wants to make them public. Why? Because authority is afraid that people may think of God knows what kind of horrible invaders. So the password still is: We have to avoid panic by all means."

"I was furthermore a witness to an extraordinary phenomenon, here on this planet Earth. It happened a few months ago in Florida. There I saw with my own eyes a defined area of ground being consumed by flames, with four indentions left by a flying object which had descended in the middle of a field. Beings had left the craft (there were other traces to prove this). They seemed to have studied topography, they had collected soil samples and, eventually, they returned to where they had come from, disappearing at enormous speed... I happen to know that authority did just about everything to keep this incident from the press and TV, in fear of a panicky reaction from the public." http://www.syti.net/UFOSightings.html

Cooper, addressing a United Nations panel discussion on UFOs and extraterrestrials in New York in 1985. The panel was chaired by then **Secretary-General Kurt Waldheim**:

"I believe that these extraterrestrial vehicles and their crews are visiting this planet from other planets, which, obviously are a little more technically advanced than we are here on Earth. I feel that we need to have a top-level, coordinated program to scientifically collect and analyze data from all over the earth concerning any type of encounter, and to determine how best to interface with these visitors in a friendly fashion. We may first have to show them that we have learned to resolve our problems by peaceful means, rather than warfare, before we are accepted as fully qualified universal team members. This acceptance would have tremendous possibilities of advancing our world in all areas. Certainly, then it would seem that the UN has a vested interest in handling this subject properly and expeditiously."

Cooper testifying to a United Nations committee:

"For many years I have lived with a secret, in a secrecy imposed on all specialists and astronauts. I can now reveal that every day, in the U.S.A., our radar instruments capture objects of form and composition unknown to us. And there are thousands of witness reports and a quantity of documents to prove this, but nobody wants to make them public."

"As far as I am concerned, there have been too many unexplained examples of UFO sightings around this Earth for us to rule out the possibilities that some form of life exists out there beyond our own world." http://www.syti.net/UFOSightings.html

"I know other astronauts share my feelings, and we know the government is sitting on hard evidence of UFOs!"

"We thought they could have been Russian. We regularly had MiG-15s overflying our base. We scrambled our Sabre jets to intercept and got to our ceiling of 45,000 feet ... and they were still way above us traveling faster than we were. These vehicles were in formation like a fighter group, but they were metallic silver and saucer-shaped. Believe me; they weren't like any MiGs I'd seen before! They had to be UFOs."

Cooper said he first encountered UFOs as a military pilot in Germany in the early 1950s when unidentified craft were spotted over an air base:

"I had a camera crew filming the installation when they spotted a saucer. They filmed it as it flew overhead, then hovered, extended three legs as landing gear, and slowly came down to land on a dry lake bed! These guys were all pro cameramen, so the picture quality was very good. The camera crew managed to get within 20 or 30 yards of it, filming all the time. It was a classic saucer, shiny silver and smooth, about 30 feet across. It was pretty clear it was an alien craft. As they approached closer it took off." When his camera crew handed over the film, Cooper followed standard procedure and contacted Washington to report the UFO and "all heck broke loose," he said. "After a while, a high-ranking officer said when the film was developed I was to put it in a pouch and send it to Washington. He didn't say anything about me not looking at the film. That's what I did when it came back from the lab and it was all there just like the camera crew reported." When the Air Force later started **Operation Blue Book** *to collate UFO evidence and reports, Cooper says he mentioned the film evidence. "But the film was never found, supposedly. Blue Book was strictly a cover-up anyway."*
http://www.syti.net/UFOSightings.html

In 1957, Cooper was one of an elite band of test pilots at Edwards Air Force Base in California, in charge of several advanced projects, including the installation of a precision landing system:

"I had a good friend at Roswell, a fellow officer. He had to be careful about what he said. But it sure wasn't a weather balloon, like the Air Force cover story. He made it clear to me what crashed was a craft of alien origin, and members of the crew were recovered."

Cooper revealed he's convinced an alien craft crashed at Roswell, New Mexico, in 1947 and aliens were discovered in the wreckage.

"It started in World War II when the government didn't want people to know about UFO reports in case they panicked," said Cooper. "They would have been fearful it was superior enemy technology that we had no defense against. Then it got worse in the **Cold War** *for the same reason. So they told one untruth, they had to tell another to cover that one, and then another, then another ... it just snowballed. And right now I'm convinced a lot of very embarrassed government officials are sitting there in Washington trying to figure a way to bring the truth out. They know it's got to come out one day, and I'm sure it will. America has a right to know!"*--- A statement when asked why has the government kept its UFO secrets for so many years? Cooper was a Mercury-9 and Gemini-5 astronaut.

126

"I believe certain reports of flying saucers to be legitimate.""

James McDivitt

"At one stage we even thought it might be necessary to take evasive action to avoid a collision."
James McDivitt commenting on an orbital encounter he and Ed White had with a "weird object" with arm-like extensions which approached their capsule. Later in the flight, they saw two similar objects over the Caribbean.

James Lovell

James Lovell was commander of both the Apollo-8 mission and the ill-fated Apollo-13 mission. He made this transmission after coming around the far side of the moon on the Apollo-8 mission on or around Christmas in 1968 with Frank Borman and William Anders. Although it was Christmas time, this statement has caused considerable controversy as "Santa Claus" was apparently a codeword used to indicate a UFO or other unusual sighting.

"Mission Control, please be informed, there is a Santa Claus."

Edgar D. Mitchell

"We all know that UFOs are real. All we need to ask is where do they come from?"
From a statement in 1971.

"I've talked with people of stature --- of military and government credentials and position --- and heard their stories, and their desire to tell their stories openly to the public. And that got my attention very, very rapidly ... the first-hand experiences of these credible witnesses that, now, in advanced years are anxious to tell their story. We can't deny that, and the evidence points to the fact that Roswell was a real incident. And that, indeed, an alien craft did crash, and that material was recovered from that crash site."

"The U.S. Government hasn't maintained secrecy regarding UFOs. It's been leaking out all over the place. But the way it's been handled is by denial, by denying the truth of the documents that have leaked. By attempting to show them as fraudulent, as bogus of some sort. There has been a very large disinformation and misinformation effort around this whole area. And one must wonder, how better to hide something out in the open than just to say, 'It isn't there. You're deceiving yourself if you think this is true.' And yet, there it is right in front of you. So it's a disinformation effort that's concerning here, not the fact that they have kept the secret. They haven't kept it. It's been getting out into the public for fifty years or more."

*"I have been, over the years, very skeptical like many others. But in the last ten years or so, I have known the late **Dr. Alan Hynek**, who I highly admire. I know and currently, work with **Dr. Jacques Vallee.** I've come to realize that the evidence is building up to make this a valid and researchable question. Further, because my personal motivation has always been to understand our universe better, and my own theoretical work has convinced me that life is everywhere in the*

universe that has been permitted to evolve, I consider this a very timely question ... by becoming more involved with the serious research field, I've seen the evidence mount towards the truth of these matters. I rely upon the testimony of contacts that I have had --- old timers --- who were involved in official positions in government and intelligence and military over the last 50 years. We cannot say that today's government is really covering it up --- I think that most of them don't know what is going on any more than the public ..."
From an interview with MSN 1998.

"The evidence points to the fact that Roswell was a real incident and that indeed an alien craft did crash and that material was recovered from that crash site," says Mitchell, who became the sixth man on the moon in the Apollo 14 mission. Mitchell doesn't say he's seen a UFO. But he says he's met with high-ranking military officers who admitted involvement with alien technology and hardware." http://www.syti.net/UFOSightings.html

Apollo 14 Dr. Edgar D. Mitchell

Lee Katchen (Former Atmospheric Physicist with NASA)

"UFO sightings are now so common, the military doesn't have time to worry about them - so they screen them out. The major defense systems have UFO filters built into them, and when a UFO appears, they simply ignore it."

Charles J. Camarda (Ph.D.) NASA Astronaut

"In my official status, I cannot comment on ET contact. However, personally, I can assure you, we are not alone!

Joseph A. Walker

On May 11, 1962, NASA pilot **Joseph A. Walker** said that one of his tasks was to detect UFOs during his X-15 flights. He had filmed five or six UFOs during his record breaking fifty-mile-high flight in April 1962. It was the second time he had filmed UFOs in flight. During a lecture at the **Second National Conference on the Peaceful Uses of Space Research** in Seattle, Washington he said:

"I don't feel like speculating about them. All I know is what appeared on the film which was developed after the flight." - Joseph Walker - to date none of those films has been released to the public for viewing.

X15 Test Pilot Joseph Walker
https://en.wikipedia.org/wiki/File:Joe_Walker,_test_pilot_and_astronaut,_X-15.jpg

Donald (Deke) Slayton

Donald (Deke) Slayton, a Mercury astronaut, revealed in an interview he had seen UFOs in 1951:

"I was testing a P-51 fighter in Minneapolis when I spotted this object. I was at about 10,000 feet on a nice, bright, sunny afternoon. I thought the object was a kite, then I realized that no kite is gonna fly that high. As I got closer it looked like a weather balloon, gray and about three feet in diameter. But as soon as I got behind the darn thing it didn't look like a balloon anymore. It looked like a saucer, a disk. About the same time, I realized that it was suddenly going away from me --- and there I was, running at about 300 miles per hour. I tracked it for a little way, and then all of a sudden the damn thing just took off. It pulled about a 45 degree climbing turn and accelerated and just flat disappeared."

Mercury astronaut Donald (Deke) Slayton

Robert White

On July 17, 1962, Major **Robert White** reported a UFO during his fifty-eight-mile high flight of an X-15. Major White reported:

"I have no idea what it could be. It was grayish in color and about thirty to forty feet away." Major White then exclaimed over the radio: *"There ARE things out there! There absolutely is!"* From a Time Magazine article

Dr. Brian O'Leary

"We have contact with alien cultures."

Storey Musgrove

"Statistically, it's a certainty there are hugely advanced civilizations, intelligences, life forms out there. I believe they're so advanced they're even doing interstellar travel. I believe it's possible they even came here."

Ed White & James McDivitt

In June 1965, astronauts **Ed White** (first American to walk in space) and **James McDivitt** were passing over Hawaii in a Gemini spacecraft when they saw a weird-looking metallic object. The UFO had long arms sticking out of it. McDivitt took pictures with a cine-camera. Those pictures have never been released.

Gemini 4 Astronauts Ed White (left) and James McDivitt
http://www.watchprosite.com/page-wf.forumpost/fi-677/ti-1011308/pi-7131583/

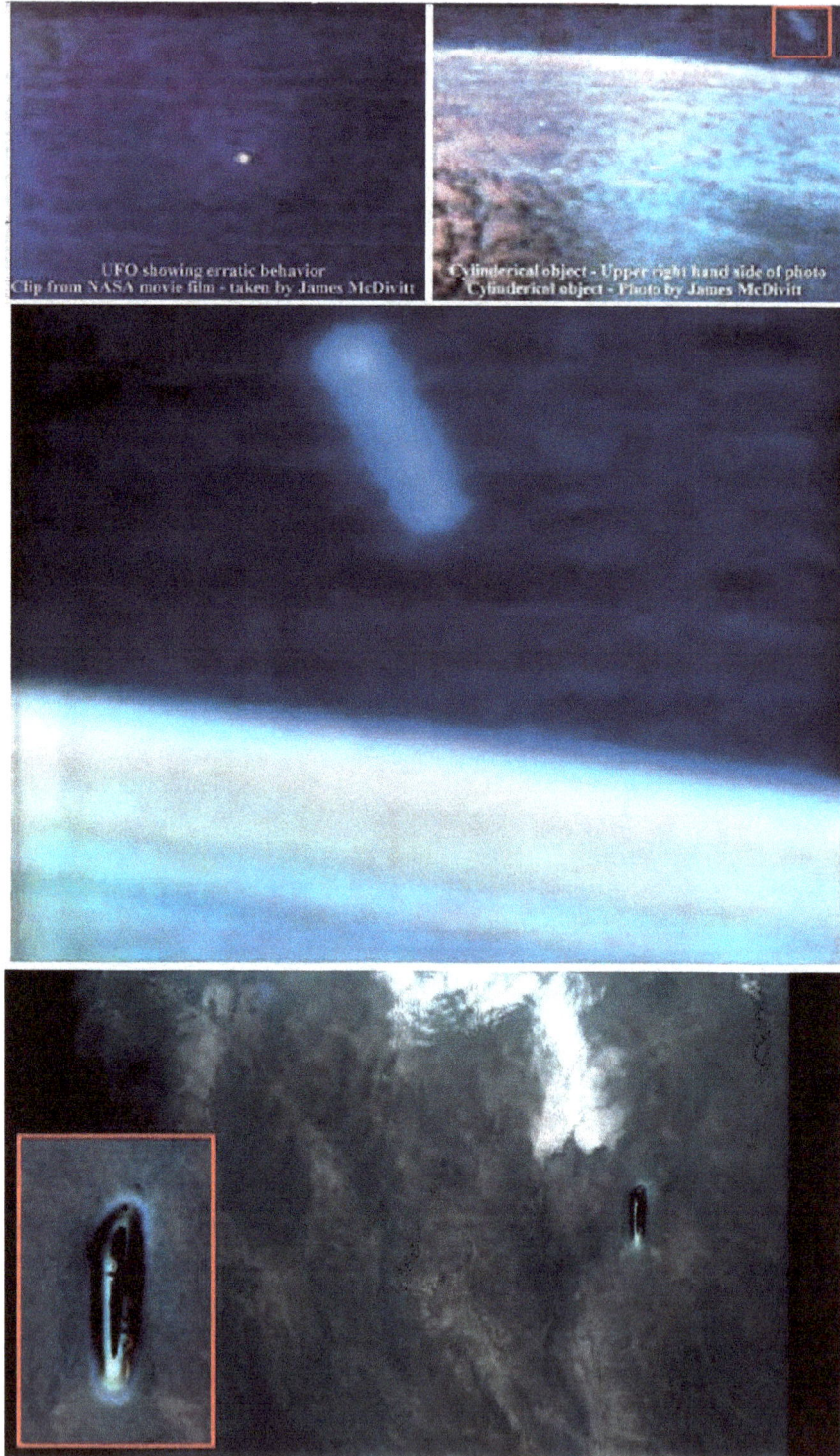

**UFOs photographed by Gemini 4 astronauts White and McDivitt
including enlarged views of unidentified cylindrical objects**

http://www.syti.net/UFOSightings.html and **(Google Image)**

http://www.educatinghumanity.com/2013/10/astronaut-ufo-photo-mcdivitt.html

132

Major Robert White

White exclaiming over the radio about a UFO encounter taking place on a 58 mile high X-15 flight on July 17, 1962:

"I have no idea what it could be. It was greyish in color and about thirty to forty feet away."
"There ARE things out there! He later reported. *"There absolutely is!"*

James Lovell and Frank Borman

In December 1965, Gemini astronauts **James Lovell** and **Frank Borman** also saw a UFO duringtheir second orbit of their record-breaking 14 day flight. Borman reported that he saw an unidentified spacecraft some distance from their capsule. Gemini Control, at Cape Kennedy, told him that he was seeing the final stage of their Titan booster rocket. Borman confirmed that he could see the booster rocket all right, but that he could also see something completely different.

Gemini 7 astronauts James Lovell and Frank Borman
https://www.pinterest.com/darlingtondon/air-space-history/?lp=true

Here is a brief communication during James Lovell's flight on Gemini 7 with Capcom:

Lovell: BOGEY AT 10 O'CLOCK HIGH.

Capcom: This is Houston. Say again 7.

Lovell: SAID WE HAVE A BOGEY AT 10 O'CLOCK HIGH.

Capcom: Roger....

(at this point the live broadcast of the conversation is interrupted by Capcom)

Capcom: Gemini 7 is that the booster or is that an actual sighting?

Lovell: WE HAVE SEVERAL .. ACTUAL SIGHTING.

Capcom .... Estimated distance or size?

Lovell: WE ALSO HAVE THE BOOSTER IN SIGHT...

Lovell is reported to have taken some magnificent photographs of two mushroom shaped UFOs on December 4. The pictures seem to show the glow of a propulsion system on the underside. The pictures were taken at a range of several hundred yards. Some believe this picture to be a blatant forgery accomplished by airbrushing a picture of light reflecting off the nose of the spacecraft.

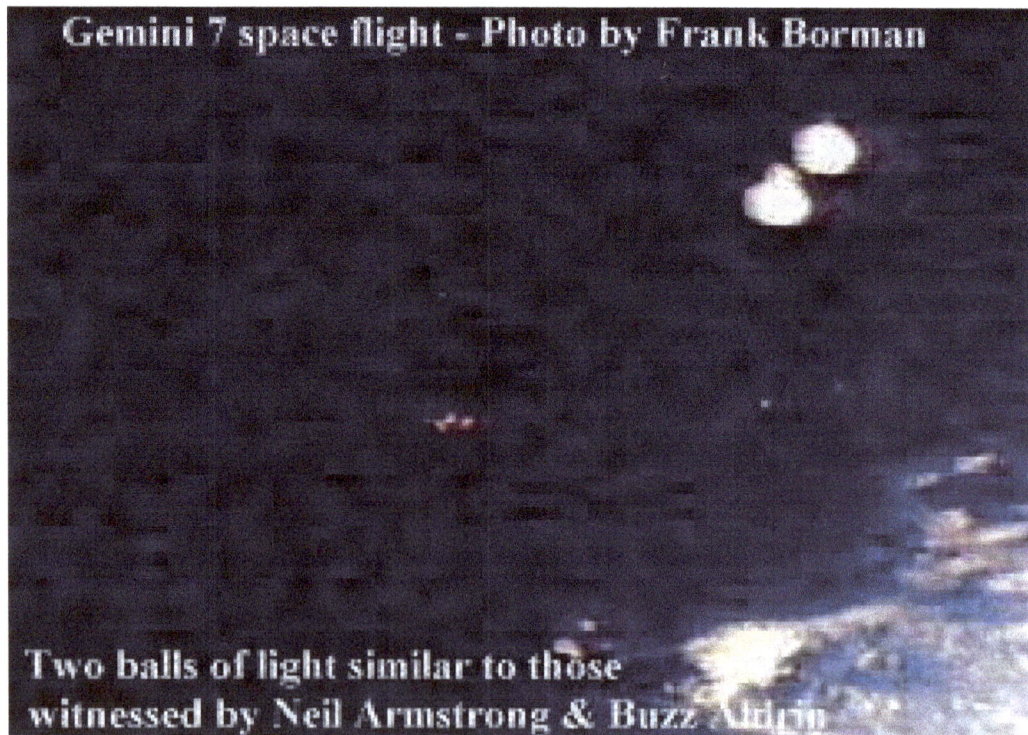

UFO photographed by Astronaut Frank Borman
http://ronrecord.com/astronauts/lovell-borman.html

Lovell also witnessed an incident during flight Gemini XII (James Lovell, Edwin Aldrin): On November 12 both astronauts are reported to have spotted 2 UFO's approximately 1/2 mile from the spacecraft. They were observed for some period of time and photographs were taken.

During the flight of Apollo 8 on Dec. 1968 (Frank Borman, James A. Lovell, and William A. Anders) as the command module emerged from behind the moon, Lovell transmits *"We have been informed that Santa Claus does exist!"* As this occurred on Christmas day 1968 it is deemed by many to have been Lovell's sense of humor for the kiddies watching the broadcast. Some speculate that it is the use of the code word originally employed by Walter Schirra on the Mercury 8 flight. On Christmas eve Borman and Lovell are reported to have seen a UFO. It is reported that an unidentified language was picked up on one of the NASA frequencies used during the mission. http://ronrecord.com/astronauts/lovell-borman.html

Neil Armstrong, Edwin Aldrin, and Michael Collins

According to NASA Astronaut Neil Armstrong, the aliens have a base on the Moon and told us in no uncertain terms to get off and stay off the Moon. According to unconfirmed reports, both Neil Armstrong and **Edwin "Buzz" Aldrin** saw UFOs shortly after that historic landing on the Moon in Apollo 11 on 21 July 1969. I remember hearing one of the astronauts refer to a "light" in or on a crater during the television transmission, followed by a request from mission control for further information. Nothing more was heard. According to a former NASA employee **Otto Binder,** unnamed radio hams with their own VHF receiving facilities that bypassed NASA's broadcasting outlets picked up the following exchange:

NASA: *What's there?*
Mission Control calling Apollo 11...

Apollo11: *These "Babies" are huge, Sir! Enormous!*
OH MY GOD! You wouldn't believe it!
I'm telling you there are other spacecraft out there,
Lined up on the far side of the crater edge!
They're on the Moon watching us!

A certain professor, who wished to remain anonymous, was engaged in a discussion with Neil Armstrong during a NASA symposium.

Professor: *What REALLY happened out there with Apollo 11?*

Armstrong: *It was incredible, of course, we had always known there was a possibility, the fact is, we were warned off! (by the Aliens). There was never any question then of a space station or a moon city.*

Professor: How *do you mean "warned off"?*

Armstrong: *I can't go into details, except to say that their ships were far superior to ours both in size and technology - Boy, were they big! and menacing! No, there is no question of a space station.* http://ronrecord.com/astronauts/lovell-borman.html

Professor: *But NASA had other missions after Apollo 11?*

**(L – R) Neil Armstrong (first man on the Moon), Michael Collins
and Edwin Aldrin (second man on the Moon)**
https://spaceflight.nasa.gov/gallery/images/apollo/apollo11/html/s69_31740.html

Armstrong: *Naturally - NASA was committed at that time, and couldn't risk panic on Earth. But it really was a quick scoop and back again.*

According to a **Dr. Vladimir Azhazha**: *"Neil Armstrong relayed the message to Mission Control that two large, mysterious objects were watching them after having landed near the moon module. But this message was never heard by the public - because NASA censored it."*
According to a **Dr. Aleksandr Kasantsev**, Buzz Aldrin took color movie film of the UFOs from inside the module and continued filming them after he and Armstrong went outside. Armstrong confirmed that the story was true but refused to go into further detail, beyond admitting that the CIA was behind the cover-up. http://ronrecord.com/astronauts/lovell-borman.html

**Edwin "Buzz" Aldrin on the Moon, and Neil Armstrong
can be seen in Aldrin's helmet visor**

https://www.theguardian.com/commentisfree/belief/2012/sep/13/buzz-aldrin-communion-moon

Maurice Chatelain

In 1979 **Maurice Chatelain**, former chief of NASA Communications Systems confirmed that Armstrong had indeed reported seeing two UFOs on the rim of a crater. Chatelain believes that some UFOs may come from our own solar system, specifically **Titan**.

"The encounter was common knowledge in NASA, but nobody has talked about it until now."
"...all Apollo and Gemini flights were followed, both at a distance and sometimes also quite closely, by space vehicles of extraterrestrial origin - flying saucers, or UFOs, if you want to call them by that name. Every time it occurred, the astronauts informed Mission Control, who then ordered absolute silence."

*"I think that Walter Schirra aboard Mercury 8 was the first of the astronauts to use the code name **'Santa Claus'** to indicate the presence of flying saucers next to space capsules.*

However, his announcements were barely noticed by the general public. It was a little different when James Lovell on board the Apollo 8 command module came out from behind the moon and said for everybody to hear: **'PLEASE BE INFORMED THAT THERE IS A SANTA CLAUS.'** Even though this happened on Christmas Day 1968, many people sensed a hidden meaning in those words." The rumors persist.

Maurice Chatelain former chief of NASA Communications Systems
http://www.leslecturesdeflorinette.fr/2018/02/maurice-chatelain-et-l-evolution-des-civilisations.html

NASA may well be a civilian agency, but many of its programs are funded by the defense budget and most of the astronauts are subject to military security regulations. Apart from the fact that the **National Security Agency** screens all films and probably radio communications as well. We have the statements by **Otto Binder, Dr. Garry Henderson,** and **Maurice Chatelain** that the astronauts were under strict orders not to discuss their sightings. And Gordon Cooper has testified to a **United Nations** committee that one of the astronauts actually witnessed a UFO on the ground. If there is no secrecy, why has this sighting not been made public?

Russian Cosmonauts' UFO Sightings

UFO stories surrounding the activities of NASA astronauts have woven a tapestry of authenticity into American Ufology for decades. A few astronauts such as **Gordon Cooper** and more recently Edgar Mitchell have been outspoken supporters of the UFO cause, but they still remain glib about their alleged encounters during the Apollo and other programs. What about Russian cosmonauts? The space program in the old USSR was formidable and full of firsts in specific accomplishments, but the secrecy behind their space programs has also been formidable, until "Glasnost" and the collapse of the Soviet Union.

A rather unusual and distressing final transmission by two unidentified Cosmonauts was reportedly monitored, perhaps from an unacknowledged military space mission) and it appears that the cosmonauts were having some spacecraft problems that needed repairs but there is also, something else outside and near to their craft that is that has caught their attention:

Female Cosmonaut: *I'll take it and hold it with my right hand. Look out the peephole! I have it!"*

Male Cosmonaut: *"There is something! If we do not get out the world will never know about this!"* --- From the final transmission of a pair of Cosmonauts whose scheduled seven-day mission was interrupted by a malfunction of unknown origins. This piece of conversation was recorded on February 24, 1961, while they were trying to repair the damage. The two Cosmonauts were never heard from again.
http://www.syti.net/UFOSightings.html

Cosmonaut Victor Afanasyev

In April of 1979, **Cosmonaut Victor Afanasyev** lifted off from Star City to dock with the Soviet **Solyut 6 space station**. But while en route, something strange happened. Cosmonaut Afanasyev saw an unidentified object turn toward his craft and begin tailing it through space.

Afanasyev commenting on a UFO sighting that occurred while en route to the Solyut 6 space station in April of 1979: *"It followed us during half of our orbit. We observed it on the light side, and when we entered the shadow side, it disappeared completely. It was an engineered structure, made from some type of metal, approximately 40 meters long with inner hulls. The object was narrow here and wider here, and inside there were openings. Some places had projections like small wings. The object stayed very close to us. We photographed it, and our photos showed it to be 23 to 28 meters away."*

In addition to photographing the UFO, Afanasyev continually reported back to Mission Control about the craft's size, its shape, and position. When the cosmonaut returned to earth he was debriefed and told never to reveal what he knew and had his cameras and film confiscated.

Those photos and his voice transmissions from space have never been released. It is only now with the collapse of the Soviet Union that Afanasyev feels that he can safely tell his story.

"It is still classified as a UFO because we have yet to identify the object." Victor did, in fact,make a drawing of the object (see below).

Cosmonaut Victor Afanasyev
http://www.aktual24.ro/dosare-strict-secrete-din-urss-cosmonaut-urmarit-de-un-obiect-ciudat-in-spatiu-vezi-cum-arata/

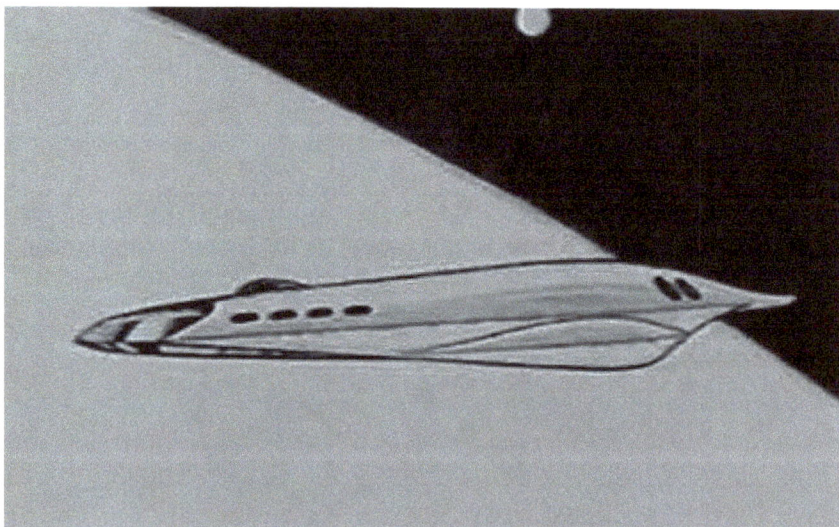

Cosmonaut Victor Afanasyev's drawing of the UFO he witnessed. Note the Similarity to some of the Roswell saucer crash descriptions and drawings
http://www.aktual24.ro/dosare-strict-secrete-din-urss-cosmonaut-urmarit-de-un-obiect-ciudat-in-spatiu-vezi-cum-arata/

140

Cosmonaut Valeri Kubasov

Open Minds Investigative Reporter **J. Antonio Huneeus** met cosmonaut **Valeri Kubasov** (veteran of Soyuz-6 in 1969, the **Apollo-Soyuz** link-up in 1975, and Salyut-6 in 1980) during a press conference in Rio de Janeiro in 1986, Huneeus asked him if it was true that some cosmonauts saw UFOs in orbit. His answer was surprising–not so much his denial, but the fact he became a NASA spokesman too.

"Both Soviet astronauts, as well as American astronauts," said Kubasov, *"have never seen until now any unidentified flying object. With regards to extra-solar life, science considers that this is possible. In our solar system there is no other planet with conditions for life; perhaps in Mars, there are some very simple bacteria, but surely there are no people on Mars."*
http://www.openminds.tv/russian-cosmonauts-ufo-sightings-and-statements/

Cosmonaut Valeri Kubasov
http://www.americaspace.com/2014/02/20/valeri-kubasov-veteran-astp-cosmonaut-dies-aged-79/

In 1980, the famous Soviet magazine **Sputnik** published a long report titled, "UFOs Through the Eyes of Cosmonauts." The article was of statements made by ten Soviet and two American astronauts (both Apollo-Soyuz crew members) regarding UFO sightings in space. Most of the cases were easily explained away as waste dumps, space boosters and accidental loss of small equipment.

Cosmonaut Yevgeni Khrunov

Cosmonaut Yevgeni Khrunov, who flew in the Soyuz-5 in 1969, publicly stated the existence of UFOs:

"As regards UFOs," said Khrunov, *"their presence cannot be denied: thousands of people have seen them. It may be that their source is optical effects but some of their properties, for instance, their ability to change course by 90 degrees at great speed, simply stagger the imagination."*
http://www.openminds.tv/russian-cosmonauts-ufo-sightings-and-statements/

Cosmonaut Yevgeni Khrunov
http://au-fil-du-ciel.eklablog.fr/ovnis-temoignages-d-astronautes-et-cosmonautes-a85767824

After the fall of communism, there was a relaxing of censorship on certain highly classified documents once regarded as top secret in the late 80's and early 90's. The Moscow newspaper **Rabochaya Tribuna** published an extraordinary account on Feb. 28, 1991, which was summarized by the **Foreign Broadcast Information Service (FBIS)**, a US government agency specialized in digesting news from the international media. Here is the pertinent account from the FBIS:

Vladimir Alexandrov

"**Vladimir Alexandrov**, the chief engineer at the **Cosmonaut Training Center**, brought a photograph of a UFO to the editorial offices of. Alexandrov claimed that the flying object in the photograph, which was published on the 28 Feb. issue of Rabochaya Tribuna (RT), was the UFO reported by cosmonauts **Valery Ryumin** and **Leonid Popov** on the night from June 14 to June 15, 1980. Alexandrov claimed that the cosmonauts' sighting had been hushed up at the time but that he was now telling RT what really happened on that night while the cosmonauts were in orbit. He said that a cluster of white, shining spots started to climb up into space from a region near Moscow and actually flew up higher than the cosmonauts' spacecraft, Salyut-6, according to Ryumin and Popov. The UFO was observed around midnight."

The FBIS digest added that the June 1980 UFO sightings were widely seen from the ground as well and that the authorities quickly explained them as a routine Cosmos satellite launch. *This was a cover-up,* according to Alexandrov, who said that *the photograph bears no similarity to that of any satellite launch,"* continued the FBIS document. *"The chief engineer told the RT editors to 'look carefully: at the tip of the luminous cloud, a dark formation in the form of a 'saucer' is noticeable. In my opinion, that is the object itself. As for the cloud–that is the plasma trail which extends back behind it. Bright flashes are also noticeable below the bottom of the saucer."* Unfortunately, we have not seen the photo in question.

George Knapp, a Las Vegas KLAS-TV journalist travelled to Moscow in 1992 to find out what the Russians knew about the UFO phenomenon and returned to the USA with a thick UFO dossier from the old **Soviet Ministry of Defense**, that contained a recent project called **"Thread-3"** from the Russian Ministry of Defense. http://www.openminds.tv/russian-cosmonauts-ufo-sightings-and-statements/

Marina Popovich, Vladamir Kovalenok and Viktor Savinykh

Marina Popovich is a retired Air Force Colonel who was a famous test pilot and also the ex-wife of **General Pavel Popovich**, a pioneer cosmonaut with a keen interest in Ufology who recently passed away. Huneeus interviewed Marina Popovich in Tokyo on Nov. 1990 together with another famous Russian Ufologist, **Dr. Vladimir Azhazha** (his report was later published in *UFO Universe,* fall 1992.) Back to the 'Thread-3' document, it mentions an audio tape of the conversation at 0:14 a.m. on June 17, 1978, between cosmonauts **Vladimir Kovalenok** and **Alexander Ivanchenko** on the **Salyut-6 space station** and Mission Control outside Moscow. The transcript of this tape was presented by Marina at a round-table talk in the editorial office of the Komsomolskaya Pravda newspaper

Here are some excerpts of my interview:

Kovalenok: *"Here is the information for you to think about. On the right, at an angle of 30degrees, there is an object flying under us."*

Control Center: *"Describe the appearance and shape of the object."*

Kovalenok: *"It is something very much like a tennis ball, bright as a flaring up star."*

Ivanchenko: *"Its rate is lower than ours."*

From left: Dr. Vladimir Azhazha, Antonio Huneeus, Marina Popovich

Salyut-6 Cosmonauts Vladamir Kovalenok and Viktor Savinykh
http://www.openminds.tv/russian-cosmonauts-ufo-sightings-and-statements/859

Interestingly, there is second UFO report involving Major-General Vladimir Kovalenok in the Salyut-6 space station in 1981. A fantastic version of it was originally published by Latvian-born reporter Henry Gris in the **National Enquirer**. It purported that an alien spaceship had flown in tandem with the Salyut-6 for two days and that ETs had even left the UFO and approached the portholes of the Russian station for a close encounter. As the National Enquirer is not a reliable source, the story was generally considered apocryphal. In 1993, however, in a videotaped interview with the Italian Giorgio Bongiovanni, Kovalenok admitted a UFO sighting took place in 1981, although not as dramatic as the Henry Gris account. We obtained a copy of this videotape from **Michael Hesemann** a while back and here is what Gen. Kovalenok stated:

*"On May 5, 1981, we were in orbit [in the Salyut-6]. I saw an object that didn't resemble any cosmic objects I'm familiar with. It was a round object which resembled a melon, round and a little bit elongated. In front of this object was something that resembled a gyrating depressed cone, I can draw it, it's difficult to describe. The object resembles a barbell. I saw it becoming transparent and like with a 'body' inside. At the other end, I saw something like gas discharging, like a reactive object. Then something happened that is very difficult for me to describe from the point of view of physics... I have to recognize that it did not have an artificial origin. It was not artificial because an artificial object couldn't attain this form. I don't know of anything that can make this movement... tightening then, expanding, pulsating. Then as I was observing something happened, two explosions. One explosion, and then 0.5 seconds later, the second part exploded. I called my colleague **Viktor Savinykh**, but he didn't arrive in time to see anything".*

*(**This UFO is a bio-engineered ship that "breathes" or pulsates and actually possesses a consciousness thus, in a sense, it is a living being but, it is also has metallic qualities. Truly, this is an example of alien technology**!) (Bold italics added by author).*

145

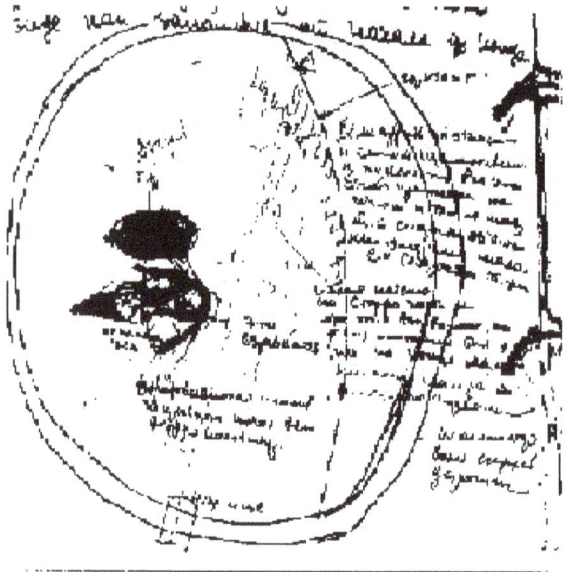

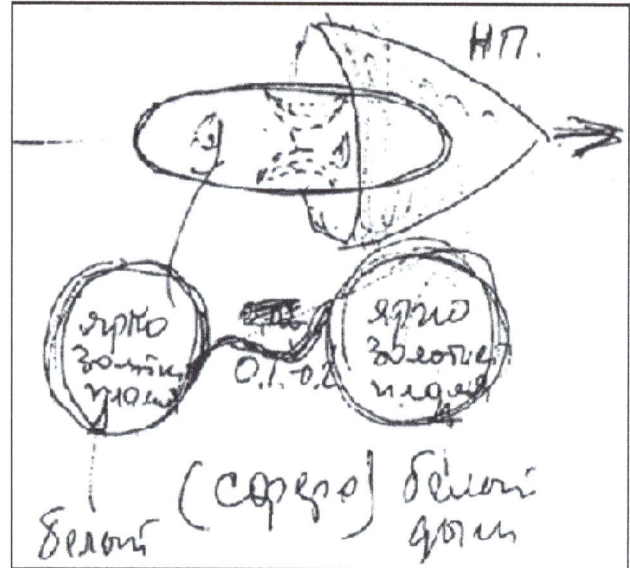

**A page from Cosmonaut Kovalenok's onboard journal, showing
his sketch of the UFO and a cleaned-up version of the drawing**

"What are the particulars? First conclusion: the object moved into a suborbital path, otherwise I wouldn't have been able to see it. There were two clouds, like smoke, that formed a barbell. It came near me and I watched it. Then we entered into the shade for two or three minutes after this happened. When we came out of the shade we didn't see anything. But during a certain time, we and the craft were moving together."

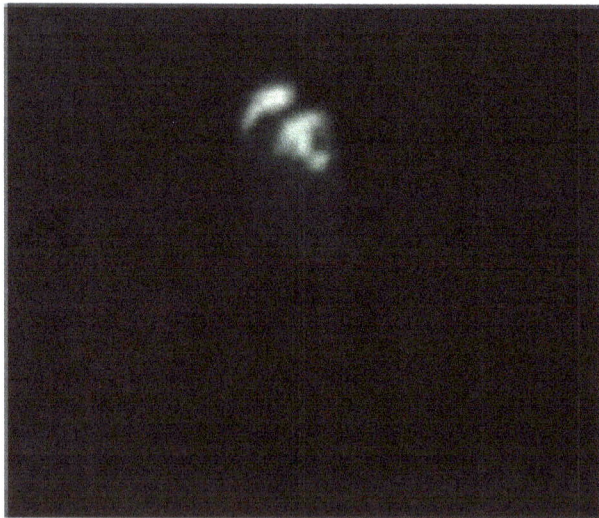

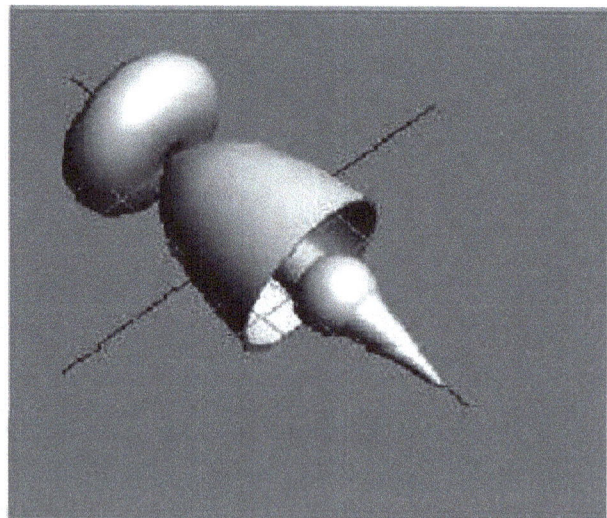

**Photo of UFO taken by astronaut Buzz Aldrin from Apollo 11 left and at right a digital
depiction of the UFO, compare this to the above Cosmonaut Kovalenok's drawing**

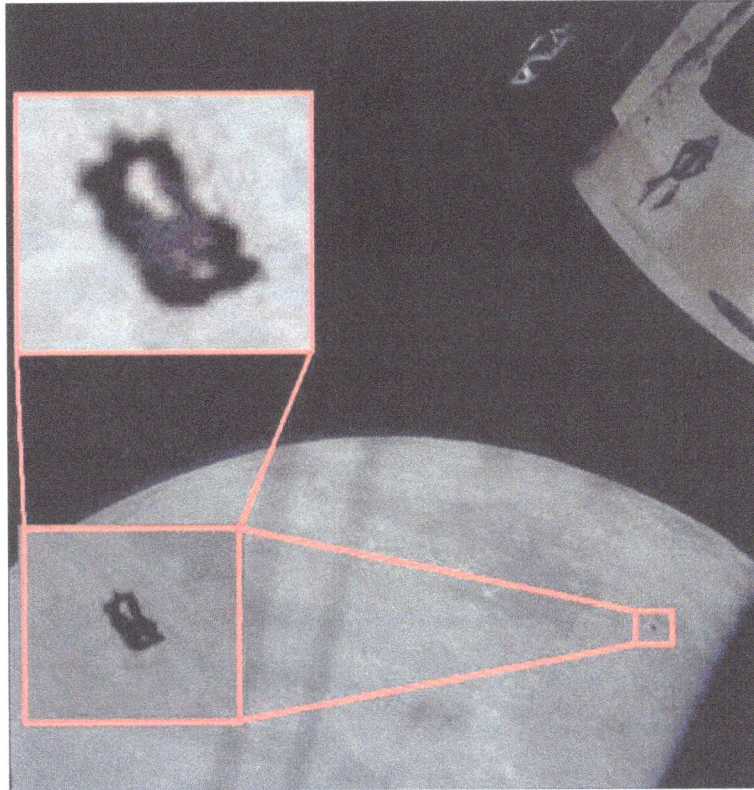

A UFO seen coming up from the Moon, compare this to the drawing seen earlier by cosmonaut Kovalenok?
(Google Image)

Pavel Popovich as a young cosmonaut and later, as a Major General of the Russian Air Force.

General Pavel Popovich

General Pavel Popovich, one of the original cosmonauts who passed away recently at a sanatorium in Crimea on September 30th, also had a UFO sighting, although not from space. He described it in a videotaped interview with St. Petersburg researcher **Valery Uvarov**: *"I saw only once something unexplainable in 1978 when we flew from Washington to Moscow at an elevation of 10,200 meters, and during the flight, we noticed through the window a flying triangle on a parallel course. We flew at the speed of 950 km/h and that triangle (which was equilateral) flew at about 1,500 km/h; he passed us, and both passengers and crew saw it and were talking without coming to any conclusion; it was a UFO and we couldn't compare it to anything."*

Gen. Popovich was also a prominent Ufologist who served for a while as President of the **All-Union Ufology Association of the Commonwealth of Independent States** in Moscow. He was somewhat skeptical of the whole Ufological scene, telling Uvarov that *"you can reject 95% of all the reports published from all sources, but 5% are definitely serious."* On the other hand, Popovich wrote in a 1992 paper that, *"the UFO sightings have become the constant component of human activity and require a serious global study."*

Gennadiy Manakov and Gennadiy Strekalov

On September 28, 1990, cosmonauts **Gennadiy Manakov** and **Gennadiy Strekalov** were interviewed by radio on the **Mir. space station** The transcript was published in the Oct. 16 edition of Rabochaya Tribuna and translated again by the American FBIS. Here are excerpts of the dialogue:

Cosmonauts Gennadiy Manakov and Gennadiy Strekalov
http://www.openminds.tv/russian-cosmonauts-ufo-sightings-and-statements/859

Question: *"Tell me, what are the most interesting natural phenomena you see on Earth?"*
Cosmonaut: *"Yesterday, for example, I saw, if one may call it that, an unidentified flying object. I call it that."*
Question: *"What was it?"*
Cosmonaut: *"Well, I don't know. It was a great, silvery sphere, it was iridescent... this was at 22:50..."*
Question: *"This was over the region of Newfoundland?"*
Cosmonaut: *"No. We had already passed over Newfoundland. There was an absolutely clean, clear sky. It is difficult to determine but the object was at a great altitude over the Earth, perhaps 20-30 kilometers. It was much larger than a huge ship."*
Question: *"Could it have been an iceberg?"*
Cosmonaut: *"No. This object had a regular shape, but what it was–I do not know. Perhaps an enormous, experimental sphere or something else..."*

Musa Manarov

A few months later, on March 31, 1991, cosmonaut **Musa Manarov** captured a cigar-shaped UFOfor over two minutes while filming the approach of a Progress cargo flight to the MIR space station. As you can see in the clip excerpted from Michael Hesemann's *UFOs: The Footage Archives* in the DVD *Ultimate UFO!* with commentary by Peter Robbins and myself, you can clearly see one end of the cylinder lit up... This is undoubtedly one of the best-known examples of UFO footage ever taken in space. http://www.openminds.tv/russian-cosmonauts-ufo-sightings-and-statements/

Cosmonaut Musa Manarov
http://www.openminds.tv/russian-cosmonauts-ufo-sightings-and-statements/859

It is evident from the above statements that astronauts from the USA and cosmonauts from Russia are seeing alien spacecraft every time they go up into space on a mission. In all probability, part of those **Space Shuttle** and **International Space Station (ISS)** missions are to monitor and track UFO activity around the Earth much in the same way that certain X15 missions were strictly UFO surveillance missions by the Air Force under the auspice of NACA before it became NASA. *(It is the author's contention that NACA never really became defunct but, instead had renewed life under another name perhaps as DARPA or the DoD, who have developed triangular spaceplanes like the TR-3, the Aurora and Senior Citizen to replace the X15 and the Dyna-soar vehicles).* [Bold italics added by author for emphasis].

NASA Tracks the Space Shuttles, the International Space Station, and UFOs!

Video and photos must exist of Extraterrestrial spacecraft that were taken by astronauts on other flight missions which have yet to surface into the public domain. A most revealing set of photos enhanced from video taken on board the ISS show an astronaut at his computer terminal working at something on the monitor screen. When the images are enlarged and contrast and brightness adjustments made, along with colour hue saturation corrections, a startling but not totally unexpected image can be seen and recognized. It is the image of two flying saucers being tracked outside of the ISS! (See photos below).

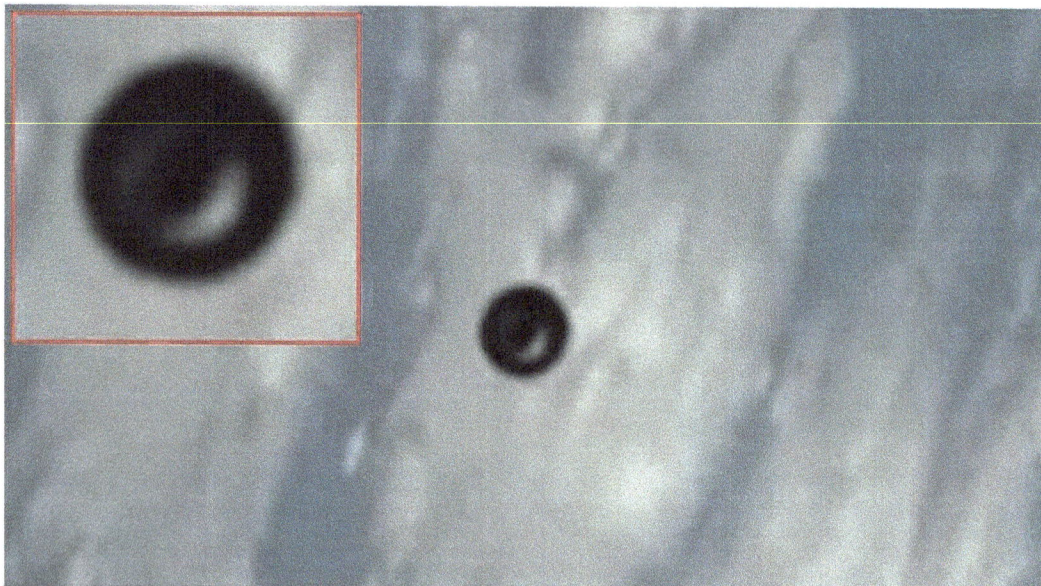

Disc shape craft photographed from Space Shuttle
http://www.taringa.net/posts/paranormal/16002239/OVNIs-en-la-mision-STS-115-del-Transbordador-Atlantis.html

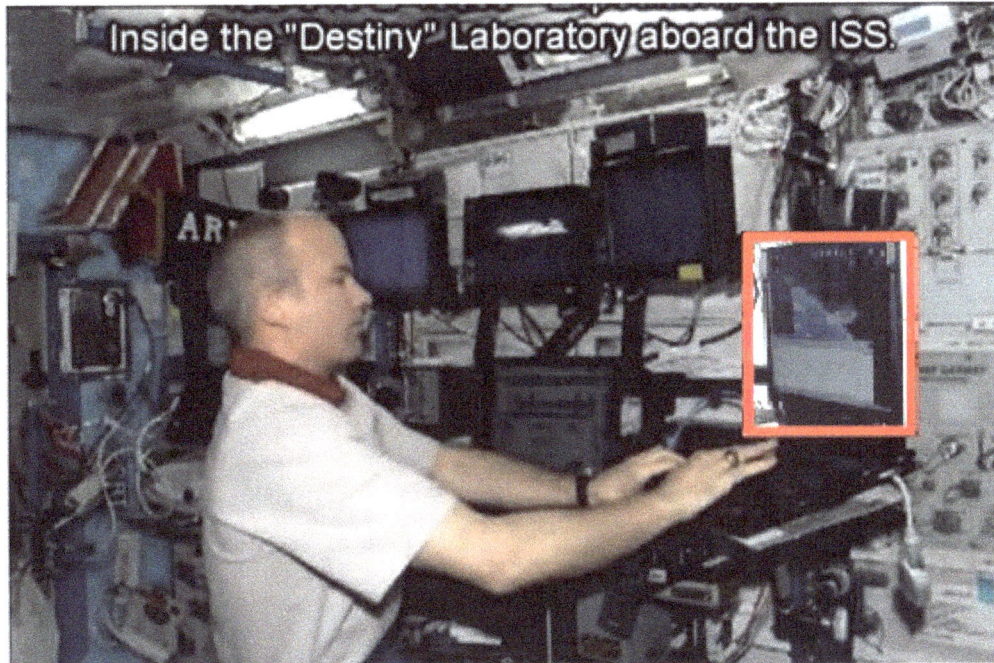

On board the International Space Station in 2006 during the Soyuz TMA-8 mission. Flight Engineer Mr. Jeffrey monitors what is outside the ISS (image enhanced)

Enlargement of computer screen showing two Flying Saucers outside of the ISS

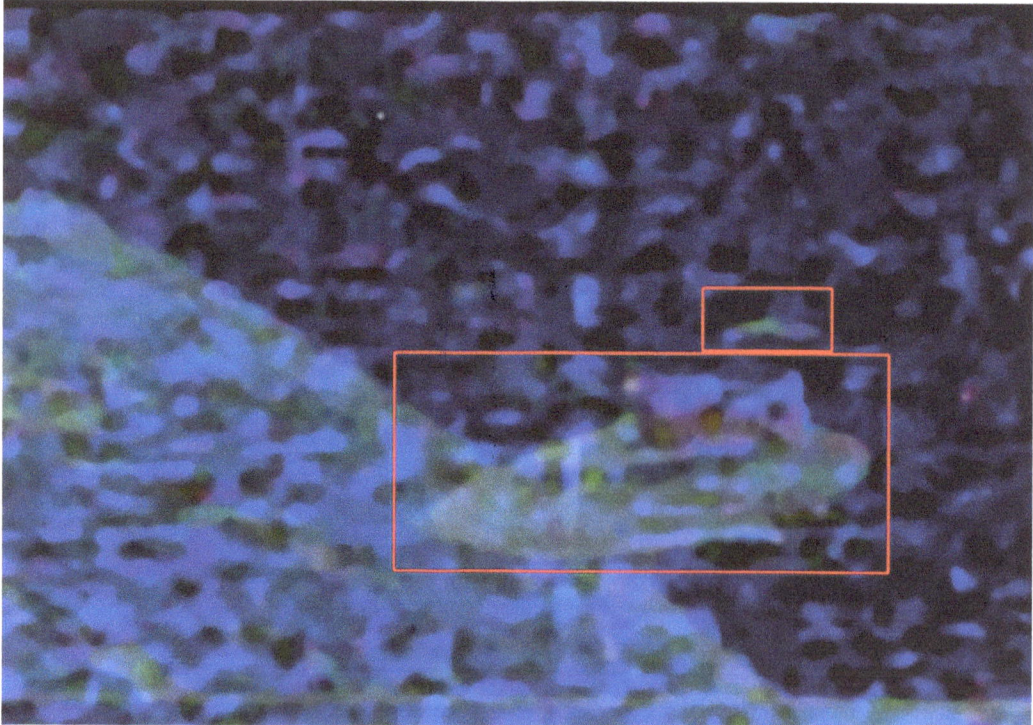

**Photo has been enlarged and enhanced to bring out detail which shows
a large craft outside the window of the space station,
a smaller ET craft can be seen above it**

**Besides monitoring the Space Shuttle, NASA also monitors and tracks UFO activity,
acommon occurrence on their jumbo screen monitors (See photos above and below)**

152

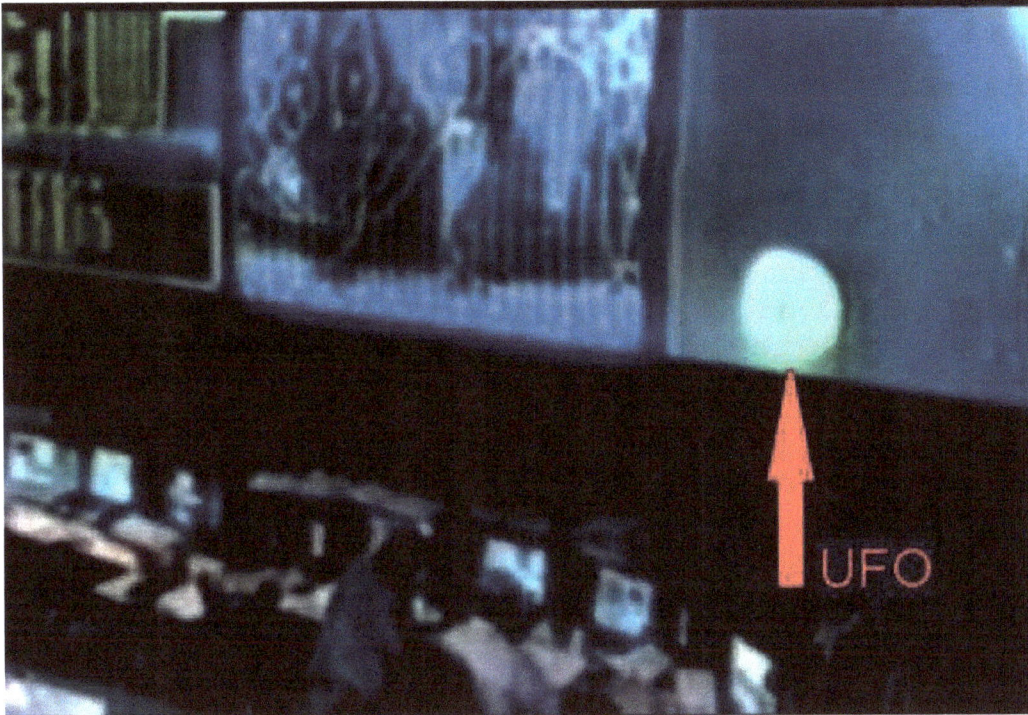

As it turns out, the suspicions and confirmations from UFO researchers is correct that NASA is not only aware of alien spacecraft that monitor our space missions but, Extraterrestrial space vehicles are routinely tracked and monitored in return response to AT surveillance. And on occasion via a military order on Earth , the nearest military missile site will target these ET craft with the expressed purpose to bring them down out of space, hopefully in a friendly country for easier crash retrieval.

Such an incident occurred in **Varginha, Brazil** in January 1996. The US military deep space radar had picked up an ET spacecraft phasing out of trans-dimensional space into physical space above the Earth *(this is the way ETI travel from one point in the universe to any other distant point in the universe. They travel beyond the crossing point of light at transluminal speeds by moving into an alter reality beyond the physical plane of existence and then pop back into physical reality at the point of destination),* recognizing the electromagnetic signature of an inbound craft about to materialize, they immediately fired a scalar particular beam weapon at thelocation point of materialization and successfully t hit the spacecraft causing it to crash to Earth. [Bold italics added by author for emphasis].

The ET spacecraft crash landed in Brazil, immediately the US military intelligence got on the phone to their Brazilian counterparts and reported to them that an alien craft was coming down intheir area and that a special crash retrieval team from the States was on its way to Brazil to collect the craft and any **EBIEs (Extraterrestrial biological entities).**

On the afternoon of 20 January, Brazilian military forces allegedly captured two live but injured aliens in the neighbouring areas of the city of **Varginha**, **Brazil** after three girls observed one of the creatures hiding in bushes near their home. Military personnel from the Army Internal Intelligence allegedly moved one of the alien bodies to a military facility in Campinas, where it underwent an autopsy at the Campinas University, some 200 miles from Varginha. (This case is well documented in the UFO literature for the reader to pursue further).

Astronaut Photos of UFOs Posted On NASA Website

Over the years NASA astronauts have seen, photographed and commented on UFOs and strangealien craft around Earth orbit or in near proximity to their own craft. This has forced NASA mission control centre to put in a 7 – 11 second (or longer) "live feed" delay out to general public viewing and newscasts. Even, with this delay, ET spacecraft and probes sometimes, just simply show up in the live transmissions as if to prove their existence and many are caught on video or photographed along with other intentional manmade objects or satellites that are a part of the flight mission. Astronauts are told to simply ignore them and to carry on with their assigned tasks but, astronauts don't always follow mission control orders. In many cases NASA will simply cut the live feed and place a message up on screen stating that they have run into a temporary transmission malfunction or will simply take remote control of the International SpaceStation cameras and point them away in some other direction or point the cameras directly at thealien spacecraft for a closer look but will still cut the live feed to the general public.

Below are images of a UFO surveillance probe that shows up behind an astronaut as he capturesa spinning satellite for repairs during his spacewalk activity.

Shuttle astronaut captures spinning satellite as orb-shape UFO flies into the area to monitor activity. Canadarm can be seen at bottom right
https://www.youtube.com/watch?v=WILN_Jcg1pc

154

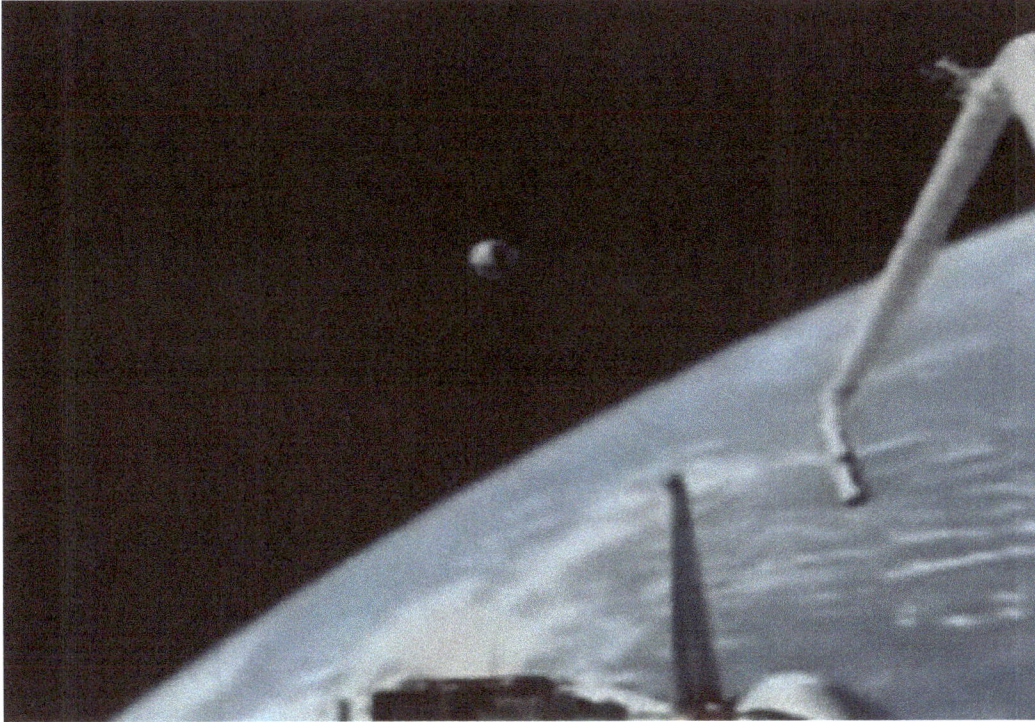

A UFO probe (similar to the Star Wars movie probe used to teach the Jedi to fight with a laser sword) flies near the Space Shuttle to monitor its activities

This close up of a UFO probe is sometimes described by NASA as one of their new surveillance probes used to monitor the Space Shuttle's exterior hull, equipment and astronauts' spacewalk activities

One of the most spectacular video footages of a UFO encounter was captured by cameras on board the **Discovery Space Shuttle Mission STS 48**, during the period of September 15 - 22, 1991. It is another case of tracking and targeting of an ET spacecraft in an apparent attempt to down a spacecraft, but the UFO managed to make a sharp turn into space to avoid being hit. The video seems to show a particle beam weapon being fired at a UFO. The shuttle was over Alice Springs, Australia at the time where a weapon was fired from this site. A six minute video of the event can be viewed at the website: http://www.youtube.com/watch?v=IiDvkB_rG-Q

What follows is official NASA photos of UFOs shot by astronauts during a mission to the International Space Station and released onto a public NASA website: http://www.thetruthbehindthescenes.org/2011/08/28/ufo-filmed-by-astronauts-flying-over-hurricane-irene-aug-26-2011/

Are these unidentifiable pieces of space debris merely junk that has been dumped by astronauts as a part of the space stations house cleaning duties or bits and pieces that have broken away from the space shuttle or space station? If either one of these possibilities is true then, someone needs to tell NASA and their astronauts not to dump their garbage in space but to recycle it instead – Go Green, Fellas! And if it is pieces of spacecraft breaking off the shuttle or ISS, then NASA better get on the case of their design engineers and tell them to nail down and screw tightly every nut and bolt on these spacecraft or there is going to be potentially more spacecraft disasters!

It is a sad commentary to discover that humans are now leaving their garbage and crap floating around the Earth in the hopes that their decaying orbit will cause them to burn up some day upon re-entry. The last thing mankind need do is to start littering one more place off planet like space, believing that infinite space is the answer to excess human waste!

If neither of the above two possibilities are happening then, whatever is being photographed needs to be seriously investigated because the only reasonable explanation left is these objects are Extraterrestrial spacecraft. If this is indeed what is being seen up in space orbiting near the space shuttle and space station then, it not merely one or two UFOs but, hundreds or maybe, even, thousands of alien spacecraft! The presence of Extraterrestrial Intelligence in orbit about the Earth in spacecraft, referred to as UFOs is real!

It matters not whether the spacecraft sent into orbit is military, quasi-civilian or merely private civilian enterprise, all space activity has been monitored by ET intelligence. With demise of any further Space Shuttle missions to the ISS in 2012 by NASA, private enterprise such as Space X, a civilian initiative to build spacecraft like the "Dragon" capable of docking with the International Space Station has picked up where NASA has left off. However, UFO activity has not abated and even, the Space-X craft have been followed and under surveillance by UFOs as indicated in the photograph below.

**Space-X "Dragon" approaches the International Space Station (ISS)
in 2012 to drop off supplies**
https://www.universetoday.com/91817/nasa-announces-feb-7-launch-for-1st-spacex-docking-to-iss/

The NASA video YouTube web link below was taken by mission **STS-75** when the Space Shuttle deployed a 40-mile cable as part of an experiment to generate electricity in space. The cable snapped and broke free of the space shuttle. Suddenly, there were hundreds of UFOs swarming like insects to a light, immediately surrounding the cable that was now drifting freely in space. These UFOs were seen as huge pulsating disc-like lights. In comparison to the length of cable, these ET spacecraft were estimated to be approximately 1.5 miles in diameter! Many crafts in the photos and in the video web link are seen swarming the cable or simply traveling to and from the Earth's atmosphere or out into outer space. The swarms of UFOs are so thick that the mission had to try hard to spot the MIR Space Station that is crowded by surrounding UFOs. This one particular video has been referred to as the best evidence of an ET presence in space and in near-Earth orbit.

http://www.youtube.com/watch?feature=player_detailpage&v=jXzH6xfgjU4 and
http://www.youtube.com/watch?feature=player_detailpage&v=X-RPWhigpQg

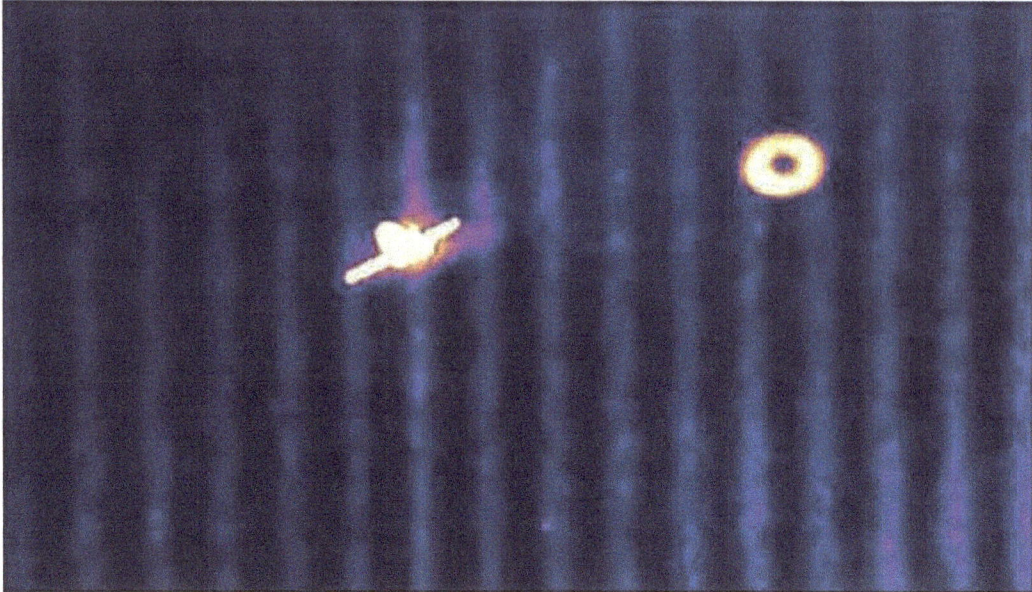

**UFO caught live during Space-X "Dragon" flight
to the International Space Station (ISS 2012 (Above and Below)**
http://www.ufosightingsdaily.com/2012/06/ufos-caught-live-during-flight-
to.html

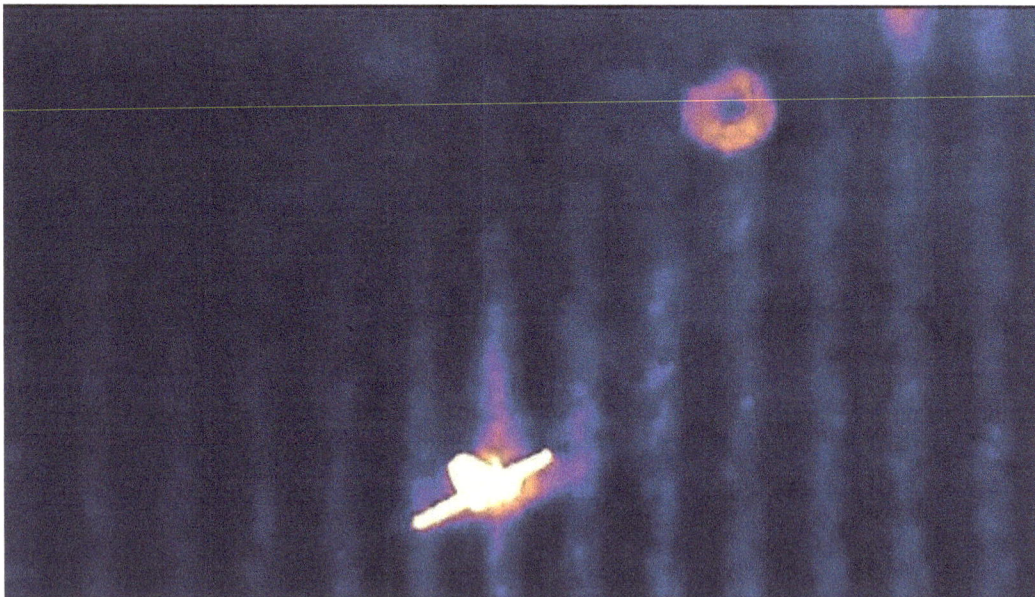

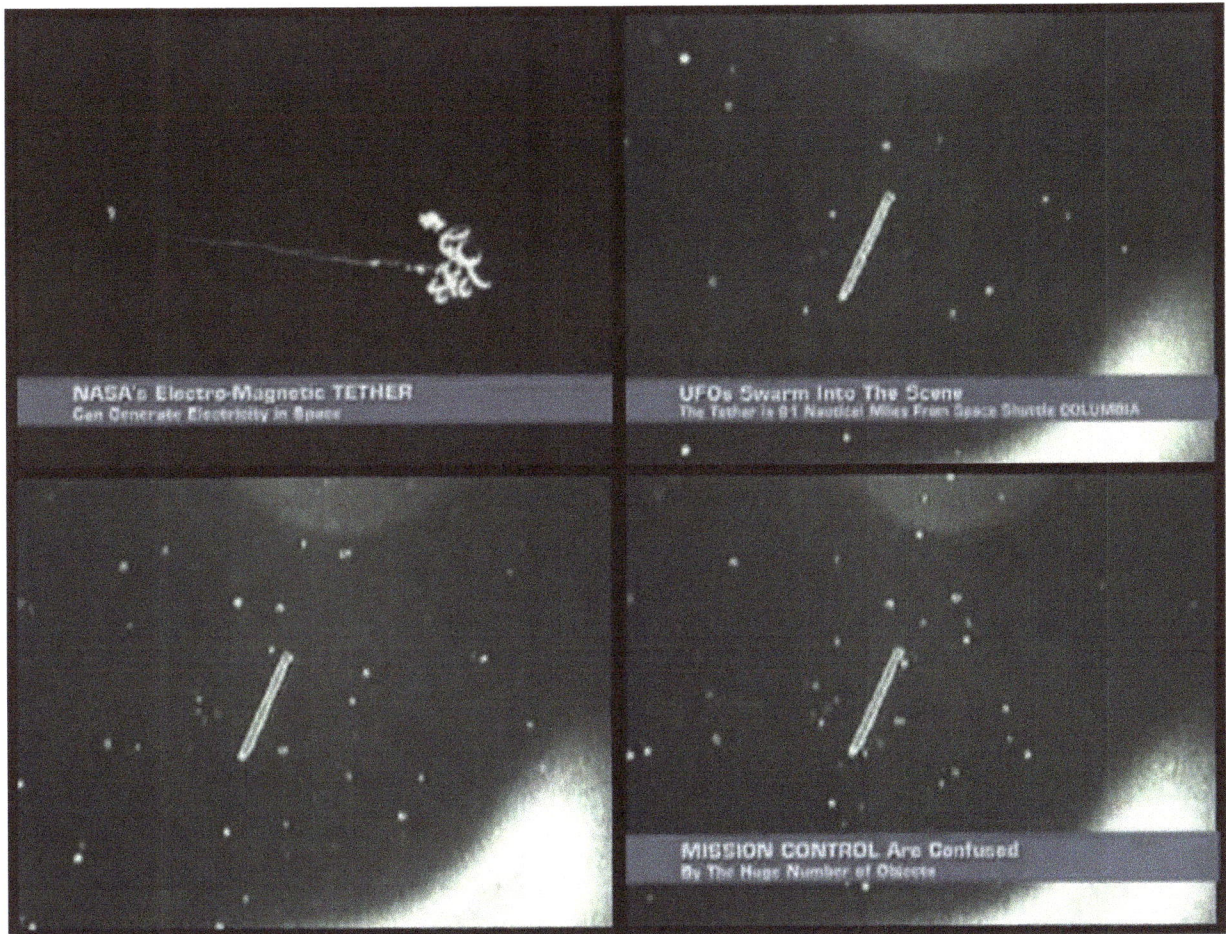

Hundreds disc-shape ET craft show up to investigate the 40-mile length of cable released by the shuttle used to generate electricity during the "Tether" Shuttle Mission STS- 75
https://www.youtube.com/watch?v=8njYpvAkMp8

Professional debunkers like **James Oberg** have dismissed this obvious proof of UFOs in space as nothing more than urine dumps from the space shuttle or the Mir Space Station or simply ice particles breaking off from these American and Russian spacecraft as they pass from the night side of Earth into the daylight warmer side of Earth. What seems odd is that Oberg never explains why the "urine droplets" or "ice flakes" appear so huge in comparison to the 40-mile length of cable or why they have a particular shape that is disc shape with a hole in the middle and a small notch in the rim. Nor does he explain the fact that they pulsate almost like living creatures as if breathing!

Author's Rant: Oberg's explanations or lack of them only serves to depict himself as a "flake" in need of some "serious de-icing" by way of a reality check and perhaps, an attitude adjustment. You simply cannot recognize astronauts on one hand as highly trained professionals with excellent observation skills and then, turn around on the other hand and dismissively marginalize their observations of UFOs in nearby space. Perhaps, Oberg needs to remove his head from out of the space shuttle's latrine or he's about to find his world of skepticism and debunking come crashing down about his head!

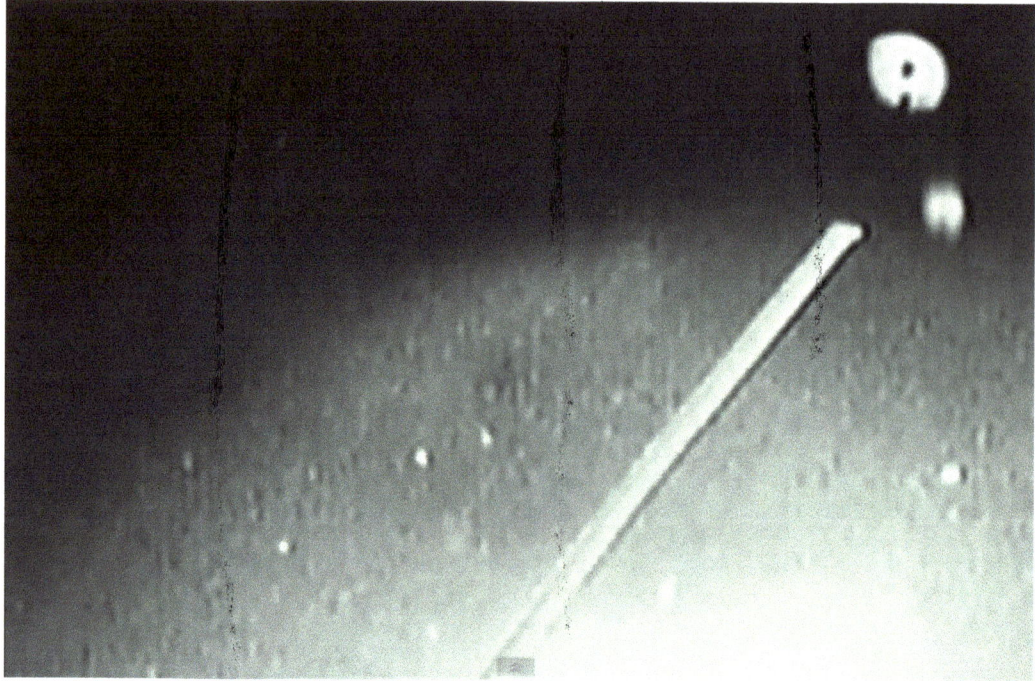

**A close up of the tether and some of the 1.5mile in diameter
ET saucer craft flying nearby**
https://www.youtube.com/watch?v=8njYpyAkMp8

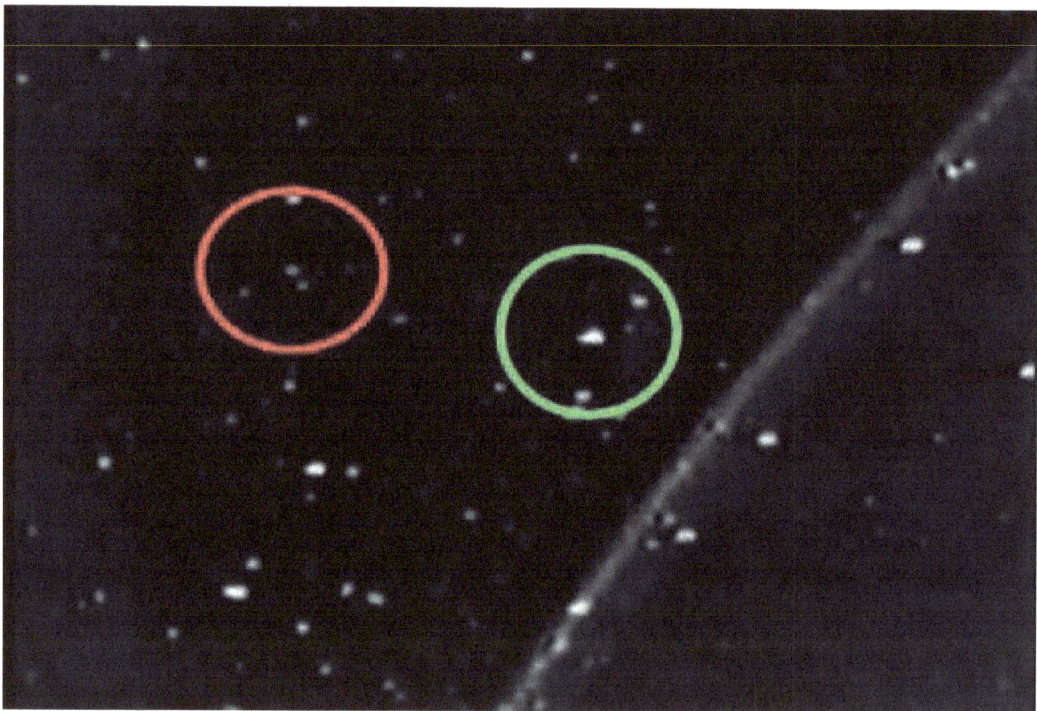

**Hundreds of UFOs near Earth. Red circle shows two UFOs travelling towards
Earth and green circle shows a UFO materializing out of nowhere**
https://www.youtube.com/watch?v=8njYpyAkMp8

Astronaut photographs clouds above Earth and captures UFO
Display Record - STS088-724-67
http://www.trinfinity8.com/the-mysterious-black-knight-satellite-who-really-owns-it/

Photo of clouds taken in 1998 from the Space Shuttle Endeavour also
shows a UFO which appears to be morphing (see next set of photos)
https://eol.jsc.nasa.gov/SearchPhotos/photo.pl?mission=STS088&roll=724&frame=69

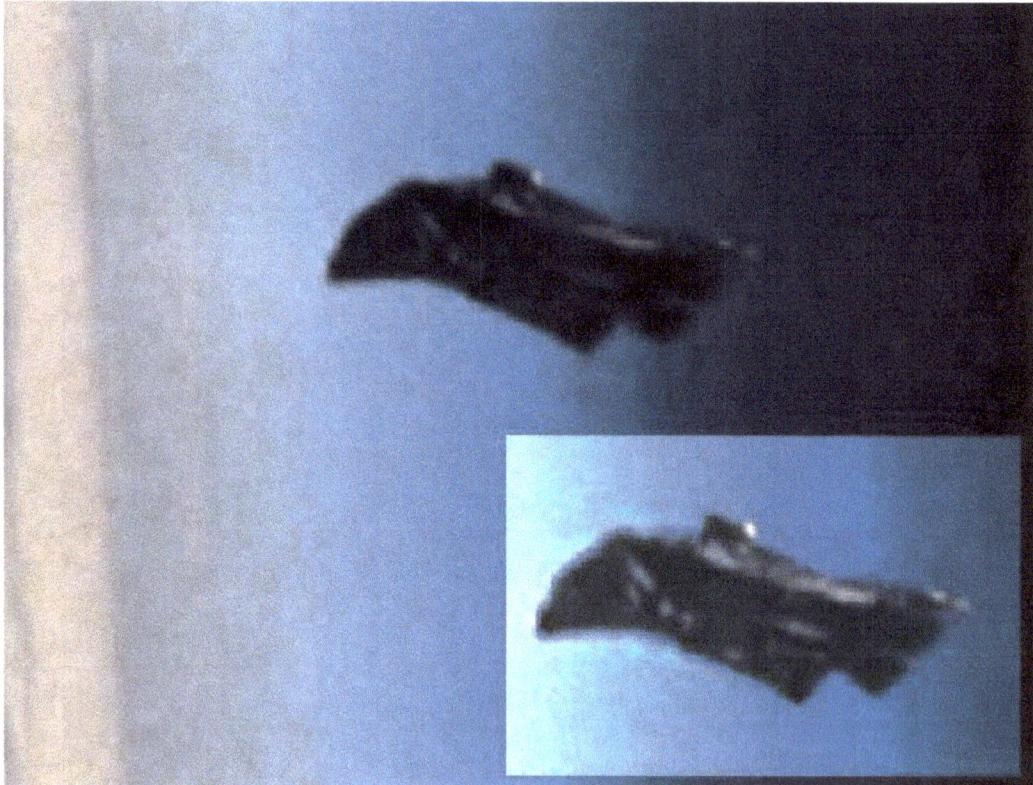

"Black Knight", a manta type UFO with enhancement inserted to bring out detail

Morphing UFO or merely another side of the same object as above

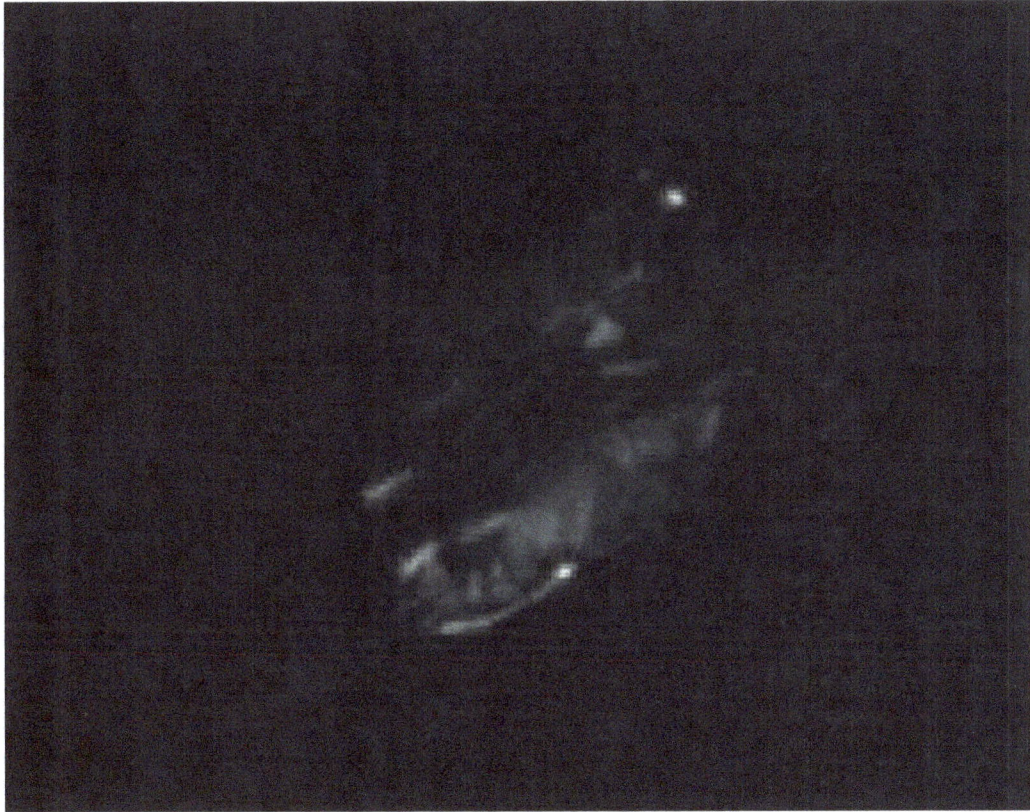

"Black Knight" UFO photographed in the night side of the Earth
https://ufoholic.com/unexplained/iss-cam-catches-glimpse-black-knight-satellite-watching-us/

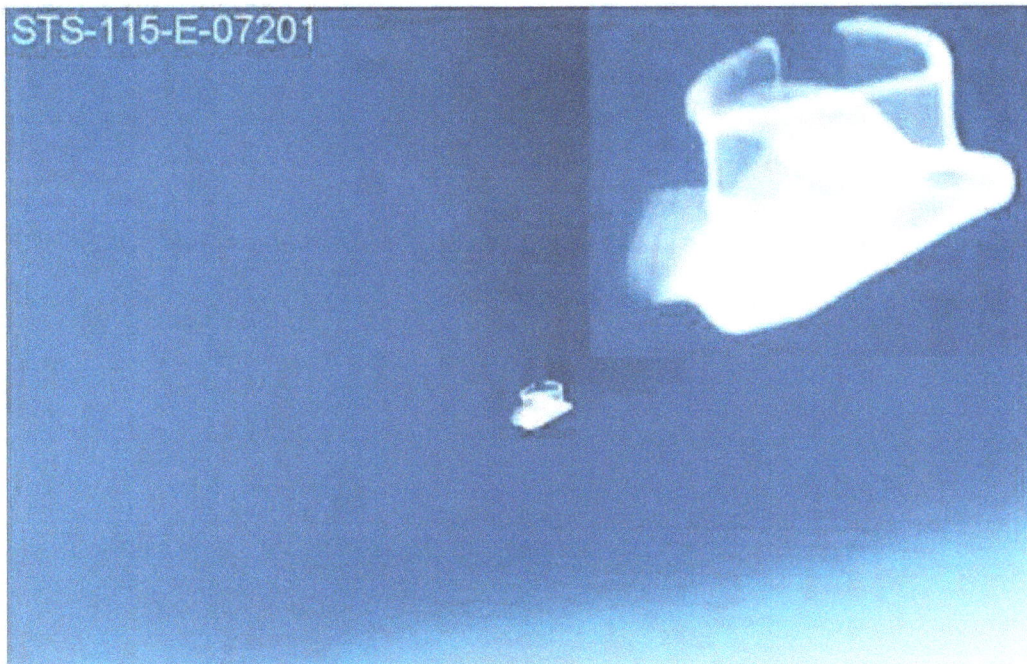

Is this ethereal-looking UFO "phasing" into Earth orbit from another dimension?
https://www.youtube.com/watch?v=oOyevZPzn1k

This ET shuttlecraft appears to be travelling up into Earth orbit at great velocity
(Google Image)

Recall the statement by astronaut **Scott Carpenter**: *"At no time, when the astronauts were in space were they alone: there was a constant surveillance by UFOs."*
https://www.youtube.com/watch?v=FF89H4S19UE

CHAPTER 68

ARE ASTRONAUTS DEBRIEFED AFTER EACH MISSION USING MIND CONTROL METHODS?

Astronauts Struggle to Remember Details of Their Missions

There has been rumours going around for some time, that early in the space programs of the US and in Russia, that whenever the astronauts or cosmonauts return to Earth after their mission in space or to the Moon, that their debriefing wasn't merely a verbal and written account of the things that went right or wrong with the mission. It entailed a very detailed mind control debriefing on what was seen, photographed or video and audio recorded, even **"black box" transcripts** were altered or sanitized of any Extraterrestrial encounters with ET spacecraft or beings!

Statements from various NASA sources, including **Maurice Chatelain,** reveal that all astronauts are briefed and debriefed, before and after every mission, and are warned not to discuss their encounters in public. Rumours of astronauts being *'threatened, 'going mad'*, or even *'losing their lives'* on their return to Earth, abound.

Every astronaut debriefing is a meticulous and intensive questioning session that could last for days. It can be very stressful time where all data and documents are thoroughly analyzed. At the end of each debriefing, astronauts are warned not to mention to anyone or among themselves or with family, friends or associates what they had seen or encountered in space that being anything that was UFO or alien related. Even the Office of inspector General (OIG) were looking at finding a way to cut the long duration period of astronaut debrief and post-mission report process. http://www.hq.nasa.gov/office/oig/hq/old/inspections_assessments/10-23-97.html

If that dire warning wasn't enough, it is believed that mind control in the form of mind-altering psychotropic drugs or psychotronic devices were used on the astronauts to ensure heir complete adherence to NASA and DOD protocols were maintained thus, this would ensure that anything of an unusual nature was forgotten for good.

It sounded far fetch but, recent astronaut behavior indicated that some of them were under extreme stress in their personal lives, the source of that stress not being readily apparent. Sometimes, because astronauts are in the public limelight that stress can erupt into public displays of irrational behavior. In radio or TV interviews, they would sometimes make statements to the effect that they couldn't remember everything that took place from their missions in space. Some astronauts, of course, have become outspoken with regard to the existence of Extraterrestrial life in space and coming to this planet but, always the language was guarded or implied or indirect, never really overt with any solid detail but, more in terms of generalizations, as if they were protecting themselves or their fellow astronauts or even, their families.

The astronauts who went to the Moon, for example, would never say, I saw artificial structures or building on the Moon's surface or "I was in contact with an Extraterrestrial Intelligence and he said so and so to me!" What the public or the Ufologist may hear would be that "UFOs exist and ETs are coming to the planet, that the Roswell saucer crash and retrieval was real; that there is a government cover-up and suppression of that knowledge!" This is considered by the **"Gatekeepers of UFO/ETI knowledge"** as safe information because it doesn't reveal any real deep secret knowledge or technical information about the subject. It may be considered as "sanitized" or as disinformation or misinformation.

One very public and humiliating display of this kind of astronaut meltdown was the attempted kidnapping of a romantic rival by NASA astronaut **Lisa Nowak** in February 2007 as reported in the China View web news site.

Lisa Nowak, who flew on a space shuttle mission seven months ago and posed for pictures next to **U.S. President Bush** soon after, was charged with attempted murder in the southeastern U.S. state of Florida.

Nowak, 43, a Navy captain and a mother of three drove 900 miles in her car from Houston to Orlando on Feb. 5 to confront a woman she thought was a rival for space shuttle pilot **Navy Cmdr. William Oefelein**'s affection. Nowak wore a wig and trench coat then, sprayed a chemical into the woman's car when she wouldn't let Nowak in.
http://news.xinhuanet.com/english/2007-02/07/content_5711503.htm

NASA astronaut Lisa Nowak , not the type of person you would expect to have a public psychotic meltdown
https://en.wikipedia.org/wiki/Lisa_Nowak

Police said Nowak followed Colleen Shipman to her car at Orlando International Airport and pepper-sprayed her. In Nowak's bag, police said they found a knife and a BB gun. In her car, they said they found latex gloves, e-mails between Shipman and Oefelein, and a letter declaring Nowak's love for him.

Shipman, 30, an Air Force officer at Cape Canaveral Air Force Station in Florida, sought a restraining order Monday against Nowak. The document Shipman filed said Nowak is "an acquaintance" of Shipman's "boyfriend" and alleged that Nowak had been stalking her for two months.

Nowak also faces other criminal charges, including attempted kidnapping and assault. If convicted of attempted murder, she faces 30 years to life in prison, authorities said. Nowak, who was ordered to wear a monitoring device, was released from jail on 25,500 U.S. dollar bail.

Nowak became the first active member of the astronaut corps to face such serious charges.

Astronaut Lisa Nowak returned to Texas from Florida Wednesday and headed to **Johnson Space Center** for a medical assessment as NASA announced it would review its psychological screening process in light of her arrest on charges of attempted murder of a romantic rival.

NASA announced Nowak had been placed on 30-day leave and removed from flight status. "We are deeply saddened by this tragic event," a statement said. "The charges against Lisa Nowak are serious ones that must be decided by the judicial system."
http://news.xinhuanet.com/english/2007-02/07/content_5711503.htm

Is it possible for NASA to cause a psychotic episode in one of its astronauts and if so, for what purpose? Was Nowak unable to abide by NASA protocols and keep her oath of loyalty in the name of national security of what she may have seen during her time in space?

NASA has been known to use mind control on children according to astronaut Gordon Cooper, one of the original seven Mercury astronauts. Cooper confirmed the existence of a mind control program administered by NASA in the 1950's and 1960's involving gifted American school children back in 2005 on the on the popular, late-night radio program, **Coast to Coast** hosted by Mike Siegel.

The children used in the mind control program had exceptional mental abilities and were put through "a kind of **"MK" (MKUltra)** program, similar to the things that are now coming out into the public domain.

Cooper described how **NASA's mind control program** emphasized cultivation of **telepathy**, **remote viewing**, and **out-of-body experiences (OBEs),** in other words, the psychic abilities.Cooper's remarks generally support the claims of a growing cadre of Americans, now in theirthirties, forties, and fifties, who are recovering memories of unusual classes that they were enrolled in as young children during the advent of the Space Age.
http://aangirfan.blogspot.ca/2012/06/nasa-brainwashing.html

It is believed that NASA's mind control program was directed at preparing children who would later be able to communicate with the non-human intelligent species that humanity might encounter in space.

This thesis is supported by the fact that one experiencer remembers being tutored in a hieroglyphic alphabet that author **Fritz Springmeier** has identified as a set of "intergalactic symbols" developed by NASA for the purpose of communicating with extraterrestrial civilizations.http://aangirfan.blogspot.ca/2012/06/nasa-brainwashing.html

It appears that NASA or some branch of NASA, possibly the DoD are not only currently engaged in mind control and deprogramming of astronauts but, are also looking towards future missions that may involve contact and communication with Extraterrestrials beings. To this mind control may be added the recruitment from a growing assemblage of pre-trained, potential astronauts that are now maturing schoolchildren entering into adulthood with the necessary psychic ability skill sets!

Buzz Aldrin is another example of an intellectually brilliant and deeply spiritual man who suffered a "nervous breakdown" very early after his return to Earth from his walk on the Moon. Public speculation from his nervous breakdown' at that time may possibly have been as a result of the government (NASA) pressure or his skyrocketing fame to sudden celebrity status. He was certainly in demand and perhaps even, somewhat deified by the news media and an adoring public. His autobiographies "Return to Earth", published in 1973, and "Magnificent Desolation", published in June 2009, both provide accounts of his struggles with clinical depression and alcoholism in the years following his NASA career. http://aangirfan.blogspot.ca/2012/06/nasa-brainwashing.html

In **Fred Steckling's**, book, "We Discovered Alien Bases on the Moon II", he says that Aldrin may have wanted to publicly proclaim the truth of what he saw on his way to Moon and on the lunar surface. Steckling feels highly certain that the Pentagon and White House officials, ever fearful that the truth on UFOs and ETs may get out to the public, would have cajoled, convinced or threatened Aldrin not to speak about such matters. http://aangirfan.blogspot.ca/2012/06/nasa-brainwashing.html

According to Steckling:

"The following list of facts I have learned from several individuals who are working or who have worked or been linked with NASA in some way over the years. For obvious reasons, I will not now use their real names. The list below represents some of the information from three people: one is currently working in DOD missions for NASA; another started working with NASA & DOD when **JSC (Johnson Space Center)** was built. The third is a scientist who has worked at NASA and other facilities (and with) **Edward Teller**."

"There are buildings on the Moon. There is mining equipment on the Moon."

Photos, NASA photos, do exist which clearly show both of these. Hundreds, but probably thousands, of NASA photos have been tampered with. Specifically, by careful use of an airbrush,

168

flying saucers, and other UFOs can be removed, and then the photo is released to the public and/or press.

Film taken by astronauts clearly show UFOs, IFOs, Alien Vehicles, etc. The NSA screens all photos before releasing them to the public. Everything that NASA has launched has been closely monitored by at least one 'alien' culture.

NASA knew about 'alien' activity on the Moon before Armstrong, Aldrin, and Collins ever set foot on it. Edwin Aldrin at one point found evidence that we were NOT THE FIRST to arrive on the moon. After first seeing and then taking photographs of footprints in the lunar soil (Aldrin) then saw the beings that made the footprints (the report and transcripts of conversations between the astronauts were not clear if Aldrin had physical and/or mental contact with the entities). **"We Discovered Alien Bases on the Moon II" by Fred Steckling; 1981; by G.A.F. International; Vista, CA. USA; ISBN 0-942176-00-6** and http://rense.com/general70/rep.htm

"Alien Vehicles flew within 50 feet of a U.S. space vehicle for one full Earth orbit and then the AV departed; again while Aldrin was present. 'Buzz' Aldrin had a nervous breakdown because of these events and the pressure not to talk. There have been 22 deaths (many 'suicides') at JSC in Houston. No astronaut who has seen AVs or ETs is allowed to talk about it, even amongst themselves. If they do and are caught, they may be fined, publicly humiliated, imprisoned, or have all pensions and future salaries taken away.'(my emphasis)

How much of the above account is fictional and how much fact? Only those who went to the moon and back know for certain. There are many second-hand accounts and alleged conversations, (Google UFO Sightings by Astronauts) where both Aldrin and Armstrong state, in no uncertain terms, they saw huge ships and other signs of alien occupation of the moon.

Return to Earth, Aldrin's autobiography, tells of his struggle with depression and alcoholism following his long and dedicated USAF and NASA career. Did NASA, CIA and the Pentagon compel Colonel Aldrin (and every other astronaut) to conceal what they saw on the moon? How much did this contribute to Aldrin's mental problems?

Some time ago in 2005, the **Science Channel** (Original Source) aired a documentary program called "First on the Moon: The Untold Story.' One segment described a UFO encounter that Apollo 11 astronauts witnessed during their flight to the moon. Aldrin told an interviewer that they saw an unidentified flying object. http://rense.com/general70/rep.htm and http://www.shockmansion.com/2012/08/27/video-buzz-aldrin-punches-reporter-for-calling-the-moon-landing-fake/

'To the best of my knowledge, this is the first time that Buzz Aldrin, an Apollo 11 astronaut, had ever publicly recounted any UFO experience associated with the Apollo 11 moon mission,' wrote **Dave Stone**.

Buzz Aldrin remarked, *"There was something out there that, uh, was close enough to be observed and what could it be?...Mike (Collins) decided he thought he could see it in the telescope and he was able to do that and when it was in one position, that had a series of ellipses, but when you made it real sharp it was sort of L shaped."*

"NASA knew very little about, um, the object reported by the Apollo 11 crew. It was obviously an unidentified flying object," said Senior NASA scientist, **Dr. David Baker**. *"But such objects were not uncommon and the history of even earth orbit space flights going back over the previous years indicated that **SEVERAL CREWS SAW OBJECTS"*** (emphasis mine).

"Now, obviously, the three of us were not going to blurt out, 'Hey Houston we got something moving along side of us and we don't know what it is," observed Aldrin. *"We weren't about to do that, cause we know that those transmissions would be heard by all sorts of people and who knows what somebody would have demanded that we turn back because of Aliens or whatever the reason is, so we didn't do that but we did decide we'd just cautiously ask Houston where, how far away was the S-IVB?'*

The S-IVB served as the third stage on the Saturn V and second stage on the Saturn IB. It had one J-2 engine. For lunar mission, it was used twice: first for the orbit insertion after second stage cutoff, and then for the **translunar insertion (TLI)**. Now NASA reported the separated stage 6,000 miles behind. Obviously, they didn't detect any UFO.

Later, in order to debunk what the Apollo 11 astronauts saw, David Morrison, a NASA Senior Scientist said that the documentary cut the crew's conclusion that they were probably seeing one of four detached spacecraft adapter panels. Their S-IVB upper stage was 6,000 miles (9,700 km) away, ***but the four panels were jettisoned before the S-IVB made its separation maneuver** so they would closely follow the Apollo 11 spacecraft **until its first midcourse correction.** (Bold italics added by author for emphasis)

Aldrin replied: *"And a few moments we decided that after a while of watching it (UFO), it was time to go to sleep and not to talk about it anymore until we came back and **(went through) debriefing.**"*(Bold italics added by author for emphasis).

NASA scientist, David Baker said, *"There were a lot of people within the program who went off later and became convinced that UFOs existed and that led to some concern on NASA's part where they got the agreement of the crew never to publicly talk about these things for fear of ridicule."* http://rense.com/general70/rep.htm and
http://www.break.com/usercontent/2011/4/20/alien-encounters-2047497

That's One Small Sucker Punch for an Astronaut and One Giant TKO for NASA!

Much later, Aldrin seemed healthier and back on the road to a full recovery with the exception of being a little bit physical in some interviews, most notably with one reporter who accused Armstrong and Aldrin of having never landed on the Moon.

On September 9, 2002, Aldrin was lured to a Beverly Hills hotel on the pretext of being interviewed for a Japanese children's television show. When he arrived, **Apollo Conspiracy** proponent **Bart Sibrel** accosted him with a film crew and demanded he swear on a Bible that the

Moon landings were not faked. After a brief confrontation, Aldrin punched Sibrel on the jaw. The police determined that Aldrin was provoked and no charges were filed. Aldrin dedicates a chapter to this incident in his autobiography "Magnificent Desolation". http://www.shockmansion.com/2012/08/27/video-buzz-aldrin-punches-reporter-for-calling- the-moon-landing-fake/ and http://www.youtube.com/watch?feature=player_embedded&v=tRBesDx1WQc

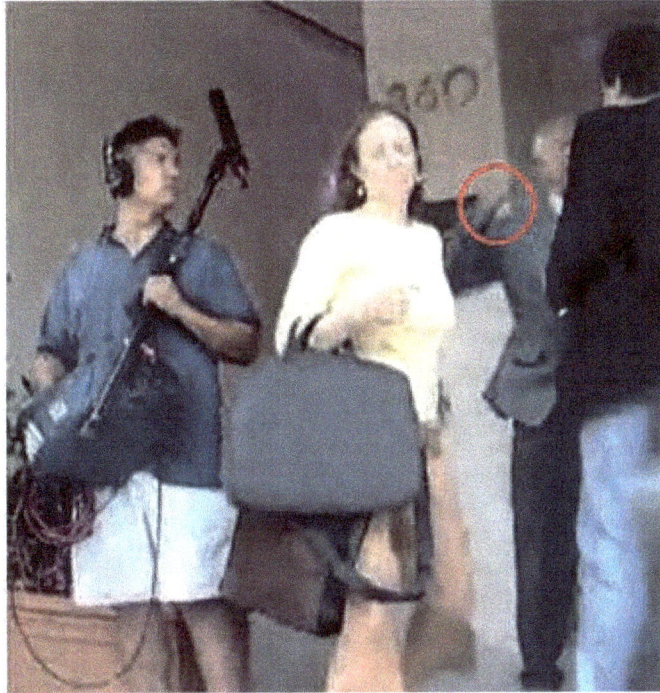

Former NASA astronaut Aldrin (in gray suit) with clenched fist lets fly a right uppercross to an antagonistic and belligerent conspiracy reporter, Bart Sibrel
http://nerdist.com/lets-all-remember-the-time-buzz-aldrin-punched-a-conspiracy-theorist-in-the-face/

It appeared that not everyone believed that the Apollo 11 Moon landing or any of the succeedingMoon landing missions were real but, hoaxed events. We will delve into this **Faked Moon Landing Conspiracy** shortly in this section.

What is evident is the fact that a lot of astronauts are under pressure to keep particular information of events and things seen in space, a secret in accordance with NASA and DoD oaths of national security on such matters. It appears that "Silence is golden" for NASA when itcomes to keeping astronauts from revealing matters of a UFO nature. Remember, NASA is an arm of the military and not a civilian agency as most people think so, it should not come as a surprise that NASA would use strong arm tactics and intelligence programs like the military.

"How is it possible to silence all astronauts who have been in space," is a question often asked with regard to this type of NASA conspiracy, when someone would have leaked that information?" This is a common method that debunkers often use to discredit conspiracy theoriesof any kind, particularly UFOs and Aliens.

NASA uses numerous methods taken from the many US intelligence agencies handbooks on keeping someone silent that range from outright threats, blunt fear tactics, blackmail, to inducing delusion and false belief systems or pathological / misguided sense of duty or false ideologies ("It's for the greater good", "God told me to do it..." etc) - this is all a core part of what sustains the growth of **pathocracy**. And then when information does come out, again there are numerous highly sophisticated methods of 'damage control', not least the global lockdown of mass media information, large scale **COINTELPRO**, and also the iron grip of control over the general population's sense of reality and their ability to perceive.

There are numerous ways of keeping silent not only the astronauts but also the people involved in all the Apollo missions and those responsible for image releasing (they get to see some nice pictures). There is some evidence that intense mind-control through hypnosis has been used on the astronauts prior and after each mission in order to conceal parts of their memory. Such techniques are commonly used in other arenas of covert intelligence by many intelligence agencies as they are very effective to manipulate and silence the people involved in such projects. https://cassiopaea.org/forum/index.php?topic=13462.25;wap2

Richard Hoagland singles out **Buzz Aldrin** as an example, who doesn't actually remember what he saw when he landed on the Moon. It's as if his memory has been in some way removed. He did hypnotic therapy (which are recorded), but even that didn't seem to work.

Mike Bara, who co-authored the book, "Dark Mission" with Richard Hoagland in an interview with **Joan d'Arc** *(yes, that really her name)* for Paranoia Magazine confirms that Buzz Aldrin is unable to describe what it was like to be on the Moon. [Italics added by author].

Joan d'Arc: Your book alleges that Buzz Aldrin is completely unable to describe what it "felt like" to be on the moon. Do you believe then that part of astronaut "debriefing" is "brainwashing"?

Mike Bara: Yes, I think there is ample evidence that the astronaut's memories have been tampered with, which we present in the book. Even the recent documentary by Ron Howard, In the Shadow of the Moon, spends tons of time on all aspects of Apollo, but virtually nothing on the actual Moonwalks themselves. Aldrin just seems to repeat the same quotes we've all *(heard before of),* what he saw on TV, not what he actually experienced. Anecdotally, Richard [Hoagland] once had a long discussion with one of the doctors who helped to hypnotize the astronauts for their debriefings, and she said she couldn't remember anything about the sessions either. But that was essentially hearsay, so we didn't put it in the book.(Italics added by author for clarity).

Joan d'Arc: So then "debriefing" is more than intelligence gathering, more than simply, OK, whatever you saw up there you're not going to talk about it. You're saying that when astronauts are "debriefed" their memories are pretty much cleaned out and maybe even replaced by screen memories or faux memories? And this is pretty much well known? And even the hypnotists have to be hypnotized? Do you have any other examples of astronauts who can't remember things?

Mike Bara: There are lots of examples in Dark Mission, but again **Ed (Edgar) Mitchell** comes to mind. He was so frustrated with not being able to recall the experience that he went through years of hypnotherapy to try to remember. That's all in his autobiography (The Way of the Explorer: An Apollo Astronaut's Journey Through the Material and Mystical Worlds). And honestly, I'm not sure about all of the Apollo astronauts having memory issues. Neil Armstrong's behavior strikes me as that of someone who's ashamed of his role in the mission of Apollo 11. That would imply that he remembers what he did and isn't too proud of it. Most of the astronauts can recite all of their scripted tasks chapter and verse. You know, "at such and such a station we deployed the **ALSEP experiment**." But when it comes to answering questions about what they saw and felt, they seem to clam up.

https://cassiopaea.org/forum/index.php?topic=13462.25;wap2

CHAPTER 69

THAT'S ONE ITTY BITTY, SMALL STEP FOR A MAN, ONE HUGE, GIGANTIC LEAP FOR WHOM… NASA?

Russia and United States Photograph Anomalies on the Moon

It's becoming clearer what the real motive was behind the Space Race between the USA and the former Soviet Union; it was a race to the Moon. It was more than just political and technological rivalry between two superpowers flexing their muscles in space, both nations knew their space efforts were being monitored not only by each other but, by an Extraterrestrial presence from somewhere else in space. Since those early days of space exploration, both nations' space agencies had foreknowledge of this Extraterrestrial presence visiting the Earth.

One of the primary objectives of these two competing nations was to determine the extent of the ET presence to our planet. Early manned missions into Earth orbit proved that humans were not alone in space but were being closely monitored and with the first satellites to the Moon, it appeared that the number of ET craft grew to armada size. The second objective of the space race was to determine if ET craft came from some hidden base on the Moon or whether they originated from within our Solar System or from beyond it? The Moon was the nearest planetary body to us and thus, was the obvious choice for unmanned and manned missions. The ultimate objective was the acquisition of ET technology in the form of artifacts, ruins, and possible alien spacecraft! The first country to discover alien technology and lay claim to it would have an incredible technological advantage over all other nations by decades!

Before man could set foot on the Moon, the lunar surface had to be mapped by orbiting satellites that would transmit back photographic data from which lunar landing sites could be selected for a manned mission.

The Soviet Lunar program had 20 successful missions to the Moon and achieved a number of notable lunar "firsts": first probe to impact the Moon, first flyby and image of the lunar farside, first soft landing, first lunar orbiter, and the first circumlunar probe to return to Earth. The two successful series of Soviet probes were the Luna (24 lunar missions) and the Zond (5 lunar missions).

The **Nation Space Science Data Centre (NSSDC)** currently holds data from the Luna 3, 9, 13, 21, and 22 and the Zond 3, 6, 7, and 8 missions. All this data is photographic in nature, except for the lunar libration data from the Luna 21 Orbiter. Lunar flyby missions (Luna 3, Zond 3, 6, 7, and 8) obtained photographs of the lunar surface, particularly the limb and farside regions. The Zond 6, 7, and 8 missions circled the Moon and returned to Earth where they were recovered, Zond 6 and 7 in Siberia and Zond 8 in the Indian Ocean. The purpose of the photography experiments on the lunar landers (Luna 9, 13, 22) was to obtain close-up images of the surface of the Moon for use in lunar studies and determination of the feasibility of manned lunar landings.

On March 3, 1959, the US launches Pioneer 4 on an Earth-Moon trajectory. It passed within 37,000 miles of the Moon before falling into a solar orbit.

On September 12, 1959, Russia launched Luna 2. It impacts the Moon on September 13, becoming the first man-made object to do so. By Oct. 4 that same year, Russia's Luna 3 orbits the Moon and photographs 70% of its surface.

Once again, in July 1964, the Russians in an ambitious program led the space race with a newly developed experimental ion rocket engine as a part of their **Zond/Mars 3MV** program. **Zond 3** was a space probe that was designed to fly to Mars by first going to the Moon, photographing the Moon's dark side and then using the Moon's gravity, sling shot the space probe toward Mars. This would be a test of satellite communications from a great distance being part of Mars 3MV project. It was unrelated to Zond spacecraft designed for manned circumlunar mission (**Soyuz 7K-L1**). Zond 3 completed a successful lunar flyby, taking a number of good quality photographs for its time. It is believed that Zond 3 was initially designed as a companion spacecraft to **Zond 2** to be launched to Mars during the 1964 launch window. The opportunity to launch was missed, and the spacecraft was launched on a Mars trajectory as a spacecraft test, even though Mars was no longer attainable.

On July 20 lunar flyby occurred at the closest approach of 9200 km. 23 photographs and 3 ultraviolet spectra of very good quality were taken of the lunar. Zond 3 proceeded on a Mars trajectory but never reached the planet. To test telemetry, the images were rewound and transmitted at succeeding greater distances, thus proving the ability of the communications system. The mission was ended and radio contact ceased when it was at a distance of 150 million km.

The first images sent back to Russia from Zond 3 were spectacular! What the Russians saw and reported to the rest of the world in newspapers and on television newscasts was a pock mark, cratered surface, unlike anything that had been seen before. For the first time, people were able to see what it was like on the **"Dark Side of the Moon"**, the side that always faces the Sun. **(The Moon rotates once in its orbit about the Earth and is one of a very few natural satellites in our solar system to do so, an oddity for sure….perhaps!)** [Bold italics added by author for emphasis].

The **"nearside"** of the lunar surface that always faces the Earth has large and small craters and very large open expanses known as **"seas or mares"** such as the **Mare Tranquillitatis** (Latin for **Sea of Tranquility)** where the first astronauts landed. What is interesting is that when you look at the Moon particularly in a waxing or waning phase, you see only the side that is sunlit (the side facing the sun) and you see the edge or the **Terminator Line** that runs from the north to south along the lunar surface dividing the nearside from the **"darkside"**. It is exactly along this Terminator Line that separates the heavily cratered dark side of the Moon from the "seaside" surface of the Moon! When the moon is full, you see it completely, except for the "darkside" but, on rather rare occasions when the sun is shining on the Earth, that same light reflects toward the Moon and the darkside of the Moon is lit by what is known as **"Earthshine"**. At these time before a **"New Moon"**, it is possible to see the "dark side of the Moon" although not very brightly lit or in any detail unless you are using some kind of telescope with night-vision and infrared capabilities. Sending a satellite to the moon would reveal what the naked eye could not see or what telescope could not discern with clarity.

**Tall lunar tower ("Tower of Babel") and structure anomalies on the dark side of the Moon.
The black dots that give the image a speckled appearance are data errors.**

Curiously, the Russian **Zond 3** spacecraft photographed and transmitted back what can only be described as lunar anomalies that really shouldn't exist at all. (See photo below). There in all its stark magnificence like a human phallus was a tower of immense, unimaginable size, along with smaller towers and what initially looked like buildings of some kind!

The Russians were ecstatic over the images and were mystified by this tall structure, sometimes referred to as the **"Tower of Babel"** among Ufologists. NASA was also puzzled by what they saw in the Russian photographs but, they were quick to point out that these anomalies were probably photographic errors or defects in the transmitted film. NASA scientists knew that they were looking at the first real evidence for a possible Extraterrestrial civilization, other than our own existing in our Solar System and so close to home!

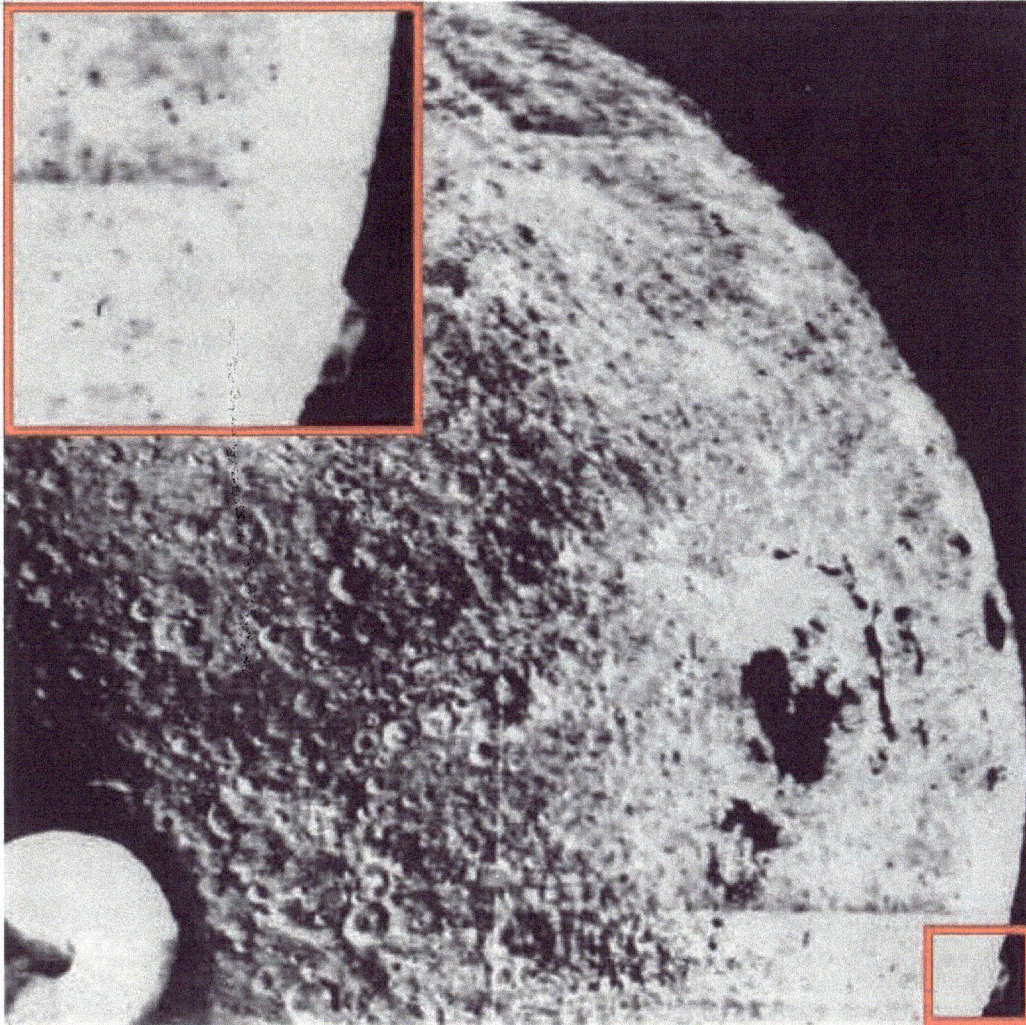

**Huge dome structure or plume anomaly on the dark side of the Moon
(See insert for detail) photograph by Russian Zond 3 spacecraft.
(Object in bottom left corner is the Zond 3 photometric target).**
http://ufocasebook.conforums.com/index.cgi?board=moon&action=print&num=1192005138

Other images revealed a large dome or possible plume of gas of some kind coming up from the lunar surface (see above photo). Was this an indication that the Moon's core was active? Was this plume, an indication of oxygen or water or some other type of gaseous geological activity? Perhaps, an atmosphere or something subterranean or was it an artificial dome shape construct?

Many people have come out to give their personal opinion as to the authenticity of the Zond 3 photographs and as on cue from some hidden source of authority, the professional naysayers weighed into the controversy with their own two sense worth of skepticism and debunking. If something appears to be a UFO, i.e. if it looks like an Extraterrestrial spacecraft or an alien being or an alien ruin then, the general public simply doesn't have a need to know about it and what better way than to dismiss the whole matter than with some professional debunking. Such matters of a sensitive nature like UFOs and ETI in space and on planetary bodies, according to the **Brookings Report**, should be the eminent investigative domain of scientists like those in

NASA officialdom! The public should be the last people to know about such truth or evidence; it may cause worldwide panic and chaos!

America had to get its own lunar satellite up there orbiting the Moon to find out for themselves what was really on the Moon's surface, to either confirm or refute what the Russian photographs had initially revealed to the world.

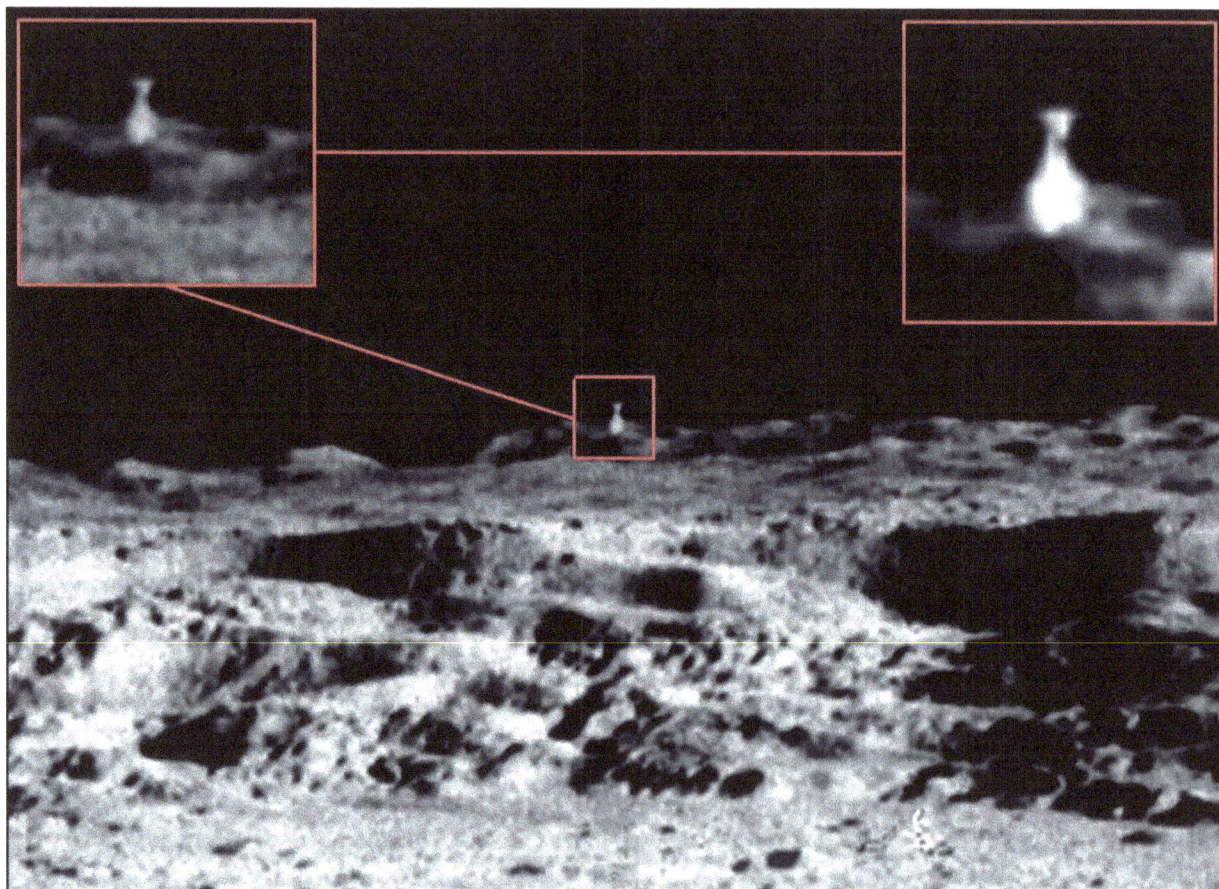

Another lunar tower with a dish shape top reminiscent of a radar tower on Earth
http://www.bibliotecapleyades.net/luna/luna_moonbases02.htm

Luna 10 became the first spacecraft from Russia to go into orbit around the Moon, and the first manmade object to orbit a body beyond the Earth. The primary objectives were to achieve the first lunar orbit, gain experience in orbital operations, presumably as a precursor to astronaut orbital missions, and study the lunar environment. The launch was timed so that the spacecraft would come around on its first orbit just as the Twenty-third Congress of the Communist Party of the Soviet Union was convening for its morning session. No doubt this was a political propaganda move on the part of the Soviet Union.

After many failed attempts with the early **Pioneer and Ranger spacecraft programs** of the late '50s and early '60s, Americans finally succeeded with **Ranger 7, 8 and 9 lunar spacecraft** in obtaining thousands of clear photographs of the Moon. These spacecraft also successfully impacted on the lunar surface. Soon, there were more American successes which followed, such

178

as the **Surveyor lunar lander** and the **Lunar Orbiter programs**. The United States had not only caught up with the Russians in the space race but was now starting to pull ahead and the US was now able to confirm with their own satellites that there were indeed lunar anomalies on the Moon, as first claimed by Russian lunar spacecraft.

It now became an imperative for America to get a man to the Moon before the Russians. There were buildings and structures on the Moon and picking the best landing sites close to these lunar structures was important for the US to understand whether ETs came from the Moon or were only based there. It also, seem strange that **President Kennedy** at that time had talked to **Premier Khrushchev** to discuss that there should be a joint effort in going to the Moon and in sharing new scientific knowledge from space exploration.

Early on, **President Dwight D. Eisenhower** pursued U.S.-Soviet cooperative space initiatives through a series of letters he sent in 1957 and 1958 to the Soviet leadership, first to **Prime Minister Nikolai Bulganin** and then to **Premier Nikita Khrushchev**. Eisenhower suggested creating a process to secure space for peaceful uses. Khrushchev, however, rejected the offer and demanded the United States eliminate its forward-based nuclear weapons in places like Turkey as a precondition for any space agreement. Feeling triumphant after Sputnik's launch, Khrushchev was certain his country was far ahead of the United States in terms of rocket technology and space launch capabilities, unlike the Soviet Union's more vulnerable geostrategic position in the nuclear arena. This would be the first of many times when space was linked with nuclear disarmament and other political issues.

Meanwhile, the United States energetically proceeded with its multinational initiative under the umbrella of the United Nations to develop a legal framework for peaceful space activities. This eventually led to the **Outer Space Treaty** and creation of the **United Nations Committee on the Peaceful Uses of Outer Space**, which a reluctant Soviet Union eventually joined.
http://www.nasa.gov/50th/50th_magazine/coldWarCoOp.html

Kennedy died two years before the first man, American Neil Armstrong walked on the Moon's surface and there was no joint space exploration agreement signed at that time with the Russians and Khrushchev was disposed of in October 1964, by the new Soviet leadership of **Leonid Brezhnev**, a Communist hardliner.

It wasn't until July 1975 that the first joint U.S.–Soviet space flight took place with the **Apollo–Soyuz Test Project** (ASTP) "Experimental flight **Soyuz-Apollo**". The last manned Moon mission by **Apollo 17** had already ended and this joint space project by the world's superpowers was the last flight of an Apollo spacecraft. Its primary purpose was as a symbol of the policy of détente that the two superpowers were pursuing at the time, and marked the end of the Space Race between them that began in 1957.

The mission included both joint and separate scientific experiments (including an engineered eclipse of the Sun by Apollo to allow Soyuz to take photographs of the solar corona), and provided useful engineering experience for future joint US–Russian space flights, such as the **Shuttle–Mir Program** and the **International Space Station**.

179

ASTP was the last manned US space mission until the first Space Shuttle flight in April 1981. It was also U.S. astronaut **Donald "Deke" Slayton's** only space flight. He was chosen as one of the original **Mercury astronauts** in April 1959 but had been grounded until 1972 for medical reasons. https://en.wikipedia.org/wiki/Apollo%E2%80%93Soyuz_Test_Project

Apollo 8 was NASA's first manned trip to the Moon with the mission of testing all facets of spacecraft operation, astronaut performance, orbiting maneuvers, more photographing of the Moon to refine maps and selected landing sites for an eventual manned landing.

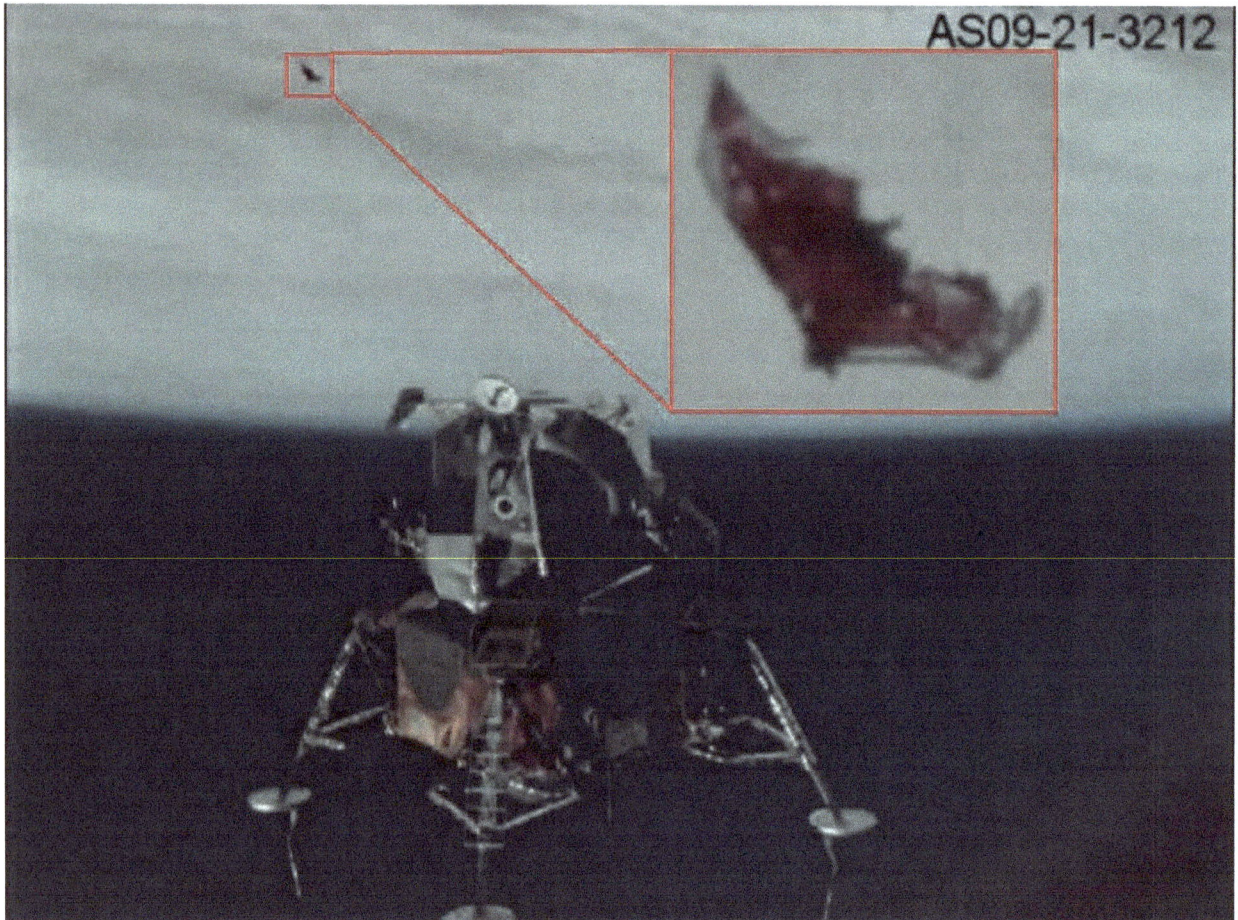

An astronaut photographs the Lunar Module separating from the Command Module while still in Earth orbit and also captures a UFO in the background
http://www.lpi.usra.edu/resources/apollo/frame/?AS09-21-3212

CHAPTER 70

ONE OF THESE DAYS ALICE, JUST ONE OF THESE DAYS --- POW! RIGHT TO THE MOON!!

Astronauts Report UFOs and Alien Bases on the Moon ("Luna")

It has been over 40 plus years since man first landed and step upon the Moon's surface, a historical event that established a new benchmark in mankind's technical accomplishments. Apollo 11, with **Neil Armstrong, Michael Collins,** and **Edwin Aldrin** was the first Apollo flight to land on the Moon, on July 20, 1969. While Collins flew in orbit around the Moon in the command module, Armstrong and Aldrin descended in the lunar module, landing in the **Sea of Tranquility** at 4:17 P.M.

Their journey to the Moon would not be without controversy, like so many other manned missions before them, Apollo 11 would have its share of UFO activity that seem to follow it right to the Moon and even down onto the lunar surface.

At one point Aldrin looking out the window of the command module could see an object that was unusual in shape. When Collins looked through a telescope, in one position the UFO had a series of ellipses, but when the focus of the telescope was sharpened, it appeared L-shaped. Not wishing to blurt out to **Huston Command Centre** back on Earth that they could see a UFO following them to the Moon. Very cleverly but cautiously, they asked Huston "how far away was the **S-IVB**?"

The S-IVB was the first stage of the rocket taking their capsule out into space; it had been jettisoned 2 days prior. Mission control had responded that the S-IVB was 6000 nautical miles away. Aldrin said that because of the distance, they had decided that the object they were looking at was not the S-IVB, but something closer. They continued to observe the object until they fell asleep, but did not discuss it over the radio with mission command again until their debriefing when they got back home.

Dr. Baker confirmed, "NASA knew very little about, um, the object reported by the Apollo 11 crew. It was obviously an unidentified flying object, but such objects were not uncommon and the history of even earth orbit space flights going back over the previous years indicated that several crews saw objects." He went on to say that others within NASA were also concerned about UFOs, "There were a lot of people within the program who went off later and became convinced that UFOs existed and that lead to some concern on NASA's part where they got the agreement of the crew never to publicly talk about these things for fear of ridicule."
http://www.openminds.tv/did-buzz-aldrin-see-a-ufo-on-his-way-to-the-moon/

Is there an "Alien Base" on the Moon? More and more people are coming forward with stories that might prove this is true. Rumors say that there is an Alien Moon Base on the far side of the moon, the side we never see from Earth.

Did you ever wonder why the Moon landings stopped and why we have not tried to build a

Moon Base? It does seem like a better and easier idea than a floating space station with no access to any raw materials or supplies? According to the NASA Astronaut Neil Armstrong, the aliens have a base on the Moon and told us in no uncertain terms to get off and stay off the Moon!

Sound farfetched? **Milton Cooper**, a Naval Intelligence Officer tells us that not only does the Alien Moon Base exist but the U.S. Naval Intelligence Community refers to the **Alien Moon Base** as **"Luna",** that there is a huge mining operation going on there, and that is where the aliens keep their huge mother ships while the trips to Earth are made in smaller "flying saucers".

Luna: *"The Alien base on the far side of the Moon. It was seen and filmed by the Apollo astronauts. A base, a mining operation using very large machines, and the very large alien craft*

described in sighting reports as mother ships exist there. " - **Milton Cooper**

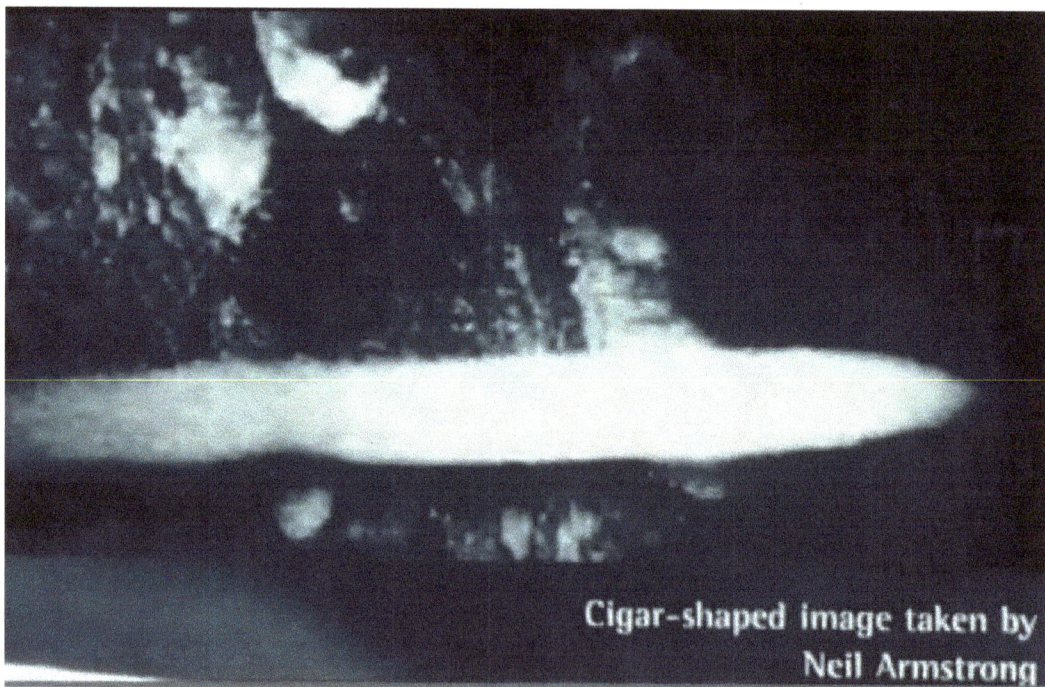

Cigar-shaped image taken by Neil Armstrong

This photo was taken above one of the Moon's crater by Neil Armstrong. This object is similar to the one photographed by the Russians over Mars
http://heavy.com/news/2015/11/astronaut-ufo-sightings-videos-pictures-photos-news-encounters-interview-testimonies-edgar-mitchell-gordon-cooper-apollo/4/

In his book, "Above Top Secret", **Timothy Good** says that there are unsubstantiated reports that both Neil Armstrong and Edwin "Buzz" Aldrin saw UFOs shortly after that historic landing on the Moon on 21 July 1969. Most of us who were alive at that time, who watched on TV the live Apollo Moonwalk, remember hearing one of the astronauts refer to a "light" in or on a crater during the television transmission, followed by a request from mission control for further information. Nothing more was heard.

According to a former NASA employee Otto Binder, unnamed radio hams with their own VHF receiving facilities that bypassed NASA's broadcasting outlets picked up the following exchange:

Armstrong: *What was it? What the hell was it? That's all I want to know!"*

NASA: What's there? .. malfunction (garble)...Mission Control calling Apollo 11...

Apollo: These "Babies" are huge, Sir! Enormous! OH MY GOD! You wouldn't believe it! I'm telling you there are other spacecraft out there, lined up on the far side of the crater edge! They're on the Moon watching us!

In 1979, **Maurice Chatelain**, former chief of NASA Communications Systems confirmed that **Armstrong** had indeed reported seeing two UFOs on the rim of a crater. "The encounter was common knowledge in NASA," he revealed, "but nobody has talked about it until now."

Soviet scientists were allegedly the first to confirm the incident. "According to our information, the encounter was reported immediately after the landing of the module," said **Dr. Vladimir Azhazha**, a physicist, and Professor of Mathematics at Moscow University. "Neil Armstrong relayed the message to Mission Control that two large, mysterious objects were watching them after having landed near the moon module. But his message was never heard by the public-because NASA censored it." Above Top Secret by Timothy Good; 1988; William Morrow and Company, Inc.; New York, N.Y. USA; ISBN 0-688-09202-0

Coloured photo of the Moon's surface with large UFO hovering above the Lunar Excursion Module (LEM) taken by Aldrin
http://www.unexplained-mysteries.com/forum/topic/195230-apollo-11-post-flight-press-conference/

According to another Soviet scientist, **Dr. Aleksandr Kazantsev, Buzz Aldrin** took color movie film of the UFOs from inside the module and continued filming them after he and Armstrong went outside. Dr. Azhazha claims that the UFOs departed minutes after the astronauts came out on to the lunar surface.

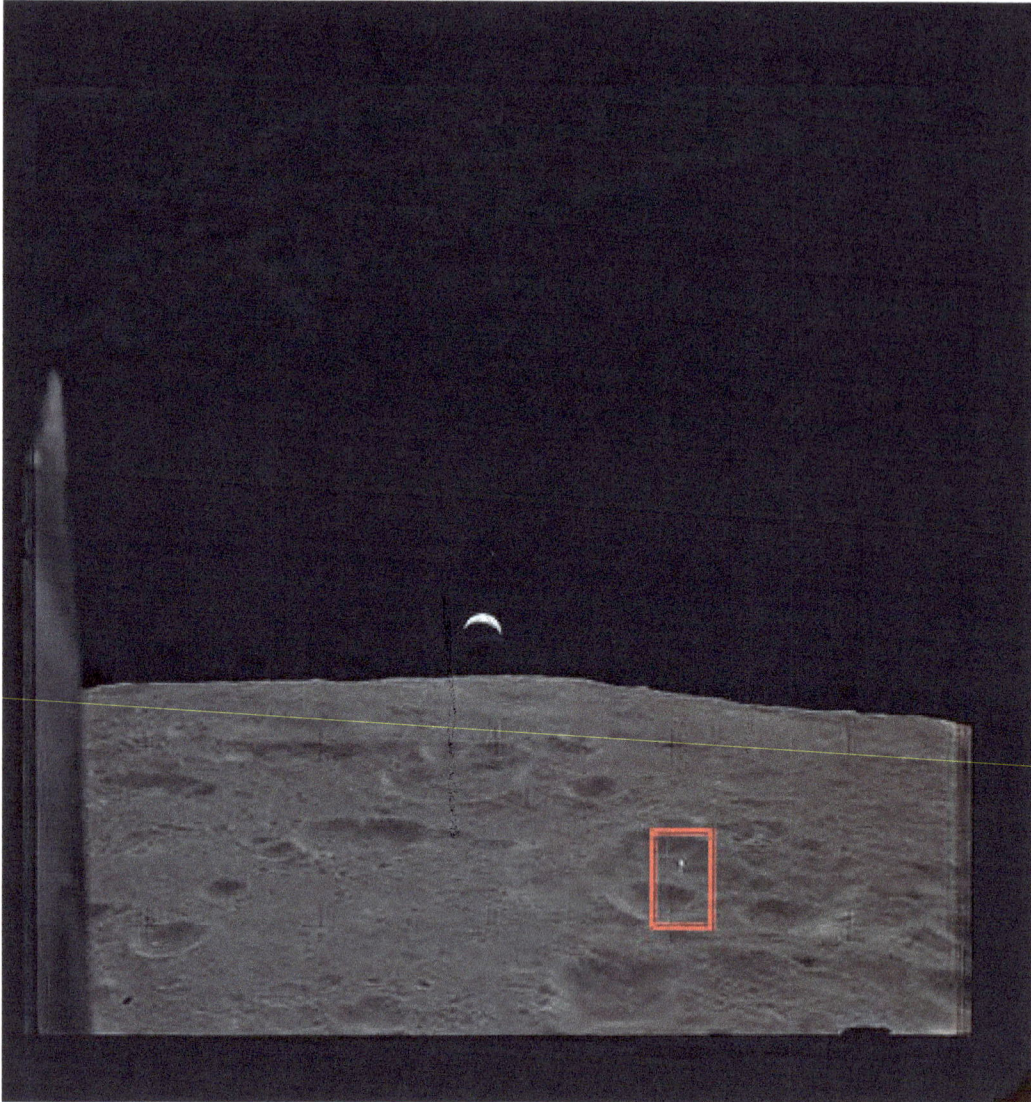

Astronaut Photography of Earth - Display Record AS12-47-6890
UFO is seen coming up from the Moon's surface (Highlighted)
http://eol.jsc.nasa.gov/scripts/sseop/photo.pl?mission=AS12&roll=47&frame=6890

Maurice Chatelain also confirmed that Apollo 11's radio transmissions were interrupted on several occasions in order to hide the news from the public. Chatelain claims that "all Apollo and Gemini flights were followed, both at a distance and sometimes also quite closely, by space vehicles of extraterrestrial origin-flying saucers, or UFOs if you want to call them by that name. Every time it occurred, the astronauts informed Mission Control, who then ordered absolute silence." He goes on to say:

"I think that Walter Schirra aboard Mercury 8 was the first of the astronauts to use the code name 'Santa Claus' to indicate the presence of flying saucers next to space capsules. However, his announcements were barely noticed by the general public.

It was a little different when James Lovell on board the Apollo 8 command module came out from behind the moon and said for everybody to hear:

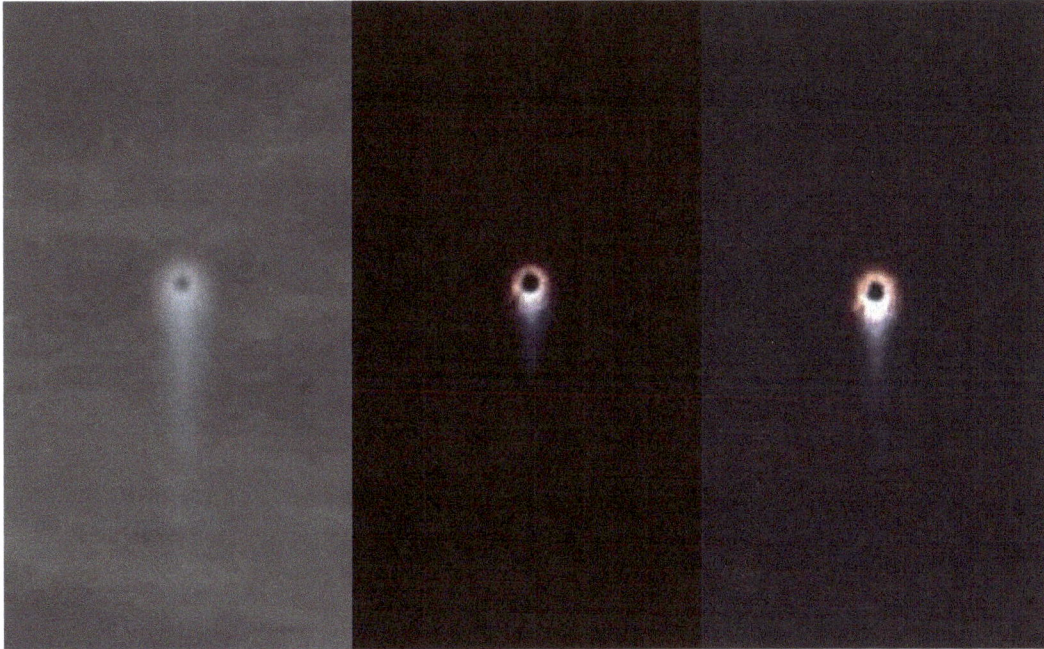

Above image has been enlarged and colour enhanced to show detail of UFO
http://eol.jsc.nasa.gov/scripts/sseop/photo.pl?mission=AS12&roll=47&frame=6890

'PLEASE BE INFORMED THAT THERE IS A SANTA CLAUS.'

Even though this happened on Christmas Day 1968, many people sensed a hidden meaning in those words."

Rumors persist. NASA may well be a civilian agency, but many of its programs are funded by the defense budget and most of the astronauts are subject to military security regulations. We have the statements by Otto Binder, Dr. Garry Henderson, and Maurice Chatelain that the astronauts were under strict orders not to discuss their sightings. And Gordon Cooper has testified to a United Nations committee that one of the astronauts actually witnessed a UFO on the ground. If there is no secrecy, why has this sighting not been made public? **Above Top Secret by Timothy Good; 1988; William Morrow and Company, Inc.; New York, N.Y. USA; ISBN 0-688-09202-0**

A certain professor, who wished to remain anonymous, was engaged in a discussion with Neil Armstrong during a NASA symposium.

Professor: What REALLY happened out there with Apollo 11?

Armstrong: It was incredible, of course, we had always known there was a possibility, the fact is, we were warned off! (by the Aliens). There was never any question then of a space station or a moon city.

Professor: How do you mean "warned off"?

Armstrong: I can't go into details, except to say that their ships were far superior to ours both in size and technology - Boy, were they big!... and menacing! No, there is no question of a space station.

Professor: But NASA had other missions after Apollo 11?

Armstrong: Naturally-NASA was committed at that time, and couldn't risk panic on Earth. But it really was a quick scoop and back again.

Armstrong confirmed that the story was true but refused to go into further detail, beyond admitting that the CIA was behind the cover-up.

Officially, Neil Armstrong was as tight-lipped and as a programmed as a robot, *"Your 'reliable sources' are unreliable. There were no objects reported, found, or seen on Apollo 11 or any other Apollo flight other than natural origin. All observation on all Apollo flights were fully reported to the public."*

NASA's debriefing protocols seem to be fully operational in ensuring Armstrong and the other astronauts were obedient to the sworn national security oaths they had taken not to divulge anything they had seen or found on the Moon. His Apollo 11 mission of being the first man on the Moon seemed to be a curse and an embarrassment to him throughout his life. The secrets of that mission weighed heavy upon him and he remained silent about them up to his death.

In the Russian newspaper Vecherny Volgograd, there was an official press-release of some of the discoveries found on the Moon that were photographed and videoed by Apollo astronauts that revealed both old and new and possibly occupied ET ruins and buildings:

"NASA scientists and engineers participating in exploration of Mars and Moon reported results of their discoveries at a briefing at the Washington national press club on March 21, 1996. It was announced for the first time that man-caused structures and objects had been discovered on the Moon." The scientists spoke rather cautiously and evasively about the functioning objects, with the exception of UFO. They always mentioned the man-caused objects as possible, and pointed out the information was still under study, and official results would be published later.

It was mentioned at the briefing as well that the Soviet Union used to own some photo materials

proving the presence of reasonable activity on the Moon. And although it wasn't identified what kind of reasonable activity it was, thousands of photo and video materials photographed from the Apollo and the Clementine space station demonstrated many parts on the lunar surface where the activity and its traces were perfectly evident. The video films and photos made by U.S.

astronauts during the Apollo program were demonstrated at the briefing. And people were extremely surprised why the materials hadn't been presented to the public earlier. And NASA specialists answered: "It was difficult to forecast the reaction of people to the information that some creatures had been or still were on the Moon. Besides, there were some other reasons to it, which were beyond NASA."

Specialist for lunar artifacts **Richard Hoagland** says that NASA is still trying to veil photo materials before they are published in public catalogues and files, they do retouching or partially refocus them while copying. Some investigators, Hoagland is among them, suppose that an extraterrestrial race had used the Moon as a terminal station during their activity on the Earth. The suggestions are confirmed by the legends and myths invented by different nations of our planet.

Ruins of lunar cities stretched along many kilometers, huge transparent domes on massive basements, numerous tunnels, and other constructions make scientists reconsider their opinions concerning the lunar problems. How the Moon appeared and principles of its revolving around the Earth still pose a great problem for scientists.

Some partially destroyed objects on the lunar surface can't be placed among natural geological formations, as they are of complex organization and geometrical structure. In the upper part of Rima Hadley, not far from the place where the Apollo-15 had landed, a construction surrounded with a tall D-shaped wall was discovered. As of now, different artifacts have been discovered in 44 regions.

The **NASA Goddard Space Flight Center**, the **Houston Planetary Institute** and specialists from the bank of space information are investigating the regions. Mysterious terrace-shaped excavations of the rock have been discovered near the Tito crater. The concentric hexahedral excavations and the tunnel entry at the terrace side can't be results of natural geological processes; instead, they look very much like open cast mines.

A transparent dome raised above the crater edge was discovered near the crater, Copernicus. The dome is unusual as it is glowing white and blue from inside. A rather unusual object, which is unusual indeed even for the Moon, was discovered in the upper part of the Factory area. A disk of about 50 meters in diameter stands on a square basement surrounded with rhombi walls. In the picture, close to the rhombi, we can also see a dark round embrasure in the ground, which resembles an entry in an underground caponier. There is a regular rectangular area between the Factory and the crater Copernicus which is 300 meters wide 400 meters long.

Apollo 10 astronauts made a unique picture **(AS10-32-4822)** of a one-mile long object called **"Castle"**, which is hanging at the height of 14 kilometers and casts a distinct shadow on the lunar surface. The object seems to be consisting of several cylindrical units and a large conjunctive unit. Internal porous structure of the **"Castle"** is clearly seen in one of the pictures, which makes an impression that some parts of the object are transparent.

As it turned out at the briefing where many NASA scientists were present, when Richard Hoagland had requested originals of the Castle pictures for the second time, no pictures were found there at all. They disappeared even from the list of pictures made by the Apollo 10 crew. Only intermediate pictures of the object were found in the archives, which unfortunately do not depict the internal structure of the object.

When Apollo-12 crew landed on the lunar surface, they saw that the landing was observed by a half-transparent pyramidal object. It was hanging just several meters above the lunar surface and shimmered with all rainbow colors against the black sky.

In 1969, when the film about astronauts traveling to the **Sea of Storms** was demonstrated (the astronauts saw the strange objects once again, which were later called "striped glasses"), NASA finally understood what consequences such kind of control could bring. Astronaut Mitchell answered the question about his feelings after a successful return the following: "My neck still aches as I had to constantly turn my head around because we felt we were not alone there. We had no choice but pray." Johnston, who worked at the Houston Space Center and studied photos and video materials done during the Apollo program, discussed the artifacts with Richard Hoagland and said, the NASA leadership was awfully annoyed with the great number of anomalous, to put it mildly, objects on the Moon. It was even said that piloted flights to the Moon could be banned in the programs network.

Investigators are especially interested in ancient structures resembling partially destroyed cities. The orbital shooting reveals an astonishingly regular geometry of square and rectangular constructions. They resemble our terrestrial cities seen from the height of 5-8 kilometers. A mission control specialist commented on the pictures: "Our guys observed ruins of the Lunar cities, transparent pyramids, domes and God knows what else, which are currently hidden deep inside the NASA safes, and felt like Robinson Crusoe when he suddenly came across prints of human bare feet on the sand of the desert island." What do geologists and scientists say after studying the pictures of lunar cities and other anomalous objects? They say such objects can't be natural formations. "We should admit they are artificial, especially the domes and pyramids." Reasonable activity of an alien civilization showed up unexpectedly close to us. We were not ready for it psychologically, and some people hardly believe they are true even now.
http://www.ufocasebook.com/moon.html

The book "Celestial Raise" by Richard Watson and ASSK records the following (continuation?) of the above remarkable dialogue of Apollo 11, which was picked up by hundreds of ham radio operators in the USA:

During the transmission of the Moon landing of Armstrong and Aldrin, two minutes of silence occurred in which the image and sound were interrupted. NASA insisted that this problem was the result of one of the television cameras which had overheated, thus interfering with the reception.

This unexpected problem surprised even the most qualified of viewers who were unable to explain how in such a costly project, one of the most essential elements could break down. Sometime after the historic Moon landing, **Christopher Craft**, director of the base in Houston, made some surprising comments when he left NASA.

The contents of these comments, which is included in the conversations [below], has been corroborated by hundreds of amateur radio operators who had connected their stations to the same frequency through which the astronauts transmitted. During the two minute interruption - which was not as it seemed, NASA, Armstrong and Aldrin with Cape Kennedy, censored both image and sound. 'I say that there were other spaceships.' **"Celestial Raise" by Richard Watson and ASSK [P.O. Box 35 Mt. Shasta CA. 96067 (916)-926-2316); 1987; page 147-148]** and **http://www.ufos-aliens.co.uk/cosmicphotos.html**

Here is reproduced completely the dialogue between the American astronauts and Control Center:

Apollo 11 Mission: July 20, 1969
Armstrong, Aldrin, and Collins landed at the Sea of Tranquility

Armstrong & Aldrin: *Those are giant things. No, no, no - this is not an optical illusion. No one is going to believe this!*

Houston (Christopher Craft): What ... what ... what? What the hell is happening? What's wrong with you?

Armstrong & Aldrin: *They're here under the surface.*

Houston: What's there? (muffled noise) Emission interrupted; interference control calling 'Apollo 11'.

Armstrong & Aldrin: *We saw some visitors. They were here for a while, observing the instruments.*

Houston: Repeat your last information!

Armstrong & Aldrin: *I say that there were other spaceships. They're lined up in the other side of the crater!*

Houston: Repeat, repeat!

Armstrong & Aldrin: *Let us sound this orbita ... in 625 to 5 ... Automatic relay connected ... My hands are shaking so badly I can't do anything. Film it? God, if these damned cameras have picked up anything - what then?*

Houston: Have you picked up anything?

Armstrong & Aldrin: *I didn't have any film at hand. Three shots of the saucers or whatever they were that were ruining the film*

Houston: Control, control here. Are you on your way? What is the uproar with the UFOs over?

Armstrong & Aldrin: *They've landed here. There they are and they're watching us.*

Houston: The mirrors, the mirrors - have you set them up?

Armstrong & Aldrin: *Yes, they're in the right place. But whoever made those spaceships surely can come tomorrow and remove them. Over and out.*
http://www.ufos-aliens.co.uk/cosmicphotos.html

The above transcript dialogue between Apollo 11 and Huston could be considered as somewhat specious, particularly when it is claimed that there were "hundreds" of amateur ham operators who tuned in on the frequency used by NASA and heard everything of the communication yet, these ham operators have never publicly been identified and their testimony has never been documented or recorded. Further investigation is required into this aspect to verify its authenticity.

However, with regard to the above questionable evidence of the ham operators, other reports from astronauts on other Apollo missions to the Moon seem to confirm Apollo 11's finding of an ET presence on the Moon. It has become widely reported in the late 1960s that NASA, meaning mankind, had been warned off to never return to the Moon again. NASA had six more manned Moon missions and landings after Apollo 11 which they were obligated to complete and then, with unusual characteristic "obedience", all further Moon landings ended. Man, since then, has never gone back to the Moon. Any plans, so we are told to set up any lunar colonies for exploration or potential mining operations were cancelled.

Many people feel that NASA has not been telling the public the complete truth regarding its space program, in particular, the Apollo Missions of the late 60's and early 70's. **Dr. Farouk El Baz**, one of NASA's foremost scientists, confirmed public suspicions when he stated: *"'not every discovery has been announced to the public"*. Is this the understatement of the millennium? Recent research has shown that conditions on the Moon could be very different from the 'official line' which NASA would lead us to believe.

Author's Rant: As a young man, who was an avid TV watcher and enthusiast of the US and Russian space programs, even I sensed that before the conclusion of the Apollo missions that the truth on these space flights and the discoveries made by these astronauts were being covered up and kept from public knowledge. It was as if my subconscious was picking up subliminal messages with each succeeding NASA moon mission. It was appearing to be that these lunar missions were truly becoming routine even; the Moon's panoramic surface was becoming all too familiar. Hills, mountains and craters appeared at times to be the same. What was really happening?

NASA has massive archives of literally millions of photographic Moon images which are supposedly in the public domain but, relatively few people have been allowed total access to them, why? Despite what NASA has photographed over the decades, the public thirst for images of the lunar landscape is miserably sated with just a few dozen "reproductions" that appear in the "official" textbooks or the "official NASA website". Something is seriously amiss and red flags are going up everywhere in a constant rhythm of inconsistencies. Many, if not all of the original photographs taken by satellites and by the astronauts are huge (32"x24") so, in order to make fit the pages of a regular book, they have been reduced and along with this reduction, clarity and

190

quality has also been reduced by the copying process which in turn make most of the images meaningless and blurry. You would think that a typical **single lens reflex (SLR) camera** or even a simple **Kodak point and shoot camera** of at that time would record better and clearer images.

In many cases, researchers are left with little more than 'smudges' and 'blurs'.

Under close scrutiny of the photos that researchers do have access to, microscopic analyze indicates that there is still telltale evidence that points to lunar anomalies on the Moon's surface. Once again, the **BIG LIE** rears one of its many ugly Hydra-like heads in the guise of NASA officialdom! Virtually everything NASA has told you about the Moon is a **LIE**!

The REAL NASA MOON PHOTOS, for example, show all kinds of structures, seemingly both old and new, such as domes, pipelines, and even pyramids. So why aren't these photos in the public domain? You can see in several of NASA's film footage, the American flag 'flapping in the wind' and yet the Moon according to NASA has no atmosphere, because it is a vacuum! One film clearly shows a desperate astronaut trying his level best to hold the flag still!

We are also told that the famous Neil Armstrong 'footprints' will remain etched on the Moon's surface forever. We are told this precisely because the Moon's 'atmosphere' is a vacuum. The laws of physics demand that dust becomes hardened and will compress in a 'vacuum' therefore ensuring the *'footprints'* remain undisturbed. And yet great plumes of dust can be seen spewing forth from underneath the 'Moon Buggy' as it travels across the lunar surface. Is this 'vacuum theory' some kind of wild hoax by NASA?

Another NASA cover-up is the small cloud formations that have been photographed above the Moon, again in a vacuum? And while we're on the subject of clouds what about the ONE HUNDRED MILE WIDE CLOUD OF VAPOR that was detected by NASA's own instruments. This embarrassing *'anomaly'* was promptly dismissed by NASA scientists as being the result of the considerable volume of urine ejected by the Apollo Mission astronauts! What were they drinking?!

For decades strange 'lights' and artificial seeming structures have been observed and recorded on the Moon by amateur astronomers. Science writer **Joseph Goodavage** observed that over two hundred white *'dome shaped'* structures had been seen and catalogued, only for them often to vanish and reappear somewhere else?. There are even colour photos from the Apollo 8 missions that clearly show evidence of green vegetation on the lunar hills.

These unusual findings, when added together with the anomalies which Richard Hoagland has shown to exist on Apollo Moon photographs, provide compelling evidence for an ongoing NASA cover-up of what the Apollo Astronauts really discovered on the Moon from 1969 to 1972.

The following are more excerpts of amazing conversations from Apollo Astronauts to Mission Control revealing what astonished Astronauts came across while on the Moon's surface; some strange artifacts, hard to explain structures and unusual sightings of unidentified craft. The

Apollo Astronaut conversations were mostly taken from the out-of-print book: **"Our Mysterious Spaceship Moon" by Don Wilson (Dell, 1975)**

Take note that the following conversations by the astronauts are sometimes in code as a means to disguise what they are referring to. Many times the excited cries of one or both astronauts seem to be out of place for the mere act of collecting of Moon rocks. A geologist might get excited…perhaps, and maybe even, a mineralogist, but an astronaut getting overly excited by Moon rock is not likely, as they would have you believe! Or did they find something much more incredible, that it was hard to contain their excitement, which was not meant for public knowledge? **home.mytelus.com/telusen/portal/index.aspx**

Apollo 12 Mission: November 14, 1969
Pete Conrad, Alan Bean, and **Richard F. Gordon** landed at the southeastern portion of the **Ocean of Storms**.

Apollo 12 carried **Charles Conrad Jr., Richard F. Gordon**, and **Alan I. Bean** through thunderclouds right at the start, experiencing an electrical discharge of short duration that did not hamper the flight. The mission, lasting 10 days, November 14-24, 1969, took Conrad and Bean to the Sea of Storms, right next t **Surveyor 3**, which had landed there two and a half years before (on April 20, 1967). Some of the more important parts from Surveyor 3 were brought back in remarkably good condition. (Italics added by author).

On the day out on the earth - moon leg of the trip, the astronauts radioed Mission Control that two flashing lights had appeared off the bow of their capsule. After rejecting the possibility that the objects could be spinning pieces of the Apollo booster rocket the CapCom suggested that they could be the jettisoned protective panels. One of the astronauts replied, *"Gee that could be, but one of those lights just shot out of here at tremendous speed".*

Despite the late Carl Sagan's assurance in his 1966 book Intelligent Life in the Universe, that *"a natural satellite cannot be a hollow object"*, there is amazing evidence *that the moon could indeed be hollow*. In 1969 the crew of Apollo Twelve, in an attempt to create an artificial moonquake sent the ascent stage of the lunar module crashing back down to the moon's surface. To everyone's surprise, the *highly sensitive seismic equipment recorded* something totally unexpected. *For more than one hour, the moon continued to reverberate like a bell.* **Dr. Frank Press** of **Massachusetts Institute of Technology (MIT)** commented, *"None of us have seen anything like this on Earth. In all our experience, it is quite an extraordinary event. That this rather small impact … produced a signal which lasted thirty minutes is quite beyond the range of our experience".* (Marrs, p.6). **http://www.greatdreams.com/moon/darkmoon.htm**

This Apollo 12 mission has more coded messages, this time, Conrad warns Beans about looking out for Indians! The original Apollo 12 Lunar Module DSEA "black box" transcript can be found on page numbers 274-275 and also, at the websites below. The following conversation is between CDR, LMP, and **CapCom Edward Gibson**:
Conrad: *One picture of that rock under the descent stage - -*

Bean: *Shall do.*

Conrad: - - *Grab the handtool carrier and head for the solar wind and grab a picture of that. In the meantime, I'll lope off to the ALSEP and check the SIDE's; I'll meet you at point 1 at head crater.*

Bean: *Okay.*

Conrad: *If you see any Indians, don't shoot until you see the whites of their eyes.*

Gibson: *Roger; we copy. And, Al, have you gotten the readings on the contrast charts?*

Bean: *Not yet and I plan to do that real quick.*
http://www.jsc.nasa.gov/history/mission_trans/AS12_LM.PDF

The conversation between Pete Conrad and Alan Bean is straightforward and matter of protocol in its dialogue, even CapCom's Ed Gibson is nonchalant in his response. The statement in its context is rather racist in its implication, unless, its intention was to imply something that may indicate a possible Extraterrestrial presence. If this statement was intended as a coded reference to ETs then, the word "Indians" may have been an error in transcript and the word "Aliens" may have been intended instead and not the stated racist appellation. What is interesting is that NASA seems to have censored the original transcript to read as two astronauts engaged in normal routine of lunar activity without the reference to "Indians". (See photocopies below).
http://spacetime.forumotion.com/t741-apollo-12-and-the-indians-on-the-moon-bizarre-astronaut-quote-reveals-censored-transcripts-audio

Official NASA Apollo 12 photograph AS12-497319 clearly shows a large UFO(the haze) hovering over an astronaut walking on the Moon. This is not a lens flare, reflection, sun glare or an extraneous light source.

¡CONFIDENTIAL

ALSEP or to head crater that you'd like polarizing
pictures taken?

05 11 58 43	LMP		Hey, hold that, would you, Pete?
05 11 58 45	CC		We'll get back to you on that. Press on now - the nominal plan right now.
05 11 58 49	CDR		Here, let me have it.
05 11 58 51	LMP		Okay, let me film the handle - -
05 11 58 52	CDR		Film the handle, yes. Hold it tight.
05 11 58 57	LMP		I got it as far as I can. Permanently mounted that baby - -
05 11 59 08	CC		Pete, we have no preference on that. Go ahead and take it as called out for in the cuff checklist.
05 11 59 12	CDR		Okay. Stick it in the ETB and we'll screw with it later.
05 11 59 14	LMP		Good idea.
05 11 59 15	CDR		Handle in.
05 11 59 16	LMP		Okay.
05 11 59 17	CDR		Drop her. Okay, Houston, one TV camera in the bag and - our plan of attack is - Al?
05 11 59 29	LMP		Go.
05 11 59 32	CDR		One picture of that rock under the descent stage - -
05 11 59 34	LMP		Shall do.
05 11 59 35	CDR		- - Grab the handtool carrier and head for the solar wind and grab a picture of that. In the meantime, I'll lope off to the ALSEP and check the SIDE's; I'll meet you at point 1 at head crater.
05 11 59 43	LMP		Okay.

CONFIDENTIAL

Apollo 12 Lunar Module DSEA "black box" transcript:
"If you see any Indians, don't shoot until you see the whites of their eyes".
(Begins at the bottom and continues on next page)
https://www.jsc.nasa.gov/history/mission_trans/AS12_LM.PDF

05 11 59 44 CDR   If you see any Indians, don't shoot until you see
                    the whites of their eyes.

05 11 59 47 CC   Roger; we copy. And, Al, have you gotten the
                    readings on the contrast charts?

05 11 59 55 LMP   Not yet and I plan to do that real quick.

05 11 59 59 CDR   Houston, Pete's on his way to the ALSEP. Okay.

05 12 00 01 CC   Roger.

05 12 00 06 CC   Roger, Al; we copy. And at ... minutes into the
                    VA - EVA, you're pretty close to the nominal time
                    line.

05 12 00 15 LMP   Okay. Very good.

05 12 00 28 CDR   Can the guy with the seismometer hear me running?

05 12 00 34 LMP   ...

05 12 01 05 CDR   Okay.

05 12 01 13 CC   Pete, we're watching you down here in the seismic
                    data - looks as though you're really thundering
                    right by it.

05 12 01 18 CDR   Yes, I - I ground to a halt there to switch to
                    intermediate cooling. I noticed that - it is
                    obviously a little bit hotter out here with the
                    higher Sun angle right now. Okay, I'm approaching
                    the infamous SIDE.

05 12 01 43 CC   Roger. And we're able to copy your rest and now
                    that you're moving again.

05 12 01 46 CDR   Okay. All right, Houston. The status is - Oops,
                    I'm going to get dust in it. The cover is off,
                    and it's pointed up at the sky at about a 60-degree
                    angle.

05 12 02 11 CC   Roger. Do not touch it right now, Pete. Which
                    way is that pointing - relative to east-west?

05 12 02 17 CDR   It is pointed down-Sun.

Apollo 12 Lunar Module DSEA "black box" transcript:
"If you see any Indians, don't shoot until you see the whites of their eyes".
(See top of the page)
https://www.jsc.nasa.gov/history/mission_trans/AS12_LM.PDF

05 11 59 17	CDR-EVA	... in.
05 11 59 18	LMP-EVA	Okay.
05 11 59 19	CDR-EVA	Drop her. Okay, Houston, one TV camera in the bag and - our plan of attack is - Al?
05 11 59 31	LMP-EVA	Go.
05 11 59 33	CDR-EVA	One picture of that rock under the descent stage - -
05 11 59 36	LMP-EVA	Will do.
05 11 59 37	CDR-EVA	- - Grab the handtool carrier and head for the sola wind and grab a picture of that; in the meantime, I'll lope off to the ALSEP and check the SIDES; I'll meet you at point 1 at head crater.
05 11 59 49	CC	Roger; we copy. And, Al, have you gotten the readings on the contrast charts?
05 11 59 57	LMP-EVA	Not yet and I plan to do that real quick.
05 12 00 00	CC	Roger.
05 12 00 01	LMP-EVA	Houston, Pete's on his way to the ALSEP.
05 12 00 08	CC	- Roger, Al; we copy. And at 30 minutes into the VA - EVA, you're pretty close to the nominal time line.
05 12 00 17	LMP-EVA	Okay. Very good.
05 12 00 30	CDR-EVA	Can the guy with the seismometer hear me running?
05 12 00 36	LMP-EVA	...
05 12 01 07	CDR-EVA	Okay.
05 12 01 11	CC	Pete, we're watching you down here on the seismic data - looks as though you're really thundering right by it.
05 12 01 20	CDR-EVA	Yes, I - I ground to a halt to switch to inter-mediate cooling. I noticed that - it is obviously a little bit hotter out here with the higher Sun angle right now. Okay. I'm approaching the Emplemus side.

Apollo 12 Lunar Module DSEA "black box" transcript (The NASA sanitized version of the Conrad – Bean conversation without the "Indian" reference)

(Google Image)

Apollo 13 Mission: April 11-17, 1970
James Lovell, Jr., Fred W. Haise, Jr., and **John L. Swigert, Jr.,** never landed on the Moon due to a damaged command module.

Apollo 13 with astronauts Lovell, Haise, and Swigert aboard, ran into trouble, seemingly confirming the superstition tied to the number 13. The mission which took place April 11-17, 1970, was halfway to the moon when one of the oxygen tanks exploded, knocking out some instruments. The question was no longer how to land on the moon, but how to get back to earth as soon as possible. It was decided to continue the flight to the moon, make a loop around it, and come back straight for splashdown, all the time saving as much oxygen as possible. The cause of the explosion was never determined, although several official explanations were given. (***It was later thought that an electrical short occurred when Huston asked Apollo 13 to stir the cryo-tanks which in turn caused the oxygen tanks to explode. This resulted in nearly one-quarter of the command module paneling being blown away revealing the ships internal structure.***) [Bold italics added by author].

There was some discussion about Apollo 13 carrying a nuclear device to the moon to be used to show seismograph recordings placed at several locations. Rumours were that the UFOs had deliberately caused the explosion on board to prevent the detonation of the atomic charge that could possibly have destroyed or endangered some moon base established by the extraterrestrials. http://www.greatdreams.com/moon/darkmoon.htm

Author's Rant: Given that NASA knew that an ET presence was monitoring American and Russian space activity, this theory is not an unrealistic premise but, I don't believe that it was the malicious or aggressive intention of ETI to cause harm to humans. It was more likely that this crippling act was specifically targeting a non-lethal component of the ship that was intended to send a clear message to humans, not to bring nuclear weapons into space or to the Moon or any other planetary body.

When Apollo Thirteen's third stage was deliberately sent hurtling into the lunar surface by radio signal, crashing with the impact of eleven tons of TNT, NASA claimed that the Moon *"reacted like a gong."* Seismic equipment as distant as one hundred and seventy-three kilometres from the crash site recorded reverberations lasting for three hours and twenty minutes and travelling to a depth of thirty-five to forty kilometres. Writer Don Wilson in "Secrets of our Spaceship Moon", claims that one NASA scientist had admitted that the United States government had conducted a series of experiments (without any public announcement) to determine if the Moon is hollow. Nobel prize winning chemist **Dr. Harold Urey** suggested that the reduced density of the Moon was due to the fact that large areas inside the Moon were "simply a cavity" and **Dr. Sean Solomin** of MIT wrote, *"the Lunar Orbiter experiments vastly improved our knowledge of the moon's gravitational field ... indicating the frightening possibility that the moon might be hollow."* http://www.greatdreams.com/moon/darkmoon.htm

It is hard to image that there are lunar cavities under the Moon's crustal surface as this would create some gravitational anomalies that may have some effect on the Moon's rotation at least that would be the logical deduction to this unusual aspect to the Moon as an orbiting body.

Apollo13 did see and photograph a very large disc or globe shape UFO (possible "mothership") and several scout ship escorting it as they travelled toward the Moon. NASA had no explanation for it other than to say it was a *"bright disc covering the Moon"*. Later, Apollo 13 astronauts photographed another UFO or possibly the same one earlier flying near the lunar surface monitoring Apollo 13's slingshot orbit of the Moon to send it back towards Earth.

Apollo 13
Official NASA caption for this image reads:
"Lunar disc with bright disc partially covering the Moon."

Hugh globe UFO measuring miles across with much smaller
UFO "scout ships" escorting the "mother ship" towards the Moon
http://ceifan.org/UFO_apollo.htm

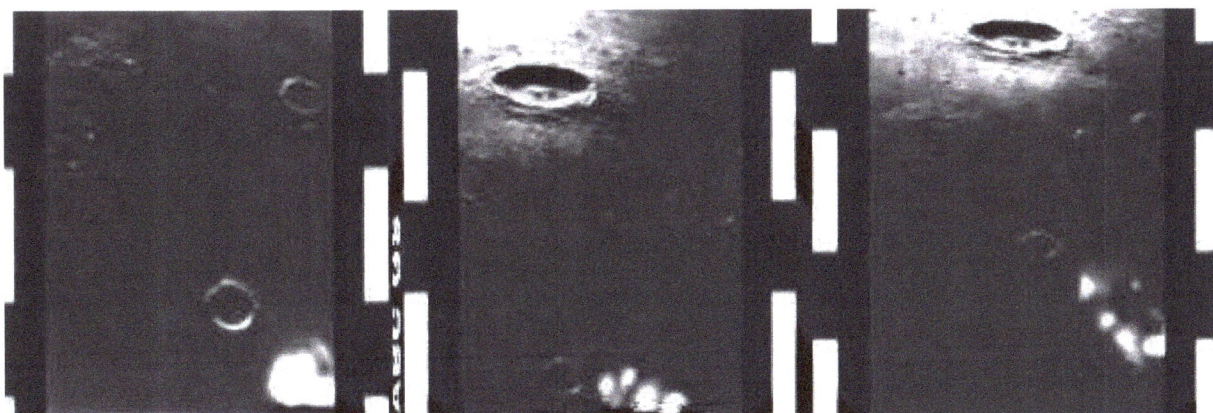

Apollo 14 photos of object flying over the lunar surface
(Bottom left corner in each frame)
(Google Image)

Apollo 14 Mission: January 31-February 9, 1971
Alan B. Shepherd, Jr., Stuart A. Roosa, and **Edgar D. Mitchell** landed in the hills of **Fra Mauro.**

Apollo 14, with **Alan B. Shepherd, Jr., Stuart A. Roosa, and Edgar D. Mitchell** aboard, went to the moon from January 31-February 9, 1971, landing in the hills of Fra Mauro, using a cart to transport the scientific instruments.

In general, the astronauts, nearly all military officers controlled by security regulations, have maintained a united front. However, there have been exceptions. **Dr. Edgar Mitchell** on the **Oprah Winfrey Show** of July 19, 1992, suggested that all information regarding UFOs had not been released, adding, *"I do believe that there is a lot more known about extraterrestrial investigation than is available to the public right now (and) has been for a long time... It's a long, long story, it goes back to World War II when all of that happened and is highly classified stuff"* **(Good, p.206).**

Colonel Gordon Cooper, in a letter to a meeting of the United Nations in 1978, to discuss UFOs, stated, *"I believe that these extraterrestrial vehicles and their crews are visiting this planet from other planets, which are obviously a little more advanced than we are here on Earth"* **(Huneeus).** http://www.greatdreams.com/moon/darkmoon.htm

Apollo 15 Mission: July 26 - Aug. 7, 1971
 David Scott, Alfred Worden, and **James Irwin**; went to the **Appenine Mountains** of the Moon. Conversation about discovering strange ***"tracks":***

Scott: *Arrowhead really runs east to west.*

Mission Control: *Roger, we copy.*

Irwin: *"Tracks" here as we go down slope.*

MC: Just follow the *"tracks"*, huh?

Irwin: *Right we're (garble). We know that's a fairly good run. We're bearing 320, hitting range for 413 ... I can't get over "those lineations", that layering on Mt. Hadley.*

Scott: *I can't either. That's really spectacular.*

Irwin: *They sure look beautiful.*

Scott: Talk about organization!

Irwin: That's the most ***"organized structure I've ever seen"!***

Scott: *It's (garble) so uniform in width.*

200

Irwin: *Nothing we've seen before this has shown such uniform thickness from the top of the tracks to the bottom.*

Wilson writes: (p. 145): "What are these tracks? Who made them? Where did they come from? Does NASA have an answer for the people?"
http://www.greatdreams.com/moon/darkmoon.htm

Apollo 15 Astronauts Davis Scott, Alfred Worden and James Irwin
https://en.wikipedia.org/wiki/Apollo_15

Another Weird Conversation About Tracks

Made by Harrison Schmitt, a trained geologist and the only civilian ever to walk on the Moon (all the rest were military men, or as **Hoagland** would say - "good soldiers", who did what they were told).

Schmidtt: I see *"tracks"* - *running right up the wall of the crater.*

Mission Control (Gene Cernan): *Your photo path runs directly between **"Pierce and Pease".** "Pierce Brava, go to Bravo, Whiskey, Whiskey, Romeo".*

Wilson writes (p. 145): "If this is not code, what is it? And why switch to the use of strange meaningless "code" words if NASA was not trying to cover up something startling, something that needed to be hidden from the public? In fact, science writer **Joseph Goodavage** maintains that "whenever something was discovered, the astronauts and CAPCOM apparently switched to a prearranged code, sometimes even on an alternate publicly unmonitorable channel".

Apollo 15, Moon, 1971

**This colour enhanced and brighten photo reveals a blue double disc
light hovering about the Apollo 15 Lunar Excursion Vehicle**
http://www.abovetopsecret.com/forum/thread482257/pg1

NASA scientist **Farouk El Baz** admitted in a magazine interview that NASA did commit itself to a secretive search for various things on the Moon. *"We're looking for something - something ...' He admitted that 'a huge bridge-like structure in Mare Crisium has been reported ... That is all I can say about it."* When asked if 'that was a bridge - that you've actually found artificial structures or some kind of intelligently placed artifact?' El Baz quickly denied it.

"No. No. I am not admitting such a thing. But when you start to think about it, almost anything is possible. There are almost no limits [to] how you can interpret the many things astronomers have been observing and reporting for several centuries. Now the astronauts are seeing many anomalies close up." **http://www.greatdreams.com/moon/darkmoon.htm**

The set of photo images below comes from a video that was only recently made public on April 5, 2015, on YouTube, possibly through a leak made by an employee of NASA. It shows one of the Apollo astronauts as he as setting up his TV camera, who inadvertently captured a disc-

202

shaped craft (possibly two craft, one smaller one is in flight) parked on a nearby crater rim shaped craft (one smaller one is in flight) parked on a nearby crater rim observing the activity of the astronauts. Such incidents have often been reported by all Apollo astronauts since the first manned moon landing made by Apollo 11 astronauts Neil Armstrong and Buzz Aldrin. Such incidents have often been reported by all Apollo astronauts, since the first manned moon landing made by Apollo 11 astronauts Neil Armstrong and Buzz Aldrin. Now there is video proof of these lunar UFO accounts! http://www.hq.nasa.gov/alsj/a15/a15.html and https://www.youtube.com/watch?v=Kjo7d5W3ic8

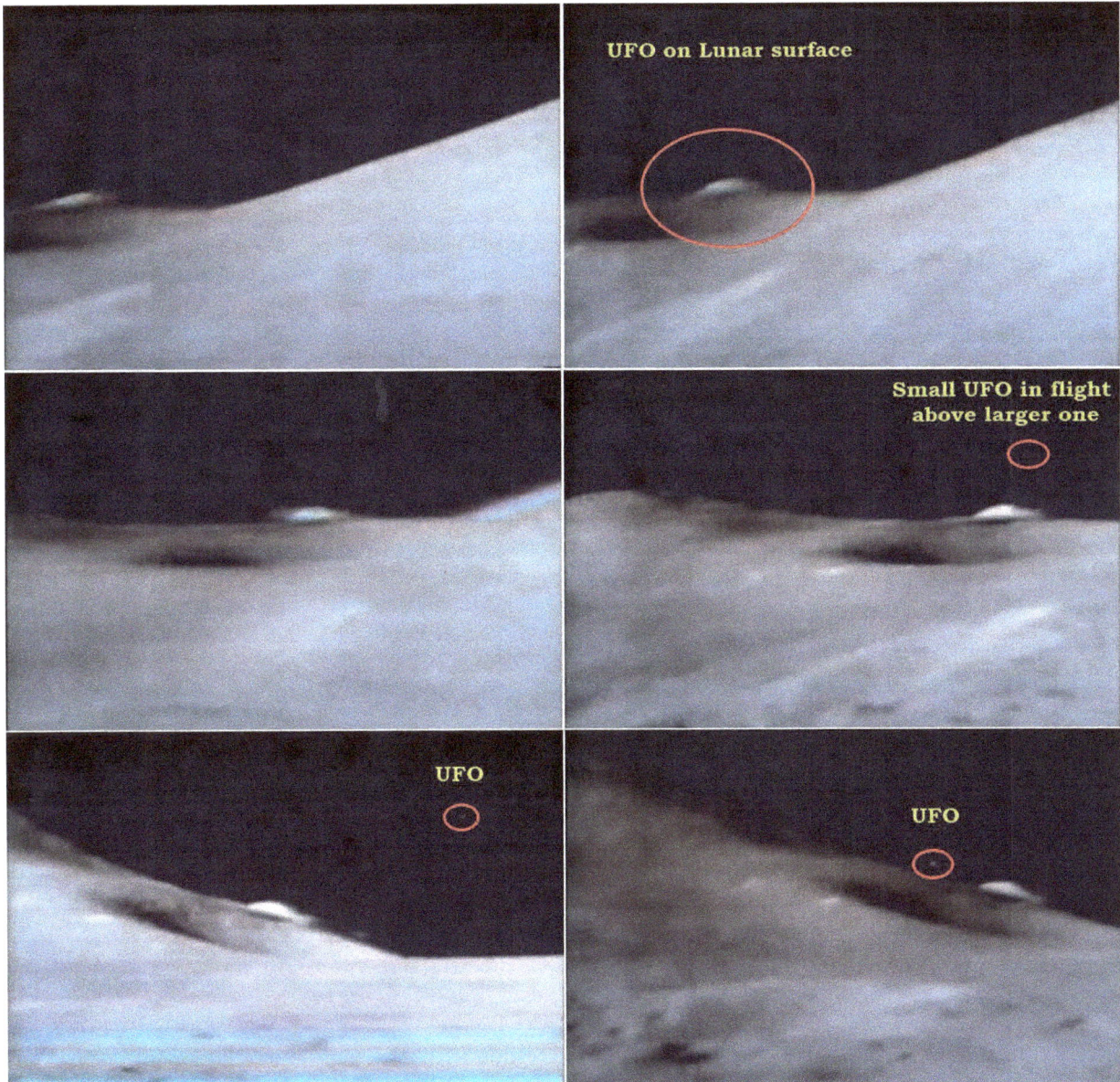

The question of whether UFOs and ETs were watching the lunar activities of Apollo astronauts is definitively answered in these pictures taken from a video as one of the astronauts unintentionally captures a disc-shaped craft parked on the edge of a nearby crater while setting up his TV camera

https://www.youtube.com/watch?v=Kjo7d5W3ic8

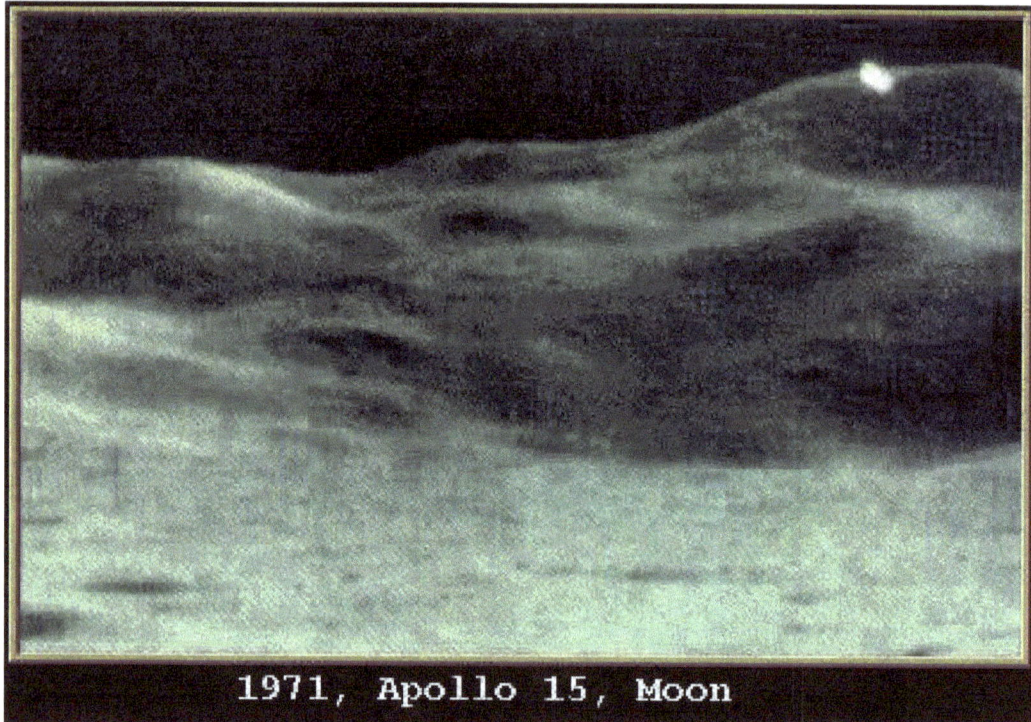

1971, Apollo 15, Moon

Bright UFO near hilltop watches Apollo 15 astronauts' activity

Apollo 15: Astronaut Sees White Objects Flying By

Capcom: *You talked about something mysterious...*

Orion*: O.K., Gordy, when we pitched around, I'd like to tell you about something we saw around the **LM (LEM or Lunar Excursion Module).** When we were coming about 30 or 40 feet out, "**there were a lot of objects - white things - flying by. It looked as if they were being propelled or ejected**", but I'm not convinced of that.*

Capcom: *We copy that Charlie.*

Wilson writes (p. 54): "What could these mysterious flying objects have been? ... Can this be considered another UFO sighting? What did the astronaut mean when he reported that these peculiar 'white things' were 'perhaps' 'being propelled or ejected'? And by whom?"
http://www.greatdreams.com/moon/darkmoon.htm

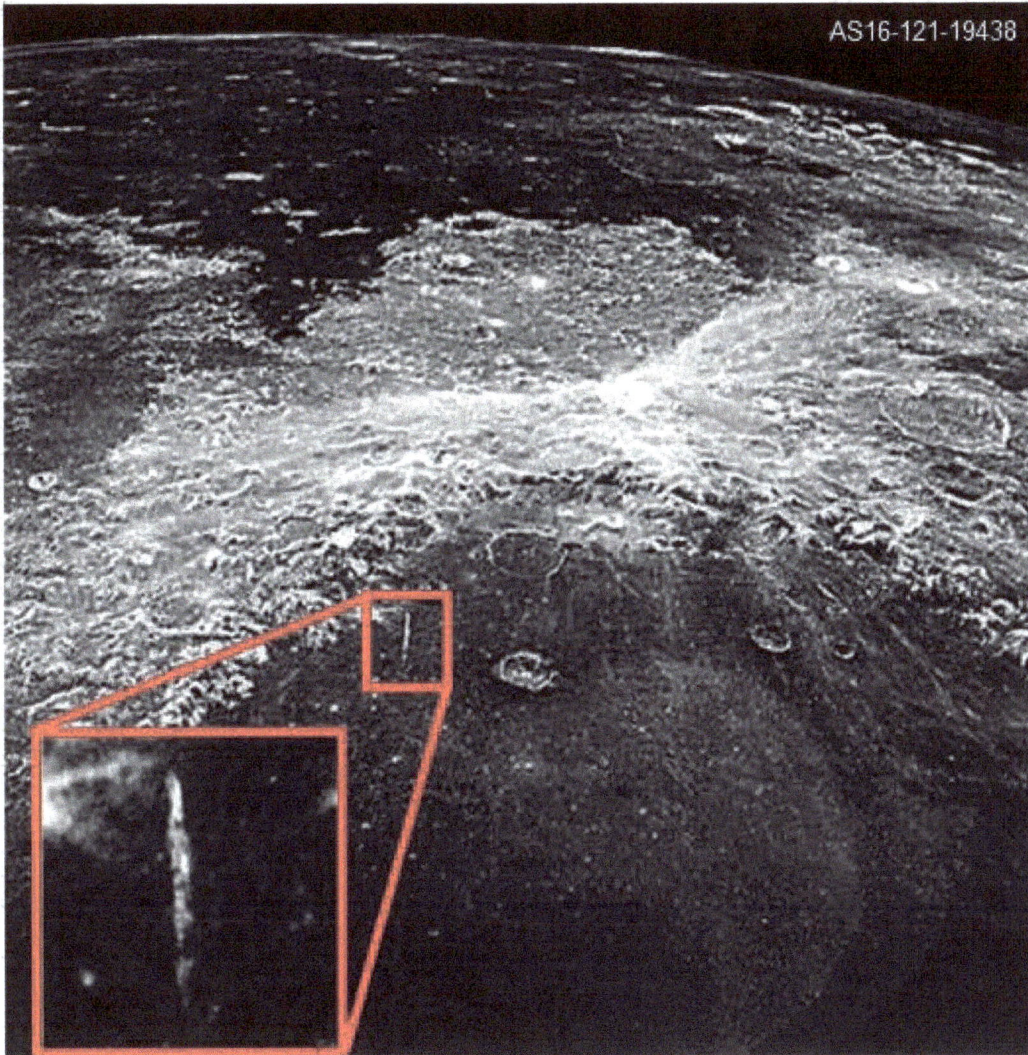

"Crisium Spire" near Picard Crater is about 20 miles in height seems to grow out of the Sea of Crises! The picture was taken in from Apollo 16.

http://m.blog.daum.net/sdigkim/254

Apollo 16 Mission: April 16 - 27, 1972
Charles Duke, Thomas Mattingly, and **John Young** landed in the **Descartes highlands**:

Duke: *These devices are unbelievable. I'm not taking a "**gnomon**" (indicator provided by the stationary arm whose shadow indicates time on the sundial) up there. [Italics added by author for clarity].*

Young: *O.K., but man, that's going to be a "**steep bridge**" to climb.*

Duke: *You got - YOWEE! Man - John, I tell you this is some sight here. Tony, the blocks in Buster are covered - the bottom is covered with blocks, five meters across. Besides the blocks seem to be in a preferred orientation, northeast to southwest. They go all the way up the wall on* *those two sides and on the other side you can only barely see the outcropping at about 5 percent. Ninety percent of the bottom is covered with blocks that are 50 centimeters and larger.*

Capcom: *Good show. Sounds like a secondary...*

Duke: *Right out here ... the blue one that I described from the lunar module window is colored because it is "**glass coated**", but "**underneath the glass, it is crystalline**" ... the same texture as the "**Genesis Rock**" ... "**Dead**" on my mark.*

Young: *Mark. It's open.*

Duke: *I can't believe it!*

Young: *And I put that beauty in dry!*

Capcom: *Dover. Dover. We'll start EVA-2 immediately.*

Duke: *You'd better send a couple more guys up here. They'll have to try (garble).*

Capcom *Sounds familiar.*

Duke: *Boy, I tell you, these **EMUs** and **PLSSs** are really super- fantastic!*
http://www.greatdreams.com/moon/darkmoon.htm

Apollo 16: Describing Domes and Tunnels on the Moon

Duke: *We felt it under our feet. It's a soft spot. Firmer. Where we stand, I tell you one thing. If this place had air, it'd sure be beautiful. It's beautiful with or without air. The scenery up on top of Stone Mountain, you'd have to be there to see this to believe it - those domes are incredible!*

Mission Control: *O.K., could you take a look at that smoky area there and see what you can see on the face?*

Duke: *Beyond the domes, the structure goes almost into the ravine that I described and one goes to the top. In the northeast wall of the ravine, you can't see the delineation. To the northeast there are tunnels, to the north they are dipping east to about 30 degrees.*
http://www.greatdreams.com/moon/darkmoon.htm

1972, Apollo 16, Moon

Mission Apollo 16 on the Moon, 1972 Astronaut Charles Duke photographed collecting lunar samples at Station 1. No explanation is given for the object seen behind the astronaut
http://www.telegraph.co.uk/news/science/space/5832435/Moon-landing-anniversary-UFOs-photographed-by-Apollo.html?image=7

Apollo 16 "Ground-to-Air" Conversation:

Capcom: *What about the albedo change in the subsurface soil? Of course, you saw it first at "Flagg" and were probably more excited about it there. Was there any difference in it there - and "Buster" and "Alsep" and "LM"?* (These terms, of course, could also refer to small rocks and/or small craters). [Author's parenthesis].

Duke: *No. Around the "" it was just in spots. At "Plum" it seemed to be everywhere. My predominant impression was that the white albedo was (garble) than the fine cover on top.*

Capcom: *O.K. Just a question for you, John. When you got halfway or even thought it was halfway, we understand you looped around south, is that right?*

Young: *That is affirm. We came upon – "Barbara".*

Wilson writes (p.140): "Joseph H. Goodavage, whom included this conversation in a Saga magazine article, comments: "Barbara? That really needs some explanation, so I made an appointment with NASA geologist Farouk El Baz at National Aeronautics and Space Museum. Here's how part of our conversation went:

Saga: *What do you suppose Young meant when he said they came upon **"Barbara"**?*

El Baz: *I can't really say. Code perhaps...*

Saga: *But **"Barbara"** is an odd name for something on the Moon, isn't it?*

El Baz: *Yes, an enigma. As I suggested, perhaps a code, but I don't really know." (Bold italics added by author for emphasis).*

Apollo 17 Mission: Dec 7 - 19, 1972
Eugene Cernan, Ronald Evans, and **Harrison Schmidt.** Landed in the **Taurus-Littrow Valley**

Eugene Cernan, commander of Apollo 17, in a Los Angeles Times article in 1973 said about UFOs: *"...I've been asked (about UFOs) and I've said publicly I thought they (UFOs) were somebody else, some other civilization."*

Check out the following coded conversation (in bold italics e.g. *"Condorcet Hotel"*) that took place:

Mission Control: *Go ahead, Ron*

Evans: *O.K., Robert, I guess the big thing I want to report from the back side is that I took another look at the - the – **"cloverleaf"** in Aitken with the binocs. And that southern **"dome"** (garble) to the east.*

Mission Control: *We copy that, Ron. Is there any difference in the color of the **"dome"** and the Mare Aitken there?*

Evans: *Yes there is... That **Condor**, **Condorsey**, or **Condorcet** or whatever you want to call it there. **Condorecet Hotel** is the one that has got the diamond shaped fill down in the uh - floor.*

Mission Control: *Robert. Understand. **"Condorcet Hotel"**.*

Evans: *"Condor". "Condorcet". Alpha. They've either caught a landslide on it or it's got a - and it doesn't look like (garble) in the other side of the wall in the northwest side.*

Mission Control: *O.K., we copy that Northwest wall of **"Condorcet A"**.*

Evans: *The area is oval or elliptical in shape. Of course, the ellipse is toward the top.*

Apollo 17 - Another Strange Conversation

Wilson writes (p. 141): "While on the Moon, did any of our astronauts see any indication of alien handiwork, such as strange constructions, disturbances or the like? Consider this strange Apollo 16 conversation:"

Orion: *Orion has landed. I can't see how fast the (garble) ... this is a "**blocked field**" we're in from the south ray - tremendous difference in the albedo. I just get the feeling that these rocks may have come from somewhere else. Everywhere we saw the ground, which is about the whole sunlit side, you had the same delineation the Apollo 15 photography showed on Hadley, Delta and Radley Mountains...*

Capcom: *O.K. Go ahead.*

Orion: *"I'm looking out here at Stone Mountain and it's got - **it looks like somebody has been out there plowing across the side of it. The beaches - the benches - look like one sort of terrace after another, right up the side".** They sort of follow the contour of it right around".*

Capcom: *Any difference in the terraces?*

Orion: *No, Tony. Not that I could tell from here. These **"terraces"** could be raised but of (garble) or something like that...*

Casper: *(**Mattingly** in lunar orbit overhead): Another strange sight over here. It looks - a **"flashing light - I think it's Annbell"**. Another crater here looks as though it's flooded except that this same material seems to run up on the outside. You can see a definite patch of this stuff that's run down inside. And **"that material lays or has been structured on top of it"**, but it lays on top of things that are outside and higher. **"It's a very strange operation".***

Apollo 17- Halo with Number One extending

They find something startling and are ordered to immediately switch to code.

LMP (lunar module pilot): *What are you learning?*

Capcom: *Hot spots on the Moon, Jack?*

LMP: *Where are your big anomalies? Can you summarize them quickly?*

Capcom: *Jack, we'll get that for you on the next pass.*

CMP (command module pilot): *Hey, I can see a **"bright spot"** down there on the landing site where they might have blown off some of that **"halo stuff"**.*

Capcom: *Roger. Interesting. Very - go to KILO. KILO.*

CMP: *Hey, **"it's gray now and the number one extends"**.*

CMP: *Mode is going to HM. Recorder is off. Lose a little communication there, huh? Okay,* *there's bravo. Bravo, select OMNI. Hey, you know you'll never believe it. I'm right over the edge of Orientale. I just looked down and saw the "**light flash**" again.*

Capcom: *Roger. Understand.*

CMP: *Right at the end of the rille.*

Capcom: *Any chances of - ?*

CMP: *That's on the east of Orientale.*

Capcom: *You don't suppose it could be "**Vostok**"? (A Russian probe).*

Wilson writes (p. 141): "The Vostok flights took place in the early sixties and were "strictly Earth orbiters". They never reached the Moon!"
http://www.greatdreams.com/moon/darkmoon.htm

Wilson writes (p.142): "And we might add that this is a very strange conversation. What are the real meanings of such terms used here as structure, blocked field, beaches, benches, terraces and the like? NASA claims that they are just metaphoric terms to describe unusual natural formations." http://www.greatdreams.com/moon/darkmoon.htm

Strange Apollo 17 Mission Conversation about "Watermarks" on the Moon

Capcom: *Roger, America, we're tracking you on the map here, "**watching it**".*

LMP: *O.K. Al Buruni has got variations on its floor. Variations in the lights and its albedo. "**It almost looks like a pattern as if the water were flowing up on a beach**". Not in great areas, but in small areas around the southern side, and the part that looks like the "**water-washing pattern**" is a much lighter albedo, although I cannot see any real source of it. The texture, however, looks the same.*

Capcom: *America, Houston. We'd like you to hold off switching to OMNI Charlie until we can cue you on that.*

DMP: *Wilco.*

LMP: *Was there any indication on the seismometers on the impact about the time I saw a "**bright flash on the surface**"?*

Capcom: *Stand by. We'll check on that, Jack.*

LMP: *A **UFO** perhaps, don't worry about it. I thought somebody was looking at it. It could have been "**one of the other flashes of light**".*

Capcom: *Roger. We copies the time and...*

LMP: *I have the place marked.*

Capcom: *Pass it on to the back room.*

LMP: *O.K. I've marked it on the map, too.*

Capcom: *Jack, just some words from the back room for you. There may have been an impact at the time you called, but the* **Moon is still ringing from the impact of the S-IVB impact**. *"So it* **would mask any other impact"**. *So they may be able to strip it out at another time, but right now they don't see anything at the time you called.*

LMP: *Just my luck. Just looking at the "**southern edge of Grimaldi**", Bob, and – "**that Graben is pre-Mare. Pre-Mare!"***

Capcom: *O.K., I copy on that, Jack. And as long as we're talking about Grimaldi we'd like to have you brief Ron exactly on the location of that "**flashing light"** you saw ... We'll probably "**ask him to take a picture of it"**. Maybe during one of his solo periods.*

Notice that the Capcom reiterates that it was a "flashing light". It was, therefore, no meteor impact that they were witnessing. Notice also that the **Lunar Command Pilot** specifically mentions the word "UFO".

Wilson writes (p. 60): "This last conversation makes it obvious that both our astronauts and NASA do not take these sightings of light or UFOs lightly. Maps were marked and photographs were taken at the sites of these occurrences."
http://www.greatdreams.com/moon/darkmoon.htm

While the Apollo 17 astronauts were discussing the "Watermarks", the sighting of the UFO occurred. The conversation then returns to the Watermarks.

LMP: *O.K. 96:03. Now we're getting some clear - looks like pretty clear high watermarks on this –*

CMP: *There's high watermarks all over the place there.*

LMP: *On the north part of Tranquillitatis. That's Maraldi there, isn't it? Are you sure we're 13 miles up?*

Capcom: *You're 14 to be exact, Ron.*

LMP: *I tell you there's some mare, ride or scarps that are very, very sinuous - just passing one. They not only cross the low planar areas but go right up the side of a crater in one place and a hill in another. It looks very much like "**a constructional ridge - a mare-like ridge that is clearly as constructional as I would want to see it"**.*
http://www.greatdreams.com/moon/darkmoon.htm

During the late 60's and early 70's, NASA's Apollo Mission astronauts all experienced close encounters with *'unidentified space vehicles'*. According to the first man on the Moon, Neil Armstrong, who took part in the Apollo 11 Mission, the reason why the US government changed their plans to build a *'Moon City'* was because they were told to change those plans by unknown extraterrestrial voyagers... *'The fact is we were warned off'* Armstrong told a NASA symposium. *'There was never any question then of a space station or Moon City.'*

Author's Rant: The real reason humans were warned off the Moon was because one or two Apollo Missions carried nuclear devices to detonate on the surface of the Moon as an act of technical superiority over Russia and China.

According to a former NASA employee Otto Binder, unnamed radio hams with their own VHF receiving facilities that bypassed NASA's broadcasting outlets picked up the following exchange:

'Buzz' Aldrin who was also with Armstrong on the Apollo 11 mission, was said to have taken colour film footage of alien craft, Armstrong later confirmed that this footage had indeed been shot by Aldrin, only to be confiscated by the CIA on their return to Earth. ***Fearing for his wellbeing,*** Armstrong refused to go in further details, ***except to confirm that the CIA were behind an extensive cover-up campaign regarding the US space program and consequent encounters with UFOs.*** In 1979, former chief of NASA Communications, Maurice Chatelain, confirmed that Armstrong and Aldrin had encountered UFOs on the Moon. To this day Chatelain vehemently protests the truth of their accounts. (Bold italics added by author for emphasis).

Chatelain also published an article in 1995 that confirmed that not only did the Apollo Moon Mission encounter UFOs, but that they also found *'several mysterious geometric structures of unnatural origin on the Moon'*.

Another former astronaut, **Dr. Brian O'Leary**, who was speaking at a science conference in 1994, confirmed the cover-up. *'For nearly fifty years, the secrecy apparatus within the United States government has kept from the public UFO and alien information, we have contact with alien cultures'*. He went on to say *'the suppression of UFO and other extraterrestrial intelligence information for at least forty-seven years, is probably being orchestrated by an elite band of men in the **CIA, NSA, NSS, DIA** and their like. This small group appears able to keep these already hard-to-believe secrets very well'*.

Author's Rant: Bingo! Give that astronaut a cigar!! He has concisely, more than anyone else up to this point apart from Dr. Steven Greer, and few other UFO researchers in this book, have the lodestone of truth on the whole subject matter of the UFO/ETI phenomenon.

Statements from various NASA sources, including Maurice Chatelain, reveal that all astronauts are briefed and debriefed, before and after every mission, and are warned not to discuss their encounters in public. Rumours of astronauts being *'threatened, 'going mad'*, or even *'losing their lives'* on their return to Earth, abound.

Donna Hare and **Sergeant Karl Wolfe** are two witnesses from **Dr. Greer's Disclosure Project** who testified at the **National Press Club** back in May 2001about the cover-up of alien lunar anomalies found on the Moon by NASA. These are credible witnesses, former NASA employees with insider knowledge of what is really going on in the supposedly "civilian agency", NASA and both have stated at the Press Club event that they would give similar testimony before the U.S. Congress. These witnesses have claimed that NASA has altered or destroyed its photos containing images of UFOs.

Donna Hare had a secret clearance while working for NASA contractor, **Philco Ford Aerospace** from 1967 to 1981 as a design illustrator and draftsman. She testifies that she was shown a photo of a picture with a distinct UFO. Her colleague explained that it was his job to routinely airbrush such evidence of UFOs out of photographs before they were released to the public. She also heard information from other **Johnson Space Center** employees that some astronauts had seen Extraterrestrial craft on the Moon and that when some of them wanted to speak out about this they were told to keep quite or threatened.

Ms. Donna Hare, former NASA employee and Disclosure Project witness say NASA photos showing UFOs was routinely airbrushed out before being sold to the public
https://www.youtube.com/watch?v=cMLrUR0_-YI

"**Sergeant Karl Wolfe** was in the Air Force for 4 and 1/2 years beginning in January 1964. He had a top-secret crypto clearance and worked with the tactical air command at Langley AFB in Virginia. In mid-1965, he was loaned to the **Lunar Orbiter Project** at NASA at Langley Field where he was required to repair a malfunctioning piece of equipment that was a photographed processing unit. An Airman 2nd Class who worked in the photographic processing lab was assisting Sgt. Wolfe showed him some photographs of a base on the back side of the Moon containing domes, towers, and spherical buildings. These photographs were taken by the Lunar Orbiter of the Moon prior to the Apollo landing in 1969.

Sergeant Karl Wolfe, another Disclosure Project witness was shown photos of a hidden base on the back side of the Moon
https://www.youtube.com/watch?v=R6QNzH4x1rY

NASA is slowly being backed into a corner that they may not be fully aware that it is happening to them, particularly when the US government was attempting to prosecute **Gary McKinnon** for hacking into Pentagon and NASA computer files. McKinnon claimed that he saw UFO-related files in NASA's computers but, NASA has denied any "cover-up". This could become a legal and public relations nightmare for NASA given their year 2000 mission statement that boasted it is both "ethical and honest" in all that they do.

Whistleblowers like former NASA employees, Hare and Wolfe and other contractors with insider knowledge of NASA's operation pose a serious challenge to NASA's claim of innocence. Wolfe has testified that NASA destroys many photos from the Apollo and other space programs which is destruction of public property and if confirmed, the alleged cost of McKinnon's hacking would be insignificant compared to NASA's annual funding of more than $17 billion.

Part of NASA's mission is to look for signs of intelligent life in outer space. So asking for more money to continue their search, after they've already destroyed evidence that they "found", is a not going to be easy particularly when they are supposed to be an agency of public disclosure and transparency.

214

CHAPTER 71

WAS THE APOLLO MOON LANDINGS HOAXED?

Before we explore further the existence of an Extraterrestrial presence on the Moon as reported by many of the NASA astronauts who actually walked upon the Moon, we need to settle the debate that has arisen and is currently circulating around the "**World Wide Web**" on the internet.

Did the Apollo 11 Moon Landing and all subsequent Apollo Moon landing missions take place or was it all a hoax as some people and particularly **conspiratorialists** have suggested?

The answer may surprise you as it is a trick question. The correct answer is **YES and NO!!!**

YES! The Apollo Moon Landings were real and did take place exactly as history recorded, *with some important exceptions* which is also the **NO** side of the answer, that is they were also, hoaxed.

How can both sides of the answer be correct? Either Apollo astronauts landed on the Moon or they didn't, right? **Wrong!** There is a third alternative which makes a lot more sense when you consider the topic that is being discussed and proven, that UFOs and ETI exist and their presence on the Moon was already an established reality among US and Russian space agencies before Apollo 11 landed!

There are four basic questions among many that that need to be asked and answered. This book will not rehash every question raised in the hoax moon landing conspiracy which is already publicly known in the literature and on the internet. The questions are:

1. Did the Apollo Moon landing missions actually take place or were they a cleverly orchestrated hoax design to act as a cover for the lack of technical capability by NASA to actually land a man on the moon?
2. Was it all a big political propaganda ploy by the American government to win international prestige in the Space Race over the Soviet Union, to basically say that America is number one among all other nations?
3. Was the hoax merely to ensure that NASA kept their lucrative budget flowing for their other space exploration programs and projects even if they couldn't succeed in a moon landing?
4. Or was the real reason for the alleged hoaxing of lunar landings a brilliant diversion and cover-up for ET artifacts and an ET presence discovered on the Moon as a result of actual Apollo Moon landings?

The **Moon landing conspiracy theories** claim that some or all elements of the Apollo and the associated Moon landings were hoaxes staged by NASA and members of other organizations. Various groups and individuals have made such conspiracy claims since the mid-1970s. The most notable claim is that the six manned landings (1969–1972) were faked and that the twelve Apollo astronauts did not walk on the Moon. **Conspiracy theorists (conspiratorialists)** base their claims on the notion NASA and others knowingly misled the public into believing the landings happened by manufacturing, destroying, or tampering with evidence; including photos, telemetry tapes, transmissions, rock samples, and even some key witnesses.

For more than 40 years conspiratorialists have managed to sustain public interest in their theories despite there being much third-party evidence for the landings and detailed rebuttals to the hoax claims. Polls taken in various countries have shown that between 6% and 20% of Americans surveyed believe that the manned landings were faked, rising to 28% in Russia...
https://en.wikipedia.org/wiki/Moon_landing_conspiracy_theories#Ionizing_radiation_and_heat

Richard Hoagland, the official science advisor to CBS News Special Events and chief correspondent **Walter Cronkite**, states in his book "Dark Mission", that he was a first-hand witness to the beginning of the "**Moon Landing Hoax**"! On February 15, 2001, Fox Television aired a widely advertised show titled "*Conspiracy Theory: Did We Land on the Moon?*" claiming NASA faked the first landing in 1969 to win the Space Race. "With this program, Fox removed the last weak link in NASA's ongoing, 40-year-old chain of overlapping cover-ups."

"It is our assertion that not only was this "Moon Hoax" tale carefully constructed as an elegant piece of professional disinformation – as a desperately- required distraction from the real lunar conspiracy documented here, which was beginning to seriously unravel as early as 1996. For, I can personally verify that I was a first-hand witness to "the Moon Hoax" true beginnings far, far earlier the 2001 Fox Special –back in 1969, and in the heart of NASA itself!"

A number of space programs were in progress in 1969 in which Hoagland was involved in, one was to oversee construction and special effects of the "Lunar Landing Day – a walk through the solar system" inside the huge North American Rockwell aircraft hangar at Downey, California. The aircraft hangar had a small miniature of the solar system and Walter Cronkite would interview Apollo engineers, project managers and special guest like Robert A. Heinlein, dean of American science fiction. After this interviews and the successful Moon landing by Apollo 11 and the start of their return back toward Earth, and Hoagland and a small CBS TV crew left the hangar to go to the Jet Propulsion Laboratory (JPL) in Pasadena for Apollo 11's "splashdown".
Dark Mission: The Secret History of NASA by Richard C. Hoagland and Mike Bara; 2007; a Feral House Book; Los Angeles, CA; ISBN: 978-1-932595-26-0

Concurrently, as the Apollo 11 Moon mission was taking place, NASA had another space program coming to fruition; the **Mariner 6** was on final approach toward Mars for a photographic fly-by on July 31st. in which television images would be sent back to Earth. It was another first not only for NASA but, for Hoagland who was suddenly put in charge of a "live" nationwide news coverage of the Martian fly-by. Just two days prior to the event, Hoagland recalls a bizarre scene at their arrival at JPL.

"It was controlled bedlam. Close to a thousand print reporters, television correspondents, technicians, special VIPs, as well as half the staff of JPL itself were all attempting to register for the limited seating in the (relatively) small **Von Karman Auditorium**" …"simultaneously –at the lobby desks specifically set up for members of the press, trying to grab the limited number of press kits on the Mission, and then nail down a seat in the Auditorium beyond.

It was at this point, as **Hoagland** was drifting around Von Karman, trying to spot where the CBS anchor desk positioned, that he noticed something strange.

At this point Hoagland saw a man who appeared to be out of place in the Auditorium "wearing jeans and a long, light-coloured raincoat", "the floppy 'great coat' that cowpunchers used to wear in old Westerns" in what was otherwise a typical hot L.A. day outside –"so, why the coat?" He had a dark leather bag slung over one shoulder and was slowly, methodically, placing "something" on each chair in Von Karman.

What was stranger to Hoagland was the fact that this man was being personally escorted around the Auditorium by another man in a white shirt and black tie; it was none other than the head of JPL press office, **Frank Bristow**!

Just as mysteriously as this appeared to be, Frank Bristow then, escorted the man in the "great coat" out of the area and into the office where space correspondents like **Walter Sullivan** (*New York Times*), **Frank Pearlman** (*San Francisco Chronicle*), **Jules Bergmann** (*ABC*), and **Bill Stout** (CBS) wrote their leads and copy after each press briefing. Bristow was now introducing the "great coat guy" to all these reporters! "Why was the official head of the JBL press office doing this?"

Hoagland had his answer as he watched Bristow's "'guest' proceeded to hand each reporter a copy of whatever he'd had been putting on the seats back in the Auditorium."

Hoagland opened up one of the handouts which contained a small aluminized mylar American flag and two mimeographed pages which he read containing an outrageous claim. The Apollo 11 astronauts who had just walked on the Moon in the "Sea of Tranquility" on July 22, 1969, weren't even halfway home toward splashdown in the Pacific Ocean. "Yet, here, someone with an obvious into JPL was handing out a mimeographed broadsheet to all reporters…claiming that NASA has just faked the entire Apollo 11 Lunar Landing… on a soundstage in Nevada!" **Dark Mission: The Secret History of NASA by Richard C. Hoagland and Mike Bara; 2007; a Feral House Book; Los Angeles, CA; ISBN: 978-1-932595-26-0**

It became clear now to Hoagland of what was really going on as he watched the other reporters throw away the two pages and keep the small flag; NASA knew what was really "out there" in the Solar System from all their space explorations to date and apparently would go to outrageous lengths to keep "the secret". He had just been a witness to a black ops operation involving Bristow, whose job was to make sure that all national reporters covering NASA at least saw what was handed out that afternoon; that shiny flag was a mnemonic device to trigger the memory of what was in the pamphlet long after it was history. The seed had been planted.

According to Hoagland, "sooner or later, a percentage of those reporters who read the pamphlet on that afternoon, would write up some quirky angle on the not-too-far-fetch official tale of apollo11. It would become a naturally-reproducing meme – "a unit of cultural information, such as a cultural practice or idea, that is transmitted verbally or by repeated action from one mind to another" – which is exactly what NASA apparently intended to plant at JPL that afternoon. To deliberately "infect" the American culture with the story that "the Moon landing was all a fake!"

The question that Hoagland asks: "Was this all some far-seeing "back-up plan" if, in some point in the future, it started to emerge why the astronauts had *really* gone to the moon?

Small wonder then, that the Fox network decided to activate the meme in 2001, the quasi-documentary special "*Did We Land on the Moon?*" in a neatly package 30 year conspiracy theory ready for those suspicious of NASA's "official transparent disclosure of accomplishments".

"An officially concocted 'inoculation' against troublemakers who would one day place before many of those same national reporters a set of embarrassing official Apollo photographs, asking the crucial question: "What did NASA *really* find during its Apollo missions to the Moon?" **Dark Mission: The Secret History of NASA by Richard C. Hoagland and Mike Bara; 2007; a Feral House Book; Los Angeles, CA; ISBN: 978-1-932595-26-0**

Here then, the reader can see that the "Moon Hoax" began back in 1969 before the safe splashdown of Apollo 11 by none other than NASA, itself!

Moon Hoax conspiracist, **Bill Kaysing** is alleged to have started the Moon Hoax based upon gut feelings or intuition even before Apollo 11 went to the Moon. Kaysing worked as a librarian at Rocketdyne, the company which built the F-1 engines used on the Saturn V rocket was not qualified as engineer of any kind and thus, did no engineering work on the Saturn V rocket engines.

According to Kaysing he worked at **Rocketdyne** starting on February 13, 1956 as senior technical writer, then on September 24, 1956 as a service analyst, September 15, 1958 he worked as a service engineer, following on October 10, 1962 as a publications analyst, and on May 31, 1963 he resigned for personal reasons. http://en.wikipedia.org/wiki/Bill_Kaysing

Could Kaysing have discovered some insider information as a librarian or service analyst, possibly he may have come across technical data that may have been of a suspicious nature but, it is unlikely, given his work position at Rocketdyne? It is more likely that Kaysing could have been a patsy for disinformation much like the information that Richard Hoagland witnessed being dispersed by an agent working for JPL/NASA.

Some people become targets by government intelligence agencies or the DOD when they get too close to the truth. These intelligence agencies will throw an individual's investigation off course into some other dead end pursuit with disinformation or they may create a hoaxed event that is contradictory to an actual event experienced by the individual. UFO researcher **Paul Bennewitz** and contactee George Adamski are two examples of this targeting of people in UFO/ET research, as are numerous people who claim alien abduction experiences after a legitimate UFO or ET encounter.

If you wondering could astronauts survive flight through the **Van Allen Radiation Belt** that surrounds the Earth, the question is easily answered.

The spacecraft moved through the belts in about four hours, and the astronauts were protected from the ionizing radiation by the aluminum hulls of the spacecraft. In addition, the orbital transfer trajectory from the Earth to the Moon through the belts was selected to minimize radiation exposure. Even **Dr. James Van Allen**, the discoverer of the Van Allen radiation belts, rebutted the claims that radiation levels were too dangerous for the Apollo missions. **Plait** cited an average dose of less than 1 rem (10mSv), which is equivalent to the ambient radiation received by living at sea level for three years. The spacecraft passed through the intense inner belt and the low-energy outer belt. The astronauts were mostly shielded from the radiation by the spacecraft. The total radiation received on the trip was about the same as allowed for workers in the nuclear energy field for a year.

The radiation is actually evidence that the astronauts went to the Moon. **Irene Schneider** reports that thirty-three of the thirty-six Apollo astronauts involved in the nine Apollo missions to leave Earth orbit have developed *early stage cataracts* that have been shown to be caused by radiation exposure to cosmic rays during their trip. However, only twenty-four astronauts left Earth orbit. At least thirty-nine former astronauts have developed cataracts. Thirty-six of those were involved in high-radiation missions such as the Apollo lunar missions.
http://en.wikipedia.org/wiki/Apollo_hoax#Ionizing_radiation_and_heat

It took literally hundreds of thousands (400,000 to be precise) of people from around the world, experts and professionals all in all fields of scientific endeavour, to help build, launch, send and land men on the Moon and then return them safely to Earth again. It took years of trial and error and a few disasters along the way to succeed and it is a major cornerstone in the evolution of human development and technology that cannot be underestimated! Mankind through NASA and the **Russian Space Agency** demonstrated without question that humans possessed the technical capability to send people to another planetary body.

The Russian space agency, the British **Jodrell Bank Observatory** and many other countries with deep space radar track every man and unmanned object that goes up into space, particularly if there are any spacecraft headed toward the Moon. If NASA faked the Moon landings, they would have been caught by one of these countries and certainly by the Russians, who would have called fouled and humiliated the Americans on the international political stage. The Russians did no such thing politically and though they may have cursed and bitten their lips in anguish over their failure to win the Space Race, they did acknowledge the Americans success. NASA could not afford failure as it would probably terminate their lucrative annual budget and perhaps, end their space program. For America, international humiliation over an alleged hoaxed event would be political suicide, not only nationally but globally.

The successful Moon landings did establish however, the USA's superior technical capability to win the **Space Race** over the Russians and unquestionably it was a political coup for Americans. It did not, however, establish US superiority in any sense of the word over its rival, much as many people would like to think but, merely that their success to failure ratio in the Space Race was much higher than the Russians. In this sense then, bragging rights go to the Americans.

The success of the Apollo lunar landings also ensured that NASA kept receiving it healthy budget infusions annually, although, since NASA was "warned off the Moon and never to return" by the ET presence there, NASA was nevertheless, committed to six lunar landing missions. In order to maintain the financial momentum flowing for its space exploration programs, NASA needed a public excuse for why it was not continuing with the Apollo program. They simply could not reveal what they had found on the Moon this would be tantamount to everything they had accomplished and discovered along the way. NASA decided that a new program of a space station and a fleet of Space Shuttles were needed for the next phase of space exploration. Besides, the Moon landings were now becoming quickly passé and the public had become bored with this type of space adventure, at least this was the implied public consensus. It was easier to say that "we've done that, got the T-shirt and we're using it as a rag" excuse. The reality, of course, was that no public opinion poll had been taken and after forty some years later, the American public and the world still wonders why the US or the Russians have never returned to the Moon.

Is it still possible that the Apollo Moon Landings were faked in a large sound stage studio and made to look like NASA had indeed actually sent men to the Moon and back?

Quite frankly, anything is possible if you throw enough money at it, in order to get a solution to the problem! Nowadays, everyone realizes that immense sound stage studios exist in Hollywood and most major film studios around the world in order to create the special effects and landscape settings required for film production. No big mystery there! Even at the time of the Apollo Moon Landings, there were immense sound stages being used by Hollywood and by NASA for the training of their astronauts. When there wasn't the right natural landscape setting like a desert or a cold arctic region or when the weather was not cooperating or the right lighting was required or just to ensure there was no public distraction or interference, sound stage studios were commonly used for astronaut training.

Stanley Kubrick Builds a Lunar Sound Stage

On Nov 15th, 2003, **Alex Strachan** news reporter for the "Vancouver Sun" newspaper of British Columbia, Canada, wrote an editorial article about the possibility of hoaxing the Apollo 11 Moon Landing, entitled "The Facts Don't Lie, But the Camera May".

"During an interview with **Stanley Kubrick's** widow, an extraordinary story came to light. She claims Kubrick and other Hollywood producers were recruited to help the U.S. win the high stakes race to the moon. In order to finance the space program through public funds, the U.S. government needed huge popular support, and that meant they couldn't afford any expensive public relations failures. Fearing that no live pictures could be transmitted from the first moon landing, **President Nixon** enlisted the creative efforts of Kubrick, whose *"2001: a Space Odyssey"* (1968) had provided much inspiration, to ensure promotional opportunities wouldn't be missed. In return, Kubrick got a special NASA lens to help him shoot *"Barry Lyndon"* (1975). A subtle blend of facts, fiction and hypothesis around the first landing on the moon, *"Dark Side Of The Moon"* illustrates how the truth can be twisted by the manipulation of images.

With use of 'hijacked' archival footage, false documents, real interviews taken out of context or transformed through voice-over or dubbing, staged interviews, as well as, interviews with astronauts like Buzz Aldrin and others, Dark Side Of The Moon navigates the viewer through lies and truth; fact and fiction. This is no ordinary documentary. Its intent is to inform and entertain the viewer, but also to shake him up - make him aware that one should always view television with a critical eye.

Dark Side Of The Moon is written and directed by **William Karel** and co-produced by Point du Jour Production and ARTE France."
http://web.archive.org/web/20070109202315/http:/www.thelastoutpost.com/site/1362/default.aspx

According to DARK SIDE OF THE MOON, the most important film of its kind since Oliver Stone's JFK - or since Rob Reiner's This is Spinal Tap, at any rate - images of Neil Armstrong's walk on the moon on July 20, 1969, were shown to the world through the lens of master filmmaker **Stanley Kubrick** and were staged on the same Borehamwood, U.K., soundstage where Kubrick made his landmark film, 2001: A Space Odyssey.

Dark Side of the Moon was written and directed last year by 63-year-old historical documentary filmmaker William Karel for France's Point du Jour Production and Arte France (the film's original, French title was Operation Lune). It uses documentary evidence, archival footage and extensive

Dark Side of the Moon was written and directed last year by 63-year-old historical documentary filmmaker William Karel for France's Point du Jour Production and Arte France (the film's original, French title was Operation Lune). It uses documentary evidence, archival footage and extensive interviews with Kubrick's widow, **Christiane Kubrick,** astronaut Buzz Aldrin and former and present-day U.S. government officials and luminaries such as **Henry Kissinger**, **Lawrence Eagleberger**, **Al Haig** and **Donald Rumsfeld**, to lay bare the lie.

And an elaborate lie it was, too, judging from the evidence. (The official **CBC** press release refers to the film's subtle blend of facts, fiction and hypothesis as a navigation through fact and fiction and asks rhetorically whether "Neil Armstrong's famous walk on the moon" was another Stanley Kubrick production. I can't tell if the misspelling of Neil Armstrong's name is meant to be ironic or incompetent.)

Dark Side of the Moon points out that, given the turmoil of the day - the Vietnam war, civil unrest, a newly elected president warily eyeing his prospects for a second term - the Nixon administration understood that it was more important that astronauts be seen to be walking on the moon than actually walk on the moon.

If the astronauts landed safely, but could not televise live images back to Earth because of some unforeseen technical glitch, then the entire expensive enterprise would have been a waste of time, from a public relations standpoint.

The Nixon administration approached Kubrick - an American ex-pat and avowed recluse, living in seclusion in a palatial estate somewhere in the suburbs of London - with a mind to stage the moon landing in advance, so that if worse came to worst, the Apollo program would still have pictures to show a doubting public.

The administration knew Kubrick would jump aboard, the film's makers suggest because it was widely known that *"**Dr. Strangelove**"*, which Kubrick directed five years earlier, in 1964, was one of Nixon's favorite films.

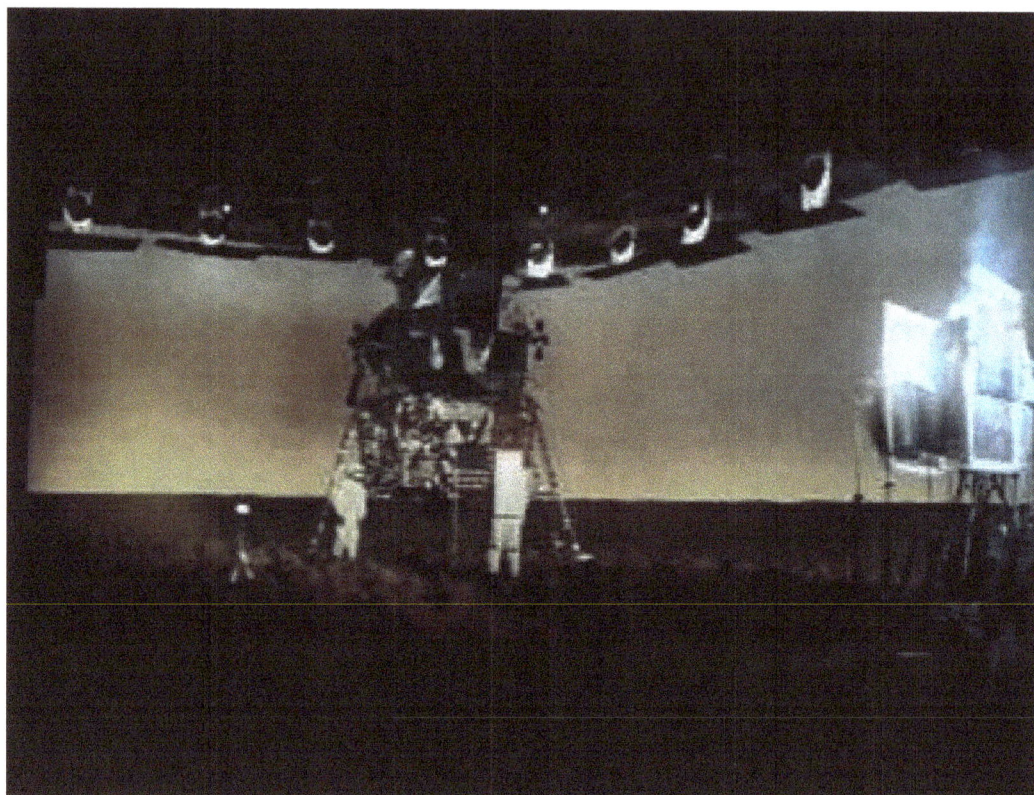

Astronaut performers on the sound stage of the lunar landscape set being filmed by the Stanley Kubrick for the upcoming Apollo 11 Moon Landing
https://cumbriansky.wordpress.com/2009/07/05/lro-and-the-apollo-hoax-believers/

The original idea was to have the CIA stage the event and film it themselves on the same soundstage where Kubrick recreated the lunar surface for 2001: A Space Odyssey. But when Kubrick - a notorious perfectionist, with a temper to match - saw how incompetent the CIA camera operators were, he demanded that he be allowed to film the scene himself.

In return, Dark Side of the Moon posits, Kubrick was allowed use of a special, one-of-a-kind Zeiss camera lens, originally designed for NASA's satellite program, to shoot his James Thackeray epic Barry Lyndon, which required a special heretofore unknown lens to depict images of life in 18th century Ireland using only available light. The film preserved the great man's vision for generations to come." **"The Facts don't Lie, But The Camera May" by Alex Strachan, Vancouver Sun, Nov 15th, 2003**

222

"But wait, there's more!

Neil Armstrong's famous line - "That's one small step for man, one giant leap for mankind" - was scripted in advance, and mangled in the translation, into "one small step for man, one great leap...who wrote this crap?" Armstrong proved to be a temperamental star. While boarding the lunar capsule prior to liftoff, for example, he was overheard to ask about the in-flight movie, about whether he was in the smoking section or non-smoking, about whether he was assigned a window seat in the back, about his requests for a kosher meal, and whether his car would be safe in the NASA parking lot.

It's the actual testimony from Kissinger, the late Vernon Walters (*speaking in Russian, and dead, under suspicious circumstances, just hours after conducting his interview for the film*), Rumsfeld (*"I'm going to tell you a fascinating story"*), Eagleberger, Haig and others - real people in real interviews, not actors playing a role - that brings Dark Side of the Moon to life. (*A cynic would point out that their comments are edited out of context, but then a cynic would already have guessed that the Apollo moon landing was staged, so why bother?*)

The decision, ultimately, was Nixon's.

"He was the president," Kissinger explains in the film, *"and he deserves the credit for having had the courage to do it."* Kissinger was awed by the sheer hubris of Nixon's actions.

"At no stage in my life could I have anticipated that this would happen," he goes on to say. *"At no stage. Not even when I was made National Security Adviser. And I think it is a great symptom of the strength of America that this was even conceivable."*

It was the right thing to do, Rumsfeld concurs, *"because we had to do something to show that we're still the United States of America...We walked out of the room and* **President Nixon** *said, 'I've decided to do that, and I need you to do this job, we're going to do it.' It was just amazing."*

Dark Side of the Moon is a mammoth undertaking. It seeks nothing less than to expose the incongruities between rhetoric and reality, by disclosing how the camera's lens can be manipulated to suit any ends, and it achieves its goal with, style and verve. It is a thoroughly entertaining and revealing film, and well worth seeing.

Oh, and one other thing. According to the final credits, any resemblance to actual living persons is purely coincidental.

That's important to know. After all, the camera lies. It's not always easy to tell." **"The Facts don't Lie, But The Camera May" by Alex Strachan, Vancouver Sun, Nov 15th, 2003**

It's at this point that all the conspiracy theorists jump out of the woodwork and proclaim, "We told you so!" then proceed to add all the "evidence" as proof of a fake lunar landing, ad nauseum. Based upon this evidence alone, one would conclude that the evidence is inescapable and it is case closed! It would seem that the conspiratorialists of the Moon Landing Hoax were right!

Not so fast! It is this author's contention, however, that this is only part of the answer and is not supportive of a total and complete "Hoaxed Moon Landing" event or theory.

What was not common knowledge and has never even been considered in the arguments for a hoaxed lunar event is that NASA had contingency plans and backup events for when things went wrong or when there was an emergency situation or when astronauts discovered something unusual that NASA did not want the public to see. That being the presence of UFOs and ET presence on the Moon.

When you send two men to land on the Moon and return them safely back to Earth, you are going to throw tens of billions of dollars or more at the program and provide for everything necessary to succeed, including the many places necessary to rehearse the planned mission. From the great outdoors to the secure indoors of an immense soundstage building where the lunar landscape and its surface can be reproduced right down to the minutest crater detail and rock.

Dry run or "dress rehearsal" practices of every aspect of the Apollo program must be gone over and over and over again and again until everyone involved in the mission gets it 100% right every time, all time whenever, a new mission comes up. Screw ups are not tolerated, they are costly and potentially life ending resulting in a failed mission and can seriously set back space exploration programs and projects by years or ending in complete termination of programs.

In this immense sound stage studio, many Moonwalk scenarios are carried out to fully train the astronauts for every contingency they may encounter for when they actually walk upon lunar surface. Every lunar aspect must be perfectly "choreographed and performed" in this sound stage. It is a highly detailed, carefully orchestrated training program following a ***carefully worded script and a set of protocols for every word spoken and every action taken***. (Bold italics added by author for emphasis).

The astronaut training is a dress rehearsal like any play, stage or movie performance that actors need to go through in order to receive a standing ovation from the audience at the conclusion of the opening night performance. In this case, NASA is the audience that must give their approval to the astronauts' performance in order for there to be a successful mission.

Anyone who follows the NASA space program knows that this type of astronaut training in a sound stage environment made to look like the Moon's surface is an integral part of the mission success. Therefore, when advocates of the hoaxed Moon landing conspiracy state that they have seen cameras or lighting equipment dangling from a ceiling or that a person has been inadvertently videotaped standing off to one side of the lunar set, that there are no footprints of astronauts leading up to where they were standing or no tire tracks from the moon buggy leading to the final resting point of destination were seen, or that there were tears in sound stage ceiling fabric or loose panelling were evident or that no dust was found on the lunar landing pads were indicative a carefully staged hoax or the reflections in the astronauts' visor show the interior ceiling structure of a sound stage then, be assured that what was seen or recorded on camera or sound equipment was indeed an event that was crafted in an immense soundstage production area!

It had to be because it was an astronaut training area for what would eventually take place for real upon the Moon! It wasn't an entirely hoaxed event!! But a carefully synchronized event!!! It was hoaxed only to the extent of covering up an Extraterrestrial presence and technology on the Moon! Between the live Moonwalk from the lunar surface when astronauts were photographing and documenting UFOs sighted on the rims of lunar craters and alien artifacts and buildings encountered in the immediate lunar environment, it became necessary to "pull away" to the lunar sound stage studio or play an appropriate film segment from the Kubrick's "faked moon landing" to avoid showing the public what was really on the Moon!

If astronauts came across anything alien in nature such as an ET spacecraft (sometimes these were unavoidable in photographs) or an alien artifact or structure and if astronauts did not pan their cameras away quickly enough unless, instructed to do otherwise, to something more natural to the lunar environment or failed to use coded speech or text then, NASA had a seven to eleven second time delay between "live feed" and "public feed" and always there was a technician or engineer with his finger on the "kill button". In such "awkward situations," the time delay was sufficient time to switch from the Moon's live feed to the sound stage lunar set where two astronaut "performers' would continue the actions of what was supposedly taking place upon the Moon! It was a carefully choreograph synchronized event!

Whenever there was an Apollo Moon mission, this was the prescribed set of NASA protocols; nothing was left to chance at any time. For a mission control specialist to miss his cue in the "lunar stage production' would be an anathema.

NASA knew before they started sending astronauts to the Moon, what they expected to find on the lunar surface, it wasn't just unusual Moon rocks and rare minerals for some later lunar mining expedition. NASA and the DOD knew astronauts would encounter alien artifacts and immense alien structures and ruins, previous unmanned satellites and spacecraft had photographed in prolific number all kinds of lunar anomalies and had carefully mapped every location where they occurred.

Every Apollo Moon mission was carefully scripted and a rigid mandatory set of protocols had to be followed, including everything that was said, seen or photographed. Astronauts were required to use coded speech or alternate transmission frequencies in their communications with mission control centre whenever they came across anything alien in nature and needed to document or photograph it. *This has been clearly shown already as indicated earlier in the above coded Apollo astronaut communications.*

This is the hoax that has been perpetrated upon the public by NASA, all in the name of national security, national self-vested interests and continued UFO/ETI secrecy!

The real reason for the alleged hoaxing of Moon landings is a brilliant diversion and cover-up for ET artifacts and an ET presence discovered on the Moon as a result of actual Apollo Moon landings.

The immense lunar sound stage was an integral part of NASA's success in the Apollo Moon mission program and for the continued cover-up of UFOs and ETI existence. The discovery of alien artifacts and huge city-like structures on the Moon's surface by the Apollo astronauts, who were sworn to keep such revelations a secret for fear that a disclosure of these truths would cause mass "public panic", established beyond all doubt, the existence of an Extraterrestrial presence on our nearest neighbour, the Moon!

The final nail in the coffin of the Moon Hoax conspiracy theorists is the many photographs taken by the **Lunar Reconnaissance Orbiter (LRO) satellite.** Since the late 2000s, high-definition photos taken by the LROC spacecraft of the Apollo landing sites have captured the lander modules and the tracks left by the astronauts. In 2012, images were released showing the Apollo flags still standing on the Moon. All the Apollo landing sites on the Moon were photographed by NASA, including all Russian Lunar Landers, laying to rest any doubt that men from Earth did indeed set foot on the Moon. The only real question left is, what did they actually find there? **"The Facts don't Lie, But The Camera May" by Alex Strachan, Vancouver Sun, Nov 15th, 2003**

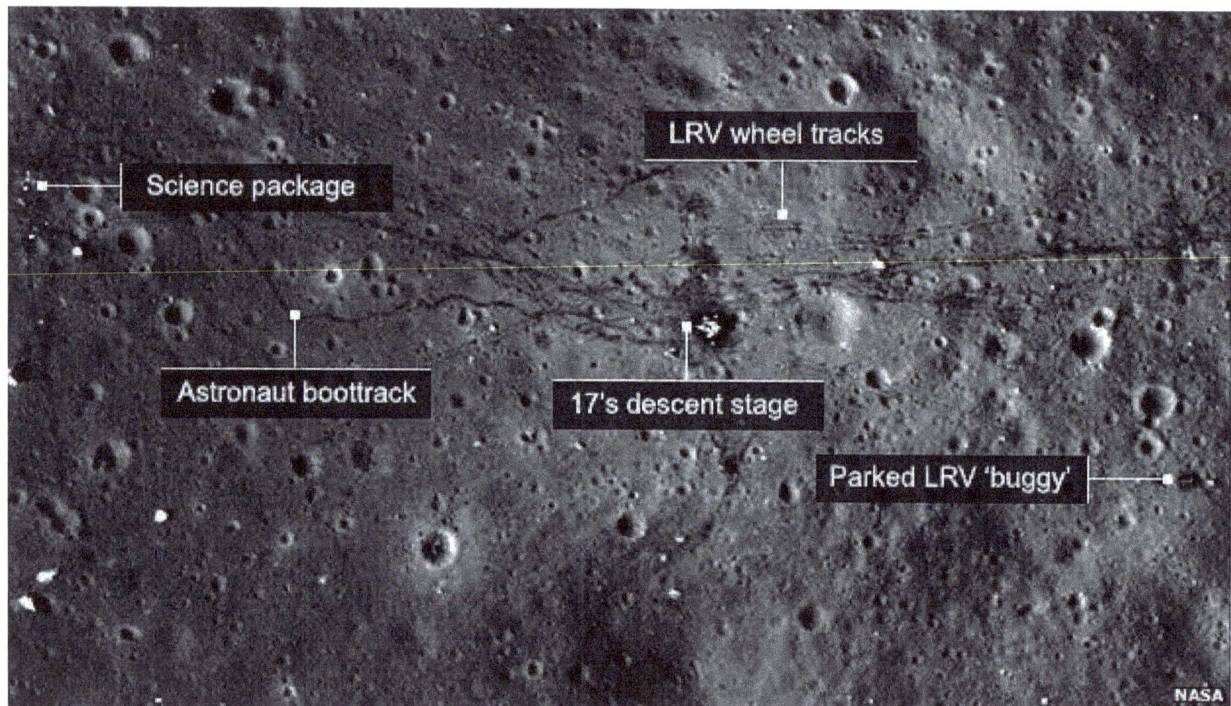

Lunar Reconnaissance Orbiter (LRO) photo of Apollo 17 landing site
http://www.hasaan.com/2012/08/debunking-moon-conspiracy-theories.html

Author's Rant: The evidence for men going to the Moon and walking upon the lunar surface is overwhelming. Its reality is an undisputable fact!

Such arguments to the contrary without full examination of all the evidence is proof of an arbitrarily contentious thought process that is neither conclusive nor logical.

People are entitled to believe whatever they want, that is a God-given right of everyone, but it doesn't alter the facts! When factual proof is offered and demonstrated to be true in counter opposing points of view then, people are duty bound by their personal sense of integrity when seeking out the truth, to acknowledge and accept the evidence. This is the process that allows for enlightenment and the advancement of civilization.

Now, it is case closed!!!

Apollo 11 landing site taken by the LRO
https://www.nasa.gov/mission_pages/apollo/revisited/#.WPKU-2e1vmE

In 2012, NASA's Space Shuttle program ended and small civilian space agencies and companies have taken up the space exploration mantle with some cooperation from NASA. The Russians have now become the sole means for American astronauts to get to the International Space Station and in true entrepreneurial fashion, they are also offering civilians a chance to ride into space to the ISS aboard a Russian Salyut spacecraft at $20 million dollars a pop. Now, that Americans have no fleet of space shuttles or alternative spacecraft to travel into space whenever they want, the question on everyone's mind is what type of next generation space shuttle will they develop. The European Space Agency continues to launch both Earthbound and deep space satellites and sends trained astronauts from different European countries to help man the International Space Station.

Currently, these are the countries that have space programs (as opposed to just space agencies): USA, Russia, China, Pakistan, Japan, S Korea, N Korea, Italy, France and Britain. Most Euro countries operate under the aegis of **European Union Space Agency,** though Iran is also fast-tracked in developing a space program.

China has quickly evolved their space program in playing catch up to the USA and Russia with several successful manned space flights. Japan and India have also successfully launched satellites into deep space to asteroids and comets and also, around the moon. Canada continues to maintain its close partnership with the US in almost every aspect of space exploration with their own satellites and Canadian astronauts are routinely sent to the ISS. China, India, and Japan all have satellites orbiting the Moon photographing its surface. They have no doubt discovered what American and Russians have already discovered on the Moon and kept suppressed from public knowledge for four decades, that the Moon has many alien structures on its surface.

It is not unexpected then, that these nations have decided to send their own men to the Moon to see what is really there. It is going to be very difficult, if not impossible to keep a lid on the lunar secrets that NASA has kept to itself for so long, secrets that have only slowly leaked out during those decades of cover-up. A new **"Space Race"** it seems is about, to begin with new players in this race. These competitors: China, Japan, and India have all stated their commitment to the challenge ahead and they are quickly showing that they have the right stuff to get to the Moon. It may happen that some of these nations will partner with existing space agencies or form new alliances with other competing nations.

This new Space Race has sent NASA to rethink its own manned space program and has forced NASA to pick up the Moon landing program again from where they left off, this time with the Orion space program. The Orion program is similar to the Apollo program except that the command module is a larger version of the Apollo Moon capsule but, completely updated and the lunar lander will also undergo drastic changes and modifications as well.

It would seem that America through NASA and the DOD has a vested interest to preserve and maintain the high ground advantage they have made on the Moon back in the late '60s and '70s. The new "Space Race" could become not only a very political driven agenda but, possibly a militaristic driven one with its own hidden agenda. The "Race" back to the Moon is no longer just a race between two countries with opposing political philosophies but, with countries all vying to get their hands on alien technology and to ultimately set up permanent moon bases. If humanity is not careful in its motives and diverse political thinking before returning to the Moon, we may find that we have brought the same problems that have plagued us for thousands of years with us, namely, our inability to get along with each other. There could well be permanent bases established on the Moon, however, these bases could be decidedly manned with a full-time military presence. Into this mix must be added a second party interest that a divisive humanity may not have factored into the space race equation that of the ET presence already established on the Moon.

CHAPTER 72

"REMOVING ONE OF TRUTH'S PROTECTIVE LAYERS" - AN ALIEN PRESENCE ON THE MOON!

Astronauts and cosmonauts have talked about it, hinted about it, photographed and documentedit, and covertly communicated it to their respective mission control centres Now, with other nations such as India and China sending their satellites to the Moon, what exactly was it that theUS and Russia found on the Moon? Every astronaut crew that landed on the Moon reported thatthere was an Extraterrestrial presence on the Moon. There were ET spacecraft which followed each Apollo spacecraft to the Moon and into lunar orbit, they hovered and monitored the astronaut lunar activities from crater rims and mountain tops. Everything humans did on the Moon from moonwalks, setting up equipment, to travelling short distance on the lunar rovers, tothe eventual blast off from the Moon's surface for the return home to Earth.

Coded communication was used by astronauts whenever they came upon something unusual or alien in nature or whenever they were photo- documenting strange alien structures on the lunar surface. In such situations, live transmission was either blacked out, terminated or re-routed to the lunar soundstage studio for a "normal, uneventful continuation of the lunar panorama" supposedly been seen by the astronauts or a "clip" from the ***Kubrick's hoaxed moon landing and lunar excursion film*** was played to the public. Everything would appear normal from a ***"dead lunar body",*** while the ***"real events"*** were unfolding behind a carefully edited veneer of cellulose, scheduled digital video blackout or a staged ***"lunar performance"*** from a soundstage.(Bold italics added by author for emphasis).

From the first Apollo 8 Moon mission through to Apollo 17, all lunar orbiting command modulepilots were photographing many lunar anomalies or re-photographing lunar surface areas with more high-powered cameras and telescopes for objects seen earlier by previous Apollo missions.There were many fantastic structures on the Moon's surface that were miles in height, tall towers, dome or mushroom shape buildings, spherical buildings and large dish shape structures, lunar bases and cities structures, things that resembled mining equipment in craters and large megalithic cubes floating above the ground! These were hardly signs of a dead world, possibly abandoned perhaps, but definitely not dead as flashing lights seen on the ground and ET craft in close proximity to Apollo spacecraft indicated that the Moon was still alive and active!

Richard Hoagland and Mike Barra have revealed with convincing arguments, facts and evidencein their book, "Dark Mission: Secret History of NASA" that NASA has a secret agenda on the Moon as a result of discovering ancient ruins left behind from an Extraterrestrial origin. They argued in their book that 40 years ago when U.S. astronauts landed on the Moon, they found ancient ruins of artificial origin as well as a previously unknown technology for the control of gravitation.
https://www.youtube.com/watch?v=u10UPBdSiV8

Has NASA a secret agenda on the Moon after the discovery of ancient ruins left behind from an Extraterrestrial origin?

http://www.bibliotecapleyades.net/antarctica/antartica38.htm

The Apollo astronauts took pictures of the objects but NASA ordered , former manager of the **Data and Photo Control Department** at NASA's Lunar receiving laboratories for the Apollo Mission, to destroy the images. Apparently, Johnston did not follow the order and leaked the information.

Many of the pictures in Hoagland's book reveal ruins of extraterrestrial buildings; dome-like objects made of glass, stone towers and castles hanging in the air. Hoagland, once a former NASA scientist, asserts that the US government's interest in the Moon exploration have been led mainly by secret society groups or sects, composed of prominent individuals. Each group apparently has their own agenda indelibly stamped on the space program that follows a prescribed secret or "occult" doctrine that traces its root beliefs back to "ancient religion and mysticism" which does not correspond with rational empiricism and has thus, promoted a veil of secrecy for over 40 years.

This bizarre obsession by NASA to ancient religious mysticism originates with three ancient **Egyptian "gods or goddesses"– Isis, Osiris,** and **Horus**, has in the current age become the iconic symbolism for NASA's celestial mechanics and astronomical research. Forearmed with the basic knowledge of the Egyptian pantheon of gods, Hoagland recognized that the "A" in the word of **Apollo** and the Apollo Program's official patch actually stands for **"Asar"** – the Egyptian designation for **"Osiris".** The Asar/Osiris link, says Hoagland, is none other than the Greek constellation of **"Orion"** – which was also the background stellar constellation on the patch itself. Dark Mission: The Secret History of NASA by Richard C. Hoagland and Mike Bara; 2007; a Feral House Book; Los Angeles, CA; ISBN: 978-1-932595-26-0

230

After Apollo 11's accidental admission of an UFO encounter en route to the Moon and **Surveyor 3's** classified photos of the Moon's **Mare Cognitum**, NASA executed its first lunar black-ops mission. The crew of **Apollo 12** conducted a dark 19-minute S-EVA at the location *Statio Cognitium*, **"THE KNOWN BASE."** Allan Bean had opened the LEM's upper escape hatch and standing through the hatch opening photographed a 360 degree panoramic view of the lunar landscape. These photographs have never been made public nor acknowledged to exist, yet it is believed that they contained evidence of alien ruins and artifacts, too large to bring back to the Earth.

Hoagland and Johnston say that a significant number and variety of lunar artifacts were successfully identified, collected and brought back to Earth by the Apollo mission crews and all further Apollo missions were terminated after the success of the secret explorations, ending with Apollo 17. The reason behind the discontinuation of further moon landings was not budgetary cuts but, rather the unique discoveries that NASA had found and the dire warning not to return to the Moon made by ETI.

The Soviet (Russian) Space Agency had also photographed the lunar surface at close range with their very successful unmanned Moon landers and robotic Moon rovers. **Lunokhod** ("Moonwalker") was a series of Soviet robotic lunar rovers designed to land on the Moon between 1969 and 1977. The 1970 **Lunokhod 1** and the 1973 **Lunokhod 2** landed on the moon were successful missions which were in operation concurrently with the Zond and Luna series of Moon flyby, orbiter and landing missions. The Lunokhods were primarily designed to support the Soviet manned moon missions and to be used as automatic remote-controlled robots to explore the surface and return pictures. The Lunokhods were transported to the lunar surface by Luna spacecraft. The moon lander part of the Luna spacecraft for Lunokhods was similar to the ones that brought Moon rocks back to Earth.
http://en.wikipedia.org/wiki/Lunokhod_programme

Below is a photo of the Lunokhod 1, moon rover (**Lunik 13**) which was the first of its kind to land on anther planetary body on December 24, 1966; America wouldn't have a similar rover until 1997 when the **"Sojourner"** rover, Mars Pathfinder landed on Mars. The Lunokhod 1 rover had a long multi-disc antenna which in the photos taken by the rover of the strange disc-like object on the Moon, one cannot help but wonder if the long antenna had simply fallen or broken off the moon rover. Yet, if the antenna had broken off then, how was the Moon rover able to transmit the photos back to Earth? A damaged antenna not able to transmit would essentially end the moon project and the other Lunokhod 2 Moon rover was several hundred miles away so, it didn't photograph the area or the supposed damaged Lunokhod 1and what about those other strange objects on the lunar surface?

The Lunokhod 1 moon rover was the first rover of its kind to land on another planetary body. Note the long antenna and compare to photos below

https://www.flickr.com/photos/71411059@N00/6098463988

| 240° | 250° | 260° | 270° | 280° |

Part of a panoramic photograph taken by the Russian Lunik 13 lunar lander. Note the obvious artificial object in the foreground

http://mentallandscape.com/C_CatalogMoon.htm

Close ups of the same artificial object (top and below) appear similar to a drilling or augur device but photographed at different times as indicated by the shadows cast by the sun

It is difficult to tell if the augur-like device has spirals or discs but it presence on the moon is very unusual with no explanation of how it got there

170°   180°   190°   200°   210°

**Another partial panoramic photo by Lunik 13 showing more
strange metal objects on the Moon's surface**
http://mentallandscape.com/C_CatalogMoon.htm

**Close up of metal objects in foreground next to Soviet Moon Lander
and another augur-like object in the background**
http://mentallandscape.com/C_CatalogMoon.htm

**A rectangular block on the rim of a crater photographed by Luna 17
that appears to have some writing on it**
https://mocksure.wordpress.com/2009/02/page/3/

The last Lunokhod 2 photo shows unusual block shapes and round objects
http://lroc.sese.asu.edu/posts/11

Are there ancient ruins on the moon? There is some very strange visual evidence that there are artificial structures and areas on the Moon's surface that are being intentionally obscured from view due to image tampering by NASA and the US Navy. Some of these structures have slipped through the censor's net and are clearly visible on photographs taken by the Apollo missions, **Clementine satellite,** and others.

The "Castle" on the Moon

This extraordinary object comes from **AS10-32-4822,** an Apollo handheld Hasselblad photo. The **"Castle"** is extremely bright and plainly visible at normal magnification, making it nearly impossible for the astronaut taking the picture to have missed it.
http://www.disclose.tv/forum/moon-anomalies-ruins-and-structures-on-the-moon-t4823.html#ixzz2IUbanD7x

On enlarging one of the small reflective specks in frame AS10-32-4822 (taken during the manned Apollo 10 mission) a stunningly enigmatic complex object is revealed. Nicknamed the "Castle" due to its shape, is almost a mile in size and this object is suspended nearly 10 miles above the lunar surface!

This side by side comparison is from 2 different versions of the same Apollo 10 photographic frame, one obtained by Hoagland and the other by another researcher. In fact, Hoagland has now identified some nine versions of this photo in various archives around the world.

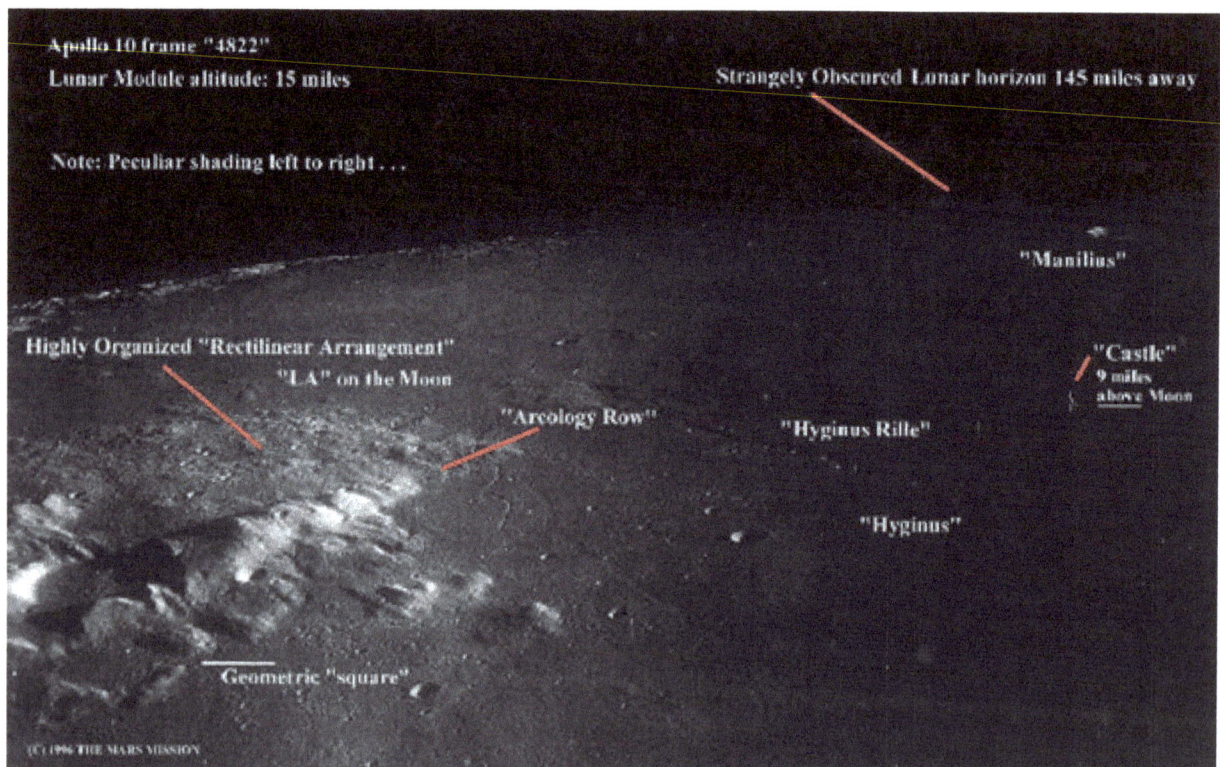

A moonscape full of anomalies including "L.A." on the Moon and the "Castle
http://www.thelivingmoon.com/43ancients/02files/Moon_Others_01.html

On the 2nd November 1966, The Washington Post ran a front page article entitled "Six Mysterious Statuesque Shadows Photographed on the Moon by Orbiter". NASA Lunar Orbiter 2 had photographed an area on the moon (**Mare Tranquilis**) of approximately 30 by 50 kilometres. Photographs showed six or seven towers, appearing in a specific geometric pattern, rising from the lunar surface. Their pointed shadows indicated that they were either conical or pyramid-shaped. NASA countered that the photographs did not show anything of any interest. The Russian magazine, Argosy, published a communication from the Russian space scientist **Alexander Abromov**. He stated that the Russian Luna 9 probe, on landing on the Moon on the 4th February 1966, had taken some strange photographs of structures that stood out of the landscape in a precisely defined pattern.

An extraordinary object, in the **Sinus Medii** region of the Moon, was photographed on one of the Apollo lunar missions using a handheld Hasselblad camera (photo number AS10-32-4822). The structure commonly referred to as the **"Castle"** is extremely bright and plainly visible at normal magnification. The photograph shows a remarkable stacking of segments that point to it being of artificial, rather than natural, construction.

The one-mile-long structure appears to be hanging some ten miles above the lunar surface. It has been speculated that it could be the remnant of a huge glass dome. A cable appears to be passing through the tip of the structure. http://blog.hallofthegods.org/2009/06/moon-anomalies-castle.html

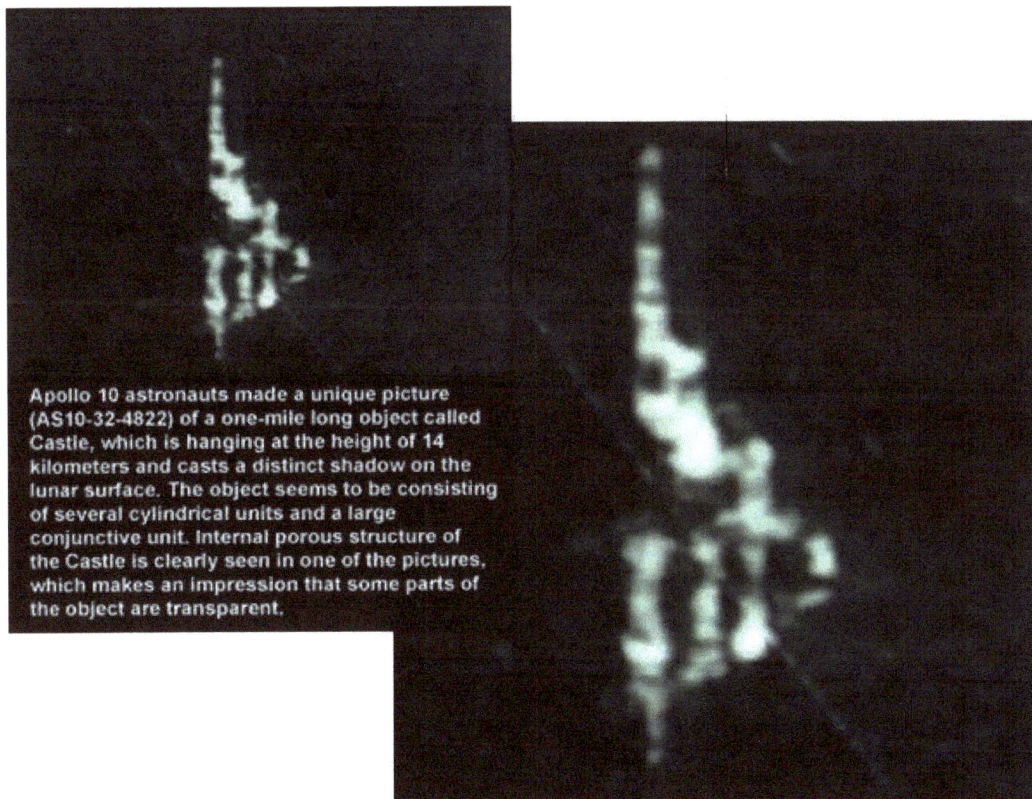

Apollo 10 astronauts made a unique picture (AS10-32-4822) of a one-mile long object called Castle, which is hanging at the height of 14 kilometers and casts a distinct shadow on the lunar surface. The object seems to be consisting of several cylindrical units and a large conjunctive unit. Internal porous structure of the Castle is clearly seen in one of the pictures, which makes an impression that some parts of the object are transparent.

These close-ups show the highly anomalous aspects of the "Castle's" composition
http://www.thelivingmoon.com/43ancients/02files/Moon_Others_01.html

Scientist and former NASA specialist, **Richard Hoagland** insists that there are other intriguing photographs of strange moon anomalies that are currently hidden deep inside NASA. Hoagland speculates that this information is being kept secret because NASA feels the public is not yet psychologically ready. http://blog.hallofthegods.org/2009/06/moon-anomalies-castle.html

This structure has been photographed by both American and Russian spacecraft and the possibility that it really exist is extremely high (if not definite). (See composite photo below).

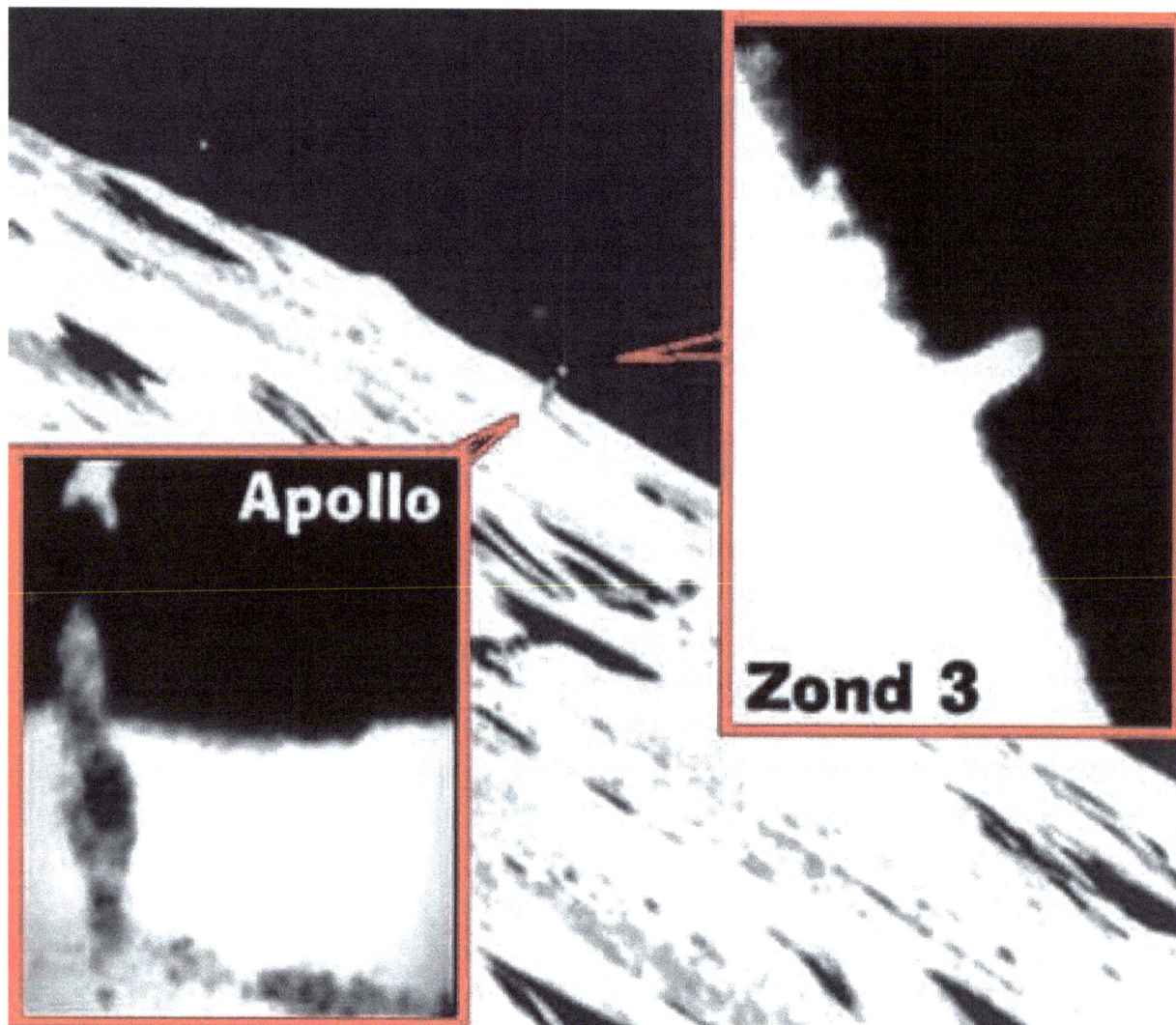

US Apollo
http://blog.sina.com.cn/s/blog_89436e570102wr7l.html

The "Shard" And "Tower"

This image is an overexposed 44x enlargement of **Lunar Orbiter frame LO-III-84-M**. Taken with the medium resolution camera at a distance of at least 250 miles, it shows an object dubbed by Hoagland the **"Shard"**. The star-like object above the "Shard" is a camera registration mark. The "Shard" has a shadow cast in the correct direction for it to be a real object on the Moon and is aligned with the local vertical rather than the grain of the film, decreasing the chance it is an emulsion abnormality. Close-ups reveal a cellular-like internal structure. Above and behind the "Shard" is the **"Tower"**, a massive 7 mile high structure with a central **"Cube"** suspended by a tripod like base. Enhancements of the "Tower" show a similar cellular construction to the "Shard", but with a distinctly hexagonal pattern.

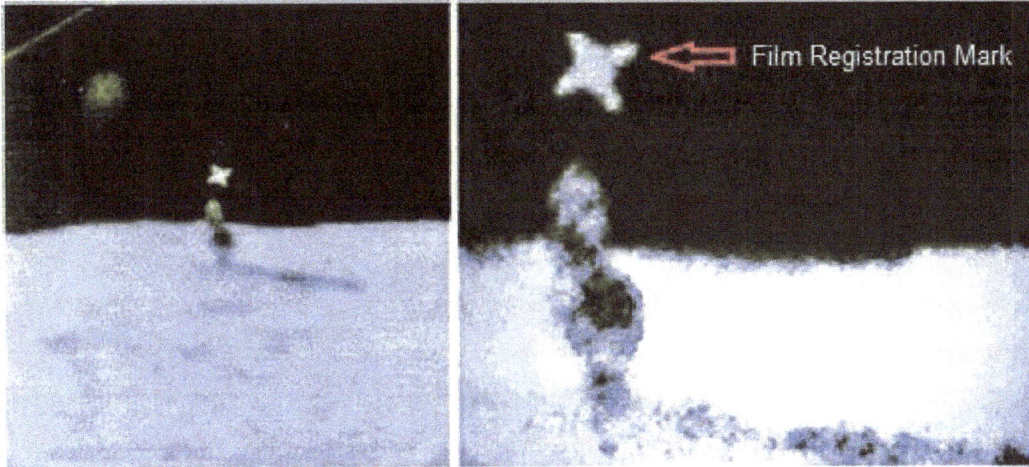

These two photos have been blue hue enhanced to show the internal lattice structure of the Shard

http://www.thelivingmoon.com/43ancients/02files/Moon_Others_01.html

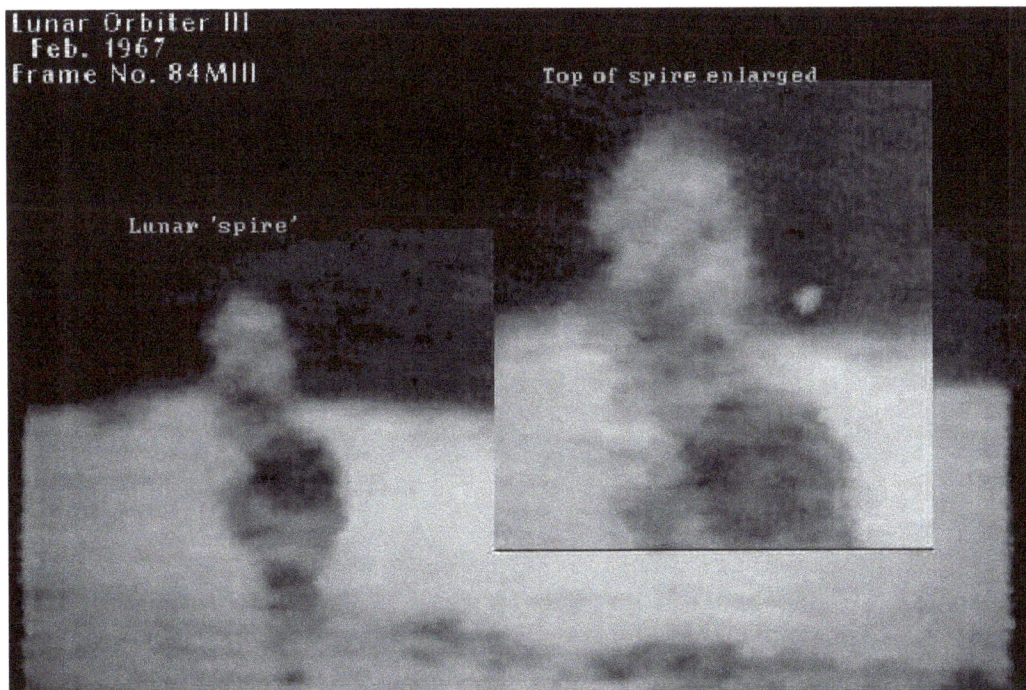

The Shard also known as the lunar "Spire" in gray scale shows internal structure

https://www.bibliotecapleyades.net/luna/esp_luna_26.htm

239

Poor resolution images like this one of the "Shard" have led some to conclude it is an ephemeral "out gassing" event. However, the enhancements reveal no "spray" or splatter which would be consistent with such a conclusion. The object appears to be solid, though badly battered by meteors.

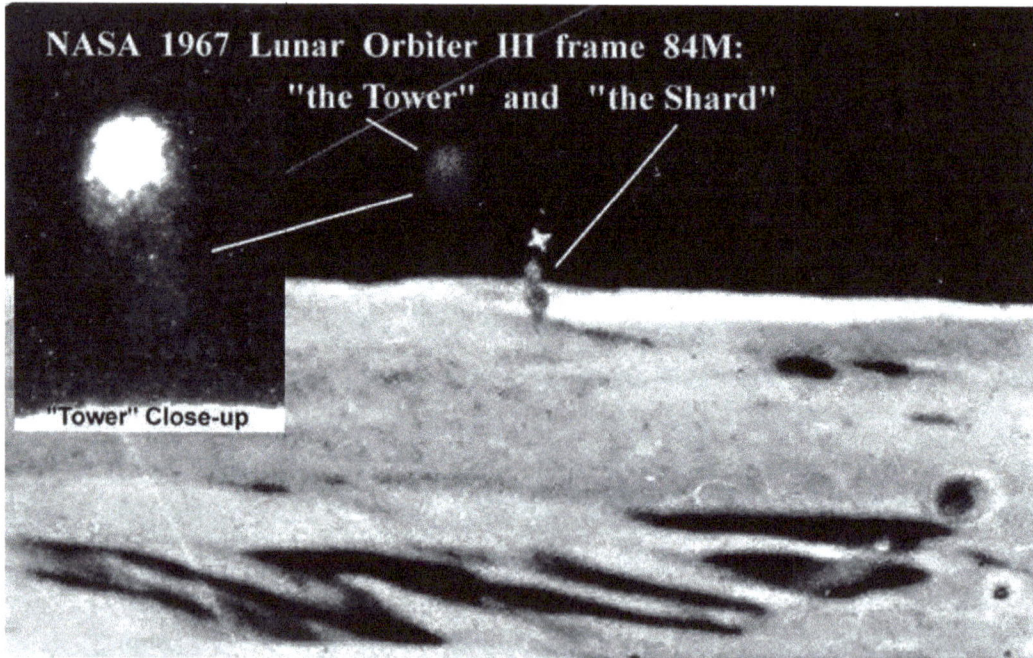

The "Shard and the "Cube"
(Cube is at top centre and "cross" object is a registration mark on film)

https://oko-planet.su/science/sciencediscussions/87626-nasa-skryvaet-pravdu-na-lune-obnaruzheny-sledy-drevney-civilizacii.html

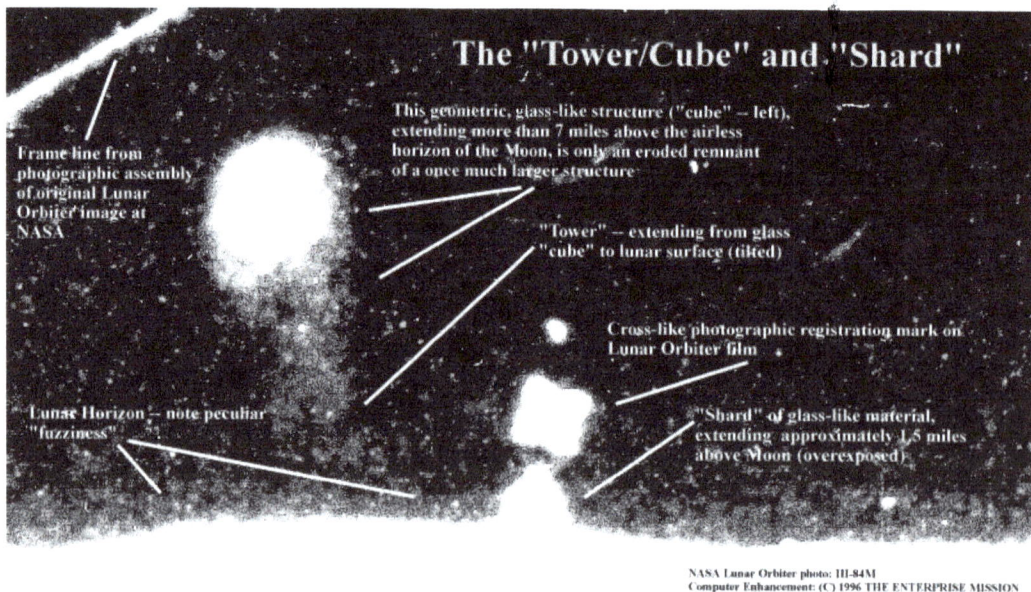

The "Tower/Cube" and the "Shard" on the moon is located in or near Mosting A.

https://www.bibliotecapleyades.net/luna/esp_luna_26.htm

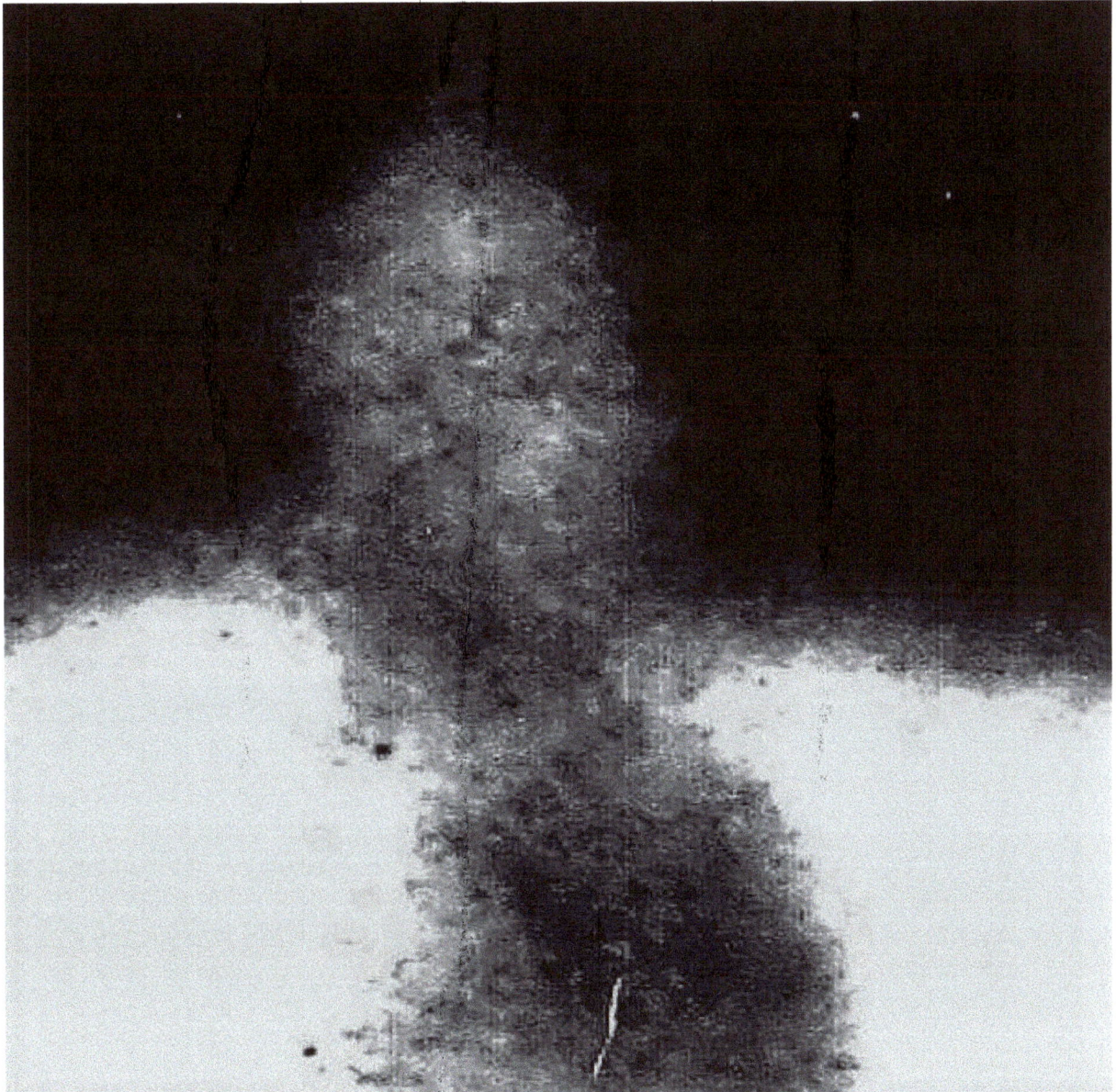

The Shard shows considerable "weathering" from meteorite impactwhich has caused degradation of the structure

Poor resolution images of the "Shard" have led some to conclude it is an ephemeral "out gassing"event. However, the enhancements reveal no "spray" or splatter which would be consistent with such a conclusion. The object appears to be solid, though badly battered by meteors. Assuming that there is no lunar weather like there is on Earth to corrode the Shard, it is hard to know how long it will remain standing, without some inevitable future outside intervention to brace or anchor the structure from falling. There are other tower structures that do appear to have collapsed or fallen over onto the lunar surface but, some are remarkably still intact from their fall. **(The readers are encouraged to search the internet for these tower structures and other Moon anomalies for themselves).**

**A colourized digital enhancement of the "Cube" on the Moon
thought to be constructed of glass and steel**
(c) Terry Tibando

CHAPTER 73
SEARCHING FOR EARTH-LIKE WORLDS
IN THE GOLDILOCKS ZONE

U.S. and Russian orbiting satellites have photographed everything in sight on the lunar ground below ranging from pyramids to massive domes miles in diameter and immense towers reaching up miles above the Moon's surface. There are floating castle-like structures and cubes, mining sites and huge mining equipment that appear to be terra-forming the lunar landscape in search of minerals of one kind or another. There is also evidence of cities of glass and steel and crystalline construction, some that are still intact and many in ruin, decay, and degradation. Logically, this should not come as a complete surprise to any inquiring mind that is searching for evidence of life off-planet. To find evidence of life on the Moon may seem unexpected, especially so close to home but, it really is not that unusual and here's why!

Astronomers and scientists consider the Earth to be in the ideal location or region of space, that being the right distance from the Sun, in order for life to have occurred on this planet. This magical region for life is referred to as the **"Goldilocks Zone"** or the **"Habitable Zone"** a reference to the children's story of "Goldilocks and the Three Bears" where a little girl – Goldilocks chooses from sets of three items, ignoring the ones that are too extreme (large or small, hot or cold, etc.), and settling on the one in the middle, which is "just right"– the **"Comfort Zone"**. Likewise, a planet following this **Goldilocks Principle** is one neither too close nor too far from a star to rule out liquid water on its surface. Habitable zones, however, arenot stable. Over the life of a star, the nature of the zone moves and changes.

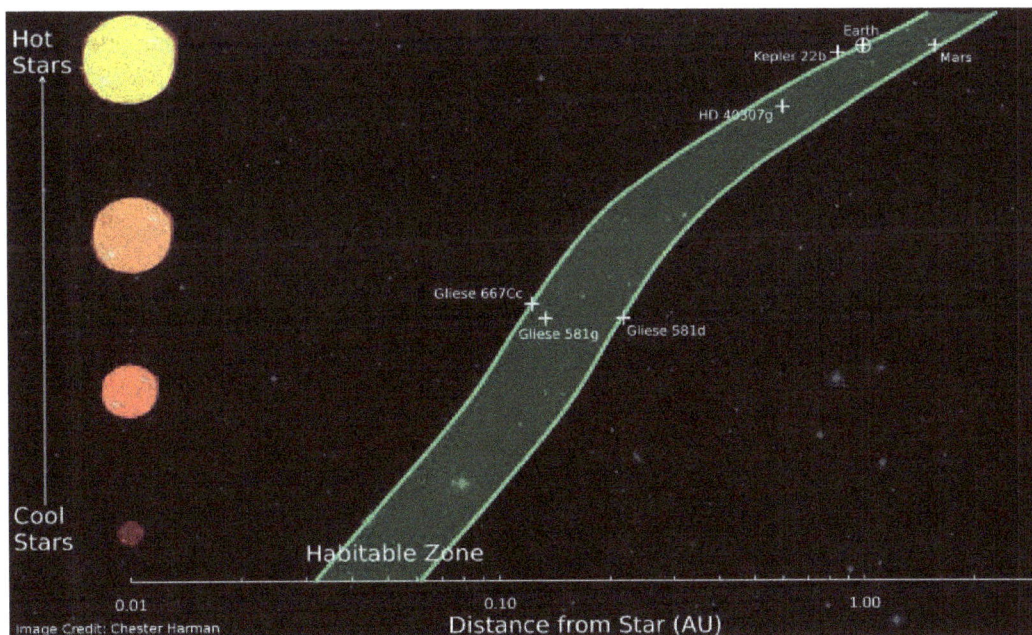

The newly revised Habitable Zone (Goldilocks Zone) where planets may support life basedon sufficient atmospheric pressure to maintain liquid water on its surface.
http://phl.upr.edu/press-releases/anewhabitablezone

A rough measurement or rule of thumb to determine the Goldilocks Zone for our Solar System is 50,000,000 to 150,000,000 miles or 0.50 A.U. to 1.50 A.U.s (Astronomical Units) from the Sun. The Earth is at 0.93 A.U.s in the Goldilocks Zone – just right! http://www.space.com/19522-alien-planet-habitable-zone-definition.html

In astronomy and astrobiology, **habitable zone** (more accurately, **circumstellar habitable zone (CHZ)** is the scientific term for the region around a star within which it is theoretically possible for a planet with sufficient atmospheric pressure to maintain liquid water on its surface.

The significance of the concept is in its inference of conditions favorable for life on Earth – since liquid water is essential for all known forms of life, planets in this zone are considered the most promising sites to host extraterrestrial life.

"Habitable zone" is sometimes used more generally to denote various regions that are considered favorable to life in some way. One prominent example is the **Galactic Habitable Zone** representing the distance of a planet from the galactic centre, based on the position of the Earth in the Milky Way. If different kinds of habitable zones are considered, their intersection is the region considered most likely to contain life. The location of planets and natural satellites (moons) within its parent star's habitable zone (and a near circular orbit) is but one of many criteria for planetary habitability and it is theoretically possible for habitable planets to exist outside the habitable zone. The term "**Goldilocks planet**" is used for any planet that is located within the **circumstellar habitable zone (CHZ)** although when used in the context of planetary habitability the term implies terrestrial planets with conditions roughly comparable to those of Earth (i.e. an Earth analog). Astronomical objects located in the zone are typically close in proximity to their parent star and as such are more exposed to adverse effects such as damaging tidal forces and solar flares. Combined with galactic habitability, these and many other exclusionary factors reinforce a contrasting theory of interstellar "dead zones" where life cannot exist, supporting the Rare Earth hypothesis.

Some planetary scientists have suggested habitable zone theory may prove limiting in scope and overly simplistic. There is growing support for equivalent zones around stars where other solvent compounds such as ammonia and methane could exist in stable liquid forms.

Astrobiologists theorize these environments could be conducive to alternative biochemistry. Additionally, there is probably an abundance of potential habitats outside of the habitable zone within subsurface oceans of extraterrestrial liquid water. It may follow for oceans consisting of ammonia or methane.

Habitable zones are used in the **Search for Extra-Terrestrial Intelligence** and is based on the assumption, should intelligent life exist elsewhere in the Universe, it would most likely be found there.

It has been stated once before and it needs to be repeated here, once again. **Baha'u'llah**, the Manifestation of God for this day and age states in one His book, *Gleanings*:

"Know thou for a certainty that every fix star has its planets and every planet it creatures which no man can compute".

By this is meant that every fix star is a stable and steady star. In other words, a star that has a constant unfluctuating equilibrium of steady radiance and not to given erratic and sudden solar changes like a nova.

Now, let's return to our example of the Earth positioned in the Goldilocks Zone, it then goes without saying that the Earth's Moon is also in the Goldilocks Zone and by the statistics of solar and planetary evolution would by necessity also have produced life on that body at the same moment that life began on this planet. All molecules, atoms, planetary gases and solar accretion that constitute our solar system were in play conservatively 4.5 to 6 billion years ago. What occurred here on Earth would ipso facto also occur on the Moon. Life should have sprung into existence on the Moon at approximately the same time as it did on the Earth. The reason that it appears so lifeless and dead is, according to many scientist and astronomers is due largely in part to bombardments by massive meteoric storms in the early evolutionary period of our solar system. This seems like a very reasonable explanation and logical conclusion, except for the fact that the Earth seems to have miraculously avoided this massive and constant meteoric bombardment to still survive and thrive and develop life but, according to our scientists the Moon was not so fortunate. Whatever, atmosphere, water and biological life forms that may have developed and got a foothold on the Moon was stripped away into space leaving a barren and dead planetary body. The result was a severely cratered surface on our lunar body. So, is the Moon a dead, lifeless and barren satellite? Well, it seems that it is not and we will give reasons why it isn't before we leave this section on the Moon. The Moon it seems is a very strange *SATELLITE* in our Solar System!

Lunar Mining or Meteor Ejecta?

For decades, people have hoped that stories of 'intelligent artifacts' and 'bases' on the far side of the Moon since the time of the Apollo Moon landings would be confirmed hopefully through official NASA or U.S. Military disclosure. This has unfortunately not happened and is not likely to ever occur given the political climate and the decades of cover-up by the Military Industrial Complex. Like most things about the UFO/ETI phenomenon, there is a truth embargo surrounding it, a suppression of knowledge to keep such lunar discoveries a closely guarded secret. This, however, has not stopped independent researchers from investigating these lunar anomalies, particularly in such areas as the **Lobachevsky Crater** as imaged by Apollo 16 and by the U.S. Navy's recent Clementine moon mapping mission.

Depending which website the reader looks with regard to the Lobachevsky Crater, he will see many photographs of this lunar crater which either show merely a crater impacted on the rim of a larger crater with "**ejecta**" (flow material) below it or a mining site with a tower like object on the rim of a crater. Below is just one example of this crater showing what appears to be a lunar anomaly. It is recommended that the reader go to the internet and bring up the images for Lobachevsky Crater and determine for himself what he sees in the many images.

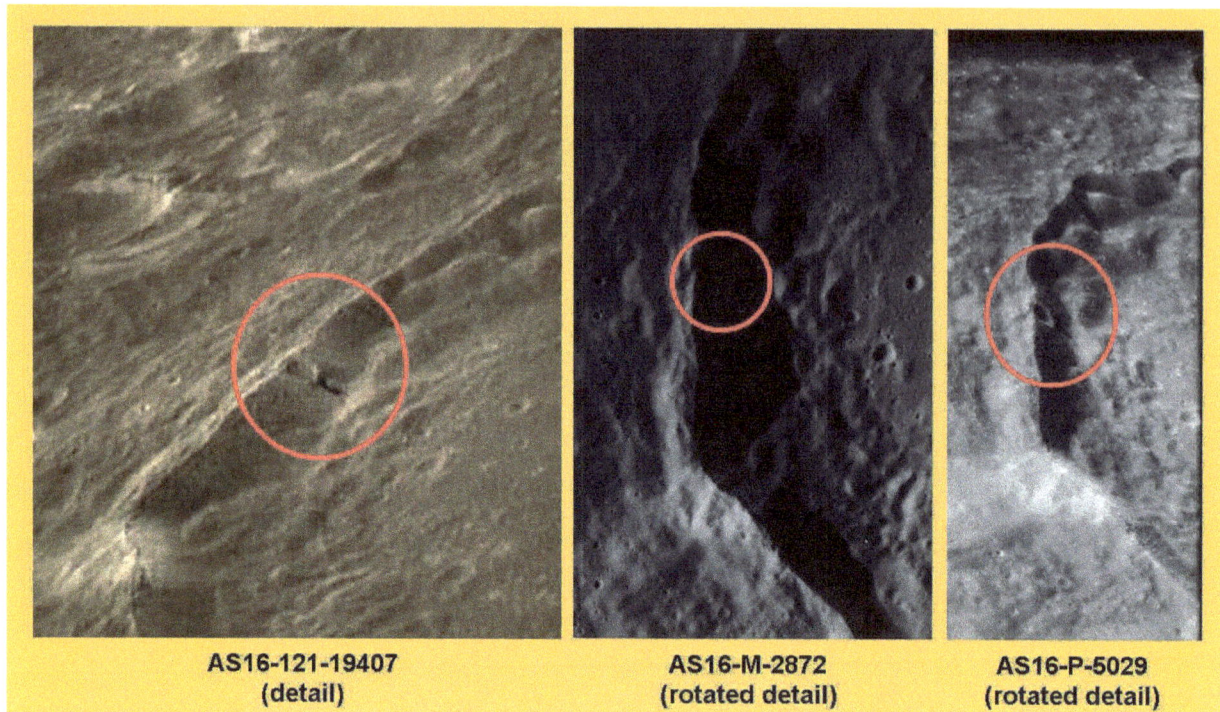

An anomaly on the rim of the Lobachevsky Crater AS16/10075825 imaged by Apollo 16

http://anomalieshunters.altervista.org/moon/mn0002.htm

American researcher **Steve Wingate** focused world attention on the Lobachevsky Crater by posting the old Apollo 16 photograph and several new views from the recent Clementine mapping mission to the **Lunascan Mailing List**. Wingate's discovery of an unusual and artificial structure on the far side of the moon was later confirmed by eminent German geologist and science writer **Dr. Johannes Fiebag** as shown in the above 40+year old NASA Apollo 16 orbital photograph, AS16/10075825.
http://www.bibliotecapleyades.net/luna/esp_luna_35.htm and
http://www.anomalies.net/archive/cni-news/CNI.0818.html

The Lobachevsky Crater clearly shows four strange objects or unusual anomalies in it:

- Item **"A"** demonstrates a strange white rectangular shape along the ridge of the crater rim and includes an unusual shadow below and to its right.
- Item **"B"** appears to be a spectacular 'spire' soaring perhaps hundreds of meters straight up from the lunar surface and standing next to what appears to be a rectangular shadowed hole or depression running from its base out to the right.
- Item **"C"** smaller spire sits in the bottom of the small valley or ravine below the crater rim. Item "C" (circled) is barely visible, even in the enlarged insert.
- Item **"D"** is set apart from the other objects and projects an extraordinary and bright reflective surface. http://www.bibliotecapleyades.net/luna/esp_luna_35.htm

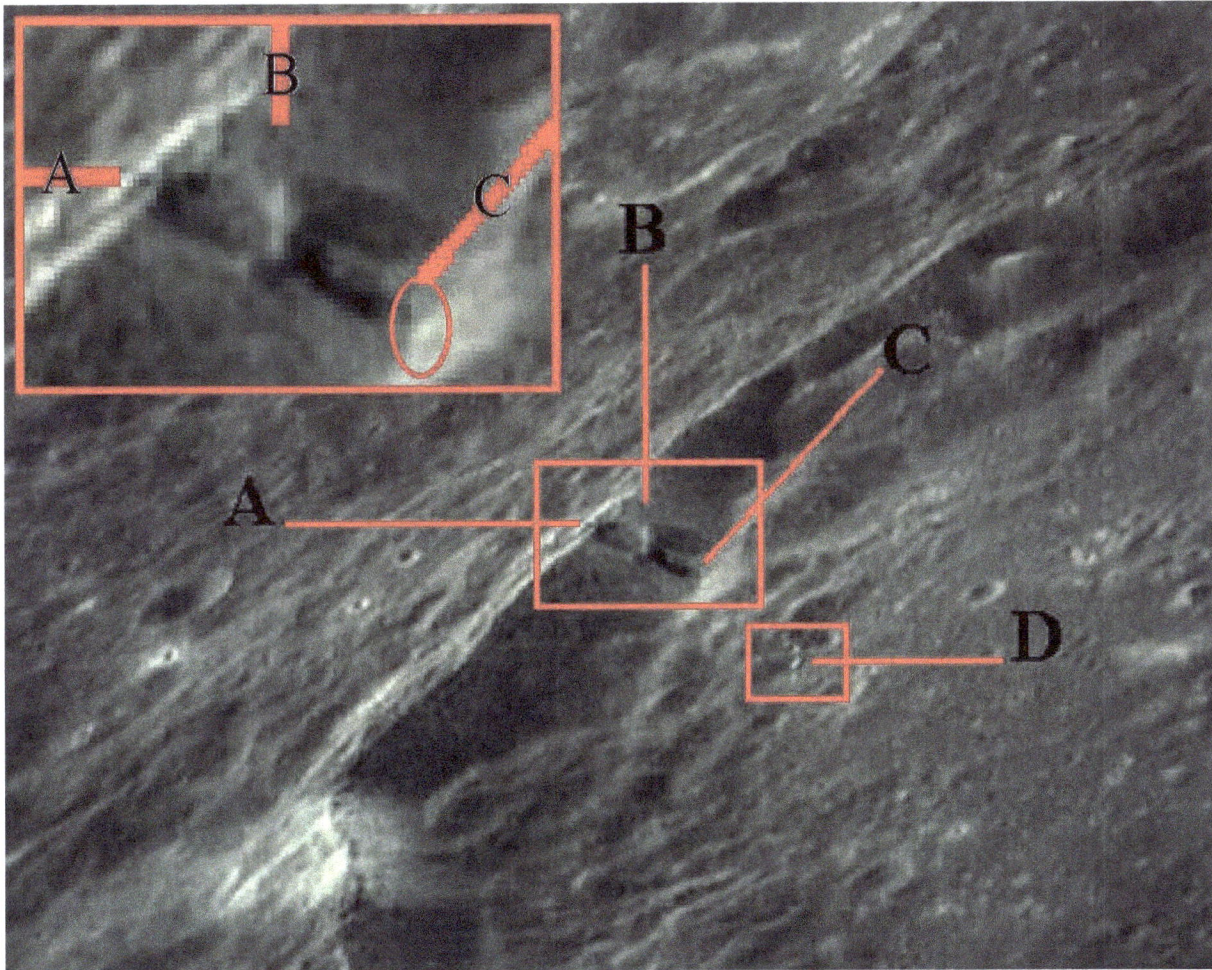

Close up of the Lobachevsky Crater (enlarged insert to highlight area of interest)
https://mocksure.wordpress.com/2009/02/page/3/

The picture was sent to graphics analyst **Liz Edwards** at *I Wonder Productions* for enhancement and further analysis. Her work quickly underscored the fact that there are indeed some bizarre anomalies in the photograph, in fact, there are several.

(Ms. Edwards' most recently exposed what appears to be a tiny missile in a photograph of a New York sunset at the exact time of the **TWA 800 tragedy**.)

Ms. Edwards made it clear that her multi-stage "enhancement process takes the image to its most critical viewing size without distorting the objects in question." Below, is her first enhancement of the **NASA** Apollo 16 photograph...
http://www.bibliotecapleyades.net/luna/esp_luna_35.htm

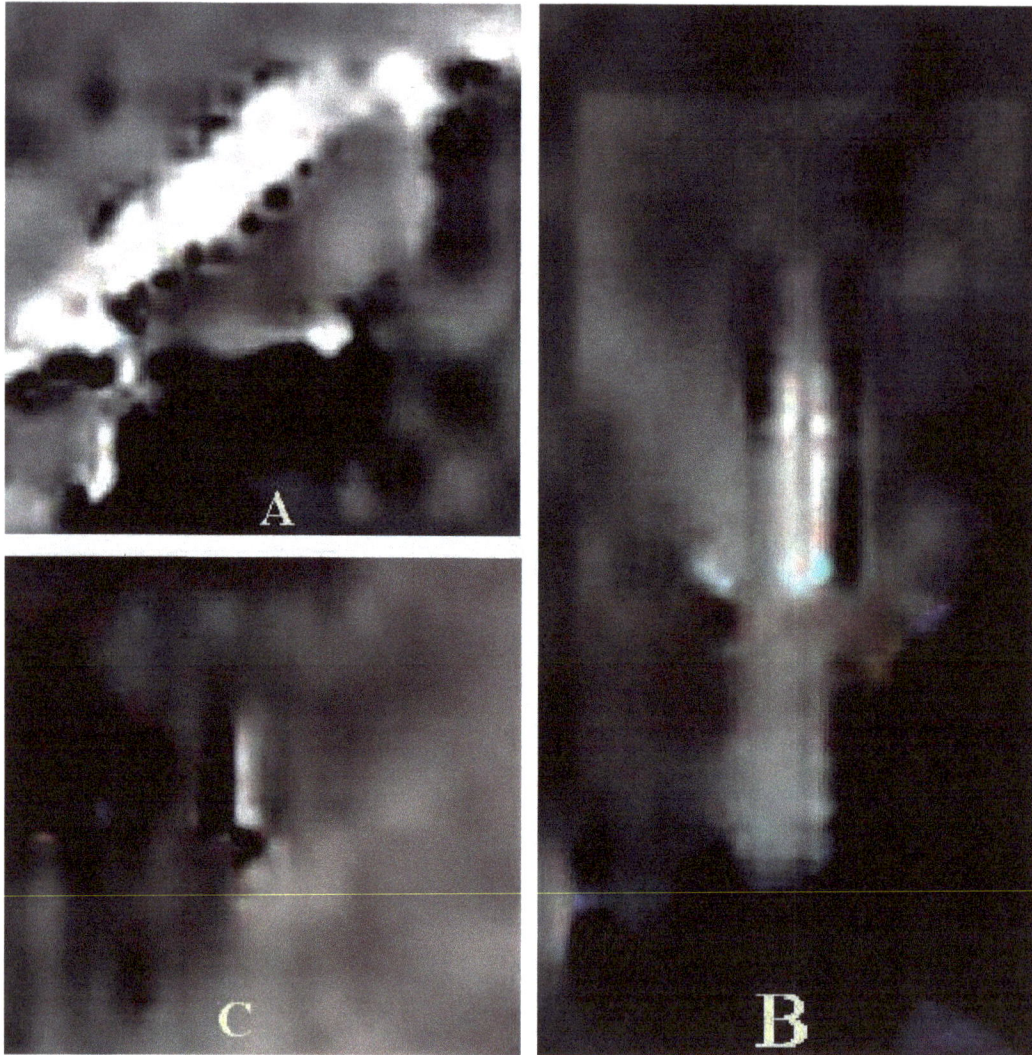

Close-up of Ms. Edwards' enhancement of three of the subject areas.

Item "A", the ridge-top on the left shows an amazing line-up of round protuberances which appears to be an integral part of the white colored cap area on top of the ridge. These black, holed objects are lined up in such a way as to resemble large 'ventilation' or exhaust stacks on an ocean liner or industrial plant, although much larger.

Item "B" appears to be a shimmering image many hundreds of feet rising straight up from the lunar surface. The sun's rays against the tower-like object appear to illuminate it indicating that its structure may be polished metal or glass crystal. This crystal lunar obelisk would be another example of an ET technology more advanced than ours which appears to be connected to the bizarre ridge-top 'vent' pipes above on the crater rim. A part of a mining or excavation operation directly related to the dark, shadowed surface depression next to it?

Item "C" at the bottom of the ravine below the ridge is a smaller spire which also rises up out of the lunar surface. http://www.bibliotecapleyades.net/luna/esp_luna_35.htm

248

NASA Airbrushes the Evidence of Lunar Anomalies Away

What is really interesting about the **Lobachevsky Crater** is that the photos from the US Navy's Clementine satellite of this same area do not reveal any unusual anomalies but natural lunar cratering and ejecta material. Is it possible that the Clementine and the NASA photos of this crater and other unusual areas on the Moon been altered or airbrushed out before being released to the public? **Donna Hare**, one of Dr. Steven Greer's **Disclosure Witnesses** has testified to that fact as being a routine practice by NASA and the DOD, as indicated by photos below of the Moon's surface.

Many times the airbrushing, whether done by hand or automatically by satellite software becomes so excessive that the photos are distorted beyond any usual visual information. In fact at times, one would question if these photos are indeed lunar photographs.

The following photos below show lunar anomalies imaged from the US Navy's Clementine Moon satellite that were discovered by **Joseph. P. Skipper** and were posted on his website: **Mars Anomaly Research** (www.marsanomalyresearch.com). The fact that so many photos of the Moon's surface containing strange alien objects were airbrushed out or altered in other ways is most telling of a deliberate cover-up within NASA and the DOD. We know that Donna Hare stated that airbrushing of alien objects out of photos was done routinely while she was in the employment of NASA but, this may have been done as a final detail for the photos the satellite camera failed to optically obscure completely as intended. A new method via technology and software is being used to obfuscate photo images intentionally after the original non-obscure photograph is taken of a target specific object or area.

The Clementine satellite had at least a two-fold mission: one, to thoroughly survey and photograph the Moon's surface for any signs of ET structures and possible technology and two, to test advanced covert optic technology designed to be target specific while being visually obfuscating which would be used on future Mars missions.

Skipper's research indicated that earlier 1960s and 1970s tampering technology was comparatively crude, layering over anomalous scenes in individual small image strips that were later joined into mosaic wider field of view images. This meant that multiple small strips showed demarcation lines between the differing tampering fields because different techniques were used and thus, showed up very strikingly in the mosaic images that were released to the public. Inefficiency by scientists to employ the rigid protocols of the scientific method permitted the US NAVY to get away with the tampering undetected.

Superfast computers utilizing artificial intelligence and advancements in mapping and evidence recognition software were first used in search of civilization evidence on the Moon. Later, it would include greater search parameters for surface water and biological surface life to be used on Mars missions. The new technology according to Skipper can deal with much larger image scenes through the mapping technology and the applications are made to target (object) specific in the scene, rather than in broad spectrum applications.

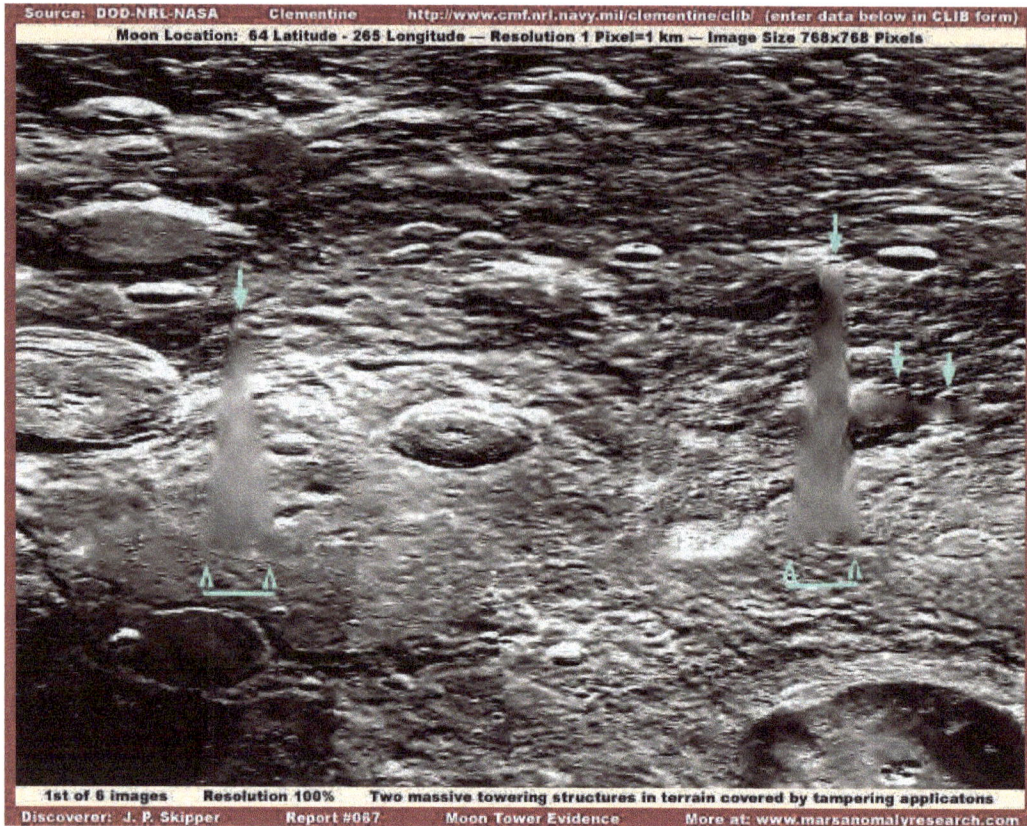

Evidence of moon towers blurred out by NASA

http://www.marsanomalyresearch.com/evidence-reports/2004/067/moon-towers.htm

The Clementine tampering work is thorough but, it isn't perfect and major mistakes still exist visually. Images are often a confused mess that brings attention to itself in that the tampering applications are literally mapped and conform too closely to the offending targeted objects. When the objects are small, it's not a big deal, however, when the objects are large, long and narrow such as a long narrow straight band like object or series of objects is still revealed by its outline still coming through. Likewise, the tall or squat massive object outlines are also still revealed as the tampering application conforms too closely to the object rather than blending out into its surrounding terrain.

So, says Skipper, "the Clementine obfuscation technology level is more effective in disguising normal familiar size objects like building structures on Earth. This would be logical to expect in the initial programming since that is what the programmers are most familiar with here as Earth evidence. On the other hand, it is also logical to expect that the programming would be less effective in dealing with massive bulky objects like the towers and squat structures and the long narrow geometric structures not familiar with here on Earth due to sheer scale".
http://www.marsanomalyresearch.com/evidence-reports/2008/153/moon-tower-shadows.htm

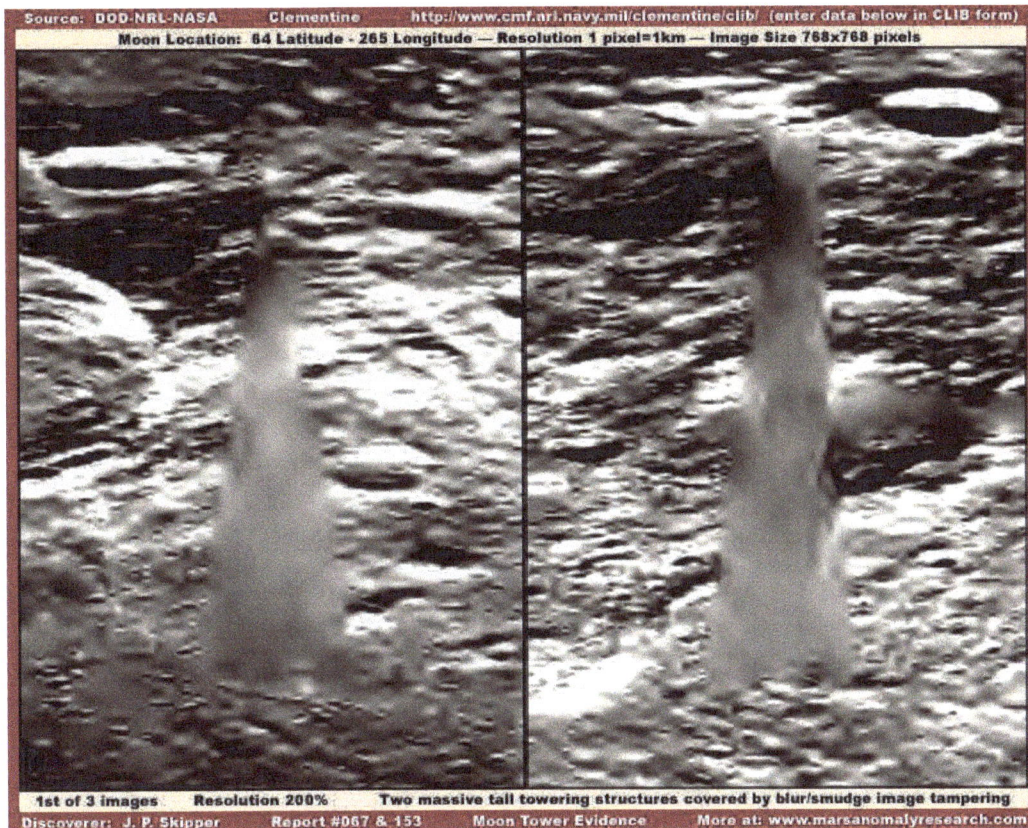

Two tall lunar towers that have been smudged, evidence of image tampering
http://www.marsanomalyresearch.com/evidence-reports/2008/153/moon-tower-shadows.htm

When one looks at the photo images below, one sees that this smudge and blur treatment conforms too closely to the dimensions of huge objects in the terrain and are not the result of camera or processing flaws? In the photo of the huge, squat object for example, in the image below, a little of the object beneath can be seen. That may be acceptable as proof, but how is itthat there is a lack of any shadows cast if these are to be consider obscured massive upright structures. So where are the shadows?

If one looks at the surface texture of the smudge and blur conforming to the objects and move your eye out into the terrain surrounding these huge objects, you'll see the same smudge and blurtexture there. It should be apparent, Skipper points out that an immense artificial structure on theMoon would not be the only solitary object in a geological terrain bare of any other civilization evidence and such image tampering applications would not be limited to just the largest most massive objects.

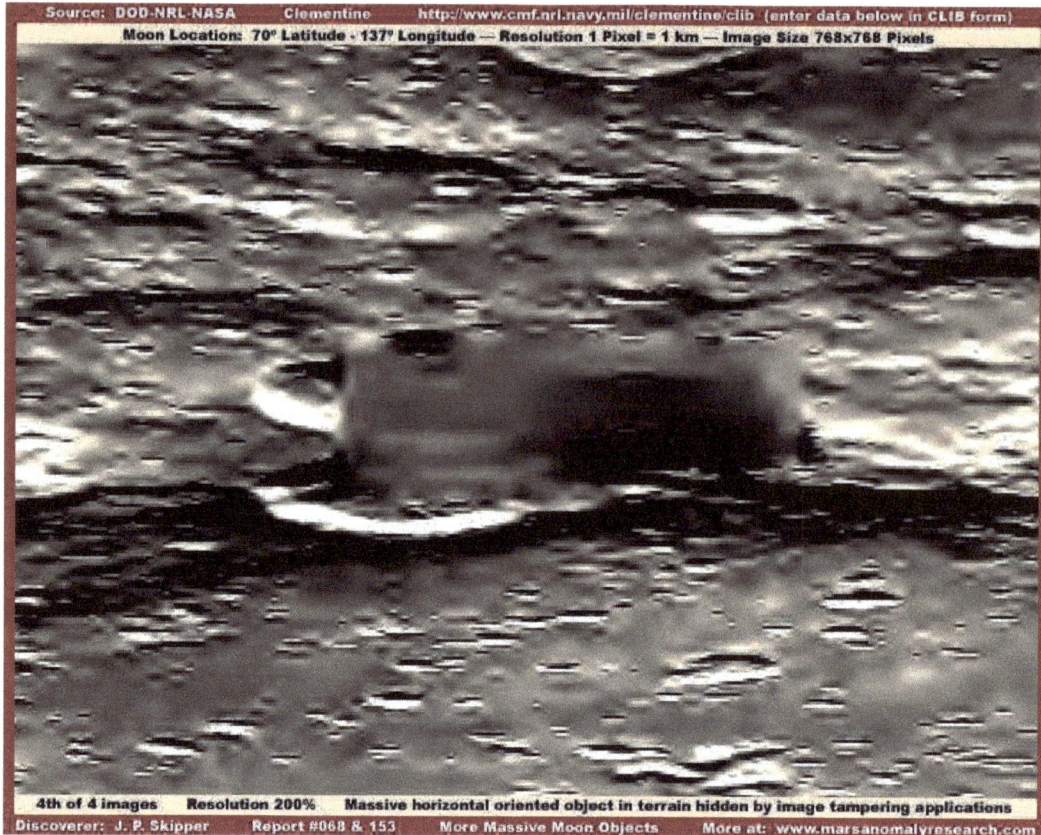

Source: DOD-NRL-NASA   Clementine   http://www.cmf.nrl.navy.mil/clementine/clib (enter data below in CLIB form)

Moon Location: 70° Latitude · 137° Longitude — Resolution 1 Pixel = 1 km — Image Size 768x768 Pixels

4th of 4 images   Resolution 200%   Massive horizontal oriented object in terrain hidden by image tampering applications
Discoverer: J. P. Skipper   Report #068 & 153   More Massive Moon Objects   More at: www.marsanomalyresearch.com

A massive lunar object has been covered up through NASA image tampering

http://www.marsanomalyresearch.com/evidence-reports/2008/153/moon-tower-shadows.htm

There are actually short shadows cast by some smaller objects that were not included in the smudge tampering. Note that the shadows are cast forward toward the viewer indicating sunlightis likely coming from behind the massive objects meaning that any shadows present for the massive objects would be thrown forward toward the viewer with some angle to it. Note that none of the shadows cast by the smaller objects are very long indicating a fairly high overhead position of the Sun and certainly not lower on the horizon. That equates to shorter size shadows.

http://www.marsanomalyresearch.com/evidence-reports/2008/153/moon-tower-shadows.htm

Inspection of the surrounding terrain displays a strange speckling that is not natural and, if you were a scientist, this should draw your attention raising questions within you as to why this is so?Each of the many small speckle spots is the "rooftop" of a relatively normal size geometric shaped structure. The tampering applies the smudge/blur application all around the geometric structure to hide it but, the software is programmed to leave a little something out and to leave a little detail behind. If it did not do this, the smooth featurelessness of the tampering smudge fieldwould be broad and blanketing, drawing attention to the field.

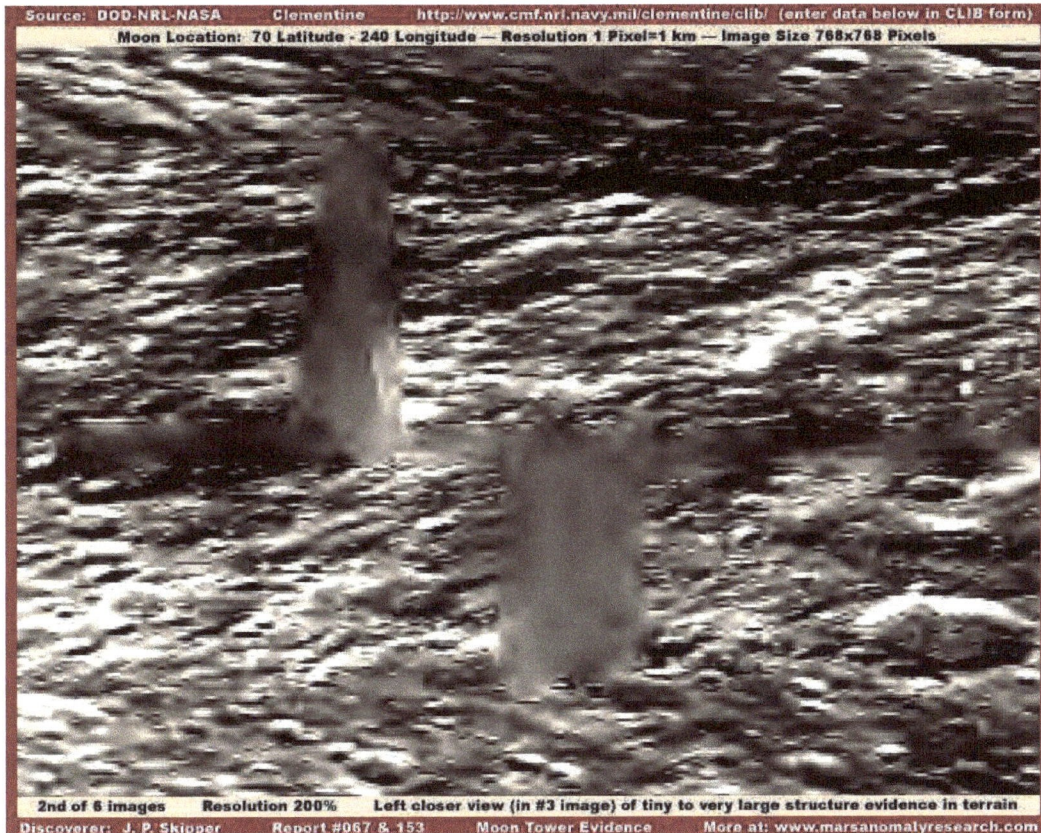

Source: DOD-NRL-NASA   Clementine   http://www.cmf.nrl.navy.mil/clementine/clib/ (enter data below in CLIB form)

Moon Location: 70 Latitude - 240 Longitude — Resolution 1 Pixel=1 km — Image Size 768x768 Pixels

2nd of 6 images   Resolution 200%   Left closer view (in #3 image) of tiny to very large structure evidence in terrain

Discoverer: J. P. Skipper   Report #067 & 153   Moon Tower Evidence   More at: www.marsanomalyresearch.com

**More evidence of heavily blurred/smudged lunar objects and towers
in an effort to deny their existence from public scrutiny**

http://www.marsanomalyresearch.com/evidence-reports/2008/153/moon-tower-shadows.htm

The broad fields of tampering are the very things the new tampering technology is supposed to prevent. Look closely at the above image, you'll see that several small white rectangular structures were actually left out of the smudge as the software no doubt got overloaded with so much detail and got confused as to what to cover and what not. Although effective at hiding the normal size structure evidence, the problem is that there are just too many "rooftops" giving the whole scene a strange speckled look, again inadvertently drawing attention to it. Note, however, that not a single speckle spot is located on any of the massive object areas.

All deficiencies stemming from the tampering obfuscation process could have proved disastrous for the US Navy and NASA that would have blown their cover-up program they nevertheless, felt confident enough to publicly release this photographic mess. They were confident of the public's ignorance to not actually question what they saw, combine this with academic pressure on scientists to keep their mouths shut, secrecy was, therefore, maintained. If this didn't work, they could always fall back on camera or processing artifact explanations.

http://www.marsanomalyresearch.com/evidence-reports/2008/153/moon-tower-shadows.htm

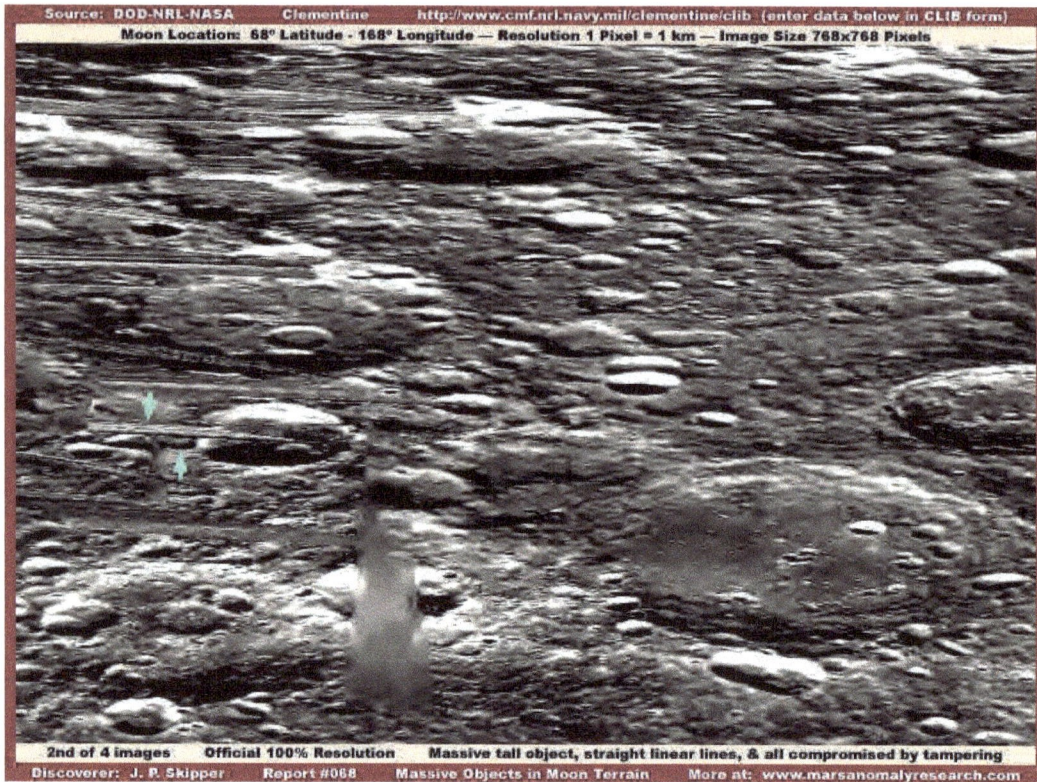

Source: DOD-NRL-NASA   Clementine   http://www.cmf.nrl.navy.mil/clementine/clib (enter data below in CLIB form)
Moon Location: 68° Latitude - 168° Longitude — Resolution 1 Pixel = 1 km — Image Size 768x768 Pixels

2nd of 4 images   Official 100% Resolution   Massive tall object, straight linear lines, & all compromised by tampering
Discoverer: J. P. Skipper   Report #068   Massive Objects in Moon Terrain   More at: www.marsanomalyresearch.com

Many lunar objects are hidden through smudging/blurring of the images top and bottom

http://www.marsanomalyresearch.com/evidence-reports/2004/068/moon-objects.htm

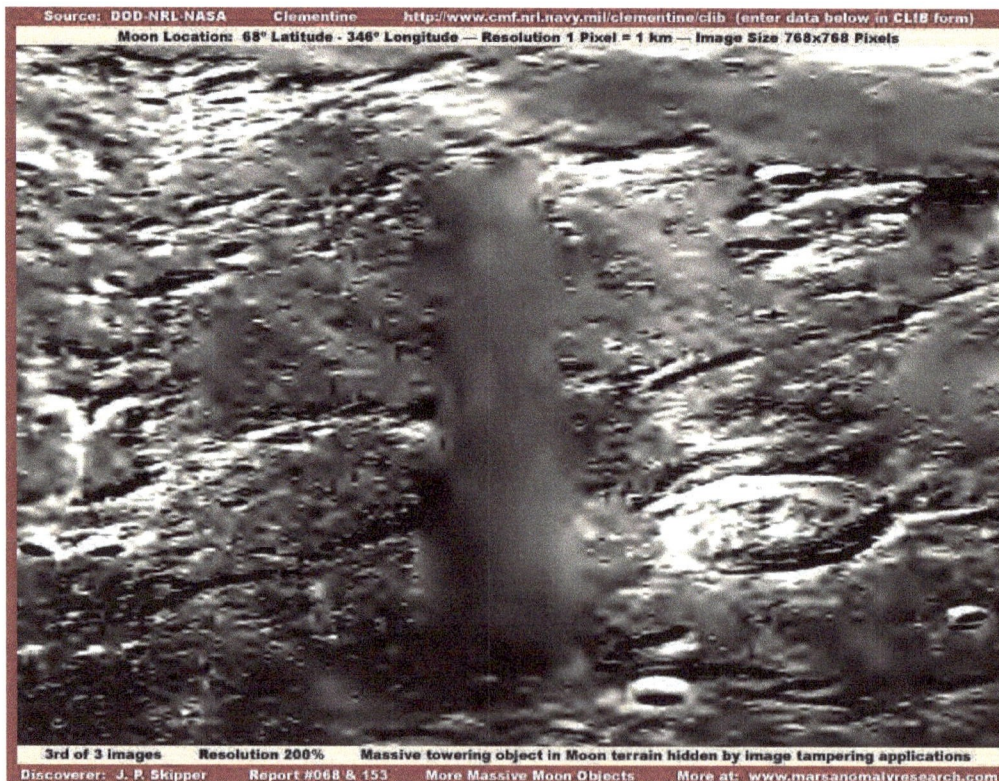

Source: DOD-NRL-NASA   Clementine   http://www.cmf.nrl.navy.mil/clementine/clib (enter data below in CLIB form)
Moon Location: 68° Latitude - 346° Longitude — Resolution 1 Pixel = 1 km — Image Size 768x768 Pixels

3rd of 3 images   Resolution 200%   Massive towering object in Moon terrain hidden by image tampering applications
Discoverer: J. P. Skipper   Report #068 & 153   More Massive Moon Objects   More at: www.marsanomalyresearch.com

254

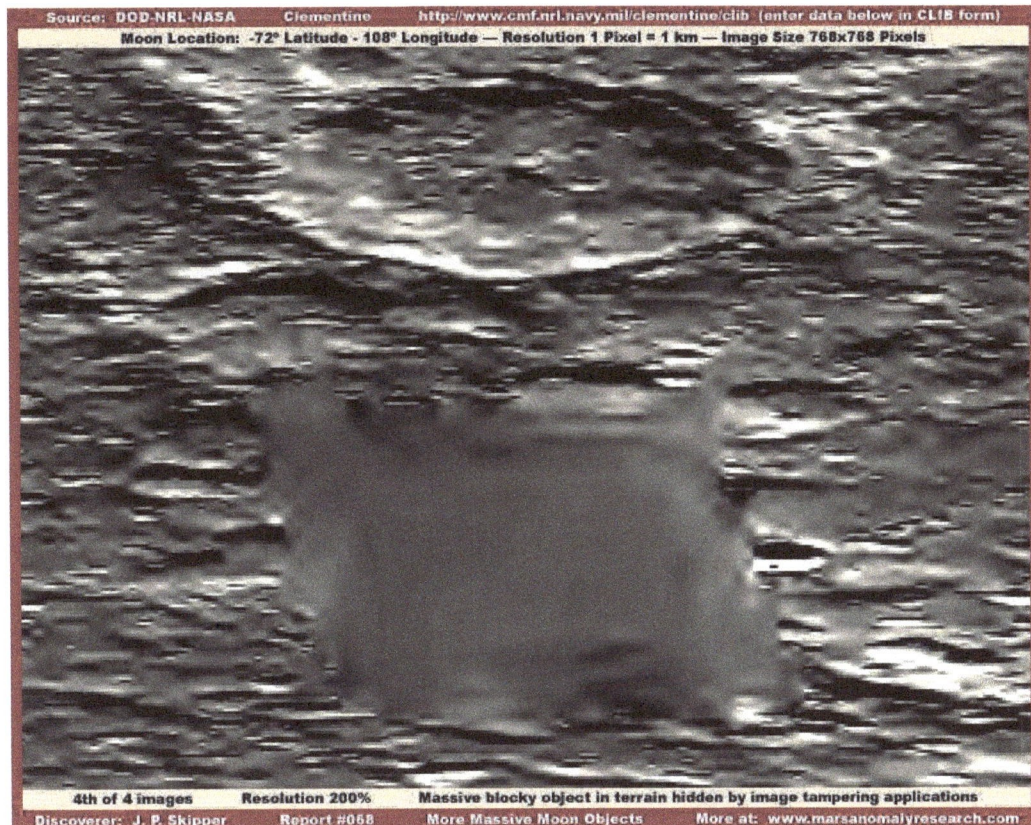

A massive blocky object is hidden through image tampering
http://www.marsanomalyresearch.com/evidence-reports/2004/068/moon-objects.htm

Is There an Alien Presence in the Form of Lunar Cities on the Moon?

Bear in mind that these tampered photos are covering up massive structures on the Moon that arein an order of miles in size, greater than anything on Earth that we have yet built. It is proof positive of a highly advanced civilization that at one time existed in our Solar System and may still be present or may have recently returned.

It is believed that because the gravity of the Moon is only one-sixth that of the Earth's, it is, therefore, possible to build structures on a massive scale, larger than can be constructed on Earth.Glass and steel composites would have a tensile strength far greater than any glass or steel manufactured here on Earth. It would be comparable to the silk thread of a spider that is spun to form a web in order to capture its prey. The tensile strength of the silk thread if constructed to a scale comparable to the diameter of steel cable used in bridge building, far fewer "silk cable" strands would be required to support the bridge. In fact that "cable of silk" could support or hold an exponential tonnage of weight that would otherwise snap a steel cable in two, even if that silkcable were stretch beyond the limits of steel. Imagine then, the ability to construct immense skyscrapers that would reach miles into the atmosphere without crushing or collapsing under its own weight through the forces of gravity. This is the type of structures that has been discovered on the Moon by NASA and the DOD.

In the photos below, note the WWII city ruins of Wesel, Germany from the aftermath of Allied bombing raids and the medieval castle of **Tahkt-i- Sulaiman** in Iran that has been passed off as an alleged "alien base" on the Moon, pay particular attention to the rectilinear aspects of both ruins and compare this to the rectilinear structure of the lunar "City".

Lunar ET ruins of "Glass and Steel"
http://www.enterprisemission.com/NPC-Russia2.htm

NASA-LRO Image of the "Glass City" – "Crystalline Ruins on the Moon"
http://www.abovetopsecret.com/forum/thread518181/pg1

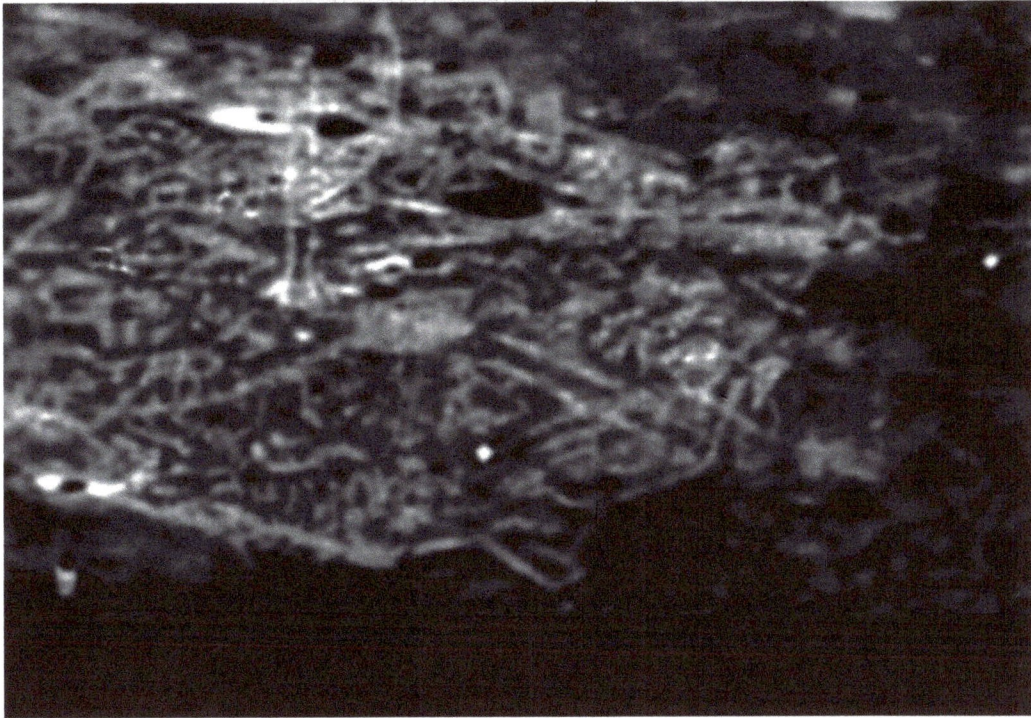

A lunar "City" on the Moon. Notice all the rectilinear structure

Aerial view of the aftermath of Allied fire-bombing of Dresden, Germany. Compare the rectilinear structure in this photo to the image of the lunar "City" above

257

**Two more views (top and bottom) of the holocaust fire-bombing of Dresden.
In both photos notice the tall "tower-like" linear aspects of the city ruins**
http://flattopshistorywarpolitics.yuku.com/topic/203/Dresden#.WPWJZmeIvmE
and http://rarehistoricalphotos.com/ruins-dresden-1945/

**The bombed-out WWII German town of Wesel after the Allied
raids on February 16, 17, 19, 1945**

The destruction to **Dresden** and **Wesel** in the above photos shows walls, foundations of buildings and a lone church steeple that now remain carpet and fire- bombing by Allied Forces. Roads can be seen with massive rubble and debris strewn over them; note sections of city block that are empty shells of their former structure, church steeples and partial ruins of apartments that still stand. Notice also, the heavily cratered landscape from Allied bombing within the city of Wesel and directly outside of it and notice also the absence of vegetation remaining anywhere.

This type of horrific bombing by the Allied Forces during the Second World War occurred repeatedly over many Germany cities that were pulverized into utter destruction and ruin. The landscape of many cities in the aftermath frequently looked barren and dead, conditions that were found repeatedly by the Apollo lunar program among the ruin cities of the Moon. Of significance is the tightly impacted cratering over much of the earth's surface that runs in linear patterns horizontally, vertically as well as in curves. How is this type of pattern or carpet bombing in the city of Wesel and in many cities throughout Germany and Europe important to what was found on the Moon? This will be discussed shortly.

Close Up Alien Moon Base

An aerial view of the Tahkt-i- Sulaiman castle in Iran, it is not an alien Moon base!
http://alien-ufo-research.com/alien_moon_bases/

Alien Moon Bases and Mining Sites

The photo above is a clever piece of disinformation and deception, designed to mislead UFO researchers by hoaxing a description of an unusual but, otherwise common medieval structure thus, providing an air of authenticity to a supposedly "alien Moon base" on the far side of the Moon, which it is not. It is, in fact, a medieval castle in Iran known as **Tahkt-i- Sulaiman.**

Author's Rant: As a student of medieval history, besides my interest in all things UFO and Extraterrestrial, I was suspicious when I first came across this photo claiming to be an alien moon base, as its appearance was very familiar to me. A brief search on the internet soon confirmed what I surmised earlier, that it was indeed a medieval castle sharing many similar features with other castle structures of its time.

Note the outside crenulated walls and towers circling the inner courtyard, the large keep or stronghold, and the village community structures beside it which are below the small lake or pond, the castles main water source. The main creek enters through the outer city wall on the

260

west side feeding into the lake; the other creek flows north (top of photo) forming a defensive moat next to the castle's main entrance where a road leads up to and through the entrance into the inner courtyard. There are a set of gatehouses just before the small lake. The whole castle community sits about 100 to 200 feet upon a hill that overlooks the countryside, truly a near impregnable medieval castle. One last thing about this alleged "Alien Moon Base" is where are all the lunar craters that typically pockmark the lunar surface? All that can be seen, where there should be moon craters are plowed and cultivated fields of crop!

UFO researchers should beware of this kind of deceptive practice if it looks too good and unusual to be true it probably is and therefore, it's a fake or hoax! The purpose behind this type of deception is to make UFO researchers appear to be poor investigators, make them look publicly foolish and ultimately discredit the whole UFO phenomenon and those who investigate it. Do your homework and save yourself some embarrassment and grief.

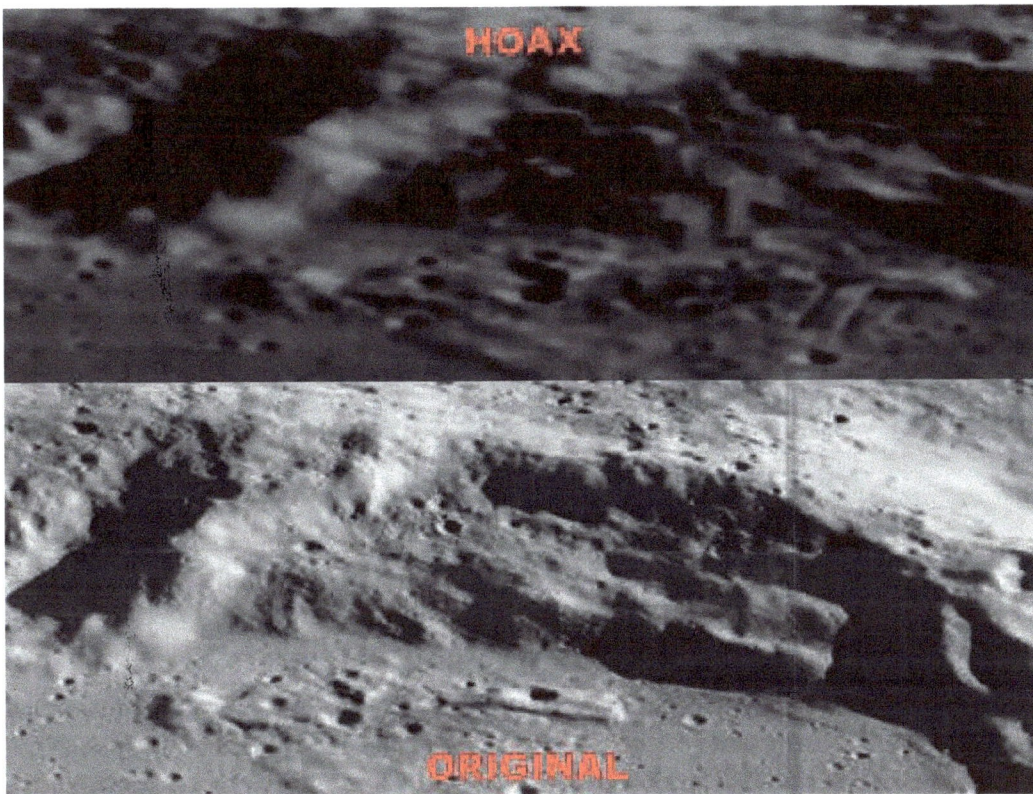

Alien Moon Base allegedly captured by China's Chang'e -2 Orbiter 2012.
This image has been altered from the original and is a fake
https://www.metabunk.org/debunked-change-2-photos-of-alien-base-on-moon-hoax.t6174/\

In recent years, the **Chinese Space agency** sent a satellite to the Moon to obtain its own photographic and scientific measurements for an eventual manned moon landing program. The photos above taken by the **Chinese Lunar Orbiter Chang'e 2** shows an alleged "alien base" or "mining site" located in a lunar crater, it is a fake! NASA's lunar orbiters (below) photographed another large crater and it too shows signs of a structure that is undeniably artificial indicating a strong similarity to manmade building structures back here on Earth. Is it possible that these buildings are secret US or Russian Moon bases for the purpose of mining or for a secret lunar colonization process? Is it possible that these are the remnants of a former ET civilization that once inhabited the Moon or are they some sort of lunar outposts designed to monitor human activity on our planet? A closer inspection of the Moon's surface is required to answer some of these questions.

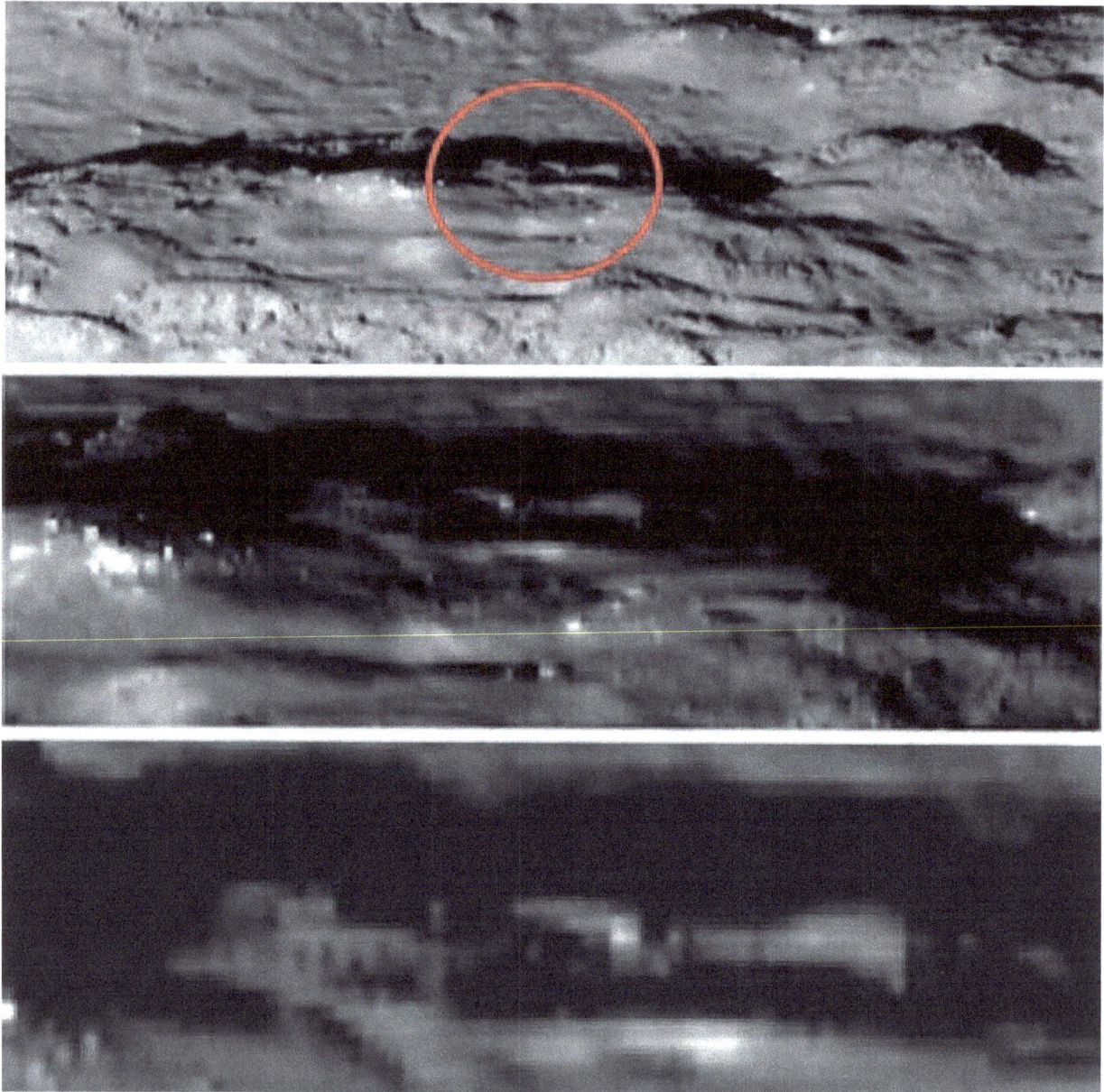

Original NASA Luna photo in top image, bottom two photos are enlargements to bring out detail that confirm the presence of lunar buildings and a tower or obelisk
http://www.dailystar.co.uk/news/latest-news/545007/ufo-alien-base-nasa-image-surface-moon

The next photo below shows what a lot of investigators believe is strong evidence for mining activity on the Moon. Many areas of interest have been circled to show mining waste till areas (top left), several buildings (top and left of ravine), and cluster of small domes or buildings (top right) and tower structures along road toward mining site (bottom middle), mining site (centre) with flow patterns beside it (centre right). Is this really a mining site or as NASA would like us all to believe, it is nothing more than a trick of light and shadow, a natural geological terrain found all over the Moon. Nothing happening here, folks! Move along, now!!

What most people and scientists would like to know, if there is or has been mining activity on the Moon by whom then, what minerals or resources were being removed from the lunar terrain? What purpose were these resources being used for? Were they used for constructing lunar buildings and cities or for fuel in their spacecraft? Answers to such remain unknown to us at this time.

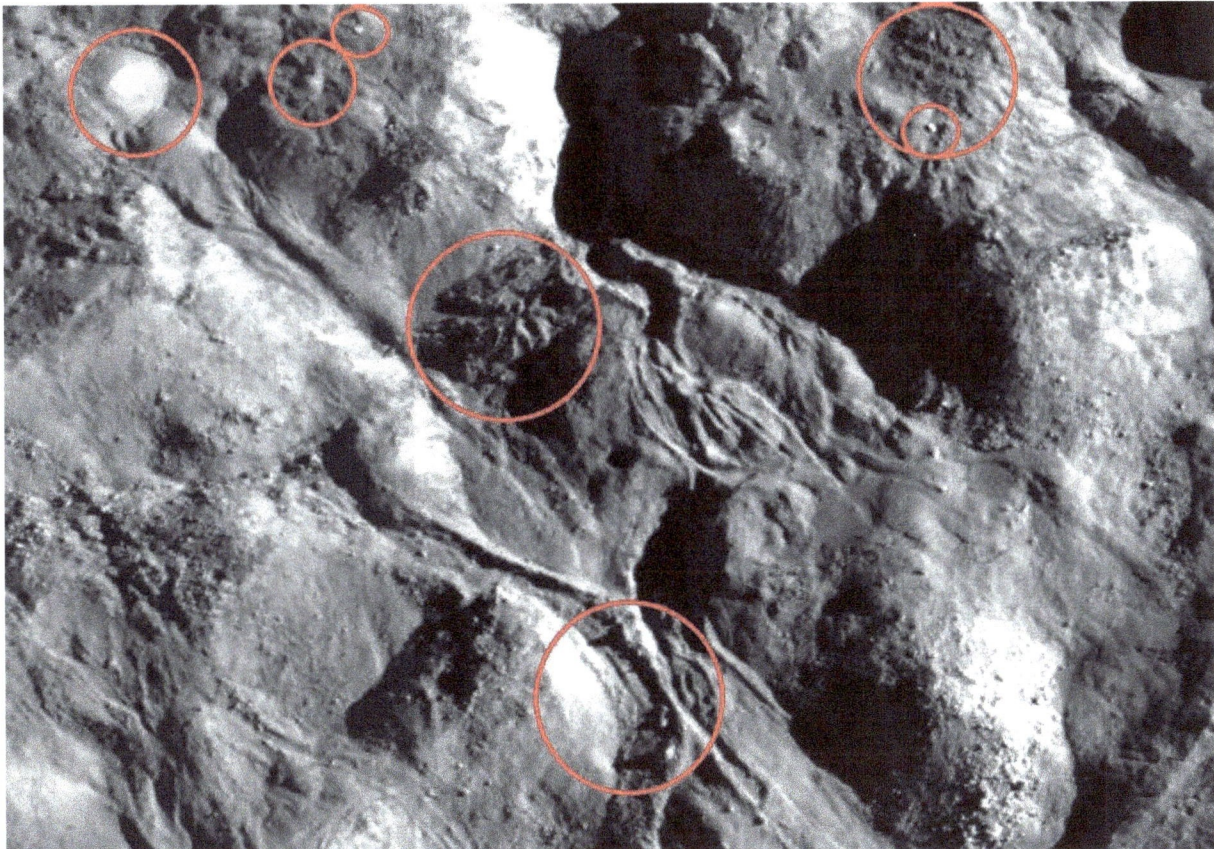

Is there mining operations going on in the Tycho Crater or are these circled areas merely natural Moon formations?
http://www.abovetopsecret.com/forum/thread514308/pg7

In the **De Moraes Crater**, another lunar anomaly was discovered. The crater is about 45-46 km across, we can assume since the alien structure covers 1/3 of that crater then, the alien base is conservatively 15-15.3 km across or 9.32 miles. Again, we are looking at an unusually immense object or structure, a structure that is beyond comprehension for size and functionality. It would appear that the photos that have been presented thus far as proof for the existence of Extraterrestrial life forms indicate that whatever is built by ETI is never done piece-meal or on a small scale. Nearly every artifact found and photographed on the Moon by Apollo astronauts and other countries lunar orbiting satellites has been a technology with gigantic proportions! Are we dealing with a race of giant Extraterrestrials?

Is this lunar anomaly in the De Moraes Crater, an alien mining machine, a spacecraft, a building or just another moon crater?

Yet, there are many buildings and structures that appear relatively small comparable to the type of manmade structures and buildings found on Earth. The fact there are immense and massive lunar structures demonstrates a technology highly advanced in comparison to our own and the reason for their construction may be simply nothing more than they have the intellect and capability to do so. There is also a power curve ratio that is required to construct such miles high structures or to suspend and float a massive city building like the "Castle" above the lunar horizon. This is also indicative of a **Type I or Type II Civilization** according to the **Kardashev Scale of Galactic Civilizations** and no doubt the acquisition and introduction of such alien artifacts and technology into Earth base technology would represent a quantum leap in our development on many levels. The growth potential for mankind would be exponential boost us into a Type I civilization almost overnight!

Consider the fact that such alien technology exists on the Moon with building structures miles high made from some new composite of steel and glass as has already been stated before in the discovery of glass or crystalline cities. ET Intelligences taking advantage of the Moon's low gravity, that is one-sixth of that of the Earth, could easily build immense tall towers and cities without any concerns for buildings collapsing or being crushed under their own weight due to the forces of gravity.

Image some type of glass – steel composite material that has the same tensile strength of a silk spider's thread that is spun into a web to capture their insect prey. The spider's thread is thin, very strong, and capable of stretching almost twice its length with breaking. Its tensile strength is far greater than steel of the same size and proportions and steel certainly, does not have the ability to stretch to any degree of more than a few inches before snapping in two. By comparison, a spider's silk thread if built to the same proportions as a steel cable used in bridge building would be able to hold an almost exponential weight of tonnage and thus fewer "silk cable" strands would be required to suspend or support a bridge. That fact that silk spider thread is used in the Kevlar vests to protect police officer and military soldiers is proof positive of its tensile strength capabilities.

Could the ETs who built the immense lunar structures have figured out the chemical formula for a composite material that has the same tensile strength as a silk spider's thread? Only landing on the Moon, by examining and gathering samples of such technological structures will that question be finally answered.

We need more evidence to support such theories and speculations that alien bases, mining sites, and other alien artifacts do exist on the Moon which would be conclusive of an ET presence. We need some artifact or technology other than a lot of fuzzy and blurry photographs. It needs to be evidence that is in your face, undeniable and conclusive beyond a shadow of a doubt. Does such evidence exist? If we accept some of the evidence that is to be found on the internet and in books written by UFO investigators then that evidence does indeed exist!

Life Is Where You Find It – Like On the Moon and On Mars!

There are at least five major pieces of evidence that may be definitive proof that life has existed on the Moon in the distant past up to and until recent recorded times. Up to this point, we have to look at other evidence on the Moon in the way of NASA and Russian satellite and astronaut images of immense and tall lunar building structures. In most other similar situations, such photographic evidence of artificial structures on Earth would be conclusive and be regarded as real and without question. The fact that artificial structures exist on the Moon is open for debate and cover-up as per Brookings Report to NASA. Although, hundreds of thousands to millions of photographic images of the Moon exist in the archives of NASA and the Russian Space Agency, what has been seen in the public domain is fuzzy and not clear or definitive thus, it is open to conjecture and interpretation.

This evidence include microbial life fossilized in Moon rock that was returned to Earth in sterilized Russian sample capsules; photographic/video evidence of astronauts actually walking through the ruins of a lunar building; photographic/video documentation of the wreckage of an immense and very ancient alien spacecraft within a lunar crater and the recovery of a well preserved Extraterrestrial corpse from that huge alien spacecraft.

If any of this evidence is factual and authentic and not some clever prefabrication planted to throw off further investigation into this subject matter then, the very physical reality of this evidence would cause us to re-evaluate life on this planet, the understanding of our own recorded history and ultimately our position within the universe! This lunar evidence apart from any evidence we have accumulated over the last 100 hundred years on the UFO/ETI phenomenon here on Earth would be enough to confirm the existence of other intelligent life in the universe and certainly within our own Solar System.

Evidence Number One: Lunar Microbial Life

In recent times, there has been some controversy over whether or not the Apollo 12 astronauts Pete Conrad and Alan Bean may have brought back some Extraterrestrial life forms from their Apollo 12 mission in the way of microbial life. The astronauts had retrieved the Surveyor 3 camera from the probe and brought it back to Earth under sterile conditions by the Apollo 12 crew. When scientists analyzed the parts in a clean room, they found evidence of a small colony of microorganisms inside the camera.

The small colony turned out to be common bacteria -- *Streptococcus Mitis* -- had stowed away on the device and did not originate from the Moon as many people had thought. NASA went back into their dusty records and archives and found video evidence that indicated a less than sterile environment where the camera was examined. The NASA technician who investigated were not dressed in full body suits and helmets, only in shirt sleeves with white caps, face masks, and rubber gloves. This eventually necessitated much higher protocols in sterile cleanliness when dealing with potential bio-hazards that could originate from space or other planetary bodies or even from more complex biological life forms, namely ETs.

**The electron microscope revealed chain structures
in meteorite fragment ALH84001**
https://www.inverse.com/article/19572-bill-clinton-alien-mars-meteor-extraterrestrial-life

This type of controversy harkens back to the Martian meteorite (**Allan Hills 84001**) that was found in Antarctica in 1996 by a research group led by **Dave McKay** of NASA's Johnson Space Center which was believed to have "little" fossilized Martians… microbes that is, embedded within it.

Such Martian meteorites are believed to have been ejected from Mars by the impact of an asteroid or comet and landed on the Earth. As of January 9, 2013, there are 114 meteorites identified as Martian. These meteorites are thought to be from Mars because they have elemental and isotopic compositions that are similar to rocks and atmosphere gases analyzed by spacecraft (**Viking Landers**) on Mars.

Not only did the shapes look like bacteria, but a form of magnetite (iron oxide) was found in the meteorite that, on Earth, is produced within the bodies of certain bacteria. The study also found tiny carbonate globules in the meteorite, which the scientists said were likely formed by living organisms in the presence of liquid water. http://en.wikipedia.org/wiki/Martian_meteorite

Since their surprising announcement, the researchers' presentations were not met with any of the excited frenzy that greeted the original 1996 announcement about the meteorite -- which led to a televised statement by President Bill Clinton in which he announced a "space summit," the formation of a commission to examine its implications and the birth of a NASA-funded astrobiology program.

Other scientists have closely examined the **Allan Hills meteorite** and concluded the microscopic shapes aren't necessarily associated with life and the different features in the meteorite all could have formed by non-biological processes. The Mars meteorite "discovery" has remained an unresolved and somewhat awkward issue. This has continued even though the team's central finding -- that Mars once had living creatures -- has gained broad acceptance among the biologists, chemists, geologists, astronomers and other scientists who make up the astrobiology community.
http://www.washingtonpost.com/wpdyn/content/article/2010/04/30/AR2010043002000.html

Well, it would seem that many scientists have not bought into the recent finds that alien microbes have hitched a ride from Mars or some other parts unknown on board a piece of meteoric rock and ended up on Earth. They do not rule out the concept of **panspermia** as a real possibility as to how life may have begun on a planet but, they also don't acknowledge the probability that intelligent life may already be visiting our planet or living on our closest neighbor, the Moon.

Russians it appears have a whole different take on whether there is microbial life on the Moon and according to two biologists at the **Russian Academy of Sciences**; they say that this is not only possible but a confirmed fact. Lunar fossilized microorganisms were returned from the Soviet's Luna and Luna 20 mission core samples and there were strong similarities to spiral filamentous microorganisms found on present day Earth.

At an astrobiology conference in Denver in July 1999, the findings of biologists **Stanislav Zhmur** and **Lyumila Gerasimenko** were presented and then, published that year. Their announcement that carbonaceous meteorites containing biological microfossils from the Moon were passed off in the usual skeptical fashion by some critics as contaminants from terrestrial examination. http://www.theforbiddenknowledge.com/hardtruth/evidence_life_moon.htm

The Russian biologists refuted the claims that the microfossils are the result of contamination stating that the Luna samples were from robotically accessed drill-cores which were then hermetically encapsulated on the Moon. The samples survived re-entry whereupon Russian biologists quickly moved them into a laboratory and there, they opened and examined the samples.

"According to **Zhmur** "The lithified remnants ...are tightly conjugated with the mineral matrix, removing the possibility that they are contaminants." In other words, the microfossils are intertwined with the rock itself, meaning that they were entombed within the rock very shortly after it was formed or ejected. Most likely, they were frozen in time and place on the Lunar surface after originating elsewhere in the solar system".
http://www.theforbiddenknowledge.com/hardtruth/evidence_life_moon.htm

Given that the lunar rocks and core samples brought back by Apollo astronauts and the two Luna Soviet missions came from only 6 regions out of 15 million square miles and the core samples from separate regions of **Mare Fecunditatis** 120 miles apart, then this would imply that from just a few samples representing less than 1 millionth of the total lunar surface area surveyed and sampled, it stands to reason then, that the lunar sub-surface must be literally teeming with fossilized life!

This raises a big question, how did it get there and what processes could have occurred to explain terrestrial-like microorganisms in a lunar drill core? Is **Tom Van Flandern's Exploding Planet Hypothesis (EPH)** occurring on Mars the explanation for the arrival of bacteria on the Moon or Earth? In his theory, blast waves propagate surface crust and organic material across that system impacting other bodies and moons like our own. Bacteria somehow survive the initial blast from the planet and the journey across space and impact on the Moon which end in their fossilization in the lunar vacuum.

The fact that the Luna fossils resembled modern terrestrial **coccoidal bacteria** could mean that our Moon and Earth might have experienced such a dynamic process. According to Richard Hoagland's the indication for the presence of microbial fossilization on the Moon is evidence of a past lunar life on a much grander scale than microbes, knowledge of which was discovered many years prior to the announcement of Russian Luna microbial fossils.

Hoagland's lunar investigations through many NASA archives of lunar photography suggest that the Moon was not always an uninhabited satellite as it is today. Hoagland discovered in many of NASA's photographs numerous examples of geometric, orthogonal **arcologies**, many exhibiting familiar earth –like design elements which were constructed by an intelligence from a past epoch. Many other site locations have reinforced the notion that the Moon to some degree was once an inhabited body. Hoagland's research of glass domes in **Mare Crisium**, organized installations in **Hortensius**, and rectilinear structures in the region of **Kepler Crater** have all revealed evidence of a vast (though ruined) civilization there.

The information we got back from the Russian Luna probes helps to verify what we already know. The Moon could have very well been inhabited by spacefarers who had physical contact with Earth and returned to the Moon and perhaps other places, bringing with them the "contaminating" terrestrial organisms as well as the technological ability to exist there. Fossilized microorganisms did not, after all, construct what Hoagland (and we) have found on the Moon. http://www.theforbiddenknowledge.com/hardtruth/evidence_life_moon.htm

In reality, the microfossil findings on the Moon support *both* the EPH/Panspermia idea, as well as Hoagland's "prior habitation" model. However, there is no way to determine which scenario of how "Earth-like" bacteria made it to the Moon. Too many millions of years have passed since the cessation of the lunar "**Copernican Crater era**" (referring to the former lunar civilization). There has been incessant meteoric rain bombarding the Moon for eons leaving immense devastation on the lunar surface that hasn't changed since the time of the dinosaurs. But if you consider the political developments both before and after the samples were brought back - over thirty years ago - you can perhaps get a clearer picture.

The exploration of the Moon by the Soviets and by NASA has always been closely shrouded in secrecy and even denied for over 40 years. We've gotten snippets of information about it here and there, like the fossil-announcement. There has been an unsettling malaise in their attitudes toward the Moon. Neither country has returned to the Moon since 1972. Russia's first planned circumlunar flight of cosmonauts in December of '68 scheduled just before Apollo 8's historic flight around the Moon got cancelled.

Historically most people believed that a combination of events such as Russia's several major catastrophic booster and rocket failures, the assassination of **President Kennedy** who wanted to partnership with the Russian space program, disposition of the political leadership of **Khrushchev** with another leader waiting the wings, Kremlin's termination of the head of the Soviet lunar program. Some believe that Russia cancelled it space program after a major USSR robotic mission returned its high- resolution films back to Earth. Why? Could they have found something on the film that "paralyzed" their lunar program? The new successor of the Russian lunar program then proceeded to destroy most of that evidence in much the same way that much of our own Apollo legacy was disposed of, including photography, blueprints, etc. We know that NASA's lunar program was abruptly and prematurely cancelled. What went wrong with our space programs 40 years ago? We never went back. Why? *It was as if both space programs had been quarantined.*
http://www.theforbiddenknowledge.com/hardtruth/evidence_life_moon.htm

If there were any laboratory results found from returned samples and cores at the Apollo landing sites as there were at the Soviet ones at **Fecunditatis**, they haven't revealed that information and nothing can be found in any of the Apollo Scientific Reports. To that extent, NASA has even denied and tried to cover up the existence of lunar artifacts and structures for the last 40 years.

Did life come to the Earth or the Moon from Mars via **panspermia** and why couldn't a Moon fossil be an indicator of previous lunar terrestrial visitation? Full disclosure that intelligent life had existed whether on Mars or the Moon may never be revealed in our time unless the political climate in the US changes or some other nation involved in space exploration steps forward and discloses what they know on the world stage. According to many, including the writers of the Brookings Institution Report commissioned by NASA in 1959, our civilization would collapse by such revelations.

Disclosure is inevitable even if it comes piecemeal, bit by bit. It's just a matter of time before all of us will know. It may happen when we make the decision to go back to the Moon to really explore then; we will discover that we really were there once before. When that happens we will have to rewrite all our history books.

Evidence Number Two: Lunar Artifacts

The lead in this second piece of hardcore evidence was spearheaded from Richard Hoagland's investigation of NASA's archives of photographs and what has been released into the public domain. Since Apollo 8 began photographing strange anomalies on the Moon's surface up to the first manned Moon landing by Apollo 11 with its reported sightings of ET spacecraft hovering over craters and hilltops, it became apparent to NASA and the DOD that all following missions required a more covert agenda, while maintaining the appearance of public openness. Apollo 12 is considered the first manned mission to the Moon with a duel agenda that of actual lunar exploration coupled with documentation and collection of alien artifacts. These missions with dual agendas concluded with Apollo 17 and from that last mission came photographic images that seem to have slipped through the cracks of NASA censors.

From Richard C. Hoagland and Michael Bara's book "Dark Mission," they tell how Apollo 17 astronauts Schmitt and Cernan had an **EVA (Extra-Vehicular Activity)** to "**Shorty Crater**, a black halo rimmed crater on the outskirts of the light mantle material from the South Massif avalanche". Along the way, Schmitt and Cernan stopped at the rim of a small crater, later named **Ballet Crater** in honour of Schmitt after he had taken a major spill there (a real no, no in lunar walking activity) to retrieve some sample bags. This short stop over at Ballet Crater ate into valuable EVA time because of Schmitt's fall nevertheless; both astronauts arrived safely at Shorty Crater. It is at this point that some interesting conversation ensues:

> **145:22:22** Schmitt: Shorty is a crater, the size of which you know (about 100 meters in diameter.) It's obviously darker rimmed, although the fragment population for most of the blanket does not seem too different than the light mantle. But inside...Whoo, whoo, whoo!

> **145:22:38** Cernan: Man, are you going to get a picture now.

> **145:22:40** Schmitt: Oh, yeah.

> **145:22:41** Parker: We can hardly wait.

Schmitt's description seemed to imply that while Shorty was relatively unspectacular on the outside, the area inside the crater was at least very interesting. Unfortunately, when the camera started up, it was pointed at the rover and the distant South Massif. It stayed positioned there as Schmitt moved away to take a panorama of the crater. Several minutes into this sequence,

Cernan oddly states "O-kaay! O-kaay." At this point, Schmitt begins to discuss something odd he noticed through his visor. Raising his filter, he suddenly absorbed what he was seeing.

145:26:25 Schmitt: Wait a minute...

145:26:26 Cernan: What?

145:26:27 Schmitt: Where are the reflections? I've been fooled once. There is orange soil!!

145:26:32 Cernan: Well, don't move it until I see it.

145:26:35 Schmitt: (Very excited) It's all over!! Orange!!!

145:26:38 Cernan: Don't move it until I see it.

145:26:40 Schmitt: I stirred it up with my feet.

145:26:42 Cernan: (Excited, too) Hey, it is!! I can see it from here!

145:26:44 Schmitt: It's orange!

145:26:46 Cernan: Wait a minute; let me put my visor up. It's still orange!

145:26:49 Schmitt: Sure it is! Crazy!

145:26:53 Cernan: Orange!

145:26:54 Schmitt: I've got to dig a trench, Houston.

Dark Mission: The Secret History of NASA by Richard C. Hoagland and Mike Bara; 2007; a Feral House Book; Los Angeles, CA; ISBN: 978-1-932595-26-0

The astronauts then begin to sample the orange soil, which was later found to be highly oxidized, a discovery which had tremendous implications for later colonization of the Moon. Extracting oxygen from the lunar surface would make the idea of a permanent lunar base much more viable. After the scooping and core samples were taken, Schmitt moves off to the side to take numerous images of the interior of the crater. In some of these images, strange objects can be seen which do not resemble the fractured, volcanic rocks which would be expected at this site.

Instead, they look like large chunks of broken machinery. Shortly before leaving for the next station, Schmitt stops and takes several high-resolution pans of the inside of Shorty crater and the area around it. Intrigued by other images of mechanical looking debris and orange soil in the area, we obtained early generation negatives of the pans and subjected them to processing and color enhancement.

In doing so, it became clear that some of the debris in Shorty was unusual, to say the least. Color enhancement showed that many of the "rocks" had highly unusual spectral qualities, reflecting

271

light more like crystals or highly polished metallic boxes than a simple "rock garden" would suggest.

As we scanned the center of Shorty, we noticed a very large and strange looking artifact that strongly resembled a pump mechanism or engine housing. Nicknamed "**the turkey**" because of its odd resemblance that terrestrial creature, this object appears to have a series of tubes and mechanical features extending from a geometric, metallic case. There are even what appear to be forged connectors or mounting points on the object.

The unusual features in the crater had Hoagland inspecting every rock and oddity of light and shadow on the crater floor and walls beyond d the "turkey" when his gaze fell upon something that did not belong in the picture. Even as he studied it, he couldn't believe what he was looking at; its appearance was all too familiar, like something out of a museum or even a movie. **Dark Mission: The Secret History of NASA by Richard C. Hoagland and Mike Bara; 2007; a Feral House Book; Los Angeles, CA; ISBN: 978-1-932595-26-0**

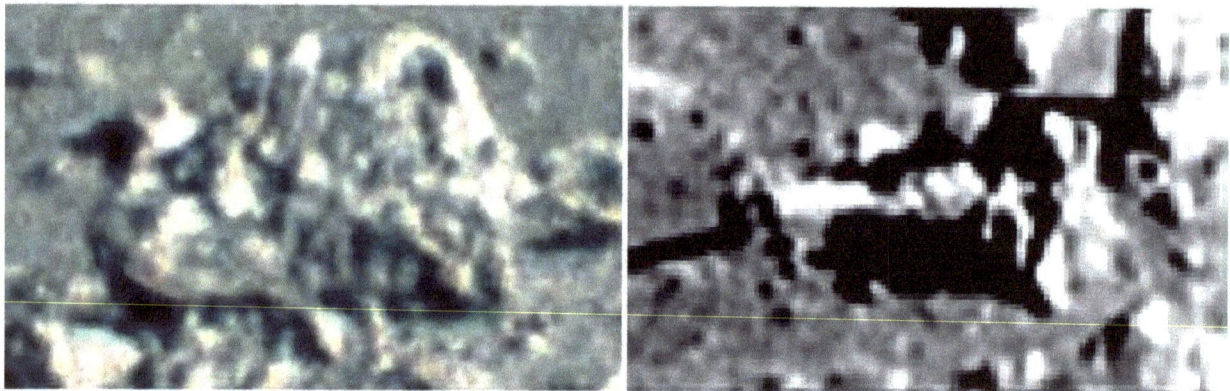

"The turkey" (left) found in Shorty Crater appears to be a mechanism similar to a pump or engine housing with tubes and connectors. Bits of Machinery (right)
http://www.enterprisemission.com/datashead.htm

Staring up from the crater floor appeared to be what looked like a human skull. How was this possible? After all, it was lying in a debris field from an impact crater, which had tossed up all manner of junk and material from just below the regolith of the valley floor. Organic matter, even a fragile fossilized bone would disintegrate from such an impact or from exposure to severe solar and cosmic radiation. Given that everything else in the crater was either mechanical in origin or moon rock, its artificiality had to be also, of mechanical origin.

The object in question was it a robot's head?

Schmitt and Cernan's entire EVA was one of astonishment; they could scarcely believe their own eyes at what they were seeing. Cernan even dubbed the entire valley as "one mysterious looking place." Did Cernan and Schmitt see a waste pit of mechanical debris when they gazed into the crater at Nansen? Were the photographs taken by them stashed away for later analyze? Was Shorty simply another example of the kind of "unbelievable" things they had seen all along on this second EVA? http://www.enterprisemission.com/datashead.htm

272

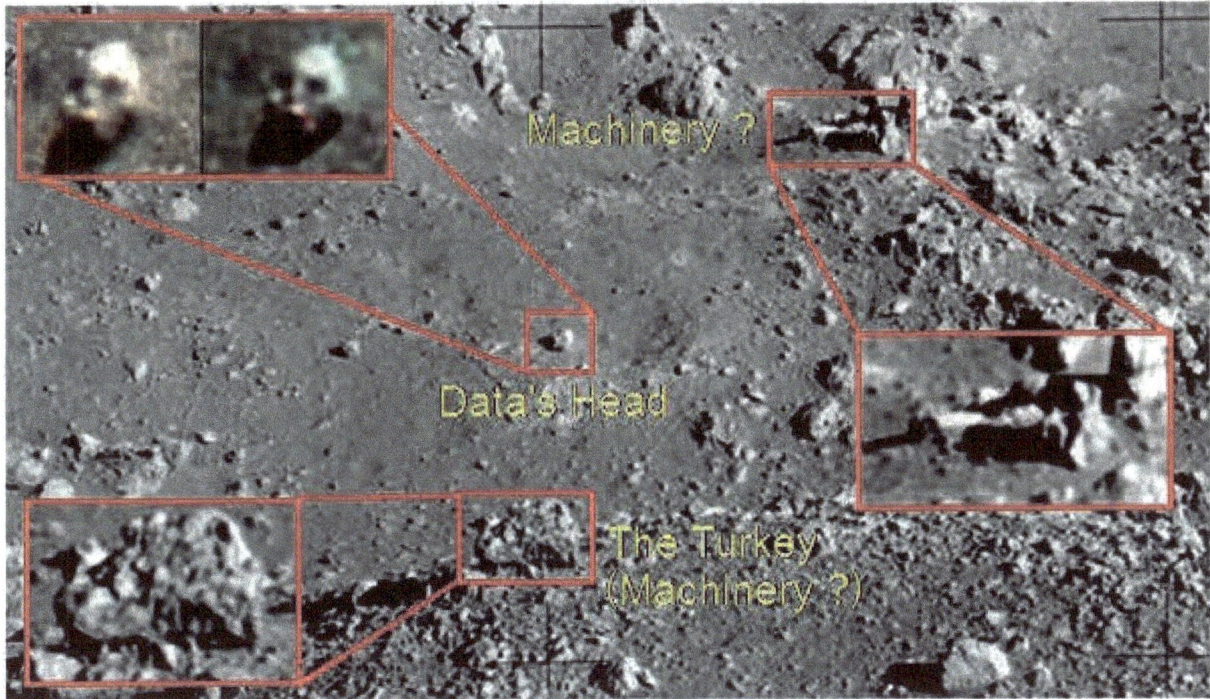

**Photograph AS17 -137 – 20997HR of Shorty Crater shows not only "the turkey"
but, also "Data's Head" and some odd bit of machinery**

https://www.hq.nasa.gov/alsj/a17/images17.html#MagC and https://www.hq.nasa.gov/alsj/a17/AS17-137-20997HR.jpg

Color enhancements showed that the "head" had a distinctive red stripe around the area where the upper lip should be, a feature that clearly appeared to be *painted or anodized* on the object. Composites of other frames showed that the head had two eye-sockets, a forehead, brow ridges, a nose with nostrils, twin cheekbones and the upper half of the jaw. The "lower jaw" seemed to be missing. Still, it was an astonishing photographic find. And the resemblance to another, even more, familiar figure did not escape the authors…

What was most striking about the C3-PO comparison – and most telling – was the *eyes*. Like C3-PO, our robots' head had indented stereoscopic, *rounded* eyes. Camera lenses. Just like C3-PO. http://www.enterprisemission.com/datashead.htm

In looking at the context panoramas from which it was taken, Hoagland was able to confirm that the head was approximately the same size as a human head, Which meant, among other things, that they could have brought it back.

The transcripts for the **Shorty EVA** show that the astronauts were certainly rushed at this station because of the time they had lost at Ballet crater. It is possible that **Schmitt** and **Cernan** never saw the object in question, or that they decided it would be too risky to try and retrieve. However, they certainly had enough off-camera time to descend the crater and retrieve it if they wanted to. http://www.enterprisemission.com/datashead.htm

273

**Color enhanced version of "Data's head " in Shorty crater.
Red stripe is not an artifact of image processing**

**From left to right, Lunar Data's Head, CP-30 from the movie
Star Wars and the actual Data's Head from Star Trek: Next Generation**

In looking at the above images, one is reminded as were Hoagland and Bara of a Star Trek, the Next Generation episode called "Time's Arrow." In it, the Enterprise is summoned to 24th century Earth to an archeological dig below San Francisco. In this dig, **Captain Picard,** and **Commander Data** – an android -- are shown a puzzling artifact. Mr. Data's disassembled robotic head.

In the course of the story, **Data's head**, and the information contained in his "positronic brain," are crucial to unlocking the mystery of Earth's past. By tapping into the memory of this ancient and damaged artifact, the crew of the Enterprise is able to stop human history from unraveling and their very existence from being threatened.
http://www.enterprisemission.com/datashead.htm

Hoagland postulates that like the story from Star Trek, a similar circumstance may have occurred in which the lunar "head" artifact may also have contained information similar to a hard drive storage unit that is commonly found in computers. This may have been the great secret of Apollo 17 therefore; it would be worth the time and effort of the astronauts to retrieve this object. Hoagland also wonders if this was the reason for Cernan's odd behavior at the NASA hero ceremony expecting that NASA would reveal the discovery to the world and but never did. Cernan was incensed that he was being asked to participate in still another cover-up on the American public and became yet one more astronaut to speak out against NASA policy to hide and suppress discoveries found on the Moon.

This alien artifact is exactly the kind of technological evidence NASA has been looking for and it is small enough and easily transportable to bring it back to the Earth for further lab studies; no doubt there must have been other similar small collectible artifacts that could also have been brought back as well. As we shall see the **lunar Data's Head ET artifact** is not the only head relic or artifact of its kind and the evidence could very well be Data's brother Lore's Head which this author has discovered on a recent photo taken by the newly arrived **Mars rover Curiosity.** The picture is much clearer image than the Data's Head artifact and appears to have all the same qualities as the lunar head artifact. Stay tune to when we visit Mars and all its alien anomalies!

Evidence Number Three: Lunar Building Ruins

This next piece of evidence is a piece of video that has circulated around the internet that gives credence to the many structures and buildings that have been seen and photographed by satellites and astronauts. It purportedly shows Apollo astronauts walking through a lunar building that appears to be a ruin with walls still standing but apparently with no roof structure. Many planetary anomaly investigators feel that this video is legitimate evidence of a former ET structure on the Moon, while few researchers acknowledge it as a clever hoax perpetrated to add further confusion to what may or may not actually exist on the Moon's surface and to further discredit the credibility of the whole UFO/Alien phenomenon.

The mere fact that such a ruse may have been used only adds to the legitimacy of UFO/ETI research whether it is found here, on Earth or on some other planetary body. This disinformation campaign actually has the opposite effect of making those who try to debunk the subject matter, even less credible in the eyes of the public and supports further the claims of an on-going cover-up and truth embargo on the UFO/ETI phenomenon.

The video essentially shows what are believed to be the Apollo 11 astronauts Neil Armstrong and Buzz Aldrin filming the interior ruins of a building as they walk through doors into rooms and peer through windows to the outside. One astronaut is seen near the end of the video pawing through rubble on the ground with a shovel or scoop, instead of simply picking up samples with it as some investigators believe he should have done which lends credence to their claims of a hoaxed scenario. Another aspect to the hoax scenario claim is the life support backpacks do not conform to the standards of Moon EVA missions. There is also the way that the astronaut who is videotaping walks through the lunar building too smoothly and appears to walk through or over rubble and rock as if it wasn't there. The video is grainy which is what you might expect from that era of video recording with non-digital cameras, but this poor quality video and audio has been identified as a VHS recording made from a television set by someone. The video was aired on a Mexican television station is also a VHS copy from someone within or who had affiliations to the black-ops, who had the original astronaut version. In other words, we are looking at duplicates or multiple generational copies, covertly made since the original was first duplicated. Each copy of a copy becomes more degraded with successive copying to the point that fidelity, colour, sharpness, and detail, as well as audio, all become lost beyond retrieval.

There is also a moment very briefly when letters appear in the upper left corner ERT1 or ET1 of the video frame that some people feel that this too supports that a hoax. In actual fact, this is a logo for a Greek television station in the upper left of the screen at one point, and *that doesn't mean it's a fake!* The Greek station had also received a copy of the lunar footage and put their logo on it; whoever was filming from their own TV apparently had stability problems with their own camera, thus the poor quality in recording and the final version with which we are left with to look on in amazement.

The original astronaut version appears to have had it own set of problems as well, there was so much brightness on the stones exposed to the Sun while recording that the camera had difficulty adjusting to light exposure, the result is the viewer has a hard time making out what's on the video. Fortunately, we get more than an eyeful to astonish us!
http://www.bibliotecapleyades.net/luna/esp_luna_32.htm and
http://www.youtube.com/watch?feature=player_embedded&v=MWkGTJEK0Mc also,
http://www.youtube.com/watch?v=NH7XPMWc8ig

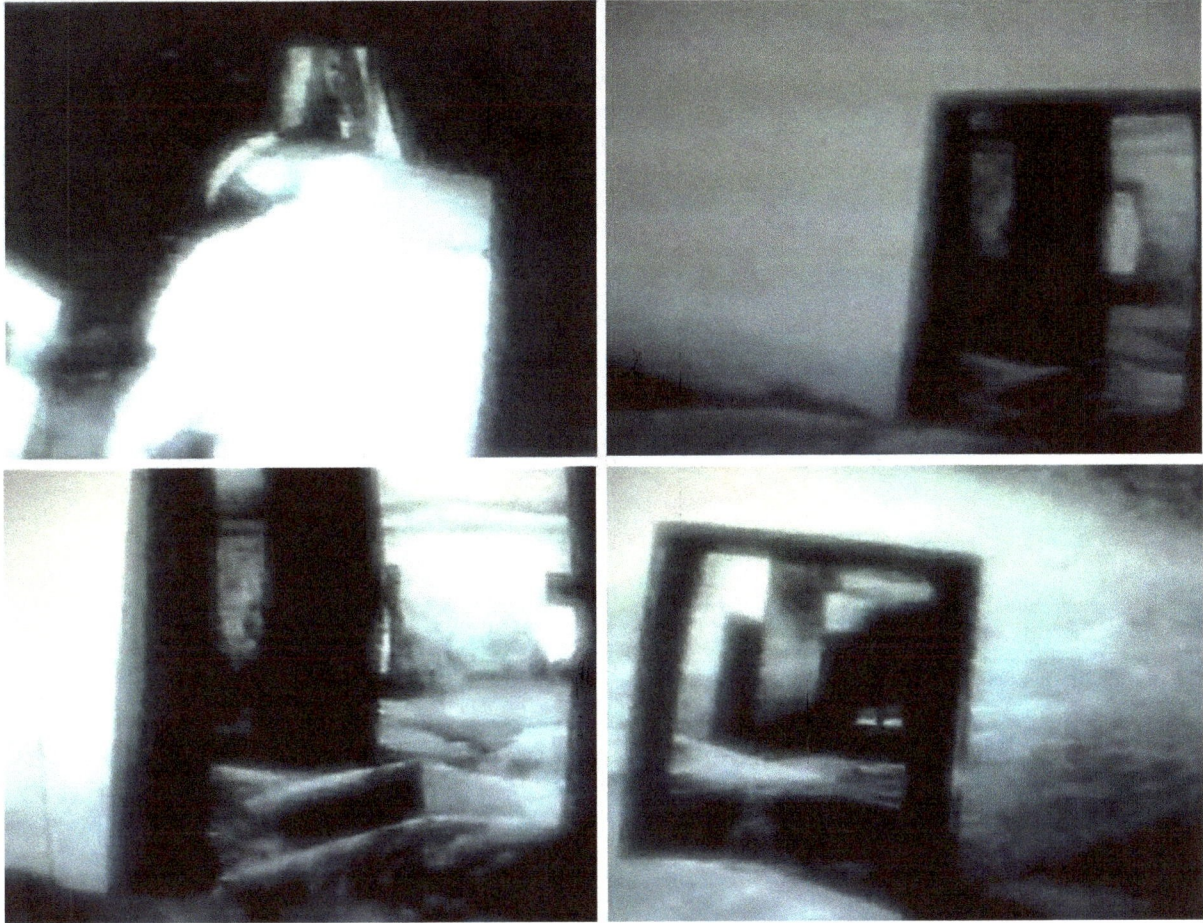

In the first image, you see a fellow astronaut exploring the ruins. In the next photos, (all of which are taken from the video), above and bottom left, show two stages of the astronaut's advance to a doorway inside the megalithic stone structure. You can see massive rectangular pillars inside and square windows looking to the outside. In the last image (right) the astronaut has now walked through the doorway, turned around and filmed the way he came in from. Notice the similarity of the structures from either side and the hugeness of the whole complex.

https://www.bibliotecapleyades.net/luna/esp_luna_32.htm

The fact that Apollo 11 astronauts filmed their tour through old lunar ruins is cause to question what was really seen by the world's general public on that eventful day on May 23, 1969. Did the public seen the actual Moonwalk by the first two men on the Moon or was there a cut away to the **"lunar sound stage"** that was discussed earlier in this section when the Apollo 11 astronauts went off to explore ancient lunar ruins? If this is the case then, astronaut actors would carry off the supposed normal, uninteresting Moonwalk that would be transmitted *"live"* for public consumption. **That folks is how it's done!**

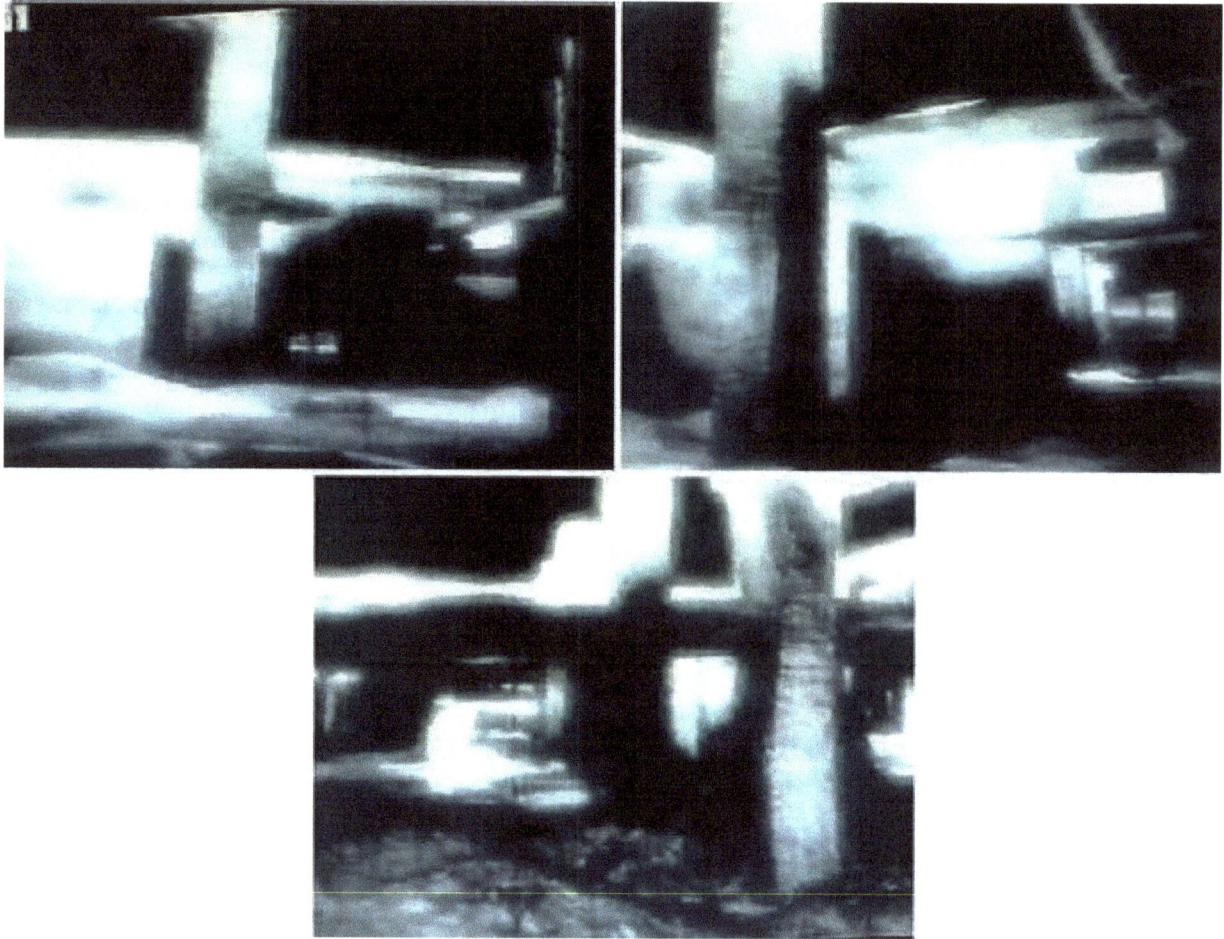

In the next frame, you see a bit of the logo (top left photo and at bottom of page), and there's an article on Rense's Sightings where you see the whole logo and a discussion about it. In the bottom photo, the astronaut is standing at the farthest point away from the complex that he can get and is filming a panoramic view.

https://www.bibliotecapleyades.net/luna/esp_luna_32.htm

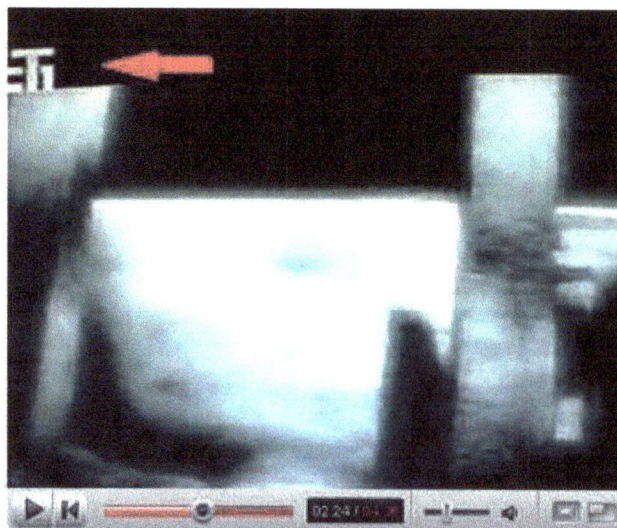

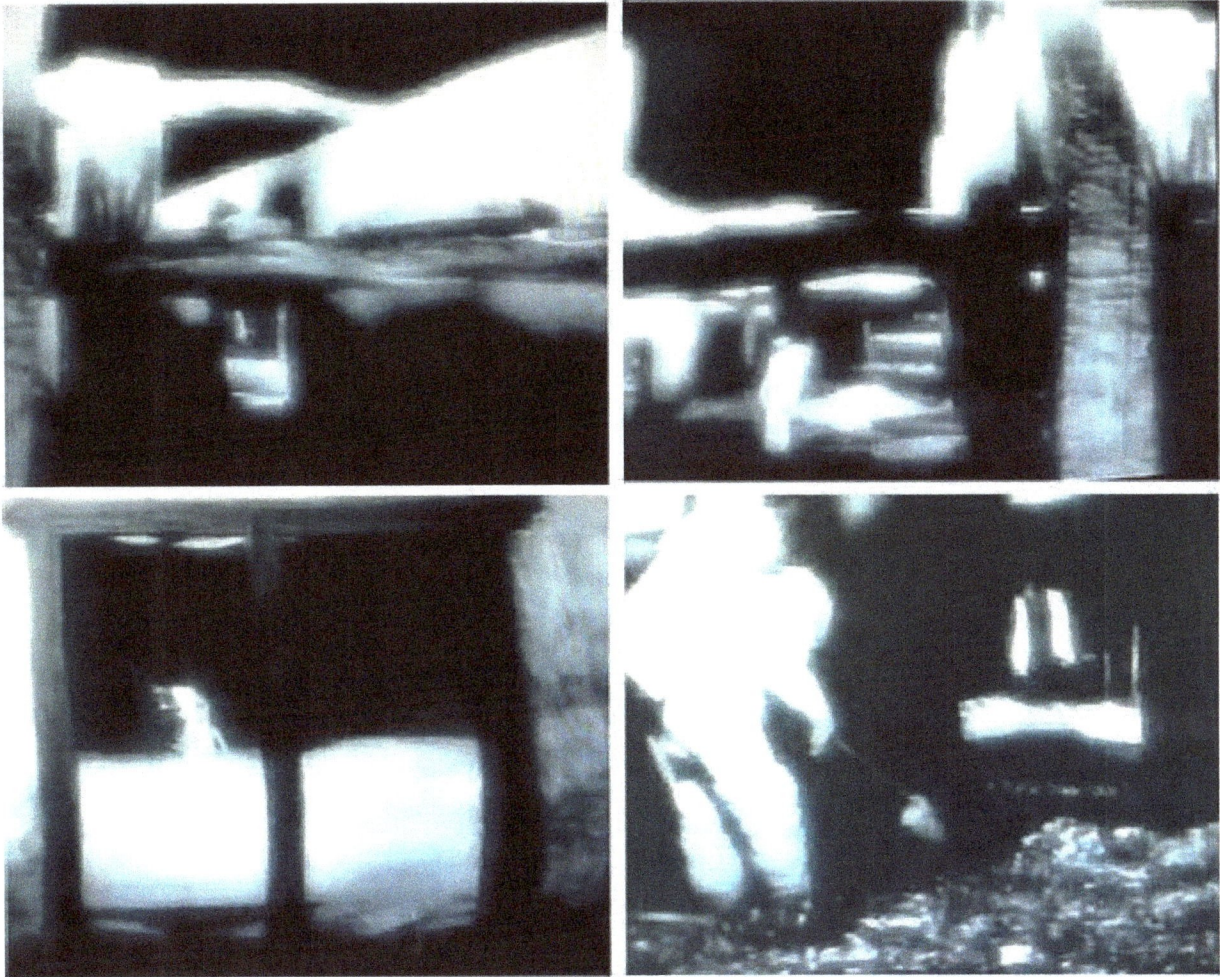

In the top left image, the astronaut has panned to the right from the last shot, and the jutting area may be a landing dock for Atlantean-era craft or merely a second-floor living area. Here you get a great look at the redundant, multi-tiered geometry to the whole complex - truly stunning. Notice the astronaut standing in the center of the frame on the opposite side of the wall for a size reference. In the bottom left image, the astronaut is filming through a stylized window, which could be made from a carved stone material, gazing back at what appears to be the lunar lander. In this last shot, near the end of the video, a fellow astronaut is taking soil samples probably to see if the area was radioactive or had any rare minerals in the surrounding soil.
https://www.bibliotecapleyades.net/luna/esp_luna_32.htm

Credit should go where credit is due and therefore, thanks goes to an individual with a high interest in the Moon anomaly phenomenon by the internet or blog name of **"NAlexo".**

Some of the comments from "NAlexo", the man who posted it, shed more light on this intriguing account:

"This structure was situated north of Moltke Crater, at the Sea of Tranquility.
NASA got its first close inspection through Surveyor 5 in 1967 collecting 19,000 photos of it. It (the lunar building) was 60m long with 12 windows on each side. The materials' consistency was 60% oxygen and 0.01% hydrogen. The prevalent wear of the walls along with the lack of atmosphere suggests an age of thousands of years. This along with other remnants were nuked for *not disturbing our social upheaval.*

Armstrong's heartbeat reached 160bpm. NASA's explanation was that he was carrying 25kgs worth of samples, yet these are no more than 4kgs on the moon. Initially, they were amazed by an array of landed discs perfectly aligned and tall beings in white suits. This is all recorded in their transmission to Houston which advises them to turn around and start working. Who do you think is going to go broke if existence of advanced noble species breaks out and we all unify on this planet?

[This footage] comes from a very credible non-US documentary, which explains *the TV logo on the top left corner* and the average resolution. I believe everyone should know [this].
http://www.bibliotecapleyades.net/luna/esp_luna_32.htm

(This also accounts for the broken English in the comments. It is not likely that the ruins were nuked; this appears to be public disinformation to stop anyone else from investigating the lunar site as there would supposedly be nothing there to see because it had been destroyed. Similar stories arose that one or more Apollo missions had carried a nuclear device on board the spacecraft to be used to nuke some area on the Moon. It would be a demonstration of the US military capability and supremacy but, their plans were thwarted by ET interference - Apollo 13 comes to mind as a near fatal mission with all its associated mechanical problems!) [Bold italics added by author].

It is believed this is but one of many such ruins on the Moon that have been explored and continue to be explored in similar fashion.

Multiple witnesses have come forward to claim that the "***secret government***" has its own large base on the moon called ***Luna***; even the Russians confirm the existence of this Moon base. It would thus, be in the realm of probability that such areas have now been extensively investigated and any usable alien technology has been mined.

Evidence Number Four: Ancient Lunar Spacecraft and Apollo 20

It should be stated here, if it is not already obvious, that each succeeding piece of lunar evidence stated thus far, admittedly stretches the boundary of credibility and increases the level of skepticism, demanding from the reader, even more, acceptance for the evidence presented before him or its outright denial.

(The reader must decide what is real and what is not by placing it within the context of everything else in this book. What is being presented here, is the best available information on the subject matter, hopefully with as little or no disinformation or misinformation creeping in, therefore, this book is a continuing work in progress, ever evolving!)

If being the first men to land on the Moon wasn't mind-blowing enough for any astronaut, exploring the interior of a lunar ruin must have sent Armstrong and Aldrin over the edge! According to "NAlexo", it did, as Armstrong's heartbeat raced with excitement and awe! Here was a building, probably one of thousands on the Moon constructed by another intelligence, built hundreds of thousands of years ago perhaps even, tens of millions of years, when mankind was supposedly still living in caves or in the trees with barely an IQ above an ice cream sandwich.

Image the surprise NASA must have had when Astronauts had photographed a small insignificant lunar crater that would normally have been overlooked by imaging experts but, something about this crater caught the eye of one photo imaging technician who immediately informed his superiors. NASA had scanned hundreds of thousands of lunar images and discovered anomalies covering much of the Moon's surface and therefore, there may have been an on-going search for anything that appeared to be out of place even in the smallest of craters. Their patience and diligence paid off when Apollo 15 astronauts, photographed an area on the far side of the Moon in the **Delporte-Izsak region**, of what appeared to be ancient wrecked spaceship not of human origin. The spacecraft was long and tubular like a submarine or missile. It appeared to be partially buried and broken, pitted with meteor craters. Its final resting place is within **Izsak D Crater**, southwest of the **Delporte Crater** where it has been accumulating moon dust for eons.

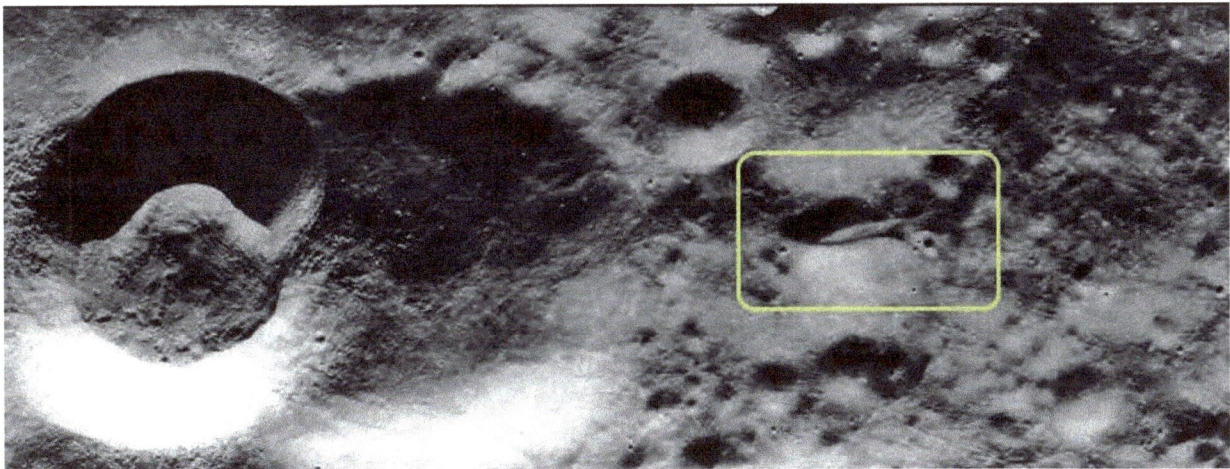

AS15-P-9625, Apollo Image Atlas (Courtesy NASA/LPI)
http://www.fallwelt.de/welten/mond/vorApollo20.htm

In Apollo 15 photos, it initially appears as a tiny, slender cigar-shaped object, its length is about the same as the crater approximately above it. From photo images, its true length is estimated to be 3370 meters long and 510 meters in height! The cigar-shaped object looks like it leans beside a crater, slightly oblique; and the visible zone in the footage is very similar to the visible region in some originals NASA photos taken by Apollo 15. The lunar coordinates where the spacecraft rests is Latitude: 10° S and Longitude: 117.5° E, Southwest of Delporte and North of Izsak. This area was also re-photographed during the Apollo 17 mission.

This is another lunar story that has circulated the internet since April 2007 and like the **Lunar Building Ruins** discussed above, it too has its supporters and detractors. The mission was a

Soviet-American partnership known as **"Apollo 20"**. Its crew consisted of **William Rutledge CDR** (from whom this story originates), formerly of **Bell Laboratories** and USAF, **Leona Snyder CSP** also from Bell Laboratories, and **Alexei Leonov**, a former Soviet cosmonaut with the **"Apollo-Soyuz"** mission, one year earlier.

The alleged photo (enlarged) taken (1976?) by covert Apollo 20 crew of the huge ET spaceship in Izsak D Crater on the dark side of the Moon
http://s289.photobucket.com/user/adventist123/media/AS20-1020.jpg.html

It has always been believed that the last official space mission to the Moon was the crew of Apollo 17 consisting of **Commander Eugene Cernan**, **Command Module Pilot Ronald Evans**, and **Lunar Module Pilot Harrison Schmitt**, which took place in December 1972, and all succeeding Moon missions including the Apollo 20 mission was cancelled by NASA in January 1970. http://en.wikipedia.org/wiki/Apollo_17

If NASA cancelled all subsequent Moon missions after Apollo 17 then, it would stand to reason that Apollo 20 and by its sequential position infers that Apollo 18 and 19 never took place either which means this may be another disinformation campaign either by NASA and/or the DOD. It could also mean that the DOD in secret cooperation with NASA and in partnership with the Russian space Agency sent its own military astronauts to the Moon to retrieve alien artifacts and exploration of some of the larger ET structures like the immense spacecraft in Izsak D Crater.

282

Side Note: Debunkers and skeptics love this type of spurious and improvable information because it plays into their public debunking agenda towards entire unidentified flying objects and the ET subject matter. This, however, does not mean that there is a disinformation campaign of a secret space program by NASA. Remember that NASA is not a civilian agency but a military agency and recall Richard Hoagland's eyewitness account of NASA propagating its own hoax lunar landings program which the media and a percentage of the public bought into, years later.

The DOD is quite capable through black ops space programs to launch their own spacecraft into space or even to the Moon, with or without NASA's approval or cooperation. The general public for the part consider Cape Canaveral and the Kennedy Space Centre in Florida as the only launch sites of moon rockets in the U.S. but, in reality, there are a number of launch facilities, all military, that can launch huge moon rockets like the Saturn V rocket.

Spaceports operate around the world, offering different capabilities and scales of operation - ranging from a basic control center, transportation infrastructure and launch platform to highly sophisticated facilities such as **Kennedy Space Center (KSC)** in Florida. KSC and the Russian Bikaner Commodore are the oldest and most advanced. But, they are not alone. Following is a list of spaceports in the U.S.

Orbital Spaceports: Launch sites for manned and unmanned missions, able to put payloads into Earth orbit and beyond; available to civilian and military government customers, commercial companies, and non-profit organizations.

- Cape Canaveral Air Force Station, Florida, USA
- Kennedy Space Center, Florida, USA
- Vandenberg Air Force Base, California, USA
- Wallops Flight Facility/Mid-Atlantic Regional Spaceport, Virginia, USA
- Kodiak Launch Complex, Alaska, USA
- Reagan Test Site, Kwajalein Atoll, Marshall Islands (USA)

Suborbital Spaceports: Launch facilities that can place objects in space but not at orbital velocities; generally lack complex launch pads and ground equipment, or have geographic restrictions such as an inadequate buffer area between their facilities and populated areas; lower costs and fewer restrictions make these facilities attractive to private investors in new space programs focused on suborbital spaceflights, such as personal spaceflight.

- California Spaceport, California, USA
- Spaceport America, New Mexico, USA
- Mojave Spaceport, California, USA http://www.spacefoundation.org/media/space-watch/spaceports-can-be-found-throughout-world

In particular, Vandenberg Air Force Base in California is considered a secret launch facility because it is military, while Edwards Air Force Base and Mojave Air and Space Port are Runway facilities for returning spacecraft, all in California. Mojave Space Port can launch manned sub-orbital vehicles but not heavy moon rockets. Area 51 is also a return Runway facility, though it is considered to be so secret as to be non-existent! (Yeah, right! It's the best kept secret military base known to the public!!)

A launch at Vandenberg Air Force Base, in California
https://militarybases.com/vandenberg-afb-air-force-base-in-lompoc-ca/

Much of the following information comes from **Luca Scantamburlo**, a well-respected writer/journalist in Italy during an internet interview with former Apollo 20 **CDR William Rutledge**, who at the time of the interview was living in Rwanda, Africa. As Scantamburlo states there are a few inconsistencies in William Rutledge's memories that you can find in reading this interview. It's understandable if you think about an old man (76 years old) who is trying to recall historical facts and events of his life.

After many years, he has become a whistleblower on NASA's secret moon missions. He was not immediately employed by NASA but, by the USAF for his work on the study of Russian foreign technology like the N1 project, AJAX plane project and the Mig Foxbat 25. Rutledge had computer navigation skills and was a volunteer for the MOL-Gemini project and chosen later for **Apollo 20** because he had a non-belief in God which was a rarity for a pilot but, essential criteria in 1976 for NASA astronauts. Hoagland also states this aspect of astronaut recruitment in his book "Dark Mission" that many of NASA's astronauts were 33rd degree Masons as was Buzz Aldrin.

 Rutledge was asked about some of the grammatical errors on the captions in the video of the flyover of the huge spaceship in Izsak D Crater and that the poor quality of the video was due to the fact that "films are not the first generation, some of them were copied in 1982..., some have a

284

blue background from the end of the 70's".
http://www.angelismarriti.it/ANGELISMARRITI-ENG/REPORTS_ARTICLES/Apollo20-InterviewWithWilliamRutledge.htm

Rutledge states that 300 people were involved in the preparation of the Apollo 20 launch which was from Vandenberg AFB and many more people witnessed departures of the **Saturn 5** or **Saturn1B rockets** into the sky. Personal cameras were forbidden all around the Vandenberg site in those days of 1976, but today a lot of Space spotters film every launch of Delta rockets from nearby towns.

Russian collaboration in the "Apollo 20" involved **W.R. James Chipman Fletcher** for USA and **Valentin Alexeiev** for Russia, **Werner Von Braun** monitored the event and **Charles Peter Conrad** and **James Irwin** were in the Capcom positions.

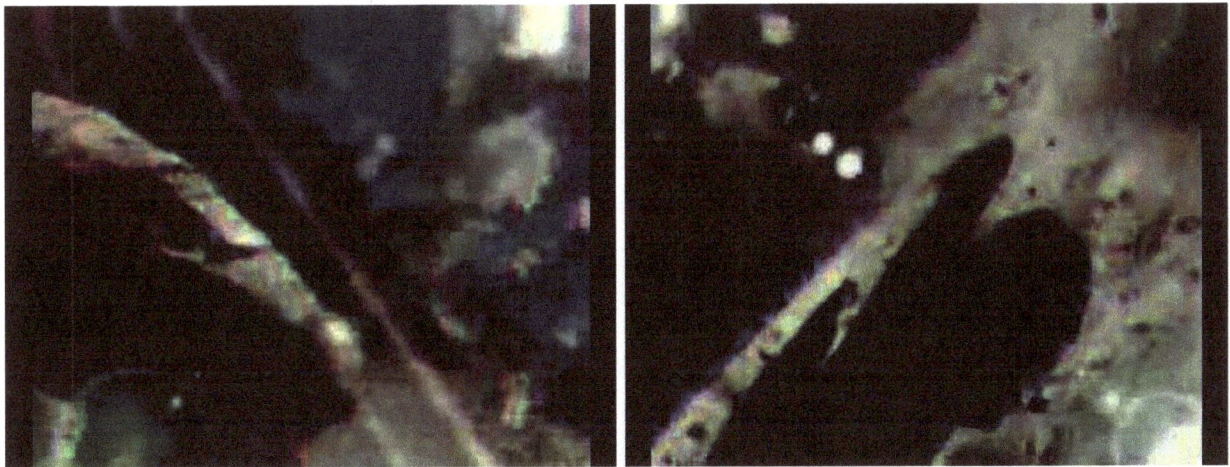

Frames from the "Apollo 20 flyover" on a presumed alien spaceship, on the backside of the Moon
http://www.angelismarriti.it/ANGELISMARRITI-ENG/REPORTS_ARTICLES/Apollo20-InterviewWithWilliamRutledge.htm

Scantamburlo asks about the **Apollo 18 and 19 missions**, especially about the last one and its failure, to which Rutledge replies:

"Apollo 18 was the Apollo-Soyuz project, the honeymoon before a moon landing mission, it was presented as a simple "shaking hands" mission in 1975. Apollo 19 and 20 were hazardous missions. On long duration flight, the helium pressure was too high on the LEM, a security disk had to burst if pressure was going high, but motor was unusable after. So it was changed on Apollo 19 and 20, but not tested in Space before. It was ok, but… in the paper. However, we got no problem with it. It was a long mission, 7 days scheduled on the Moon; every ray of light was used 'till ascent.

Apollo 19 had a loss of telemetry, a brutal end of mission without data. Now the truth is unknown but it seems that it was a natural phenomenon, a collision with a "quasi-satellite", like Cruithne, or a meteor (the probability is higher I think). The goal was the same, the landing site was the same, the exploration program was different, they had a big job to do with the rover, exploring the roof of the ship by climbing on the "Monaco hill", (I'll have to put a lunar map online). NO American astronaut is listed; I discovered since May that many people find many William Rutledge in NASA. I can be found in the list of the Test pilots of Chance Vought, on the consultant list of the James Forrestal Center, I was involved in fluid mechanics. My boss was Bogdanoff (nothing in common with the Bogdanoff scientists)."

"Russian collaborations; I don't know how, but Russian were informed of the presence of a ship on the far side. Luna 15 in July 1969 crashed just at the South of the nose of the ship. It was a probe similar to Ranger or Lunar Orbiter. They provided maps, precise charts of this area. The center of decision was located in the Ural, in the town of Sverdlovsk. The chief of the program was Professor Valentin Alekseiev, who became later president of the academy of science in Ural. Leonov was chosen because of his popularity in the communist leading staff and secondary only because he was on Apollo-Soyuz. In 1994, I met again Valentin Alekseiev in Ural, Yekaterinburg, and he had a model of the spaceship made of malachite with incrustations of gold on his desk." (Bold italics added by author for emphasis).

http://www.angelismarriti.it/ANGELISMARRITI-ENG/REPORTS_ARTICLES/Apollo20-InterviewWithWilliamRutledge.htm

The balance of the interviewed is given here in toto as it appears on the above website:

16) L.S. *Now we can discuss the ancient "alien spaceship" and "the City" on the farside of the Moon. Did you go inside the spaceship? How big was it and what did you find inside?*

W.R. *We went inside the big spaceship, also into a triangular one. The major parts of the exploration was; it was a mother ship, very old, who crossed the universe at least milliard of years ago (1.5 estimated).* **(A milliard = 1 billion, therefore, the estimated age of the spaceship based on meteorite impacts and age of the moon rock in the general vicinity of the spaceship is between 1 to 1.5 billion years!)** *There were many signs of biology inside, old remains of vegetation in a "motor" section, special triangular rocks that emitted "tears" of a yellow liquid which has some special medical properties, and of course signs of extrasolar creatures. We found remains of little bodies (10cm) living in a network of glass tubes all along the ship, but the major discovery was two bodies, one intact.* (Bold italics added by author).

17) L.S. *Did you visit "the City" on the Moon? Where was it? Did you understand if there was a connection with the Spaceship? Are "the City" and "the Ship" still there?*

W.R. *The "City" was named on Earth and scheduled as station one, but it appeared to be a real space garbage, full of scrap, gold parts, only one construction seemed intact(we named it the Cathedral). We made shots of pieces of metal, of every part wearing calligraphy, exposed to the sun. The "City" seem to be as old as the ship, but it is a very tiny part. On the rover video, the telephoto lens make the artifacts greater.*

18) L.S. *What about the "Mona Lisa EBE"? [The correct Italian name is "Monna Lisa"] How does she look like and where was she at that time, when you found out her on the Moon. Where do you think she is now?*

W.R. *Mona Lisa – I don't remember who named the girl, Leonov or me - was the intact EBE. Humanoid, female, 1.65 meter. Genitalized, haired, six fingers (we guess that mathematics are based on a dozen). Function; pilot, piloting device fixed to fingers and eyes, no clothes, we had to cut two cables connected to the nose. No nostril. Leonov unfixed the eyes device (you'll see that in the video). Concretions of blood or bio liquid erupted and froze from the mouth, nose, eyes and some parts of the body. Some parts of the body were in unusual good condition, (hair) and the skin was protected by a thin transparent protection layer. As we told to mission control, condition seemed not dead not alive. We had no medical background or experience, but Leonov and I used a test, we fixed our bio equipment on the EBE, and telemetry received by surgeon (Mission Control meds) was positive. That's another story. Some parts could be unbelievable now, I prefer tell the whole story when other videos will be online. This experience has been filmed in the LM. We found a second body, destroyed; we brought the head on board. Color of the skin was blue gray, a pastel blue. Skin had some strange details above the eyes and the front, a strap around the head, wearing no inscription. The "cockpit" was full of calligraphy and formed of long semi hexagonal tubes. She is on Earth and she is not dead, but I prefer to post other videos before telling what happened after.*
http://www.angelismarriti.it/ANGELISMARRITI-ENG/REPORTS_ARTICLES/Apollo20-InterviewWithWilliamRutledge.htm

19) L.S. *Were you able to understand the origin of the spacecraft and how old was it?*

W.R. *The age was estimated to 1.5 milliards of years, it was confirmed during exploration, we found ejections from the original crust, anorthosite, spirals in feldspathoids, coming from the impact which formed Izsak D; The density of meteor impacts on the ship validated the age, also little white impacts on the Monaco hill at the West of the ship...*

20-21) L.S. *Can you give me the technical details for every material you disclosed on YouTube? I mean, can you distinguish among the TV transmissions from the lunar rover and the camera footages, during the flyovers? I would need to know the details of shooting for every video you spread on the Web. What is the meaning of the strange numbers visible on the videos, which sometimes slowly stream over the frames, in the flyover of the Moon?*

W.R. *I have answered so much time on this, especially to an ESA astronaut. The transfer was made in Rwanda, [...] with codec and sound recuperation is not good, but it becomes better. The subtitles are not genuine but put on the videos after transfer. I asked to remove the voices sometimes to protect one person from mission control.*

We used three video cameras in Apollo, one on the rover, called GTCA; it is not the name of a company (a commenter made a mistake on this) but a Westinghouse color camera. All three color cameras has a color wheel that produced a time frame delay when transmitting to the earth. I think it could be possible for a company to restore a good TV picture. The CSM camera had a black and white monitor and produced stable pictures, sharp because focusing was visible on the monitor. The LM camera had a glass visor. The CSM camera was used one time on the AGC Visor, using the coordinates I've transmitted during the flyover video. The Flyover video was made in zero gravity. I was located on the left window, attitude horizontal, legs around the hammock, lens on the polycarbonate glass. The cameras had a Vidicon tube sensible with light, a large quantity of light, or changing the diaphragms put dropouts during transmission. The markings, numbers are used to perform a good landing. During program 64, when [...] in almost in vertical attitude, we had to put the "60" number on the landing site and hold it on the target minutes before landing. These marks are on the both [...] windows; you can verify it on a NASA site. Please verify on a genuine NASA site, (I got a flame by somebody who verified on the Apollo 13 movie) the markings had a special angle inclination. If you check it, you'll have an idea of my precise position during this sequence.

22) **L.S**. *How did you get years ago the copies of the footages of the mission?*

W.R. *About the footages [...] one day, someone I know told me he was charged to maintain security around a container. A building had to be destroyed, and archives had to be burn [sic] by a plasma torch. The nuclear power plant didn't deliver energy at the right price, so the container was plenty of interest things during some days. As (a) human is naturally curious, people (in) charged of security went inside... My friend took video films, a couple of 16 mm plates, boxes of B/W paper, two enlargers... He contacted me for selling the unused paper, and that's how I discovered the other things. I've already seen some picture before, 11*16 pictures were violet/blue, old RC photography, I watched the tapes, it was not a business affair, I put them in security, the only important thing for me were the BW sheets of paper. It was 15 years ago.* http://www.angelismarriti.it/ANGELISMARRITI-ENG/REPORTS_ARTICLES/Apollo20-InterviewWithWilliamRutledge.htm

23) **L.S.** *Have you ever met Clark McClelland, former NASA engineer who lost his job years ago because of what he discovered at the KSC (I suppose alien bodies or alien objects from the Space)?*

W.R. *But you can give me the links. Documents can be at KSC, but no bodies or alien craft I think.*

24) **L.S.** *You mentioned in a former letter C. M. and M. Who are they?*

W.R. *C. M. is the website officer on Oceans NASA site [http://oceancolor.gsfc.nasa.gov]; M. also. A. M. is 508 Coordinator. I only have mail exchange with Johnson Space Center for the moment. There is a moment of panic I think since May 18. Check 508 coordinator NASA on a search engine. Statement 508 is the way of pushing NASA to declassify material. I expected a reaction.*

Luca... it is a part of my strategy; NASA has the right to block me if I download unauthorized information. If they explain why they block me, they recognize that the videos can be obtained from them. If I sue them for violation of Statement 508, they will be forced to prove I downloaded unauthorized material, and it's not the case, I never go on a NASA site.

Since May 18, I have not a precise answer; I have to wait for a decision from the headquarters. Even as Italian citizen, you have the right to ask for material from a federal agency like NASA, see 508 Statement: http://www.section508.nasa.gov/

25-26) L.S. *Aren't you afraid of the U.S. Government's reaction and why did you talk about the date of September 2007, when NASA and USAF (according to you) "will be forced to tell the whole story before September 2007" What does it mean and who is your "deep throat"? In another recent your communication to me, you talked about the "2012" year. You said: "In 2012, the weaker will die, and governments preserve the only bit of their heritage [...] everybody has to be prepared for 2012". Is there any connection with the "Planet X" return (the ancient Nibiru, adored by the Sumerians in Mesopotamia)? What did you know about it?*

25 W.R. *I'm the deep throat. What can NASA USAF do now? Blocking or suing me would be an acknowledgement. They can speak of hoax or fiction. I'm just afraid they could open a site or another account with my name or putting almost perfect false videos with voluntary errors to disinform. Fortunately, bureaucracy and time works for me. It's a race.*

That's why the idea of putting the Leonov files is a good idea, no controversy anymore; there are no Leonov footage, no videos of this period of Leonov in a LM or on a USA USAF base. It is unthinkable related to the official version.

26 W.R. *I am a passionate of the Sumerian period, of the Genesis as related by Sumerian. They clearly explain how gods created man. But I have no indications on Sumerian cosmogony, send me some links.*

There is a question you didn't ask for and I'm always surprised that nobody does. This could be your question 27 - why is it necessary to hide UFOs, why disinformation, why putting all this under the carpet? It's question of economics. All currencies on Earth are based on the value of gold. Not many citizens know that but gold is an extraterrestrial metal coming from the death of a star. When a star is dying, its mass is growing, atoms are compressed and when the star explodes, it spreads large amounts of gold in young solar systems. That's why gold is not a mineral to treat but a perfect, carbon-free metal. This mean that it is the most common substance in the universe, no more value than a piece of plastic.

That's enough to put down all world currencies. Imagine also that an EBE says: "coffee has a good taste, rare in this galaxy", the only perspective of trading coffee through universe would displace the economic power to countries of the South in one day. You see, not a problem of panic, but simply a problem of economy. http://www.angelismarriti.it/ANGELISMARRITI-ENG/REPORTS_ARTICLES/Apollo20-InterviewWithWilliamRutledge.htm

As other UFO investigators like Dr, Michael Salla started looking into this amazing story, inconsistencies began to appear from the upload of data, video, and audio by W. Rutledge which were not all original material of the alleged covert Apollo 20 Moon mission. One inconsistency is that Rutledge used audio from the Apollo 11 and 15 missions and inserted it into the video footage of the alleged Apollo 20 mission. This does not bode well in validating the authenticity of Rutledge's Apollo 20 claim, although the photo images and video do remain valid of something seen and photographed on the dark side of the Moon. Another inconsistency is the ***"flight insignia"*** that Rutledge used was wrong for a joint US-USSR mission, compare with the Apollo-Soyuz 1975. (See below).

Flight Insignias of Alleged Apollo 20 and Apollo 18 – Soyuz
http://blog.altpov.tv/was-there-an-apollo-20-mission/ and https://www.bibliotecapleyades.net/luna/esp_luna_36a.htm

Flight insignias are, however, not a basis of proof or disproof when you are part of a covert space or lunar mission where secrecy and security are of primary importance. If you are operating within the DOD outside of NASA's knowledge and cooperation, you launch your own spacecraft with your own rockets. You are not going to worry about what type of patch you're wearing on your spacesuit as you head off on a secret mission to the moon. And, you are not going to be concerned with any warnings that some Extraterrestrials told to some other astronauts, about staying away and not coming back to the Moon. You are the United States Air Force, the US Military and you don't take orders from anyone except from your commanding officers and certainly not from any ETs! You give orders and you kick butt! That is the name of the game!! This simple point seems to be lost on a lot of UFO and ETI investigators. The US Military pretty much does what it wants, when it wants, particularly when orders come from the top brass or from the Military industrial Complex.

According to Rutledge's report supported by some outstanding videos uploaded on YouTube since April 2007, after the Apollo 17 (December 1972) and the "Apollo18-Soyuz" mission had taken place in July 1975, there were other *two missions on the Moon:*
1. the Apollo 19 (failed because of "a loss of telemetry, a brutal end of mission without data",(see the interview with W. Rutledge)
2. the Apollo 20 (August 1976), which were both classified Space missions launched from the Vandenberg Air Force Base (California)

The goal of these two presumed secret joint space mission, result of American-Soviet collaboration, was to reach the backside of the Moon (the Delporte-Izsak region, close to the well-known **Tsiolkovsky crater**) and to explore a huge object found during the Apollo 15 mission. What the Apollo 20 crew found, it was a huge and ancient alien spaceship, "approximately 4 kilometers long" (W. Rutledge).

Initial estimates of the alien spacecraft's length appear to be incorrect and recent measurements were made with comparison to individual pixel size of the Izsak Crater. The revised length of the cigar-shaped craft is a whopping 4.80 kilometers (4800 meters or nearly 5.0 kilometers) long!

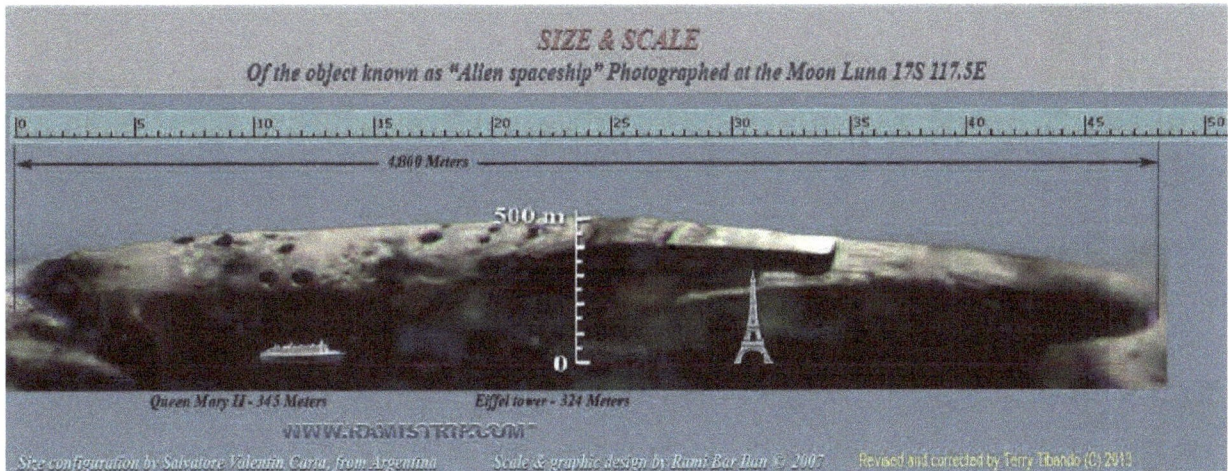

The Ancient Lunar Spacecraft to size and scale (above). The revised spaceship length is nearly 5.0 km long making it longer than the New York or Toronto Skyline (below)

https://www.pinterest.com/henrytaly/ancient-alien-encounters/?lp=true

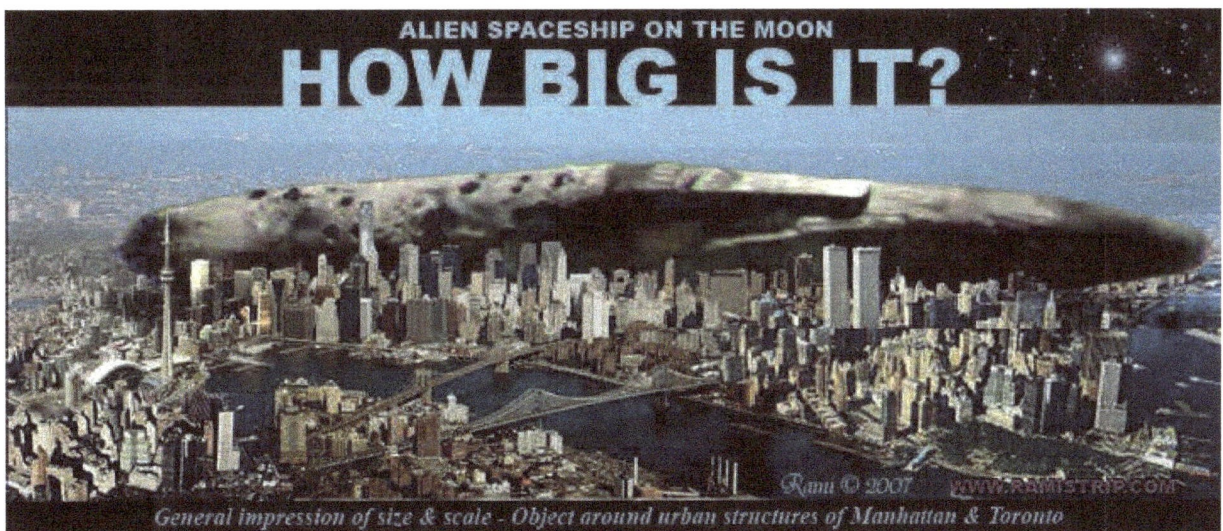

And as a matter of fact, some official NASA pictures archived by the **LPI (Lunar and Planetary Institute** in Houston), which is "a research institute that provides support services to NASA and the planetary science community" (http://www.lpi.usra.edu/lpi/about.shtml), show a

strange and big object on the far side of the Moon. LPI is "managed by the **Universities Space Research Association (USRA)**".

http://www.bibliotecapleyades.net/luna/esp_luna_36a.htm#Did%20the%20USA/USSR%20Fly%20a%20Secret%20Joint%20Mission%20to%20the%20Moon

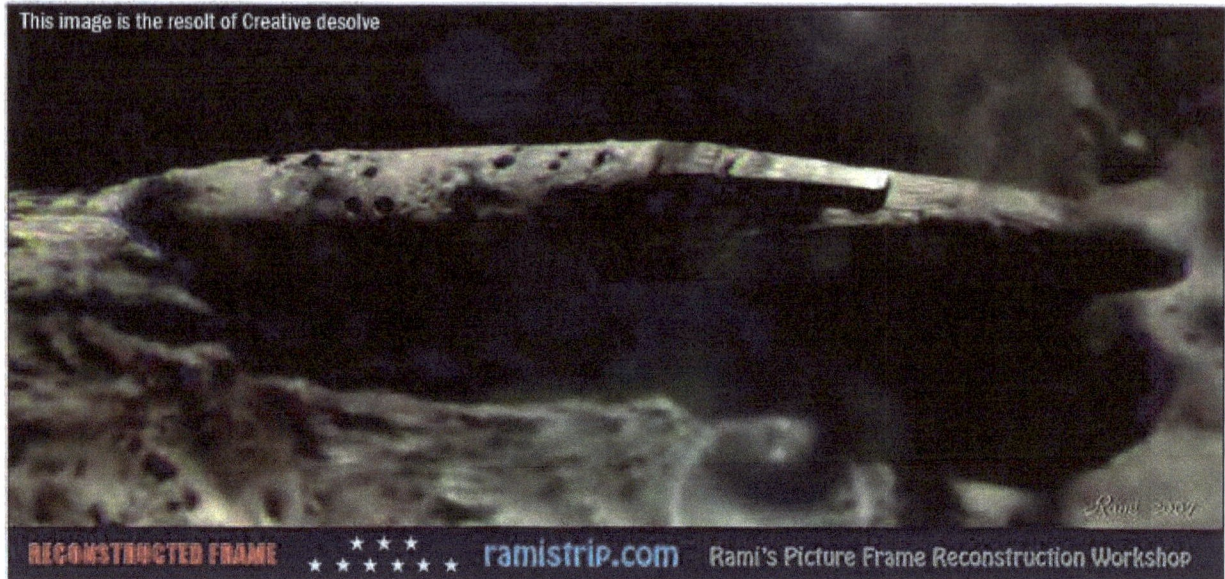

The photo is a composite of video images spliced and digitally smoothed, sharpen and enhanced to bring out detail that is hard to see in the original video without losing the integrity of the original image of the ancient lunar spacecraft

https://pics-about-space.com/apollo-20-alien-spaceship?p=1#

Salla stated that he had to re-evaluate his original position to the Apollo 20 mission story as a possible hoax or as some disinformation mixed with factual information as an attempt by Rutledge to have other whistleblowers come forward and disclose the whole secret mission of that really happened. He views Rutledge's testimony as a fabrication. However, Salla still believes that there was a covert mission that did occur in 1976 but it probably didn't use a Saturn V rocket nor was **NASA** involved. He suspects that the **USAF's fleet of antigravity vehicles** built through reverse engineering of extraterrestrial technologies in cooperation with certain corporations would obviate the need for any NASA personnel or launch vehicles that use outdated propulsion technologies and navigation principles. This is a big presumption to be sure but, certainly not beyond the realm of possibility.

According to Salla, that part of Rutledge's story doesn't make much sense given that deep black covert antigravity technology was developed in the 1950s/1960s "**Alien Reproduction Vehicles**" (**ARVs**) have operated since that time. If a crashed *extraterrestrial mothership* was located on the moon, surely a joint mission would use the most advanced ARVs, rather than Apollo hardware.

Sources for the existence of ARVs include:
1. **Mark McCandlish** who claimed he saw in a secure Air Force facility at an air-show
2. **Gary McKinnon** who learned about a Naval space fleet during his hacking ventures

3. **Nick Cook** who describes the covert nature of antigravity research in "The Hunt for Zero Point."

We cannot disqualify all the videos or the testimony that Rutledge is uploading to the internet (YouTube). Videos of the crashed ETV appear to be authentic originating from an actual covert mission, the details of which still need further investigation. We can suspect this was a joint US-USSR mission to explore a huge ancient alien spacecraft on the Moon and much of Rutledge's information appears on the surface to be accurate but, there is an inherent plausible deniability built into that information. We will just have to wait and see what else develops around this piece of lunar evidence.

http://www.bibliotecapleyades.net/luna/esp_luna_36a.htm#Did%20the%20USA/USSR%20Fly%20a%20Secret%20Joint%20Mission%20to%20the%20Moon

Evidence Number Five: The Lunar "Mona Lisa" and ET Calligraphy

This next particular piece of lunar evidence is really part and parcel of the above ancient alien spacecraft evidence already mentioned. According to alleged Apollo 20 astronaut, William Rutledge, there was another spacecraft, triangular in shape, nearly buried below the lunar surface but, still visible and accessible; located near the nose end of the long cigar-shaped alien ship. In his internet interview with Luca Scantamburlo, Rutledge was asked if he entered into the large spacecraft and he stated that he did. Rutledge exact words:

We went inside the big spaceship, also into a triangular one. The major parts of the exploration was; it was a mother ship, very old, who crossed the universe at least milliard of years ago (1.5 estimated). There were many signs of biology inside, old remains of a vegetation in a "motor" section, special triangular rocks that emitted "tears" of a yellow liquid which has some special medical properties, and of course signs of extrasolar creatures. We found remains of little bodies (10cm) living in a network of glass tubes all along the ship, but the major discovery was two bodies, one intact.

(Repeated from above): W.R. *Mona Lisa – I don't remember who named the girl, Leonov or me - was the intact EBE. Humanoid, female, 1.65 meter. Genitalized, haired, six fingers (we guess that mathematics are based on a dozen). Function; pilot, piloting device fixed to fingers and eyes, no clothes, we had to cut two cables connected to the nose. No nostril. Leonov unfixed the eyes device (you'll see that in the video). Concretions of blood or bio liquid erupted and froze from the mouth, nose, eyes and some parts of the body. Some parts of the body were in unusual good condition, (hair) and the skin was protected by a thin transparent protection layer. As we told to mission control, condition seemed not dead not alive. We had no medical background or experience, but Leonov and I used a test, we fixed our bio equipment on the EBE, and telemetry received by surgeon (Mission Control meds) was positive. That's another story. Some parts could be unbelievable now, I prefer tell the whole story when other videos will be online. This experience has been filmed in the LM. We found a second body, destroyed; we brought the head on board. Color of the skin was blue gray, a pastel blue. Skin had some strange details above the eyes and the front, a strap around the head, wearing no inscription. The "cockpit" was full of calligraphy and formed of long semi-hexagonal tubes. She is on Earth and she is not dead, but I*

prefer to post other videos before telling what happened after.
http://www.angelismarriti.it/ANGELISMARRITI-ENG/REPORTS_ARTICLES/Apollo20-InterviewWithWilliamRutledge.htm and http://www.viewzone.com/monalisa.html

**Lunar Calligraphy found in one of the triangular spaceships
or is it just some scribble and doodle?**
http://conspiraciones1040.blogspot.com/2012/06/apolo-20-nave-espacial-extraterrestre-y.html

Once again, with this type of story, inconsistencies arise with the testimony of W. Rutledge, who said in another interview (below) that he did not enter the long spaceship but instead, entered one of two triangular craft where the bodies were found. The problem here is that from the official NASA photo strips of the Delporte – Izsak crater area, only the huge cigar-shaped ship is visible yet, we are told by Rutledge that there were two smaller triangular spacecraft (buried?) in which the two ET bodies were retrieved from one of these craft. These kinds of errors and inconsistencies have even forced Scantamburlo to call into question, as to whether Rutledge is really an imposter. Scantamburlo asks him for further proof to support the legitimacy of his claim of being an astronaut and to re-establish the authenticity of his testimony which somehow, Rutledge comes up with the necessary information and proof.

The reader would easily be forgiven for their suspension of disbelief up to this point and that if they fell down laughing with incredulity, it would be understandable. Nothing does more damage to the credibility of an account than when it is filled with inaccuracies and inconsistencies, whether unintentional or deliberate. But, we will plod onward because there appears to be a payoff at the conclusion of this evidence.

L. S. About triangles spacecrafts on the far side of the Moon, and about the reasons for (why) he decided to speak out, Rutledge said to me:

W. R. *[...] about the spacecraft, one triangle was accessible) in a trisangle **(triangle)** craft, we found two bodies, one in bad condition, a meteor cut the body at the neck level, we tok **(took)** the littl **(little)** skull on board. The other body was strange, NDNA, not dead not alive, but crusted with impacts, stalagmites of blood coming out of hemmoragias **(hemorrhages)** zones. One body was on **(Apollo)** 20, fixed on a hammock, and we passed hours watching the hands, the strange hair, not the kind you see a **(sci-fi)** movie. The hair was in good condition, we can say alive,*

294

(Alexei) tested it. The ship was not explored on the 4 kilometers, but no place detected for weapons. (May 23, 2007, 07:42 AM, from retiredafb's (retired AFB's) message to my YouTube Account).

[...] -no military craft, exploring one, Crew of 300, two female (pilots) on (triangles). [...] I (choose) to (do) it know (now) because all (Apollo) program had to be definitely locked with (the) outgoing of 'the marvel of it all' presented (April) 20, presenting the 12 astronauts alive who stayed on the moon.

It was a trahison (treason) for me, for (Alexei), and for the 3 dead astronauts of amollo(Apollo) 19. My girlfriend Stephanie (Ellis), first (American) woman in sace (space), fisrt (first) afro maerican (American) woman was killed during this mission, (I) have no place to pray for her, her remains are still in orbit around earth. (May 23, 2007, 07:48 AM, from retiredafb's (retired AFB's) message to my YouTube account). **(Bold italics were added by author because the spelling was atrociously poor and certain words lacked capitalization).**
http://transition888.heavenforum.org/t568-mona-lisa-ebe

When we sort through the language difficulties of translation and the grammatical errors in script, we learn that there are possibly two huge triangular spacecraft beside the immense tubular spaceship. These Triangle craft are not apparent in the aerial photo images when the LEM flew over the debris site. With computer enhancement, details in the photographs (see below) are brought out and the triangle spacecraft appears from the shadows.

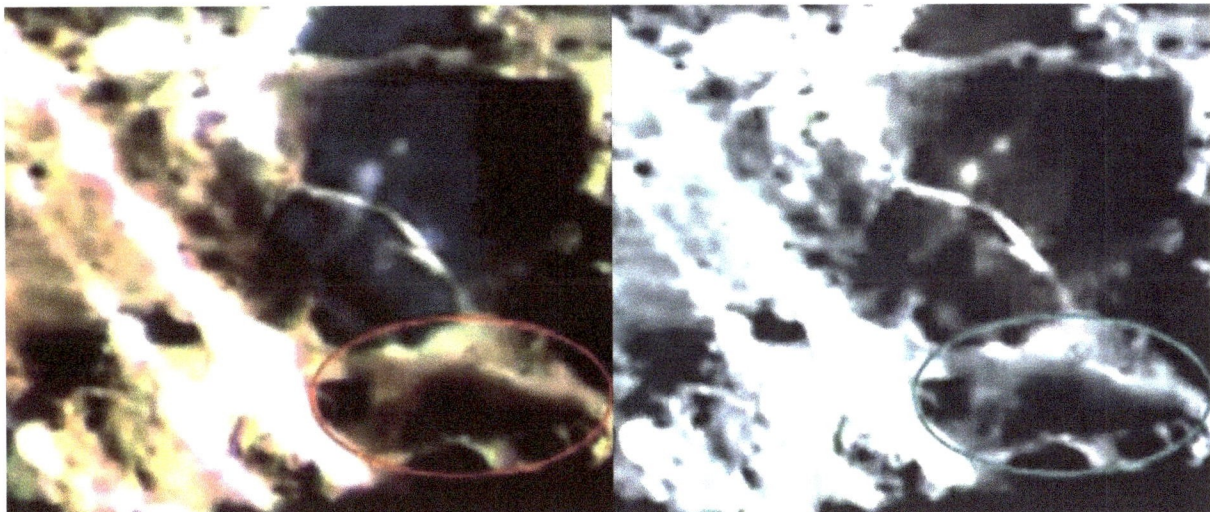

One of two triangle alien spacecraft in front of the nose end of the cigar shape spaceship Black and white photo shows contrast detail of craft not easily seen above lunar surface
(Google Image)

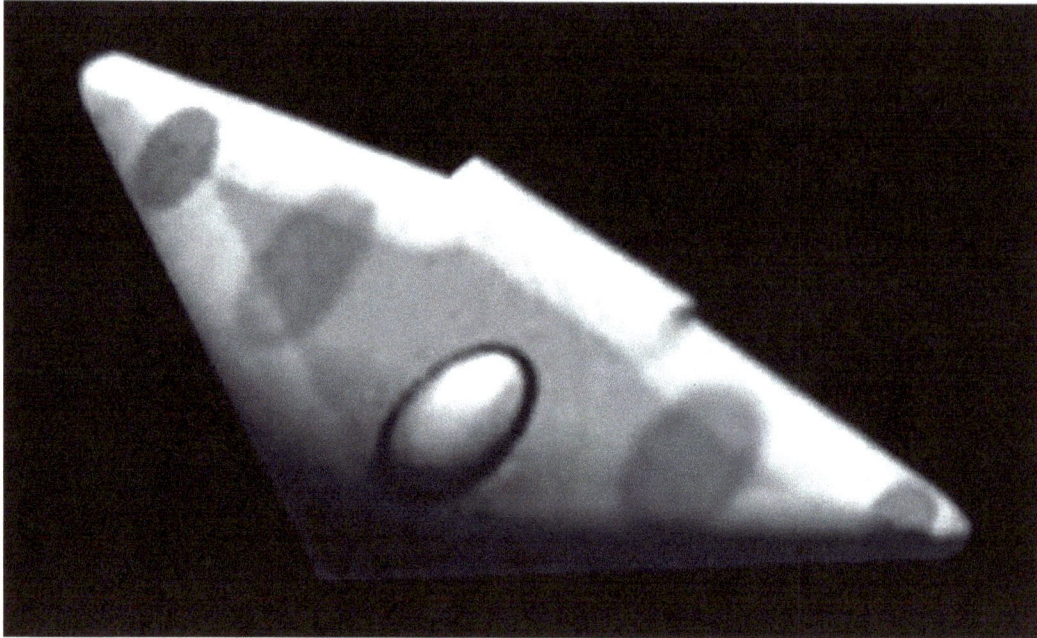

Digital image of triangular craft that contained a crew of 300 with two female pilots

A female **Extraterrestrial Biological Entity (EBE)** completely intact, neither dead or alive also, the head of a second ET with a pale blue-gray complexion (the rest of its body was badly deteriorated) were found inside one of these triangle spacecraft and both were retrieved and brought back to the **LEM (Lunar Excursion Module)** for detailed examination and photo documentation.

Once inside the LEM, Cdr. W. Rutledge and A. Leonov place the female ET body in the hammock and started video recording the ETI. What was immediately apparent to both astronauts was that the female ET was in remarkably good shape as if not really dead at all. Another obvious aspect was the similarity in taxonomical structure of the EBE to human beings, though slightly shorter in stature and with the exception of six fingers on each of her hands. The EBE was nicknamed **"Mona Lisa"** by one of the astronauts as she reminded them of Da Vinci's Mona Lisa painting. "Mona Lisa" apparently was found with no clothes and it was, therefore, easy to determine she was female as she had breasts and female genitalia.

Most strange of all was that the Mona Lisa's face and hands was covered with short metal cables (which in the video looked like "chicken bones") which were attached with some type of skin adhesive compound from the centre of the forehead to both eyes and from the eyes down to the sides of the mouth which held the mouth open. Cables to the nose had been cut and almost immediately concretions of blood or bio liquid erupted and froze from the mouth, nose, eyes and some parts of the body. All cable attachments were removed from the face and from inside the nostrils and around the eyes. Some parts of the body were in unusually good condition, Mona Lisa's hair was styled in a way that was reminiscent of the women of ancient Greek and Minoan cultures (see photos below) and the skin was protected by a thin transparent protection layer.

This thin transparent protective layer looks like plastic and some investigators have said that this gave the appearance of the body of being a cheap rubber mannikin and thus, some people have said that the whole video was a hoax. This is a legitimate statement but may be premature, if we assume that this **Lunar Mona Lisa** was in some type of **homeostasis** or **suspended animation** which would account for her appearance of neither being dead or alive. In fact, when Leonov connected their bio equipment on the EBE, telemetry was received by a surgeon at Mission Control and the signals were positive indicating there were still life signs!! The cables or metal tubes and the transparent protective layer may have been part of some equipment to keep Mona Lisa in a protective **homeostasis sleep**! This would mean however a sleep-like condition that has been maintained somehow for the last *"billion and half years"*!! This in itself would be one hell of a remarkable discovery over-shadowing almost every other technology yet found! The implications, if this EBE is still alive are profound beyond words!

This is where the absurdity of the situation takes hold and one is left thinking on the side of the debunkers and skeptics. Another problem is transporting the female ET from her spaceship to the lunar lander without the protection of a spacesuit or a self-contained portable environment to endure the Moon's cold, near-vacuum atmosphere. If she was still somewhat alive before the astronauts transported her, she certainly would have been dead upon arrival to the LEM.

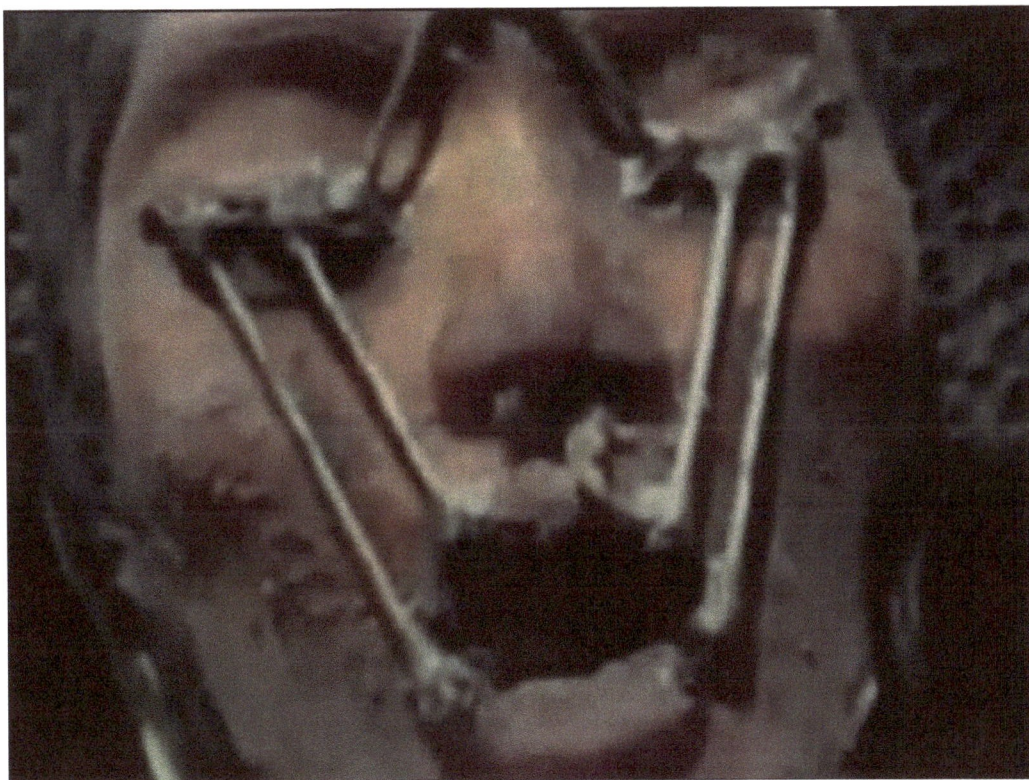

Cables or tubes attached to the face of the Lunar Mona Lisa\ may be part of some homeostasis equipment also, not the EBE's hair style

http://www.viewzone.com/monalisa.html

**Note the hair styles on these statues of Greek and Minoan women
and compare them with the "Lunar Mona Lisa" above**
Google Image and http://www.wikiwand.com/es/Kore_del_Peplo

**The Lunar Mona Lisa without the facial prosthesis and the protective
dermis layer removed by Apollo 20 astronauts**
http://alien-ufo-sightings.com/2017/03/mona-lisa-alien-girl-apollo-20-found-moon/

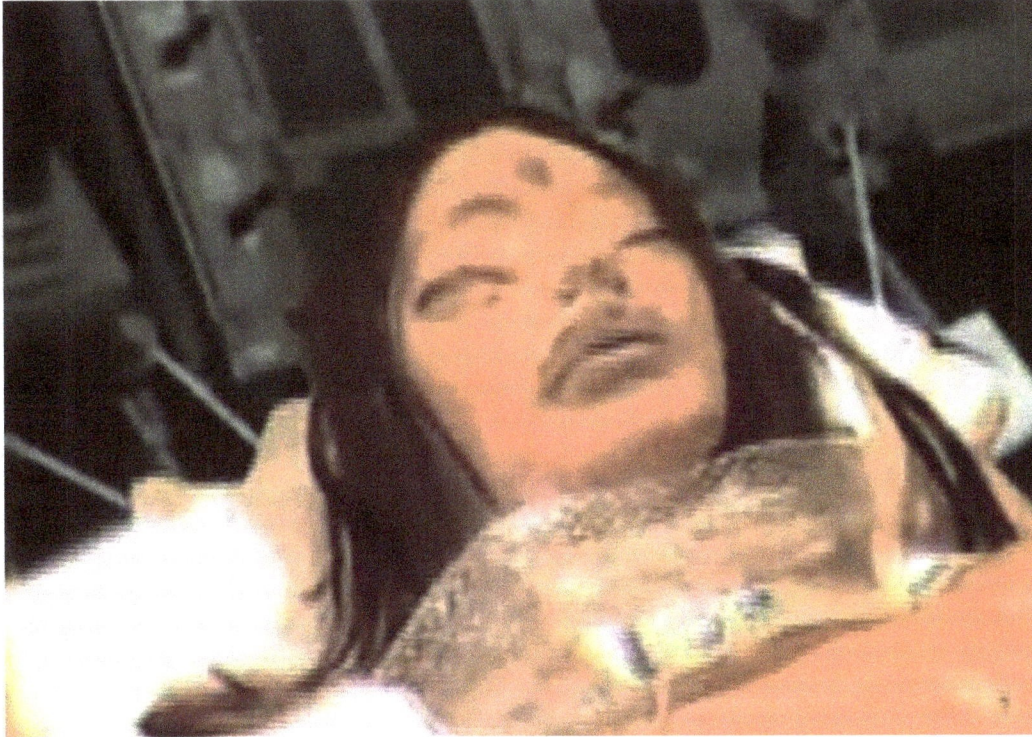

The female EBE named Mona Lisa in a hibernated sleep
https://pics-about-space.com/apollo-20-alien-spacecraft-with-woman?p=2

The Moon's surface atmosphere on average is a frigid instant iceberg temperature of ***minus 77°C below zero*** with temperature fluctuations of 280 degrees ranging from 90°C above zero to nearly 200°C below zero, depending where on the Moon you are located. http://en.wikipedia.org/wiki/Moon and http://wattsupwiththat.com/2012/01/08/the-moon-is-a-cold-mistress/

Yet, if we can get pass these absurdities, there are some other fascinating oddities about Mona Lisa.

It was mentioned earlier that Mona has six fingers on each hand which may be the basis of Earth's measurements of 12 inches in a foot, 12 months in a solar year, 12 major divisions in a circle or compass points, 12 zodiac signs, etc., etc. Once all the facial apparatus was removed and a general cleaning of Mona Lisa's face, we see that she has a Middle-Eastern appearance even, a somewhat Oriental appearance with eyes slightly larger in size than most humans. As mentioned the Lunar Mona Lisa has a hair style similar to women of ancient Greece yet, she has a raised **gibbosity** on her forehead that is also similar to the beauty mark on the foreheads of women of India. Mona Lisa beauty seems to incorporate fashion styles of many major Earth cultures yet, she literally transcends all of them. The Mona Lisa of the Moon is indeed, a rare beauty!!

A Wellspring of Lunar Evidence Emerges

As of this writing, **William Rutledge** has put many more short videos up on the internet, shows a pan shot of a lunar "City" with many arches, buildings, and cathedral-like structures all connected together which looks very alien except, for one thing, it a fake! It has proven to be a clever digital manipulation of an existing lunar landscape shot with a computerized image of an alien-like city in the background.

Another video made from the LEM before landing shows the flyover of the **Izsak D Crater** with the cigar-shaped spacecraft in the crater. This video black and white has an air of being genuineness to it.

One more video in colour shows the lunar land rover and its antenna dish coursing along the Moon's surface toward the site of the immense alien spaceship.

Still another black and white video shows the lunar landscape as the lunar rover travels pass many conical towers or spires seen not too far in the distant background and at one point, a round globe-shaped UFO flies overhead, past one of the very tall towers and comes down close to the rover within mere feet! It size can be extrapolated as approximately 10 feet in diameter and surface detail on the globular UFO can be discerned. This could be a CGI hoax as the globe UFO that flies by a large rock is the same rock that appears in the Apollo 17 photo with the moon rover parked beside it!

Another B&W video indicates that **"Apollo 20"** astronauts had a 3rd EVA to the spacecraft wreckage site where they pan the camera over the external view of one of the triangular shaped spacecraft set against the lunar background. Again, this could be real or another clever digital hoax.

There is also a B&W video with audio of the blast off of the **Lunar Command Module (LCM)** from the Moon's surface; this looks real as well.
http://www.youtube.com/user/moonwalker1966delta

Finally, there is a colour video of the LCM approaching the **Command Service Module** (piloted by **Snyder**) for the final docking maneuvering, when suddenly a glowing globe UFO comes into the picture and circles around the LCM and then, positions itself momentarily between the **Lunar Module** and the **Command Modular** before flying off. Is this real? It's possible. But with Rutledge many things seem real and yet, some are obvious hoaxes designed to mislead and create confusion in the mind of the viewer or reader. Yet, when he is put on the spot by **Scantamburlo**, to prove himself and his evidence, he comes back with convincing information that only an astronaut would know!

We are left with many video images not knowing for sure what is real and what is faked. There is an overall message being stated here, and Rutledge is operating an information campaign or possibly a disinformation campaign, without coming out personally into the public limelight to take some real hard-hitting questions from many UFO investigators as well as some of his supporters and detractors, all seeking answers.

300

As **Michael Salla** has stated, **Rutledge** main purpose may not necessarily be to present accurate documented evidence, but rather to instigate public inquiry into NASA's and the DOD's cover-up of an Extraterrestrial presence on the Moon. By creating enough factual information and evidence sandwiched between smaller lies of disinformation, Rutledge spurs further research into the lunar alien phenomenon. Admittedly, this gives Rutledge's current agenda the benefit of the doubt but, this type of information dispersal ploy does have a limited degree of succeeding and may, in the end, be detrimental to setting UFO/ETI investigation spinning in its tracks, instead of moving forward. Only time will tell and evidence of reality is everything! Stay tuned.

CHAPTER 74

SOME LUNAR CURIOSITIES, ODDITIES, AND ANOMALIES – IS THE MOON A GIANT SPACESHIP?

Neil Armstrong, *"First Man on the Moon";* July 20th, 1994*: **"There are great ideas, undiscovered breakthroughs available, to those who can remove one of truth's protective layers!"***

Robert Jastrow, First Chairman, NASA Lunar Exploration Committee
"The moon is the Rosetta stone of the planets."

After hundreds of years of detailed observation and study, our closest companion in the vast universe, Earth's moon, remains an enigma. Six moon landings and hundreds of experiments have resulted in more questions being asked than answered. Among them:

The Moon is considered as Earth's natural satellite and is one of the five largest satellites in the Solar System. After the Sun, the Moon is the brightest object in the night sky. In fact, there are many interesting and unusual moon facts including information about its origin and important orbital and physical data. It has a ratio of diameter to the Earth of 0.25 or one quarter the size of the Earth which is very unusual, whereas, all other natural satellites (moons) in our Solar System,when compare to their parent planet have a ratio diameter of 0.025 or much less, making the Moon a true oddity for a "natural satellite". But, there are many other oddities to the Moon. (Author's statements in bold italics are added for emphasis).

1. Moon's Age: The moon is far older than previously expected. Maybe even older than the Earth or the Sun. The oldest age for the Earth is estimated to be 4.6 billion years old; moon rocks were dated at 5.3 billion years old, and the dust upon which they were resting was at least another billion years older. ***This shouldn't be the case if the stellar nursery in which the Sun and all the planets were born should have come into existence before any of the planets' moons. The Sun would have formed first then, the planets next; the moons would have formed last or at least at the same time as the planets, not before the planets.***

2. Rock's Origin: The chemical composition of the dust upon which the rocks sat differed remarkably from the rocks themselves, contrary to accepted theories that the dust resulted from weathering and breakup of the rocks themselves. The rocks had to have come from somewhere else. ***This dust on the Moon's surface couldn't have come from the rain storm of meteor impacts of the past, although, scientists and astronomers would like you to believe that this was the historical conditions after the planets had formed. If this was true then, the Earth should have suffered a similar fate as did the Moon.***

3. Heavier Elements on Surface: Normal planetary composition results in heavier elements in the core and lighter materials at the surface; not so with the moon. According to **Wilson**, *"The abundance of refractory elements like titanium in the surface areas is so pronounced that several geologists proposed the refractory compounds were brought to the moon's surface in great quantity in some unknown way. They don't know how, but that it was done cannot be questioned."* ***This speaks for itself!***

4. Water Vapor: On March 7, 1971, lunar instruments placed by the astronauts recorded a vapor cloud of water passing across the surface of the moon. The cloud lasted 14 hours and covered an area of about 100 square miles. ***This is truly an oddity given that scientists believe that no atmosphere should even exist in what is thought to be a surface environment in a near vacuum.***

5. Magnetic Rocks: Moon rocks were magnetized. This is odd because there is no magnetic field on the moon itself. This could not have originated from a "close call" with Earth—such an encounter would have ripped the moon apart. ***This by itself would indicate artificiality as all natural satellites and planets have a magnetic field emanating from within and surrounding them.***

6. No Volcanoes: Some of the moon's craters originated internally, yet there is no indication that the moon was ever hot enough to produce volcanic eruptions. ***Without an active magnetic spinning core, there would be no volcanic activity, this would indeed then, make the Moon a dead world, a real anomalous phenomena in our Solar System.***

7. Moon Mascons: Mascons, which are large, dense, circular masses lying twenty to forty miles beneath the centers of the moon's *maria*, "are broad, 'disk-shaped objects' that could be possibly some kind of artificial construction. For huge circular disks are not likely to be beneath each huge *maria*, centered like bull's-eyes in the middle of each, by coincidence or accident." ***Again, scientists will say that such mascons are the result of magma solidifying as the Moon lost it magnetic field, this resulting from the core mysteriously slowing down to a complete stop.*** http://www.bibliotecapleyades.net/luna/esp_luna_16.htm and
"The Alien Chaser" by Ronald Regehr from an article by Don Ecker "Long Saga of Lunar Anomalies"; UFO magazine, Vol. 10, No. 1 2 (March/April 1995),

8. Seismic Activity: Hundreds of "moonquakes" are recorded each year that cannot be attributed to meteor strikes. In November 1958, Soviet astronomer **Nikolay A. Kozyrev** of the **Crimean Astrophysical Observatory** photographed a gaseous eruption of the Moon near the crater Alphonsus. He also detected a reddish glow that lasted for about an hour. In 1963, astronomers at the Lowell Observatory also saw reddish glows on the crests of ridges in the Aristarchus region. These observations have proved to be precisely identical and periodical, repeating themselves as the moon moves closer to the Earth. These are probably not natural phenomena. ***If there is no volcanic activity within the Moon then, how are there gas eruptions from some of the craters on a regular periodical basis, be possible?***

9. Hollow Moon: The moon's mean density is 3.34 gm/cm^3 or 3.34 times an equal volume of water, whereas the Earth's is 5.5. What does this mean? In 1962, NASA scientist **Dr. Gordon MacDonald** stated, *"If the astronomical data are reduced, it is found that the data require that the interior of the moon is more like a hollow than a homogeneous sphere."* ***Here again, this indicates a possible artificiality of the Moon, unless, there is some natural phenomena and process that we still as yet do not understand.***

Nobel chemist **Dr. Harold Urey** suggested the moon's reduced density is because of large areas inside the moon where it is "simply a cavity."

MIT's **Dr. Sean C. Solomon** wrote, *"the Lunar Orbiter experiments vastly improved our knowledge of the moon's gravitational field... indicating the frightening possibility that **the moon might be hollow**."*

In **Carl Sagan**'s treatise, Intelligent Life in the Universe, the famous astronomer stated, *"A natural satellite cannot be a hollow object."* ***How is this possible? Is this a natural phenomena or one of artificial construction in which case, we must seriously consider the Artificial MoonHypothesis that the Moon is an intelligently constructed spaceship, no matter how absurd thatmay sound.***

10. Moon Echoes: On November 20, 1969, the Apollo 12 crew jettisoned the lunar module ascent stage causing it to crash onto the moon. The LM's impact (about 40 miles from the Apollo 12 landing site) created an artificial moonquake with startling characteristics—the moon reverberated like a bell for more than an hour.

This phenomenon was repeated with Apollo 13 (intentionally commanding the third stage to impact the moon), with even more startling results. Seismic instruments recorded that the ***reverberations*** lasted for three hours and twenty minutes and traveled to a depth of twenty-five miles, leading to the conclusion that the moon has an unusually light—or even no—core. ***If by now, the growing preponderance of evidence hasn't yet sunk into the mind of the reader that the Moon is indeed hallowed and very probably artificial then, read on!***
http://www.bibliotecapleyades.net/luna/esp_luna_16.htm and
"The Alien Chaser" by Ronald Regehr from an article by Don Ecker "Long Saga of Lunar Anomalies"; UFO magazine, Vol. 10, No. l 2 (March/April 1995),

11. Unusual Metals: The moon's crust is much harder than presumed. Remember the extreme difficulty the astronauts encountered when they tried to drill into the maria? Surprise! The maria is composed primarily illeminite, a mineral containing large amounts of titanium, the same metal used to fabricate the hulls of deep-diving submarines and the skin of the **SR-71 "Blackbird".** Uranium 236 and neptunium 237 (elements not found in nature on Earth) were discovered in lunar rocks, as were rustproof iron particles. ***At this point, one would have to seriously consider the probability that the Moon is an intelligently alien construct! Marias of illeminite containing titanium in great quantities and rustproof iron particles in Moon rocks suggests a process of manufacturing. The discovery of Uranium 236 and neptunium 237 in lunar rocks is very unusual. It suggests that the debris particle fields may be the result of a nuclear holocaust on the Moon's surface; this may indicate why there are so many cities and buildings in ruins on the Moon!***

12. Moon's Origin: Before the astronauts' moon rocks conclusively disproved the theory, the moon was believed to have originated when a chunk of Earth broke off eons ago (who knows from where?). Another theory was that the moon was created from leftover "space dust" remaining after the Earth was created. Analysis of the composition of moon rocks disproved this theory also.

Another popular theory is that the moon was somehow "captured" by the Earth's gravitational attraction. But no evidence exists to support this theory. **Isaac Asimov**, stated, *"It's too big to have been captured by the Earth. The chances of such a capture having been affected and the moon then having taken up nearly circular orbit around our Earth are too small to make such aneventuality credible."* ***If the Moon is not a natural satellite of the Earth, either through naturalplanetary*** *processes and not a captured body due to the gravity of the Earth then, how did it come into being and into Earth orbit?*

13. Weird Orbit: Our moon is the only moon in the solar system that has a stationary, near-perfect circular orbit. Stranger still, the moon's center of mass is about 6000 feet closer to the Earth than its geometric center (which should cause wobbling), but the moon's bulge is on the far side of the moon, away from the Earth. "Something" had to put the moon in orbit with its precise altitude, course, and speed. ***This appears to be an intelligently controlled, self-correcting orbit, one which you would expect from a manmade spacecraft orbiting the Earth!***

14. Moon Diameter: How does one explain the "coincidence" that the moon is just the right distance, coupled with just the right diameter, to completely cover the sun during an eclipse? Again, Isaac Asimov responds, *"There is no astronomical reason why the moon and the sun should fit so well. It is the sheerest of coincidences, and only the Earth among all the planets is blessed in this fashion." **This is a true curiosity of the Moon in its relationship to the Earth unless this was an intentional design if the Moon is indeed, artificial.***
http://www.bibliotecapleyades.net/luna/esp_luna_16.htm and
"The Alien Chaser" by Ronald Regehr from an article by Don Ecker "Long Saga of Lunar Anomalies"; UFO magazine, Vol. 10, No. l 2 (March/April 1995),

15. Spaceship Moon: As outrageous as the ***Moon-Is-a-Spaceship Theory*** is, all of the above items are resolved if one assumes that the Moon is a gigantic extraterrestrial craft, brought here eons ago by intelligent beings. This is the only theory that is supported by all of the data, and there are no data that contradict this theory.

Greek authors **Aristotle** and **Plutarch** and Roman authors **Apollonius Rhodius** and **Ovid** all wrote of a group of people called the **Proselenes** who lived in the central mountainous area of Greece called Arcadia. The *Proselenes* claimed title to this area because their forebears were there ***"before there was a moon in the heavens."***

This claim is substantiated by symbols on the wall of the **Courtyard of Kalasasaya**, near the city of **Tiahuanaco, Bolivia**, which record that the moon came into orbit around the Earth between 11,500 and 13, 000 years ago, long before recorded history. ***Here, we must be careful that the history from all human cultures be considered and whether their traditions and recorded history tell of a time when the Moon did or did not exist around the Earth. This would be conclusive evidence for the Artificial Moon Hypothesis.***

Could the Moon be an artificial satellite? As we explore other moons or satellites within our Solar System, we will find that many moons have a "cookie cutter" similarity to each other, strongly suggesting that they were also created by an ancient highly advanced intelligence. The implications of these similarities imply that much of our Solar System may in fact, not be the creation of natural origin, but be in some part of deliberate artificial design!!!

1. Ages of Flashes: Aristarchus, Plato, Eratosthenes, Biela, Rabbi Levi, and Posidonius all reported anomalous lights on the moon. NASA, one year before the first lunar landing, reported 570+ lights and flashes were observed on the moon from 1540 to 1967. *Now, this is real and historical fact. Some may be dismissed as meteorite impacts but many are not and these historians knew the difference between comets, meteorites and things that were truly anomalous.*

2. Operation Moon Blink: NASA's *Operation Moon Blink* detected 28 lunar events in a relatively short period of time. *Admittedly more information is required as to what these "blinking events" may be.*

3. Lunar Bridge: On July 29, 1953, **John J. O'Neill** observed a 12-mile-long bridge straddling the crater Mare Crisium. In August, British astronomer **Dr. H.P. Wilkens** verified its presence, *"It looks artificial. It's almost incredible that such a thing could have been formed in the first instance, or if it was formed, could have lasted during the ages in which the moon has been in existence."*

4. The Shard: The Shard, an *obelisk-shaped object* that towers 1½ miles from the Ukert area of the moon's surface, was discovered by Orbiter 3 in 1968. **Dr. Bruce Cornet**, who studied the amazing photographs, stated, *"No known natural process can explain such a structure."*

5. The Tower: One of the most curious features ever photographed on the Lunar surface (Lunar Orbiter photograph III-84M) is an amazing *spire that rises more than 5 miles* from the Sinus Medii region of the lunar surface.

6. The Obelisks: Lunar Orbiter II took several photographs in November 1966 that showed several obelisks, one of which was more than 150 feet tall. ". . . the spires were arranged in precisely the same way as the apices of the three great pyramids." *This lunar bridge, the Shard, the Tower and the Obelisks are all well documented and photographed factual evidence.* http://www.bibliotecapleyades.net/luna/esp_luna_16.htm and **"The Alien Chaser" by Ronald Regehr from an article by Don Ecker "Long Saga of Lunar Anomalies"; UFO magazine, Vol. 10, Nol 2 (March/April 1995),** also from Informant News website.

If the Moon is indeed an artificial body, an intelligently engineered and constructed spaceship which is literally 2160 miles (3,476 kilometers) across then, we need to re-evaluate everything we know about physics and the engineering of structures on a planetary scale. When we contemplate megastructures on a planetary scale, what immediately comes to mind is a **Dyson's sphere**.

A Dyson Sphere is a hypothetical megastructure originally described by **Freeman Dyson**. Such a "sphere" would be a system of orbiting solar power satellites meant to completely encompass a star and capture most or all of its energy output. Dyson speculated that such structures would be the logical consequence of the long-term survival and escalating energy needs of a technological civilization, and proposed that searching for evidence of the existence of such structures might lead to the detection of advanced intelligent extraterrestrial life. n Dyson's original paper, he speculated that sufficiently advanced extraterrestrial civilizations would likely follow a similar power consumption pattern as humans, and would eventually build their own sphere of collectors. Constructing such a system would make such a civilization a **Type II Kardashev Civilization**. http://en.wikipedia.org/wiki/Dyson_sphere

A Dyson's sphere has always been considered an intellectual exercise of the imagination and not something that is even remotely feasible, until now!

According to **Sgt Major Robert Dean**, who carried the highest security clearance there is while serving at **SHAPE (Supreme Headquarters Allied Powers Europe)** in the 1960's, the US intelligence community is in possession of a top-secret dossier entitled *The Assessment,* detailingthe history of extraterrestrial visitation, and an assessment of their purposes for visiting our planet. For the past twenty years, Sgt Major Dean had fought to make this information public knowledge.

In 2009, Robert Dean gave a lecture in Barcelona, Spain at a UFO conference attended by many UFO researchers, investigators, and the general public. In his lecture, Deans tells the audience how NASA destroyed 40 rolls of official Moon mission photographs among other photographs which reveal incredible objects and structures not only on the Moon but also, out in space. The officials in NASA were so frightened by what they had found in the way of lunar structures and flying objects that they destroyed the evidence and ensured that the public would never find out about this evidence either. If NASA was fearful and perplexed as to what to do then, they assumed that the public would also be fearful as per Brooking Institute Report to NASA. The fullvideo of Dean's lecture can be found at http://www.youtube.com/watch?v=_ngvIP0Za9M.

Fortunately, not all space agencies around the world think like NASA but, act more independently in their exploration of space. The **Japanese Space Agency** was one such space agency and they decided to send Robert Dean some very interesting photographs taken by the Apollo 13 astronauts on their way to the Moon. NASA at one time sent photos out to many spaceagencies of their space exploration endeavours particularly the Apollo Moon missions. However these agencies didn't destroy their photocopies like NASA but, instead decided to send some to Dean which he shared with the people of Barcelona. (See below).

The UFOs below (A, B, and C) were of immense size, saucer-shaped objects B and C were estimated to be approximately 2miles in diameter and the cigar-shaped object C is believed to be5 miles in length! Here were ET spacecraft beyond anything that humans could build.

These giant Extraterrestrial spacecraft are real, they ply the vast interstellar reaches of space and now they are flying through our solar System. These huge craft represent an ET presence in our Solar System, a fact which is undeniable to all those who investigate its reality. However, from the viewpoint of astronomers, they are still in search for the elusive signs of ET's presence in the universe. Scientifically, the percentage of certainty that life exists elsewhere increases daily as NASA discovers new planets orbiting in other star systems with the aid of high-powered telescopes like Hubble and Kepler. We then must ask ourselves, do they exist on any of the planets within our Solar system or do they live many light years away from us? And, just how large are these Extraterrestrial spaceships?

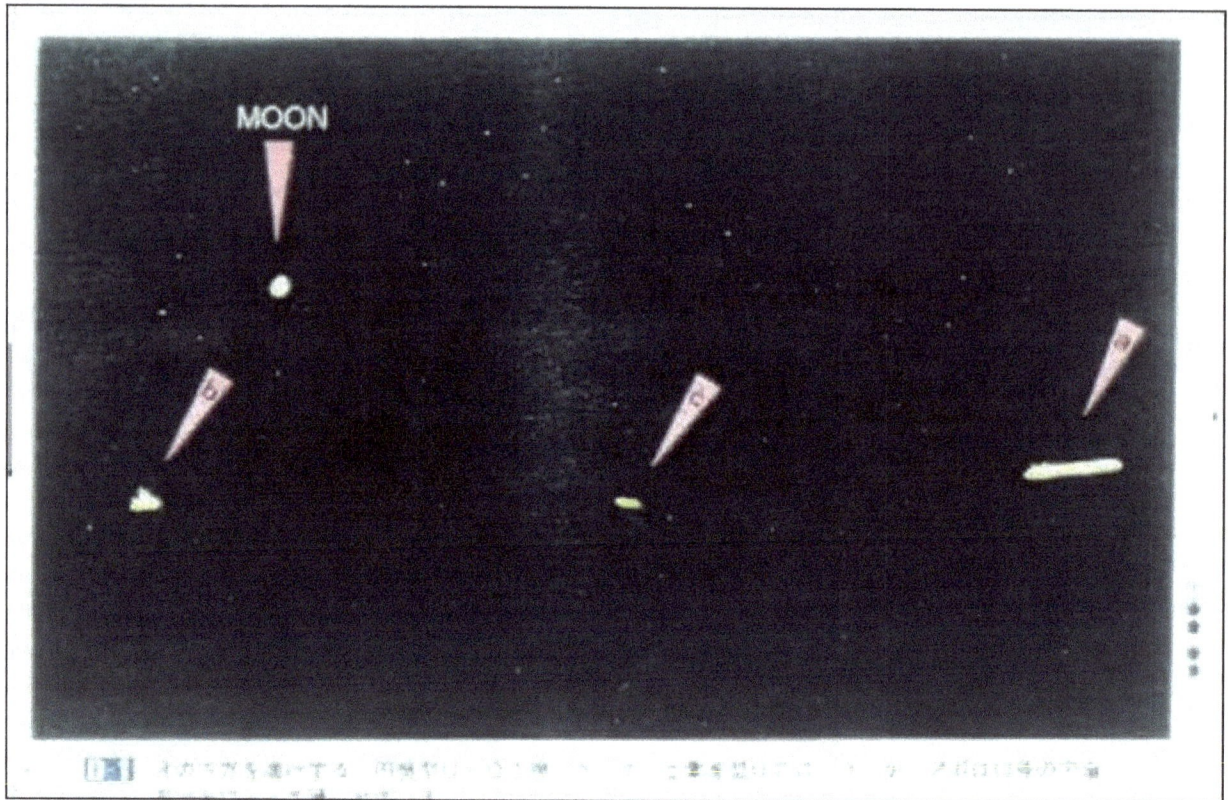

**A Japanese duplicate of an American NASA photo taken by astronauts aboard
the ill-fated Apollo13 command capsule.
The Moon is marked and three UFOs are labelled as A, B, and C.**
https://gregdougall.wordpress.com/tag/leak/

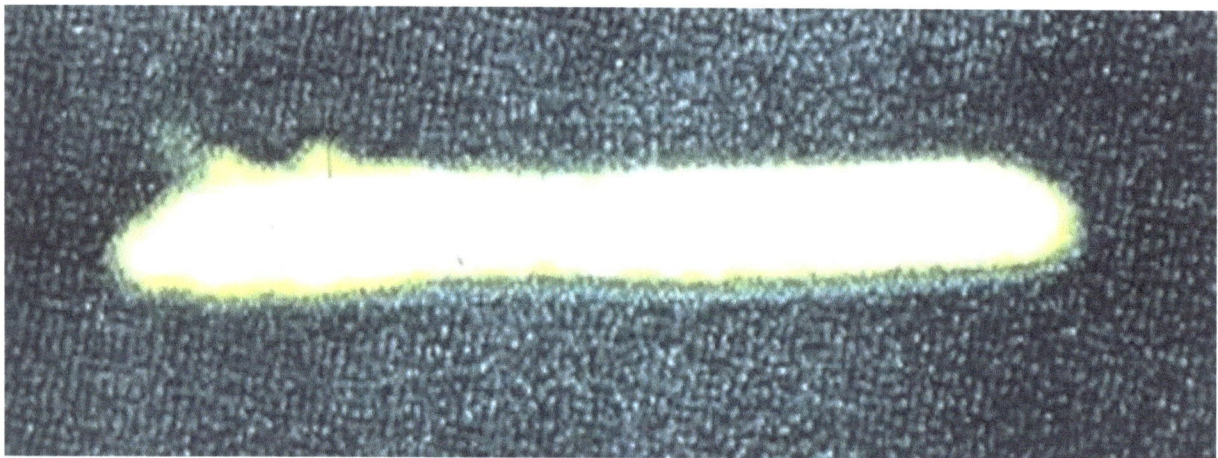

**The cigar shape UFO - positive image above and the same cigar shape UFO - negative
image below is estimated to be 5 miles in length!**
https://gregdougall.wordpress.com/tag/leak/

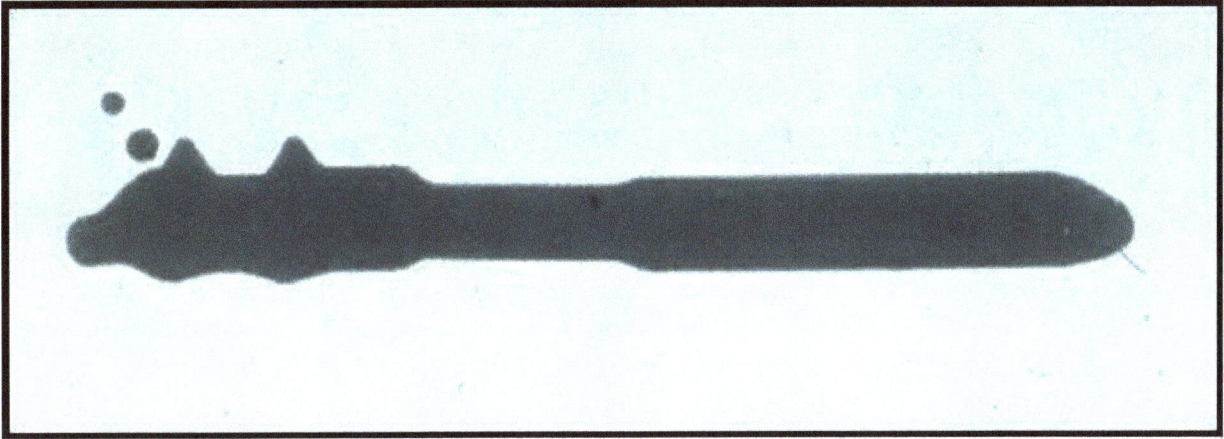

The negative image of the 5 mile long mothership, note the two smaller spacecraft at the rear of the ship
https://gregdougall.wordpress.com/tag/leak/

CHAPTER 75

TO INFINITY AND BEYOND: A VIRTUAL SPACE TOUR TO THE PLANETS IN THE SOLAR SYSTEM

Intelligent, sentient life forms are coming to our planet on a regular basis perhaps, to monitor andassess the direction of humanity. Like NASA, we need to explore the universe by taking a virtualjourney into space. Since almost everyone here on Earth will never have the opportunity to actually leave the planet physically as an astronaut at least in our lifetimes, we will have to settle for the technological efforts of other countries' space programs and learn through the eyes and the experiences of astronauts and cosmonauts. In similar fashion to the late astronomer, Carl Sagan and his popular television Cosmos, we must take a virtual voyage not only of the mind but, through robotic eyes of the onboard cameras of orbiting satellites, planetary landing vehiclesand robotic excursion rovers, as we explore the planets and their moons in our Solar System. Weare in search of life beyond Earth and beyond the Moon.

We've have already looked at the space around in Earth orbit, we have gone to the Moon with the astronauts of the Apollo program and through their testimony and photographic evidence, we've established and proven beyond any doubt that ET life once existed on the Moon and may still exist there today, possibly below its surface. Now, we are searching to see if life may exist elsewhere in our solar neighbourhood. Our cosmic voyage will take us to every planetary body inour Solar System, every moon, asteroid, comet and meteorite that flies into our corner of the galaxy.

Keep in mind the old adage that **"life is where you find it"** and our Solar System promises tolive up to that adage with many surprises that we will find along the way in our journey.

Our first stop is to our solar system's star, the **Sun.**

Fasten your seat belts and harnesses, it going to be an amazing journey!

CHAPTER 76

THE SUN – ("SOL / HELIOS" – THE GIVER OF LIFE)

Our first visit is to venture inward toward **Sol**, our **Sun**, and we behold that radiant and luminescent body, that giver of life within our **Solar System**, in a way not ever imagined before. Through the various special spectral filters and lenses aboard solar observatory spacecraft, we behold the Sun's surface stripped of it blinding brilliance and see for the first time, with crystal clarity, an immense nuclear blast furnace, a turbid surface pockmarked with the occasional, but predictable dark sunspot activity. At the same time, we witness spasmodic ribbons of solar prominences or solar matter being ejected hundreds of millions of miles into the far reaches of our Solar System. One such flick of a **Solar Flare** would be enough to end our world were it notfor the protective magnetic field surrounding our planet. We exist in our star system through a divine choreographed dance, between a precarious balance of life and death!

SOHO

SOHO (Solar and Heliospheric Observatory) is a project of international collaboration between **ESA** and **NASA** to study the Sun from its deep core to the outer corona and the solar wind. The SOHO spacecraft was built in Europe was launched on December 2, 1995. The twelveinstruments on board SOHO were provided by European and American scientists. Nine of the international instrument consortia are led by European **Principal Investigators (PI's)**, three by PI's from the US. Large engineering teams and more than 200 co-investigators from many institutions supported the PI's in the development of the instruments and in the preparation of their operations and data analysis. NASA was responsible for the launch and is now responsible for mission operations. Large radio dishes around the world which form NASA's **Deep Space Network** are used for data downlink and commanding. Mission control is based at **Goddard Space Flight Center** in Maryland. http://sohowww.nascom.nasa.gov/about/about.html

STEREO

STEREO (Solar Terrestrial Relations Observatory) is the third mission in NASA's **Solar Terrestrial Probes program (STP)**. The mission launched in October 2006, has provided a unique and revolutionary view of the Sun-Earth System. The two nearly identical observatories -one ahead of Earth in its orbit, the other trailing behind - have traced the flow of energy and matter from the Sun to Earth. STEREO has revealed the 3D structure of coronal mass ejections; violent eruptions of matter from the sun that can disrupt satellites and power grids and help us understand why they happen. STEREO is a key addition to the fleet of space weather detection satellites by providing more accurate alerts for the arrival time of Earth-directed solar ejections with its unique side-viewing perspective. http://www.nasa.gov/mission_pages/stereo/main/index.html

After the first flights on captured V-2 rockets had paved the way with early observations in the far ultra-violet (UV) starting in 1946, NASA initiated the **Orbiting Solar Observatory (OSO)** series of dedicated unmanned solar satellites and undertook the first major manned solar missionfrom space called **Skylab**. Together, these early solar programs provided many outstanding

observations in the UV, EUV, X-rays, and gamma-rays that provided the foundation of solar space science. Several other nations followed with their own solar missions, either alone as in the case of the Soviet Union or in collaboration with the United States. What follows is a summary of the major solar missions that have made or are expected to make major contributions to our understanding of solar physics through remote sensing.
http://www.scholarpedia.org/article/Solar_Satellites

Table 1. Spacecraft from the NSSDC Master Catalog (NMC) and other missions with significant solar observations

Name	Country	Launch Dates		
1 AE-C, D, E	USA	12/16/1973	10/6/1975	11/20/1975
2 Aeros-A , B	Germany/USA	12/16/1972	7/16/1974	
3 Alouette 1, 2	Canada/USA	9/29/1962	11/29/1965	
4 Apollo 16 CSM	USA	4/16/1972		
5 Ariel 1	UK/USA	4/26/1962		
6 Atlas 1 , 2, 3	USA	3/24/1992	4/8/1993	11/3/1994
7 ATS 2, 5	USA	4/6/1967	8/12/1969	
8 Azur	Germany/USA	11/8/1969		
9 CGRO	USA	4/5/1991		
10 Coriolis - SMEI	USA	1/6/2003		
11 CORONAS-F, I, Photon	Russia	7/31/2001	3/2/1994	1/30/2009
12 Cosmos 262	Russia	12/26/1968		
13 D2B	France	9/27/1975		
14 DMSP 5D-2/F06	USA	12/21/1982		
15 EOM-A	Multiple	11/1/1985		
16 EOS-CHEM1	USA	1/1/2002		
17 ERBS	USA	10/5/1984		
18 ERS 17, 27	USA	7/20/1965	4/28/1967	
19 ESRO 2, 2A	USA/ESA	5/17/1968	5/29/1967	
20 EURECA 1	ESA	8/2/1992		
21 Explorer 33, 35	USA	7/1/1966	7/19/1967	
22 Galileo Orbiter	USA/Germany	10/18/1989		
23 Genesis	USA	8/8/2001		
24 GLAST (renamed Fermi Gamma-ray Space Telescope)	USA	6/11/2008		
25 GOES 1-12	USA	10/16/1975	-	7/23/2001
26 GOES-G, K, M, X	USA	5/3/1986	-	7/22/2001
27 Helios-A, B	Germany	12/10/1974	1/15/1976	
28 Hinode	Japan	9/22/2006		

Table 1. Spacecraft from the NSSDC Master Catalog (NMC) and other missions with significant solar observations

Name	Country	Launch Dates		
29 Hinotori	Japan	2/21/1981		
30 IMP-H	USA	9/23/1972		
31 Injun 3	USA	12/13/1962		
32 Interball Tail Probe	Russia	8/2/1995		
33 Intercosmos 4, 7, 9, 11, 16	Russia	10/14/1970	-	7/27/1976
34 ISEE 3/ICE	USA	8/12/1978		
35 KH-5 9034A	USA	5/15/1962		
36 Lambda 4S-1, 2, 3, 4	Japan	9/26/1966	-	9/22/1969
37 Mariner 4, 6, 7	USA	11/28/1964	2/24/1969	3/27/1969
38 Nimbus B, 3, 4, 7	USA	5/18/1968	-	10/24/1978
39 NOAA 11, 13, 14, 15, 17, 18	USA	9/24/1988	-	5/20/2005
40 OAST Flyer	USA	1/11/1996		
41 OGO 2, 3, 4, 5, 6	USA	10/14/1965	-	6/5/1969
42 Ohsumi	Japan	2/11/1970		
43 OSO 1 - 8	USA	3/7/1962	-	6/21/1975
44 OV1- 1, 10, 11, 15, 17, OV5-9	USA	1/21/1965	-	5/23/1969
45 P78-1	USA	2/24/1979		
46 Phobos 1, 2	Russia	7/7/1988	7/12/1988	
47 Pioneer 6	USA	12/16/1965		
48 Pioneer Venus Orbiter	USA	5/20/1978		
49 Prognoz 1, 2, 4 - 10	Russia	4/14/1972	-	4/26/1985
50 RAE-A, B	USA	7/4/1968	6/10/1973	
51 Ranger 2, 3	USA	11/18/1961	1/26/1962	
52 RHESSI	USA	2/5/2002		
53 S3-1	USA	10/29/1974		
54 SAC-B	USA	11/4/1996		
55 SAMPEX	USA	7/3/1992		
56 San Marco-D/L	Italy/USA	3/25/1988		
57 Shinsei	Japan	9/28/1971		
58 SIGNE 3	France/Russia	6/17/1977		
59 Skylab	USA	5/14/1973		
60 SME	USA	10/6/1981		
61 SMM	USA	2/14/1980		
62 SMS 2	USA	2/6/1975		

Table 1. Spacecraft from the <u>NSSDC Master Catalog (NMC)</u> and other missions with significant solar observations

Name	Country	Launch Dates		
63 SNOE	USA	2/26/1998		
64 SOHO	ESA	12/2/1995		
65 SOLRAD 1, 2, 4B, 6, 7A, 7B, 8, 10, 11A, 11B	USA	6/22/1960		3/15/1976
66 SORCE	USA	1/25/2003		
67 Spartan 1, 201-1, 201-3, 201-5	USA	9/13/1994	-	11/1/1998
68 Spartan-A, B, C	USA	6/17/1985	7/9/1992	12/1/1986
69 Sputnik 2	Russia	11/3/1957		
70 SR 4/GREB 4	USA	1/24/1962		
71 SSBUV04, 05, 06	USA	3/24/1992	4/8/1993	11/3/1994
72 STEREO A, B	USA	10/26/2006	10/26/2006	
73 STP P78-1, P80-2	USA	2/24/1979	1/1/1981	
74 STS 3, 9, 34, 41, 43, 45, 51B, 41F	USA	3/22/1982		3/24/1992
75 STS Sunlab-A	USA	7/1/1986		
76 Taiyo	Japan	2/24/1975		
77 TD 1A	Netherlands	3/12/1972		
78 TERRIERS	USA	5/18/1999		
79 TIMED	USA	12/7/2001		
80 TRACE	USA	4/2/1998		
81 UARS, UARS 2	USA	9/12/1991	10/1/1989	
82 Ulysses	ESA	10/6/1990		
83 Vela 5A/B, 6A/B	USA	5/23/1969	4/8/1970	
84 Wind	USA	11/1/1994		
85 WRESAT	Australia	11/29/1967		
86 Yohkoh	Japan	8/30/1991		

<u>http://www.scholarpedia.org/article/Solar_Satellites</u>

Planetary Size UFO Craft Around the Sun

All of these above satellites listed have provided remarkable data and/or photographs of the Sun and its **heliosphere**. The photos transmitted back to Earth have frequently shown not only amazing surface images of sun activity but, quite unexpectedly reveal the presence of anomalous objects orbiting and interacting with the Sun. These objects are of planetary size, as small as Mercury or a moon and as large as Jupiter yet, they are not planets but artificial constructions! They are not defects in photographic imaging from the orbiting satellites or flaws in laboratory processing. They often appear as disc-shaped or linear structures or as immense globes or a combination of cylinders, discs, and balls similar to the "Star Trek Enterprise" or as a huge "X-Wing fighter from Star Wars"!

In some cases, there are **"Borg-like Cubes"** photographed near the Sun but, these and a few of the orbs could possibly be digital flaws where pixel information is missing but, this is to be expected, just not on so many photographs or from so many different satellites. This would be indicative of poor quality engineering of multi-million dollar satellites that don't function properly and this can hardly be the case in all situations where these strange objects have been imaged. Neither are these objects merely distant stars or galaxies shining through solar flares and prominences and neither are they chunks of solar mass ejecta which usually dissipate in a very short time period. They are also, not digital imaging remnants of a planet photographed the day before, like Mercury because, these objects are within the orbit of Mercury and are therefore closer to the Sun! In fact, some have been video recorded coming out of the Sun and then zipping away into other reaches of space!!!

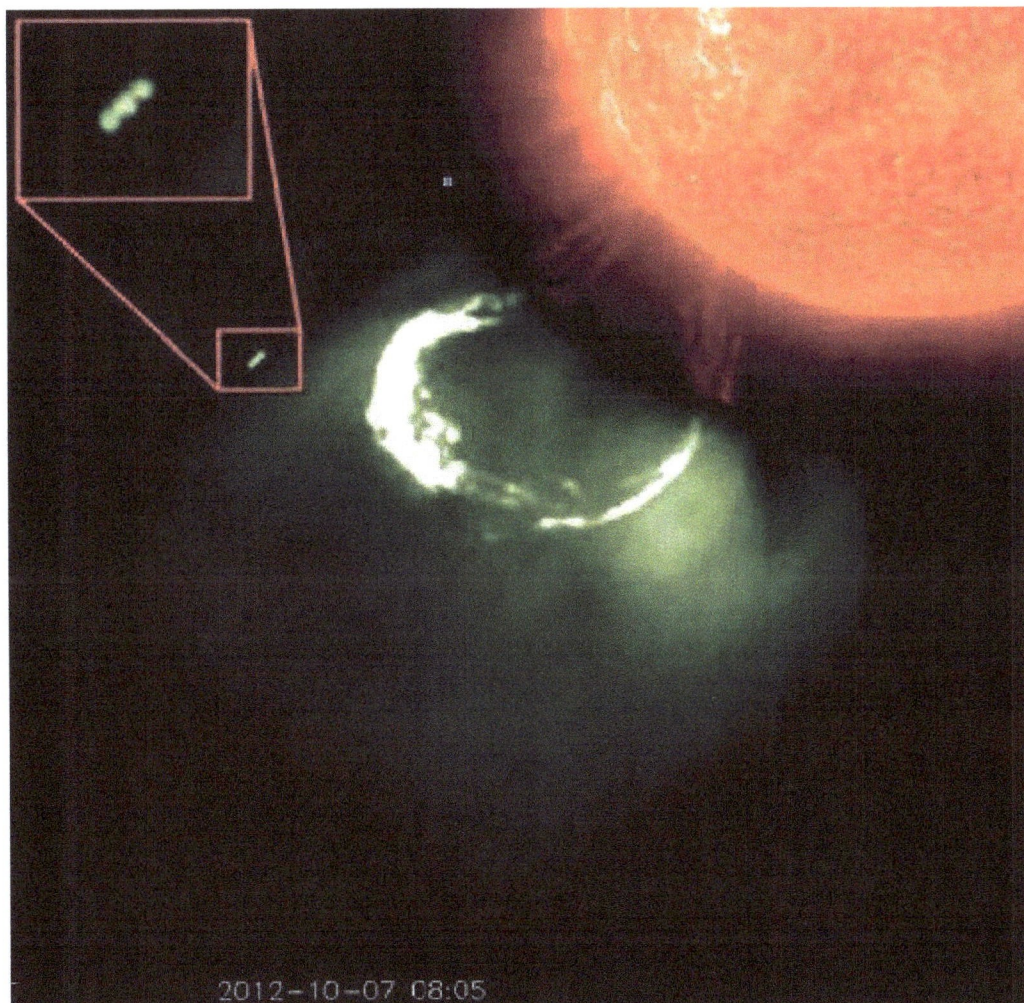

The STEREO spacecraft photographs the Sun in extreme UV light and the corona in white light with its COR1 instrument. Earth is represented by the blue dot just above the solar flare for size comparison!
https://www.youtube.com/watch?v=WB-_Li7rACA

If these mysterious objects are artificial, it, therefore, implies an Extraterrestrial presence that is very active within our Solar System, whether originally from here or from another star system. If this is indeed the case, we are dealing with such highly advanced Extraterrestrial civilizations as to be considered as either, a **Type I or Type II Civilization** maybe even, a **Type III** based upon the **Kardeshev Scale of Galactic Civilizations**, a concept promoted by American theoretical physicist **Michio Kaku**. In comparison to their technology, humanity is not even a Type 1 Civilization, we are still in our infancy playing with chemical rockets.

The above photo shows a strange shape like 3 cylinders or globes joined together below a solar prominence. **Prominences** are clouds of cooler plasma that hover above the Sun, tethered by unstable magnetic forces, which often break away or fade within days or weeks. This planetary size object is too big to be space debris, yet its structure appears to have equal proportions that in a balance to each other which eliminates anything being emitted from the Sun.
http://www.thetruthbehindthescenes.org/2012/10/18/giant-ufo-near-sun-when-prominence-breaks-away-from-the-sun-october-2012/

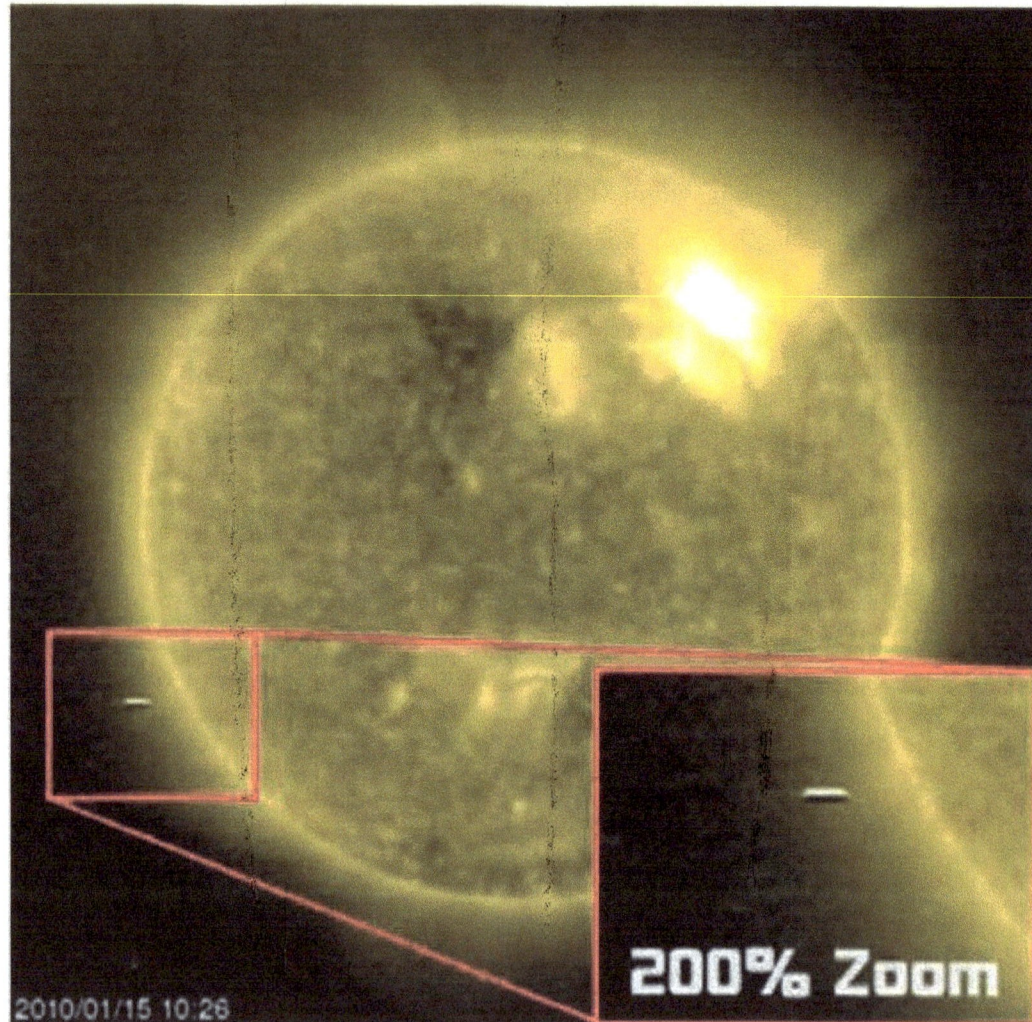

A disc-shaped craft is photographed the SOHO satellite on January 15, 2010
http://www.ufohypotheses.com/sun.htm

316

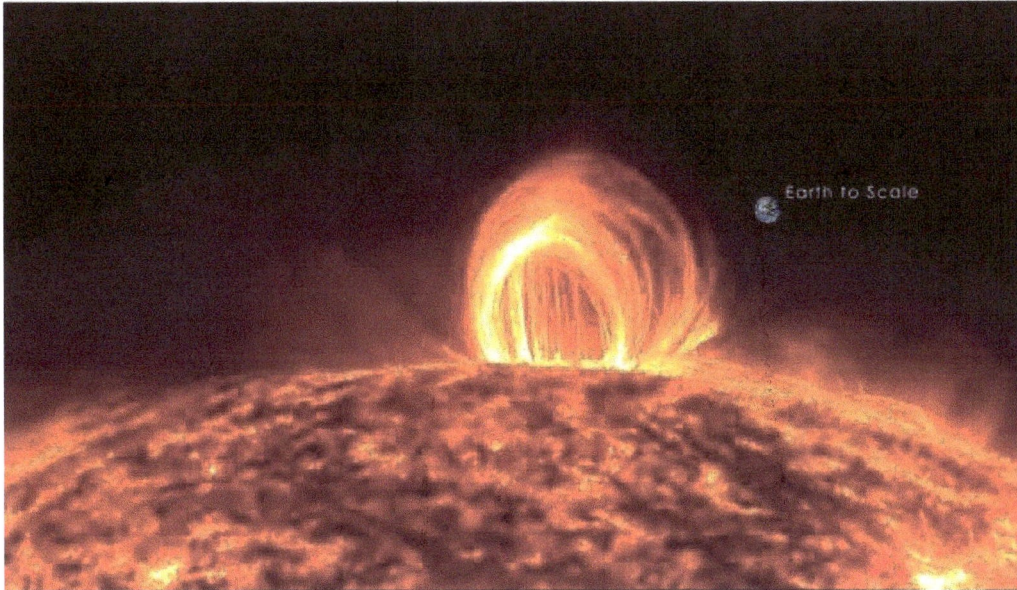

The Earth in size comparison to a Solar Prominence

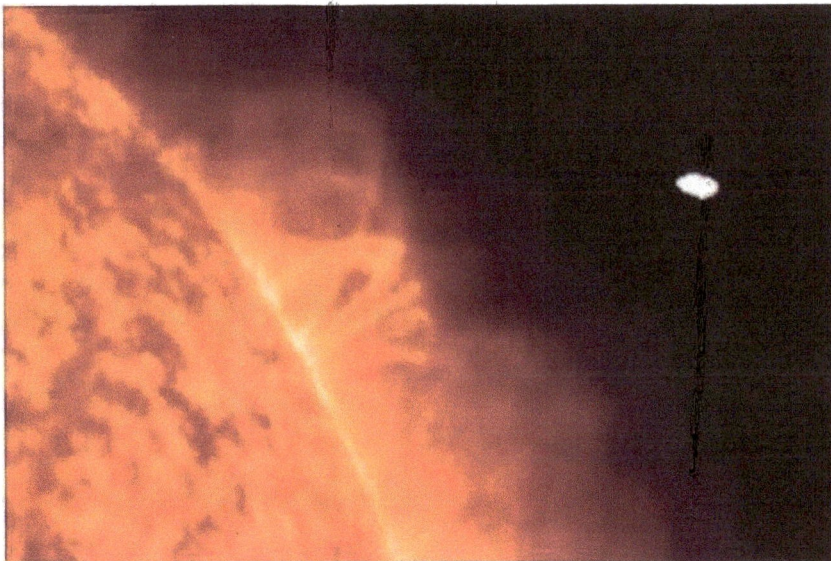

**This SOHO image captured a brilliant reflective UFO on January 28, 2012
and its size would indicate that it is many times larger than the Earth**

Many Ufologists have become amateur astronomers who view the night sky with a computer instead of a telescope by scanning the many photographic images and videos taken by the NASA and ESA solar observatory satellites. One such image showed a giant object, football-like in shape, many times larger than Earth and seems to be reflecting the sun's light which may indicate a metallic exterior. It can be argued that an ET spacecraft constructed on a planetary scale would be impossible, but if their technology is millions of years ahead of ours, then an ET civilization may be using utilizing other engineering methods that would probably be beyond our concept of understanding and science. The science of consciousness may be that method by which planetary scale structures are engineered. Suffice it to say, that no robots are needed; no longevity of its artisans and engineers is necessary, and no extended time periods of hundreds of years for the creation and construction of the ships. Merely the unified desire of a world civilization at peace with itself is all that is needed.

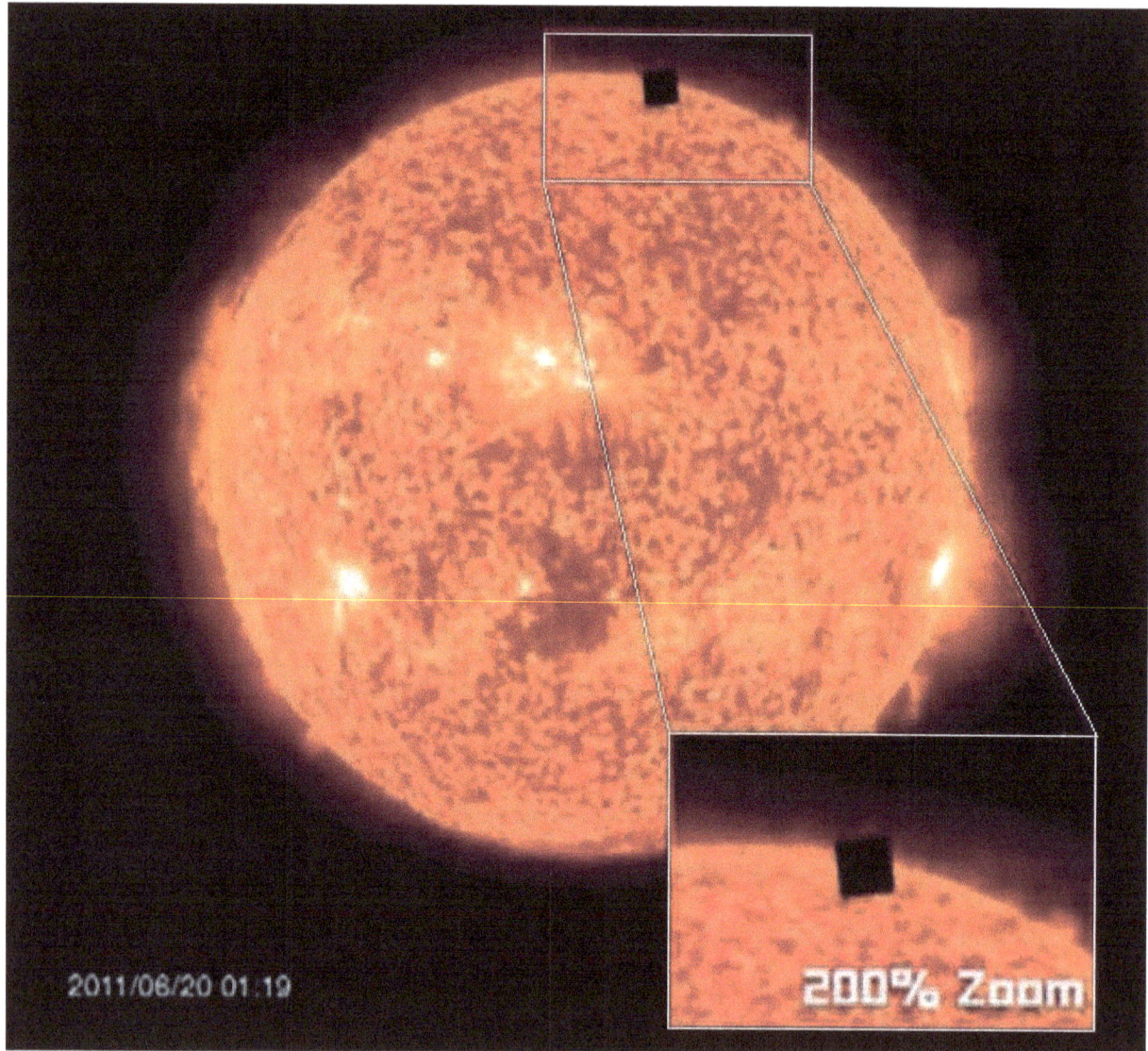

Is this an immense "Borg" ship orbiting our Sun or a data flaw in pixel information transmission or is it a deliberate elimination of data of something more unusual?
http://www.dailymail.co.uk/sciencetech/article-3573998/What-mysterious-giant-black-cube-spotted-near-sun-Alien-hunters-claim-images-accidentally-reveal-Nasa-plot-hide-orbiting-UFOs.html

Is it be possible that over enthusiastic UFO hunters are seeing and interpreting every photo blemish transmitted back from the solar observatories like SOHO and STEREO as UFOs and justifying their claims with the available **photoshop** software to enhance what they think they are seeing? This can be a problem when someone is an untrained or amateur photographic imaging

318

specialist who doesn't understand the mechanics behind photos that contain important information from those that contain errors in data. Once again, this enthusiasm to be the first to discover something important in Ufology may, in fact, hinder its progress. If you are not sure about what you are looking at, ask an expert in that field to assist you in your investigations, rather than publishing or sharing it on YouTube.

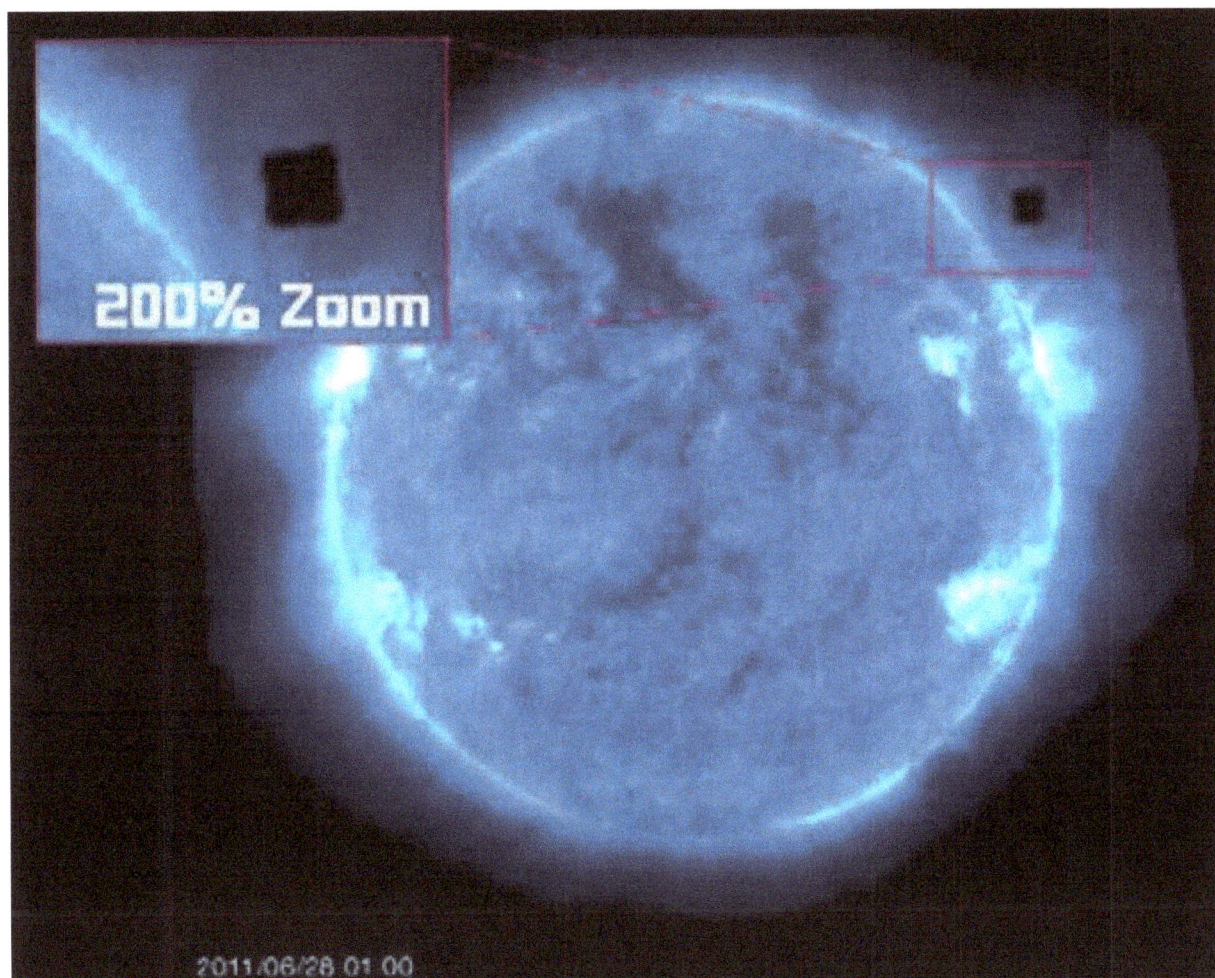

Another pixel flaw in data transmission or some unusual deliberately removed by NASA?
http://www.ufosightingsdaily.com/2011/06/giant-black-cube-orbiting-earth-sun.html

The photograph below with inserts is from Helioviewer.org an open-source project for the visualization of solar and heliospheric data. The project is funded by ESA and NASA.

In the screenshot, you can see a huge UFO emitting light, which can't be a sun reflection due to its odd light emitting angle. UFO has more lights on its horizon, which also clear suggest that UFO has its own light source. Interestingly NASA didn't upload the above image on their official SDO website "http://sdo.gsfc.nasa.gov/data/aiahmi/"
http://erigia.blogspot.be/2012/07/incroyable-ovni-geant-de-lumiere-sur-le.html

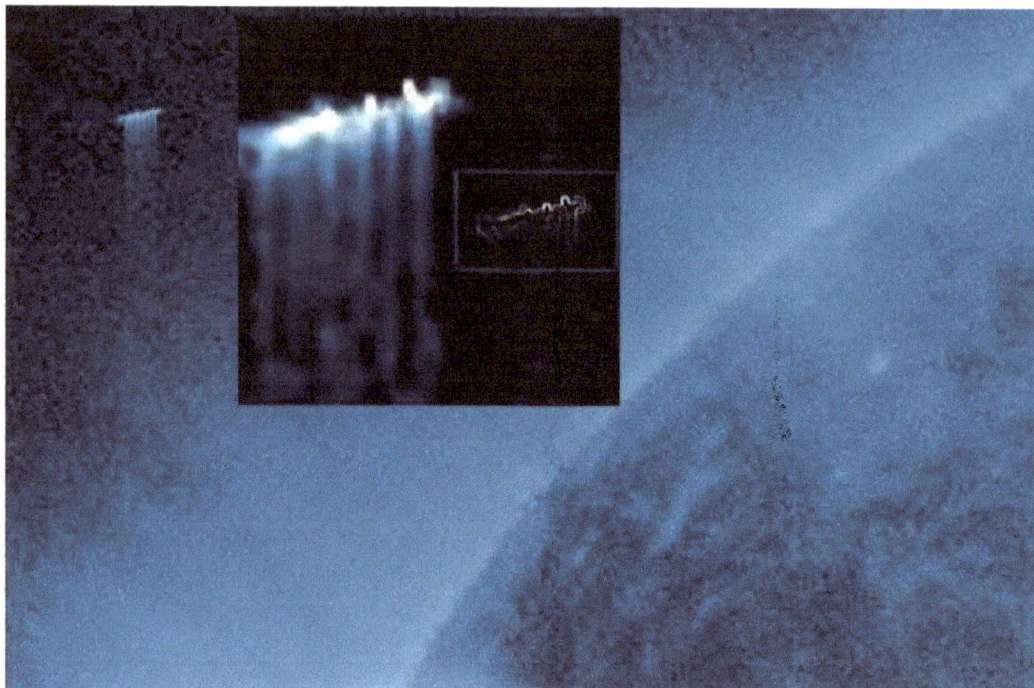

**On July 22, 2012, at 14:40:46 UT NASA SDO observatory
photographed a gigantic UFO Mothership near Sun.**
https://pics-about-space.com/nasa-ufos-near-the-sun?p=1#

**Could this be an armada of planet size UFOs in orbit around the Sun
or is it nothing more than data transmission flaws in image pixelation?**
http://truedemocracyparty.net/2016/07/ufo-fleets-massive-near-sun-again/

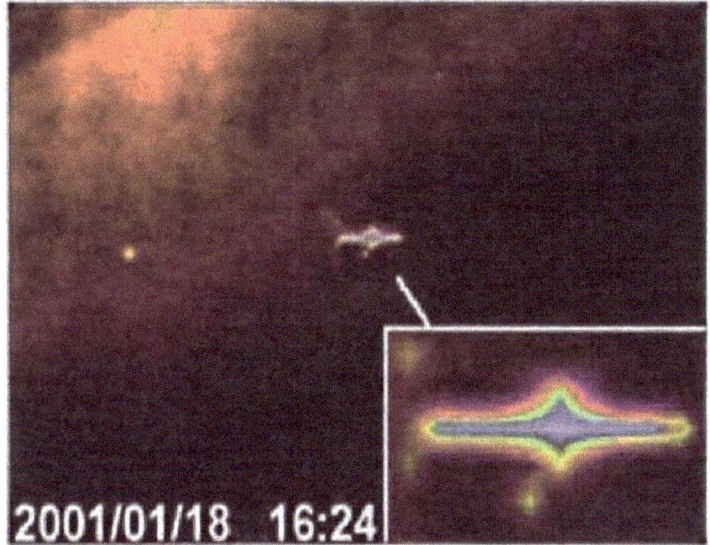

'UFO' on NASA camera

By TIM UPTON

WASHINGTON: The object is certainly unidentified and appears to be flying.

Whether this enlarged image really shows a UFO piloted by aliens remains to be seen. But according to the people who released this photo and hundreds like it are the best evidence yet of the existence of spacecraft from other worlds.

UFO investigators say the image was captured by the Solar and Heliospheric Observatory (SOHO), a NASA satellite that was launched in 1996 to observe the sun. Since then, it is said, SOHO has captured hundreds of images of UFOs moving along a kind of alien superhighway.

SOHO is more than 1.5 million kilometres from Earth, with its camera trained towards the sun. Experts say the photographed objects are likely to be only hundreds of kilometres from its lenses.

Graham Birdsall, editor of UFO magazine, said: "The images are irrefutable in that they are from official satellites owned by NASA. They resemble the kind of spacecraft we used to see in sci-fi films like Star Trek."

2001/01/18 16:24

UTTERLY ALIEN: The image investigators say shows a UFO.

A newspaper clipping of a saucer shape UFO that NASA says is an enhanced pixel error (See explanation below)
http://www.filosofix.com.br/blogramiro/?p=37

"What are Those Flying Saucer-Shaped Objects in the LASCO Images?"

"The "funny-looking spheroid" is a typical response of the SOHO LASCO coronagraph CCD detector to an object (planet or bright star) of small angular extent but so bright that it saturates the CCD camera so that "bleeding" occurs along pixel rows. There is a bright horizontal streak on either side of the image because the charge leaks easier along the direction in which the CCD image is read out by the associated electronics.

CCD stands for charge-coupled detector and refers to a silicon chip, usually a centimeter or two across, divided into a grid of cells, each of which acts like a small photomultiplier in that an incoming photon knocks loose one or more electrons. The electrons are "read out" by row (fast direction) and column (slow direction), the current converted to a digital signal, and each cell or picture element ("pixel") thus assigned a digital value proportional to the number of incoming photons in that pixel (the brightness of the part of the image falling on that pixel). This is the same kind of detector as is used in a hand-held video camera, though until recently, the analog-to-digital conversion was left out in consumer devices.

If you point a video camera at a very bright source (say, the Sun), the image "blooms" or brightens all over --- there are so many electrons produced in the pixels corresponding to the bright source that they spill over into adjacent rows and column, perhaps over the entire detector.

Better CCD's will "bleed" only along the fast readout direction (a single row), and perhaps a few adjacent rows.

The LASCO and EIT CCD cameras include "anti-bleed" electronics which limit the pixel bleeding around bright sources to less than the full row (and usually no adjacent rows). In the case of a marginally too-bright object, the pixel bleeding will be only a few pixels in either direction along the fast readout direction. Thus, the "flying saucer" images.

A few of the LASCO images that have appeared on the "extraterrestrial" Web sites show much larger and brighter, but still saucer-like features. These images are in fact obtained with the instrument door closed, but with an incorrectly long exposure. The big "saucers" result from massive pixel bleeding along every row of the detector containing part of the image of the "opal," or small diffusing lens, in the instrument door, that is used for obtaining calibration data.

If your correspondents still prefer to believe that the pixel-bled images of planets or bright stars are something else, ask them why the extended part of the "saucers" (i.e., the pixel bleeding) always occurs in the same direction relative to the image --- even when the spacecraft is rolled relative to its normal orientation relative to the Sun."
http://sohowww.nascom.nasa.gov/explore/faq.html

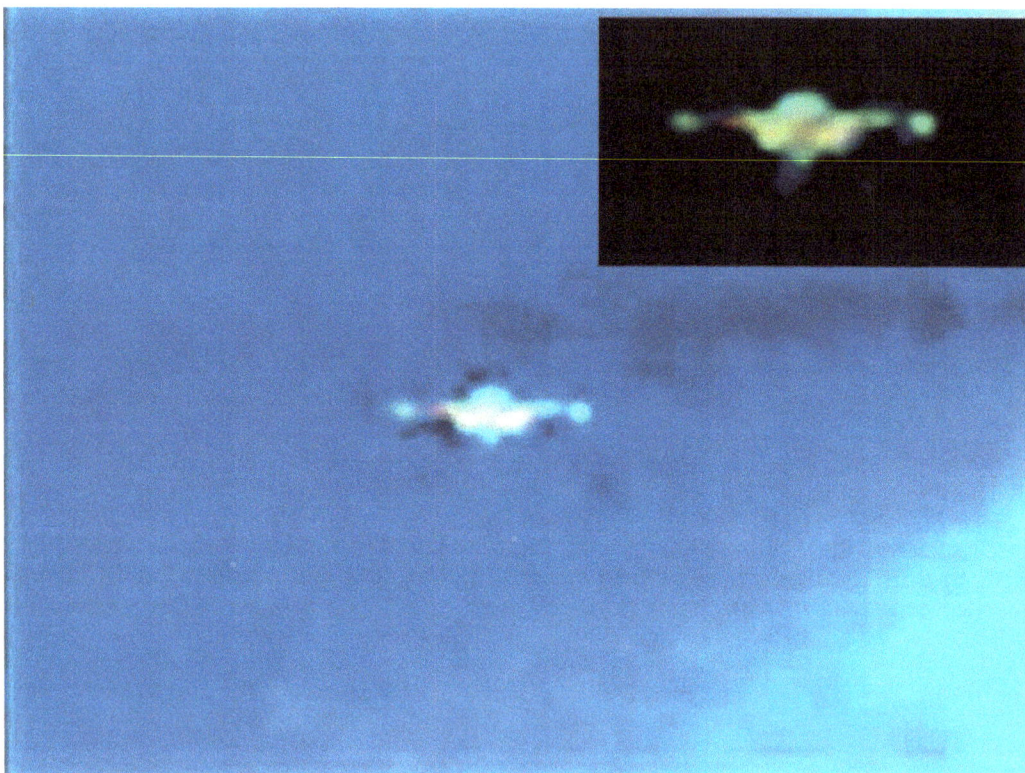

The "Suncruiser," is a flying saucer-like anomaly that has appeared frequently on SOHO images. Such explanations from SOHO officials vary and only increase the suspicions of those following the solar observatory developments

http://www.ancientmoons.com/beamships.htm

322

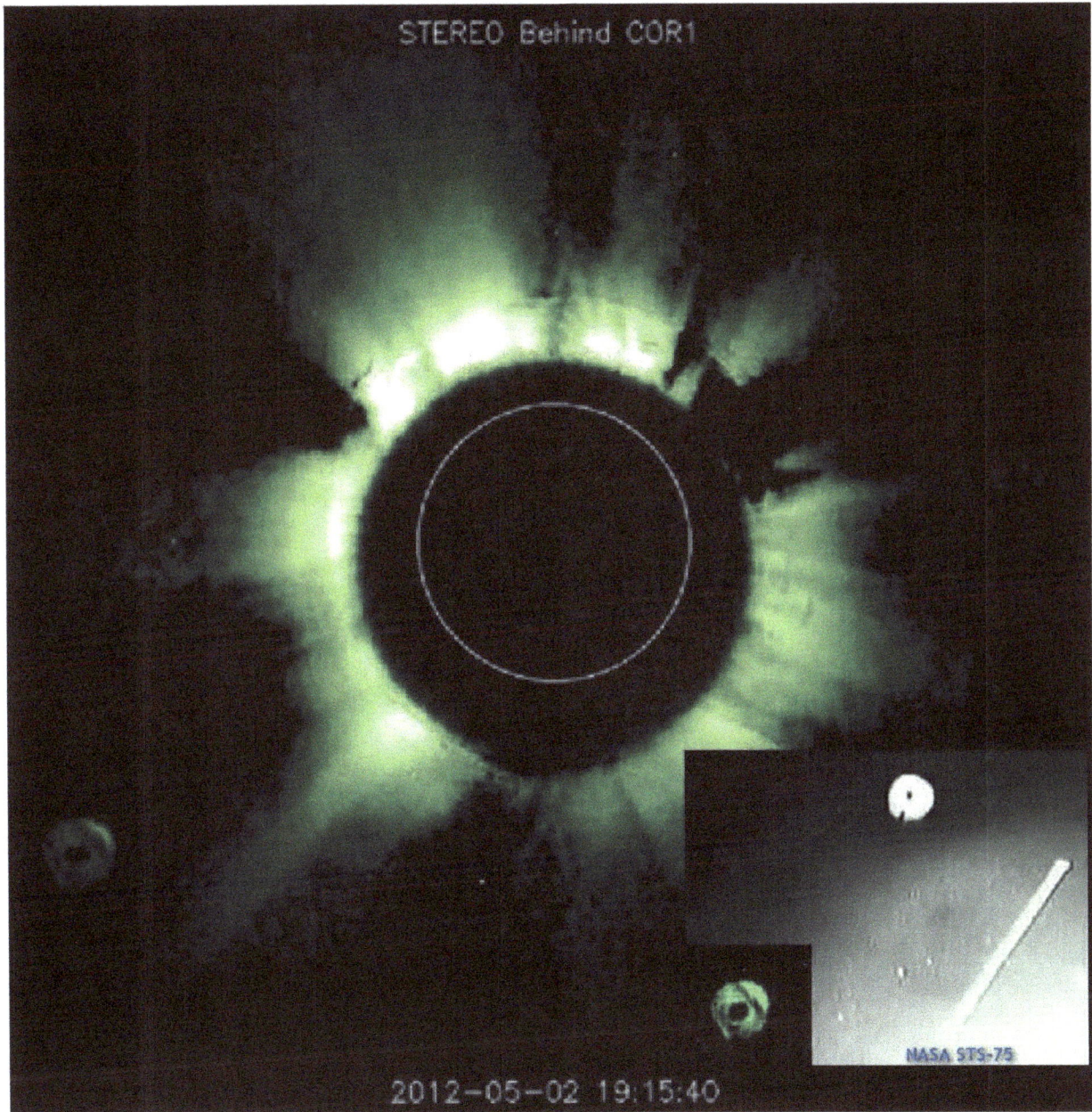

A planet size "donut" shape UFO taken by the STEREO solar satellite close to the Sun is similar to the UFO seen beside the cable of NASASTS-75 "tether experiment"

A UFO that was seen on SOHO/NASA photos of the sun on May 2nd. This UFO is similar to the NASA STS-75 photo as seen in the above photo insert. It is the same exact shape as the UFO that cut the NASA cable that the space shuttle had released into space. The space tether experiment was a joint venture of the US and Italy, it was a scientific experiment in which a large, spherical satellite was to be deployed from the US space shuttle at the end of a conducting cable *(tether)* 20 km (12.5 miles) long. The idea was to let the shuttle drag the tether across the Earth's magnetic field, producing one part of a dynamo circuit. However, the tether was broken in two when one of these UFOs flew through it cutting it either unintentionally or on purpose.

These round UFOs actually look like giant living creatures which seem to throb or pulsate. The reason for this is that these objects actually do have a biometric sentient quality to their structure. They are in fact a living machine! These objects are piloted by ETs and are able to approach near the sun's magnetic field area perhaps, "feeding" or interacting in some special way on the magnetic fields much in the same way as they did with NASA's tether experiment to interact with its magnetic field.

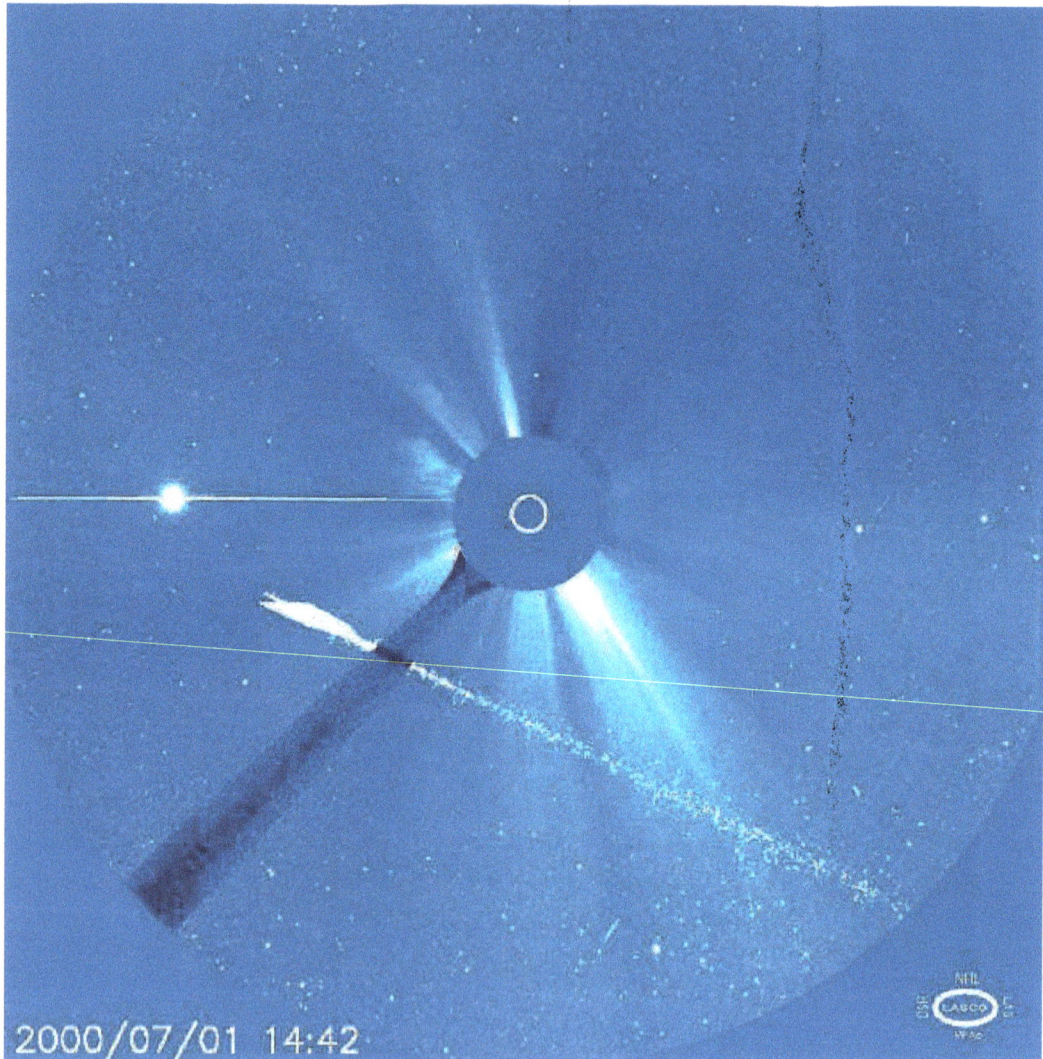

**Is this yet another disc-shaped craft passing in front of SOHO's camera
or is it shooting some kind of Star Wars particle beam weapon?
(Mars is seen to the left as a bright globe with a line through it)**
http://www.enterprisemission.com/soho.htm

Once again, **Richard Hoagland** and his team from *'The Enterprise Mission'* were paying attention to some of NASA's releases of photographic images to the public when "on July 1st, 2000 SOHO recorded yet another extraordinary event. What seemed to be a wedge or disk shaped object and a trail of some kind could be seen passing right in front of the SOHO imager. Now internally, we've had some disagreement as to what this image actually shows. Is it a large wedge-shaped object, firing a beam of some kind? Or is it two distinct objects, one disk-shaped and in the foreground, and the other having passed behind the "disk" from SOHO's perspective?

The implications of this second option are somewhat staggering. If the second, background object is a projectile, captured by the camera just as it passed by the "disk" in a near miss, are we seeing some kind of "Star War" taking place in front of SOHO? It has certain characteristics of a rail gun type projectile, similar to what was seen in the STS-48 video. And if indeed somebody is shooting at somebody else out there, just who is doing the shooting? And who owns the "target?"
http://www.enterprisemission.com/soho.htm

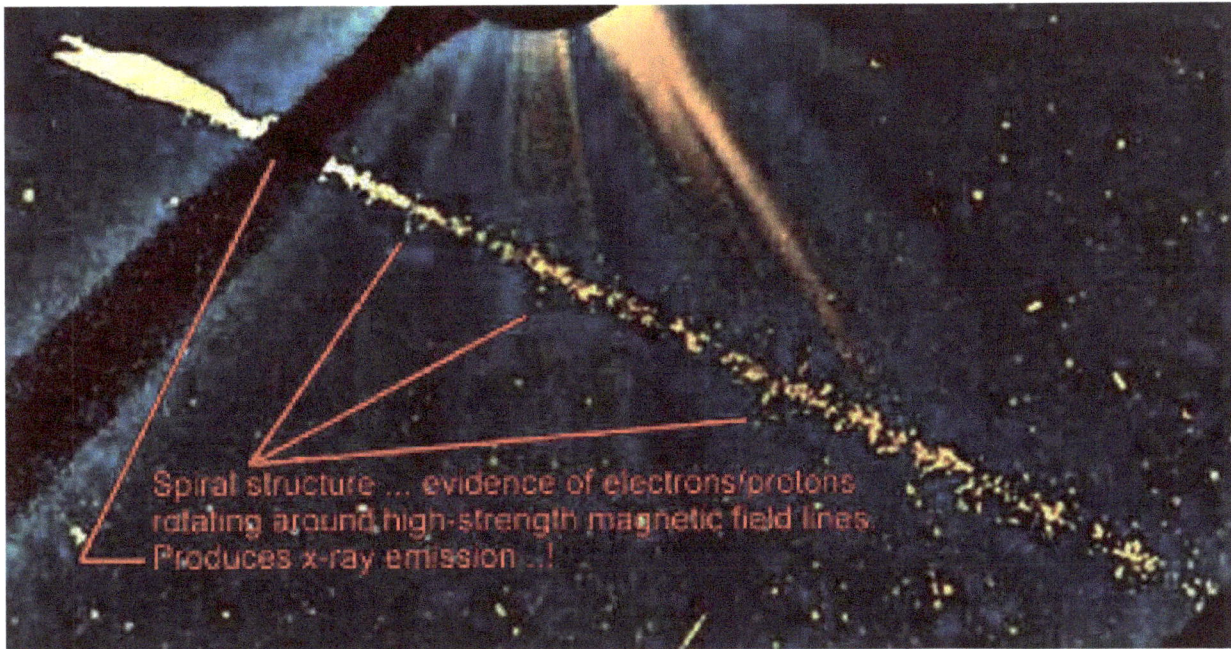

A colour enhanced image showing the ET craft firing a high energy particle beam weapon
http://www.enterprisemission.com/soho.htm

"But there is still a better case to be made that this beam is not a dissipating trail of the "projectile," but a coherent beam of energy.

And make no mistake; this event was staged specifically for the **SOHO** cameras. SOHO has a battery of instruments that take images on a fairly infrequent basis, about twice an hour. Even at conventional spacecraft speeds, a near-SOHO object would have passed the camera's field of vision in moments. And if the object (or objects) were far away, they would have to be immense -- on the order of *miles* across -- to even be visible. Either way, SOHO had virtually no chance of seeing this event "by accident." The question now is: who staged it? And assuming that NASA did not, why are we being allowed to see it? NASA has total control over what SOHO image are released to the public. Why let this one out ... and why now?"
http://www.enterprisemission.com/soho.htm

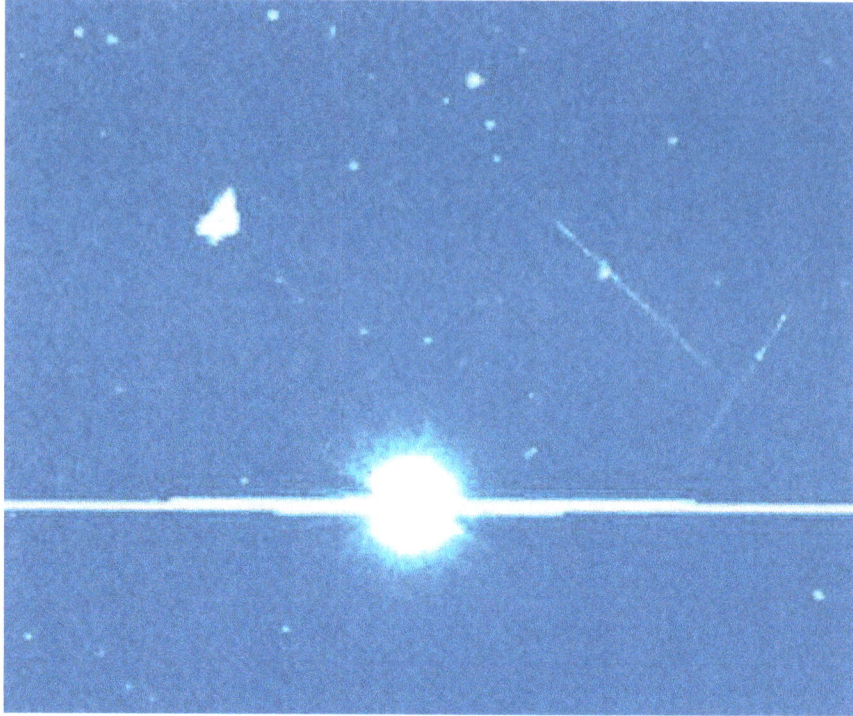

A delta-winged craft was seen hovering the next day, in the line of sight of Mars
http://www.enterprisemission.com/soho.htm

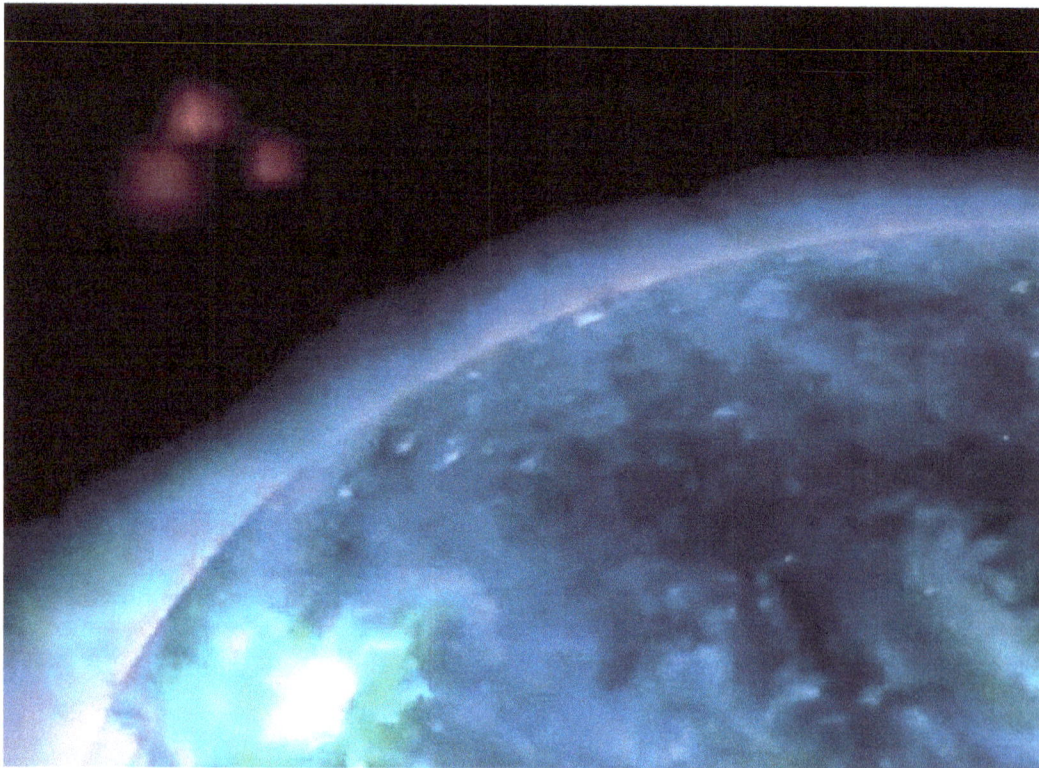

A triangle UFO near the Sun possibly interacting with the Sun's magnetic fields
https://www.youtube.com/watch?v=-W27zzVYuPk

It would seem that, if NASA and the US government are slow toward official disclosure of the Extraterrestrial presence in the Solar System then, the ETI would not be as dismissive of the opportunity to let their presence be known to the general public. The takeaway message from ETI is that they are present, they are very real and that they do want to interact with humans. Perhaps, NASA knows this to be true, but will not allow this information to be disclosed to the general public for fear, not of any social upheaval and panic as per **Brookings Report** but, rather the displacement or removal of power from the officialdom of the government, the **Department of Defence (DoD)** and by the corporate elite held for so long over the public masses.

We need to turn our gaze away from the Sun not because it is too bright to behold but, to see into the darkness that has obscured the truth of our search for Extraterrestrial life in our Solar System. We must now look outward, away from the Sun toward Mars, the one planet in the Solar System that is very much like our own Earth in so many ways.
https://www.youtube.com/watch?v=Vw6Hcbw3XOY and
https://www.youtube.com/watch?v=aZUGuex5XYY

Is there life on the Sun and on Stars?

Before leaving our star Sol, the giver of life to all the planets in our Solar System, there is a Baha'i quote from **Abdu'l-Baha**, son of **Baha'u'llah**, the Prophet and Founder of the **Baha'i Faith** who talks about the possible life that may exist on stars! Based on discoveries made by scientists, such as S. A. Mitchell, Abdu'l-Baha states:

"The earth has its inhabitants, the water and the air contain many living beings and all the elements have their nature spirits, then how is it possible to conceive that these stupendous stellar bodies are not inhabited? Verily, they are peopled, but let it be known that the dwellers accord with the elements of their respective spheres. These living beings do not have states of consciousness like unto those who live on the surface of this globe: the power of adaptation and environment moulds their bodies and states of consciousness, just as our bodies and minds are suited to our planet. For example, we have birds that live in the air, those that live on the earth and those that live in the sea... The components of the sun differ from those of this earth, for there are certain light and life-giving elements radiating from the sun. Exactly the same elements may exist in two bodies, but in varying quantities. For instance, there is fire and air in water, but the allotted measure is small in proportion. They have discovered that there is a great quantity of radium in the sun; the same element is found on the earth, but in a much smaller degree. Beings who inhabit those distant luminous bodies are attuned to the elements that have gone into their composition of their respective spheres."
"Divine Philosophy," compiled by I.F. Chamberlain (Boston: The Tudor Press, 1918), pp. 114-15

As incredible as this may seem to the rational mind, it now appears to be not an outburst of wild speculation, but a factual reality! There could indeed be some form of life on the Sun and by that, on other stars. We may not even recognize it as a lifeform or it may possibly have similar aspects of life found on Earth and those found on or near the radioactive element Radium (Ra). This is worth exploring further.

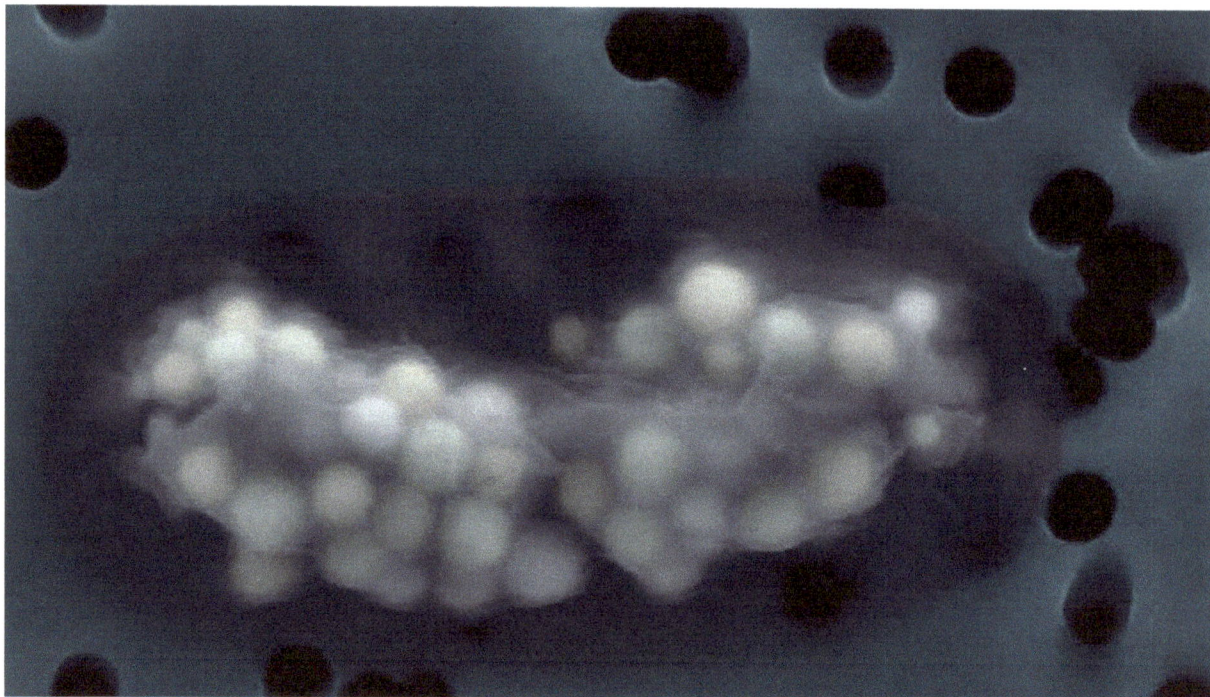

A *Gloeomargarita lithophora* bacterium containing carbonate granules (white spheres).
The microbe's ability to form such granules internally might explain how
it draws radioactive isotopes from its surroundings.
Credit: Karim Benzerara and Stefan Borensztajn/CNRS
https://www.nature.com/articles/d41586-019-03087-1

Photosynthetic bacterium discovered in a Mexican lake accumulate two radioactive isotopes in their cells and could help to soak up radioactive contaminants in polluted waterways.

Benjamin Kocar at the **Massachusetts Institute of Technology (M.I.T.)** in Cambridge, Massachusetts, Karim Benzerara at the Sorbonne University in Paris and their colleagues found that the bacterium *Gloeomargarita lithophora* is particularly adept at sucking up the radioactive isotopes, **radium-226** and **strontium-90**. Ra-226 has a half-life of 1,600 years, can be found in mining and power-plant runoff, and is one of the most common radioisotopes in groundwater. By contrast, Sr-90 mainly reaches waterways from nuclear tests and accidents and has a half-life of 29 years. Both pose risks to the environment and human health, and prior research has explored the ability of various microbes, fungi and other organisms to gobble them up.

Compared with organisms studied previously, *G. lithophora* showed the highest uptake of both ^{90}Sr and ^{226}Ra. This ability is probably related to a process by which *G. lithophora* draws in substances from its environment to form internal clumps of calcium carbonate.
https://www.nature.com/articles/d41586-019-03087-1 and
https://pubs.acs.org/doi/10.1021/acs.est.9b039821

Extremophiles

Extremophiles are **microorganisms** that have adapted so that they can survive and even thrivein extreme environments that are normally fatal to most life-forms. **Thermophiles** and **hyperthermophiles** thrive in high temperatures. **Psychrophiles** thrive in extremely low temperatures. – Temperatures as high as 130 °C (266 °F), as low as −17 °C (1 °F **Halophiles** such as *Halobacterium salinarum* (an archaean) thrive in high salt conditions, up to saturation. **Alkaliphiles** thrive in an alkaline pH of about 8.5–11. **Acidophiles** can thrive in a pH of 2.0 or less. **Piezophiles** *thrive at very high pressures: up to 1,000–2,000 atm, down to 0 atm as in a vacuum of space*. A few extremophiles such as *Deinococcus radiodurans* are **radioresistant**, resisting radiation exposure of up to 5k Gy. *Extremophiles* are significant in different ways.

They extend terrestrial life into much of the Earth's hydrosphere, crust and atmosphere, their specific evolutionary adaptation mechanisms to their extreme environment can be exploited inbiotechnology, and their very existence under such extreme conditions *increases the potentialfor extraterrestrial life.*

https://en.wikipedia.org/wiki/Microorganism

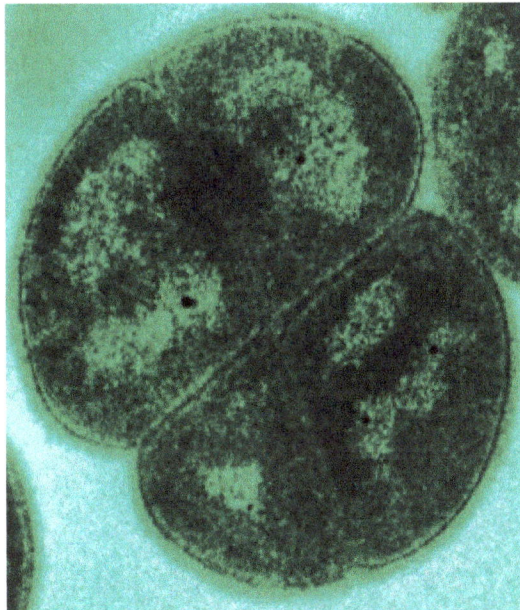

A tetrad of Deinococcus radiodurans, a radioresistant extremophile bacterium
https://en.wikipedia.org/wiki/Microorganism

Microorganisms have been detected in a variety of extreme environments (see first picture below), virtually in any location where liquid water is available for life to use. This demonstratesthat life can adapt to a wide range of parameters (see second picture below). It is therefore imperative to determine the minima and maxima for each parameter, and even more importantly,to understand their combined effects, in order to evaluate the limits of Earth's life and advance our understanding of the potential for life elsewhere.
https://www.ncbi.nlm.nih.gov/pmc/articles/PMC6476344/

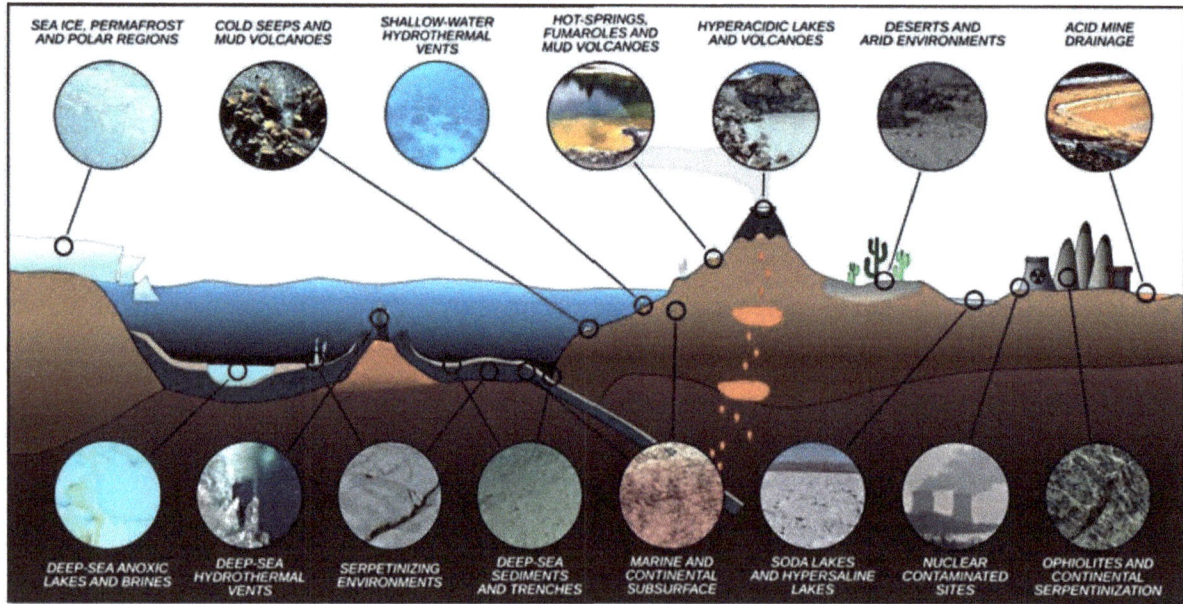

Representative idealized cross section of Earth's crust showing the diversity of extreme environments and their approximate location.

https://www.ncbi.nlm.nih.gov/pmc/articles/PMC6476344/

(**Courtesy of National Library of Medicine (NLM)**)

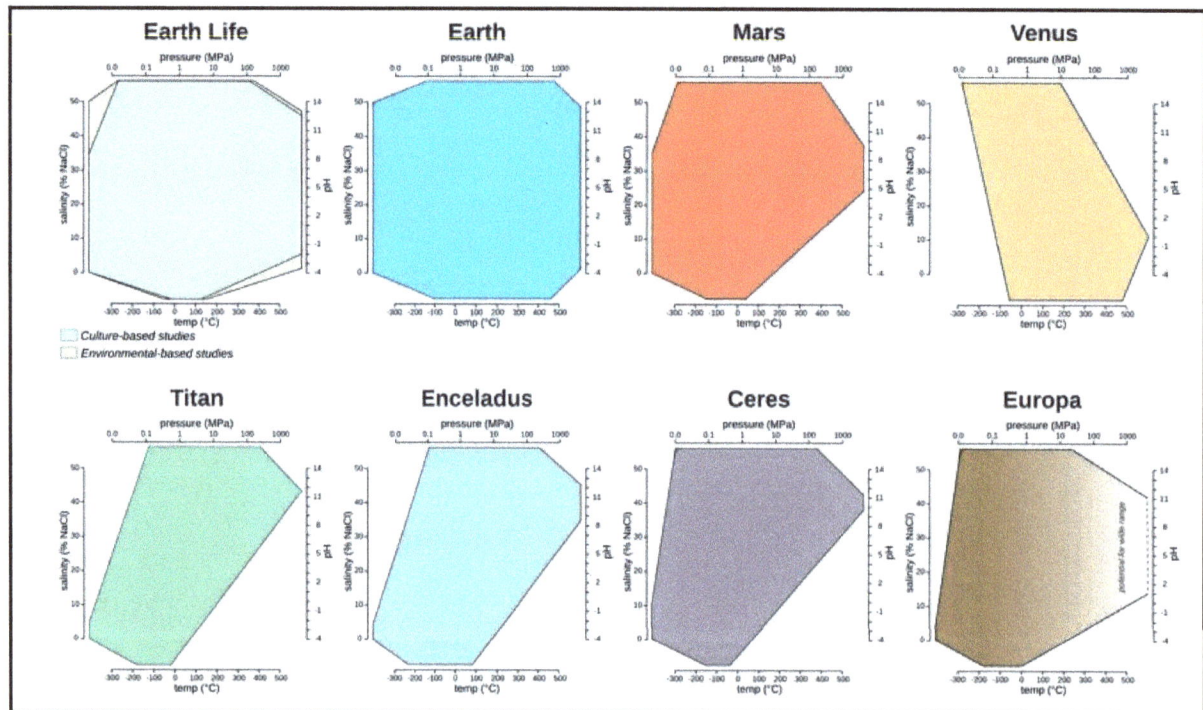

The temperature, pressure, pH, and salinity boundaries observed for life on Earth compared to the phase space observed on planetary bodies discussed in the main text.

https://www.ncbi.nlm.nih.gov/pmc/articles/PMC6476344/ (**Courtesy of National Library of Medicine (NLM)**)

In the above diagram, Polygon charts are designed to represent ranges in multidimensional space. Each edge represents the range for the specific variables. Single values (e.g., when min =max) are represented by a single vertex on an axis, while missing values (e.g., NA or NR) are represented by the absence of the corresponding polygon edge on the corresponding axis. https://www.ncbi.nlm.nih.gov/pmc/articles/PMC6476344/

An excellent and fascinating science paper to read is the **Frontiers of Microbiology is "Livingat the Extremes: Extremophiles and the limits of Life in a Planetary Context".**

In the **Popular Astronomy** magazine Vol. XXI, No. 6 dated June-July 1913, an eleven-page article that addressed the question: **"Is Radium in the Sun?"** written by **S. A. Mitchell.** In the article, spectrographic analysis of the Sun's light spectra (a series of lines) indicated the presenceof **Helium (He)** before it was discovered on Earth and that **Radium (Ra)** being discovered on Earth in small quantities, may in turn, be present in the Sun's atmosphere but in vast quantities.
 http://articles.adsabs.harvard.edu/cgi-bin/nph-iarticle_query?bibcode=1913PA21..321M&db_key=AST&page_ind=0&data_type=GIF&type=SCREEN_VIEW&classic=YES (courtesy of SOA/NASA Astrophysics Data System)

Is radium in the stars? And if so, is it also in our sun as our concepts of cosmic evolution makes us believe that if radium is in the Earth , then it must also be in the sun and at least in some starsas we and all the stars and the universe are made up of the same star dust! Or as **Dr. Carl Sagan**was so famously fond of saying in his television series "Cosmos", *"We are the stuff of stars!"*

Since Radium is and gives off **radioactive substances** which manifest themselves as exciting phosphorescence and fluorescence, by causing air near them to become conductors of electricityand with the continuous generation of light and heat. These radioactive effects owe their origin to the emission of rays called **Alpha, Beta, and Gamma rays** which manifest spontaneous and continuous source of energy.

Through various processes of radium division into radium emanations and **α (alpha) particles,** and disintegration one atom of **Helium (He)** is produced and one radium emanation and this appears in the spectrum from radium, since emanation is produced from radium are the lines dueto helium, showing conclusively that helium is produced from radium. Helium is also produced from not only from radium, but from other radioactive elements such as actinium, uranium, thorium.

Eclipse spectra since 1868 have told us that helium is present in the sun. If it has had its origin as the result of radioactive changes, then must radium be in the sun, and likewise the parent of radium, uranium. No helium lines are found in the ordinary or Fraunhofer spectrum of the sun. Helium is however, readily seen in the chromosphere, and its lines are found in the spectrum of the chromosphere at the time of an eclipse.

This was later confirmed by the **Astronomer Royal of England, Dr. F. W. Dyson**, made a comparison with the chromosphere lines obtained by him from observations of the eclipses of 1900, 1901 and 1905, and with the chromosphere lines obtained by **Sir Norman Lockyer** at the eclipse of 1898, and drew the following conclusion, *"It seems to me that these lines in the **chromosphere** may reasonably be attributed wholly or partially to radium."* http://articles.adsabs.harvard.edu/cgi-bin/nph-iarticle_query?bibcode=1913PA

21..321M&db_key=AST&page_ind=0&data_type=GIF
&type=SCREEN_VIEW&classic=YES

With the scientific understanding that microbes and even, some fungi and other organisms can consume radioactive isotopes, it stands to reason that they can live directly on the element of radium and survive without any problems to its existence. By deductive reasoning, radium is alsofound within the Sun and probably other stars, therefore, it is reasonable to consider that microbial life of some kind exists on our Sun!

Now, that we know that there is conclusive proof of radium in the Sun as on the Earth and no doubt probably throughout the universe. This similarity is also indication that the microbial lifefound on radium on Earth which scientists say can consume the radioactivity emitted by radiumis also possible on the Sun and other stars. It sounds like science fiction but we are discovering that it may no longer be fiction but science fact!

Therefore, could the strange planet size UFOs moving around in orbit of the Sun be other ETI coming to investigate the properties of the star and the probable life that may inhabit its surface or in the **Chromosphere** (outer atmosphere of the Sun) or is it possible these craft originate fromthe Sun, itself?

Has the life on the Sun advanced to the point of space travel? Are theses beings microbial in sizeor are they incredibly gigantic size beings? Do they pilot planetary size spacecraft or do they even require spacecraft to travel through space and is it even possible? Sheer speculation to be sure!

Out virtual journey through the Solar System may reveal that answer.

CHAPTER 77

PLANET MERCURY (THE "MESSENGER")

Mercury is the innermost planet in the Solar System. It is also the smallest, and its orbit is the most eccentric (that is, the least perfectly circular) of the eight planets. It orbits the Sun once in about 88 Earth days, completing three rotations about its axis for every two orbits. The planet isnamed after the Roman god Mercury, the messenger to the gods.

Mercury photographed by the NASA Messenger orbiting space probein the blue and near-infrared spectrum
https://www.universetoday.com/13943/mercury/

Mercury's surface is heavily cratered and similar in appearance to Earth's Moon, indicating that it has been geologically inactive for billions of years. Due to its near lack of an atmosphere to retain heat, Mercury's surface experiences the steepest temperature gradient of all the planets, ranging from a very cold 100 K at night to a very hot 700 K during the day. Mercury's axis has the smallest tilt of any of the Solar System's planets, but Mercury's orbital eccentricity is the largest. The seasons on the planet's surface are caused by the variation of its distance from the Sun rather than by the axial tilt, which is the main cause of seasons on Earth and other planets. At perihelion, the intensity of sunlight on Mercury's surface is more than twice the intensity at aphelion. Because the seasons of the planet are produced by the orbital eccentricity instead of the axial tilt, the season does not differ between its two hemispheres.

Because Mercury's orbit lies within Earth's orbit, it can appear in Earth's sky either as a morning star or an evening star. While Mercury can appear as a very bright object when viewed from Earth, its proximity to the Sun makes it more difficult to see than Venus.

While there doesn't appear to be any life on Mercury, new observations by the **Messenger spacecraft** provide compelling support for the long-held hypothesis that Mercury harbors abundant water ice and other frozen volatile materials in its permanently shadowed polar craters. If water may exist on the Mercurian surface, then **extremophile** life may also exist.
http://www.nasa.gov/mission_pages/messenger/main/index.html

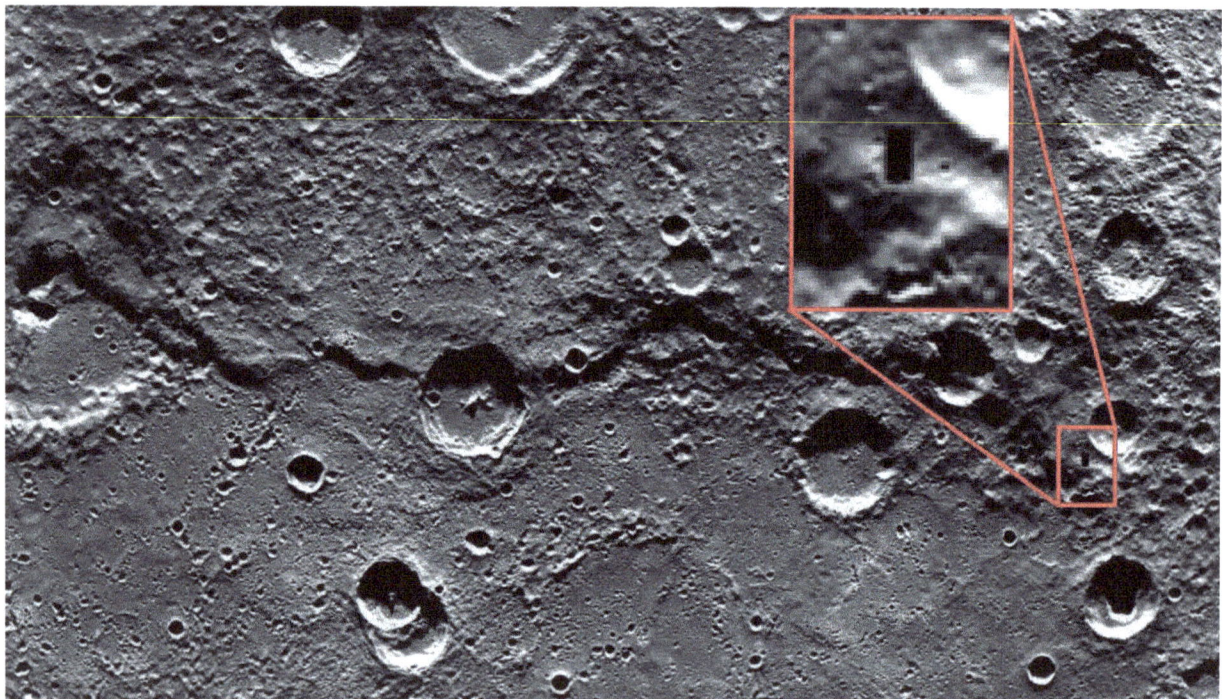

A large rectangle opening or monument 5.6 Km or 3.5 miles in size discovered on the surface of Mercury. Is it a pixel omission flaw or a true artificial anomaly?
http://www.dailymail.co.uk/sciencetech/article-3926832/A-doorway-world-Alien-hunters-say-monolith-entrance-Mercury-latest-bizarre-claim.html

Mercury, like Venus, appears to be a dead lifeless world, however, there is at least one anomaly on its surface that does appear to be artificial, it is a large rectangular opening that is estimated to be 3.5 miles in size and reminiscent of an Ionic Square! Is it an opening or simply a photographic flaw of missing pixels? (See photo above).

On Dec. 1, 2011, a camera onboard NASA's **STEREO** spacecraft recorded a wave of electrically charged material shooting out from the sun and blasting Mercury. Footage of this **"coronal mass ejection" (CME)** or solar flare has caught the attention of alien-hunters, who say it has unveiled a giant, "cloaked" spaceship parked near the solar system's innermost planet.

In the footage, one sees a huge spurt of plasma and other solar ejecta washing over Mercury; peculiarly, the material seems to flare up as it hits another nearby object, too. "It's cylindrical on either side and has a shape in the middle. It definitely looks like a ship to me, and very obviously, it's cloaked," YouTube-user **siniXster** said in his video commentary on the footage, which has quickly spread across the Web. http://www.lifeslittlemysteries.com/1966-cloaked-mothership-mercury-ufo.html

While watching a coronal mass ejection traveling from the sun, one of NASA's solar observatories appears to have discovered a massive alien spacecraft docked in orbit next to the tiny planet Mercury! The video of this alien object appears on YouTube for anyone to view.

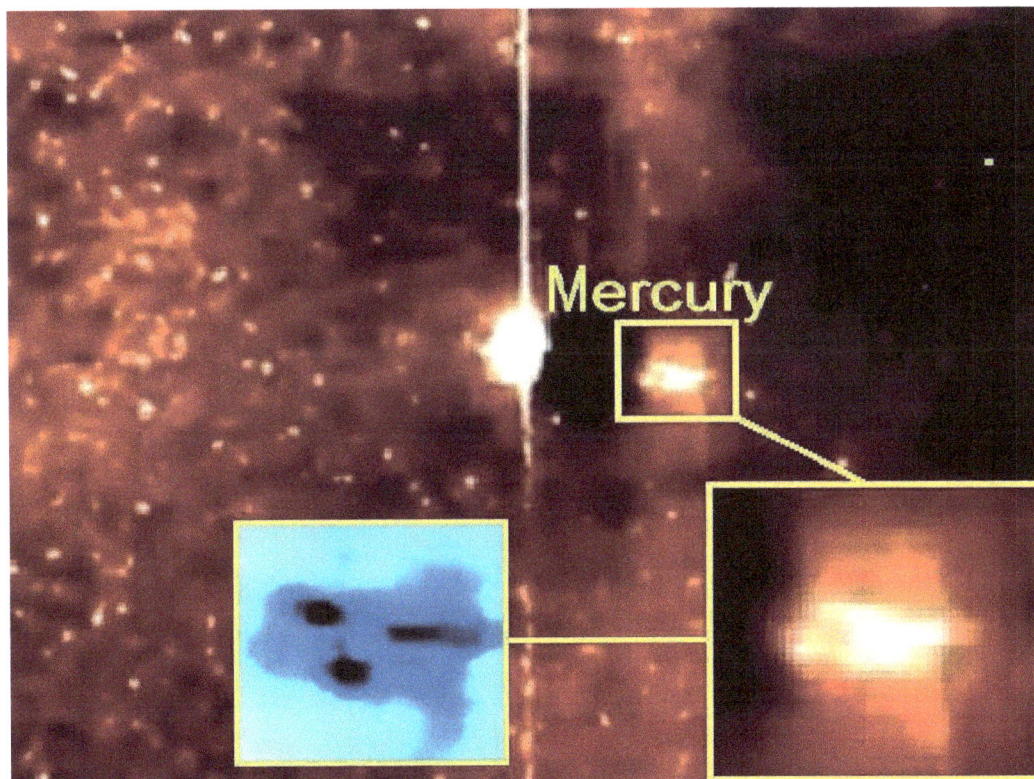

Giant cloaked UFO next to Mercury (enlarged and reversed colour enhanced) on SECCHI HI1-A on 12/01/11, appears when a CME (Corona Mass Ejection) hits it
http://www.livescience.com/17339-cloaked-mothership-mercury-ufo.html

335

It has cylindrical and linear dimensions to its overall shape. It looks artificial and most unusual of all was that it had been invisible or "cloaked" until hit by a massive coronal ejection. Its sudden appearance suggests some sort of ship.

The question was meant rhetorically, but nonetheless, the video is curious, so we've put it to scientists in the solar physics branch at the **United States Naval Research Laboratory (NRL)** — the group that analyzes data from the **Heliospheric Imager-1 (HI-1)**, the telescopic camera that shot the new footage.

As you might suspect, there *is* a non-UFO explanation of the apparent flare-up near Mercury. According to **Russ Howard**, head scientist of the NRL group, and Nathan Rich, lead ground systems engineer, it is simply an artifact left over from the way raw HI-1 telescope data gets processed. Rather than a UFO mothership parked near Mercury, the bright spot is "where the planet was on the previous day," Rich told Life's Little Mysteries.
http://www.lifeslittlemysteries.com/1966-cloaked-mothership-mercury-ufo.html

This seems like a reasonable, logical explanation and it is probably the correct solution to this astronomical mystery until you start looking at other photographic images of even stranger, more unusual object captured by the same solar orbiting satellites. These UFOs are also immense, on a planetary scale and are even closer to the Sun where there are no other orbiting planets.

Therefore, this explanation becomes no longer tenable when viewed within the big picture of the many strange and planetary size objects in orbit around the Sun.

In this vast universe, we are finding that almost daily that we are not alone. There is more and more evidence that extraterrestrial beings do exist and their intelligence is far superior to ours. We are discovering that not only is our planet is being visited on a regular basis but, ETs also, seem to be interested in the other planets within our solar System. The YouTube video shows a huge alien spaceship which inadvertently became de-cloaked by Solar Flare while it was flying close to Mercury. The big flying object's cloaking device was probably altered by the Solar Flares causing it to reveal itself rather fortuitously to photographic recording. Mercury has a diameter of 4,879 kilometers, so this object is truly massive, around 2500 km in diameter. So far, NASA has offered no official explanation for this "Mercurial object".

CHAPTER 78

PLANET VENUS (THE "GODDESS OF LOVE")

Venus is the second planet from the Sun, orbiting it every 224.7 Earth days and named after the Roman goddess of love and beauty. It is the brightest natural object in the night sky after the Moon. Venus reaches its maximum brightness shortly before sunrise or shortly after sunset, for which reason it is sometimes known as the **Morning Star** or **Evening Star**.

Venus can cast shadows and can be, on rare occasion, visible to the naked eye in broad daylight Venus lies within Earth's orbit, and so never appears to venture far from the Sun, either setting in the west just after dusk or rising in the east a little while before dawn. Venus orbits the Sun every 224.7 Earth days. With a **rotation (sideral) period** of 243 Earth days, it takes longer to rotate about its axis than any other planet in the Solar System by far, and does so in the opposite **(retrograde)** direction to all but Uranus (meaning the Sun rises in the west and sets in the east). Venus does not have any moons, a distinction it shares only with Mercury among the planets in the Solar System.

It has the densest atmosphere of the four terrestrial planets, consisting of more than 96% carbon dioxide. The atmospheric pressure at the planet's surface is about 92 times the sea level pressure of Earth, or roughly the pressure at 900 m (3,000 ft) underwater on Earth. Venus has, by far, the hottest surface of any planet in the **Solar System**, with a mean temperature of 737 K (464 °C; 867 °F), even though Mercury is closer to the Sun. Venus is shrouded by an opaque layer of highly reflective **clouds of sulfuric acid**, preventing its surface from being seen from space in visible light. It may have had water oceans in the past but these would have vaporized as the temperature rose due to a runaway greenhouse effect. The CO_2-rich atmosphere generates the strongest **greenhouse effect** in the Solar System. It is a planet that is considered extremely hot, hellish, and volcanic, not really a place for humas survivability!

In 1995, the ***Magellan* spacecraft** imaged a highly reflective substance at the tops of the highest mountain peaks that bore a strong resemblance to terrestrial snow. This substance likely formed from a similar process to snow, albeit at a far higher temperature. Too volatile to condense on the surface, it rose in gaseous form to higher elevations, where it is cooler and could precipitate. The identity of this substance is not known with certainty, but speculation has ranged from **elemental tellurium** to **lead sulfide (galena).**

The existence of lightning in the atmosphere of Venus has been controversial since the first suspected bursts were detected by the Soviet ***Venera probes.*** In 2006–07, ***Venus Express*** clearly detected whistler mode waves, the signatures of lightning.
In 2007, *Venus Express* discovered that a huge double atmospheric vortex exists at the south pole *Venus Express* also discovered, in 2011, that an ozone layer exists high in the atmosphere of Venus. On 29 January 2013, **ESA** scientists reported that the ionosphere of Venus streams outwards in a manner similar to "the ion tail seen streaming from a comet under similar conditions." https://en.wikipedia.org/wiki/Venus and
https://www.nytimes.com/2020/09/14/science/venus-life-clouds.html

Global radar imaging view of the surface of Venus taken by Magellan during 1990–1994

In December 2015, and to a lesser extent in April and May 2016, researchers working on Japan's *Akatsuki* **mission** observed bow shapes in the atmosphere of Venus. This was considered direct evidence of the existence of perhaps the largest stationary gravity waves in the solar system

The possibility of life on Venus and in its distant past, has long been a topic of speculation, andin recent years has received active research. Following a 2019 observation that the light absorbance of the upper cloud layers was consistent with the presence of microorganisms, a September 2020 article in *Nature Astronomy* announced the detection of **phosphine gas**,

biomarker, in concentrations higher than can be explained by any known abiotic source.
https://en.wikipedia.org/wiki/Venus

Venus' atmosphere consists mainly of carbon dioxide, with clouds of sulfuric acid droplets. The thick atmosphere traps the Sun's heat, resulting in surface temperatures higher than 880 degrees Fahrenheit (470 degrees Celsius). The atmosphere has many layers with different temperatures. *At the level where the clouds are, about 30 miles up from the surface, it's about the same temperature as on the surface of the Earth!* The possibility of life in the top cloud layers of Venus' atmosphere, where the temperatures are less extreme is also speculated. (Bold italics added by author for emphasis).

No human has visited Venus, but the Russian **Venera spacecraft** has landed on the surface of Venus, but due to the extreme Venusian surface heat and the sulfuric atmosphere the craft's overheated electronics failed within a few hours of its landing. It seems unlikely that a person could survive for long on the Venusian surface given it hostile environment and atmosphere.
https://solarsystem.nasa.gov/planets/venus/in-depth/

Life on Venus

Because of the extreme adverse physical conditions on Venus, life has not taken hold and it would be hard to imagine any Extraterrestrial intelligence wanting to set up a base on this planet Because of the extreme adverse physical conditions on Venus, life has not taken hold and it would be hard to imagine any Extraterrestrial intelligence wanting to set up a base on this planet. But let's re-evaluate the concept that Venus is too hostile, too hellish for life to exist.

From our investigation, we established the case for radium being a constituent of the sun's environment and that bacteria can consume and live off the radiation of radium and live on radium as it is found on Earth. We have found by deductive reasoning that some form of **extremophile life** must also exist on the Sun and this is supported by spiritual knowledge found in the writings of the **Baha'i Faith.**

Extremophiles exist not only in the most hostile environments on Earth, i.e. in hot deserts, in artic cold, in the depths of the ground and the ocean and around sulfuric lakes and vents but, just about anywhere you can imagine an inhospitable location. So, if life could exist on the Sun and probably on other stars, then it is not a stretch of the imagination to look at Venus as a larger hostile environment when compared to the pockets of hostile regions on Earth.

Venus is classified as a terrestrial planet and is sometimes called Earth's "sister planet" owing to their similar size, gravity, and bulk composition (Venus is both the closest planet to Earth and the planet closest in size to Earth). However, it has been shown to be very different from Earth in other respects. Venus is shrouded by an opaque layer of highly reflective clouds of sulfuric acid, preventing its surface from being seen from space in visible light. It has the densest atmosphere of the four terrestrial planets, consisting mostly of carbon dioxide. The atmospheric pressure at the planet's surface is 92 times that of Earth's. With a mean surface temperature of 735 K (462 °C; 863 °F), Venus is by far the hottest planet in the Solar System, if there is a Hell then, Venus comes closest to that physical reality. It has no carbon cycle to lock carbon back into rocks and surface features, nor does it seem to have any organic life to absorb it in biomass. Venus may have possessed oceans in the past, but these would have vaporized as the temperature rose due to the runaway greenhouse effect. Venus's surface is a dry desert landscape interspersed with slab-like rocks and periodically refreshed by volcanism. http://en.wikipedia.org/wiki/Venus and https://www.nytimes.com/2020/09/14/science/venus-life-clouds.html

The possibility of life on Venus is a subject of interest in astrobiology due to its proximity and similarities to Earth. To date, no definitive proof has been found of past or present life on Venus. Theories have decreased significantly since the early 1960s, when spacecraft began studying the planet and it became clear that its environment is extreme compared to Earth's. However, there is ongoing study as to whether life could have existed on the Venusian surface before a runaway greenhouse effect took hold, and related study as to whether a relict biosphere could persist high in the modern Venusian atmosphere.

With extreme surface temperatures reaching nearly 735 K (462 °C; 863 °F) and an atmospheric pressure 90 times that of Earth, the conditions on Venus make water-based life as we know it unlikely on the surface of the planet. However, a few scientists have speculated that **thermoacidophilic extremophile microorganisms** might exist in the temperate, acidic upper layers of the Venusian atmosphere. In September 2020, research was published that reported the presence of phosphine in the planet's atmosphere, a potential biosignature. However, doubts have been cast on these observations

As of 8 February 2021, an updated status of studies considering the possible detection of lifeforms on Venus (via of phosphine) and Mars (via methane) was reported.

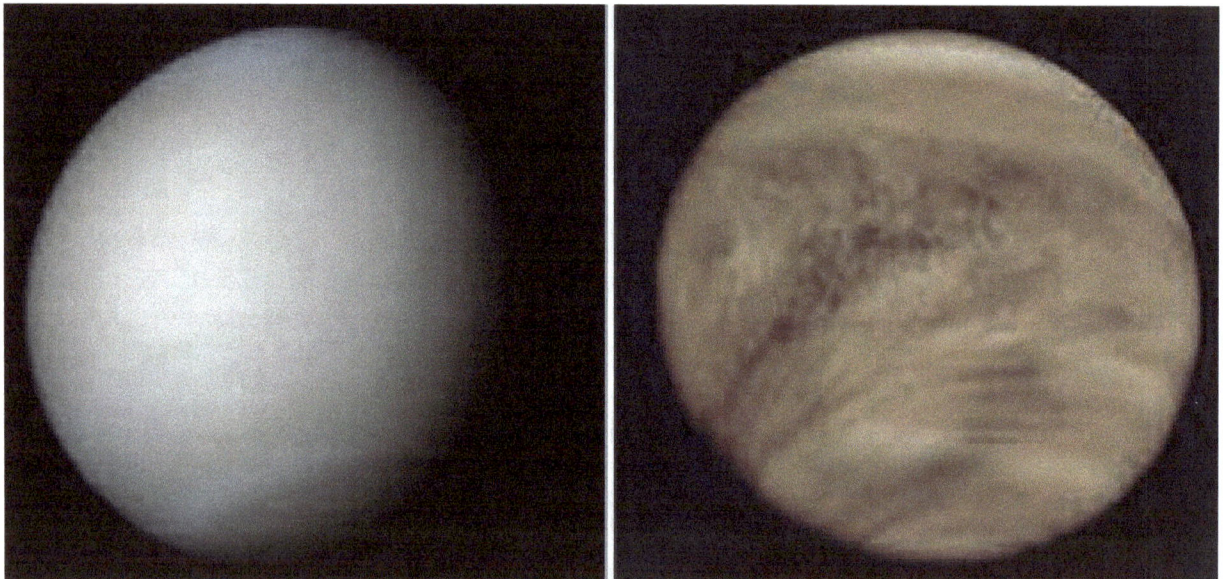

**Venus in true colour (left). The surface is obscured by a thick blanket of clouds.
Cloud structure in the Venusian atmosphere (right) revealed
by ultraviolet observations by Pioneer Venus Orbiter**
https://en.wikipedia.org/wiki/Venus

All stars in the galaxy eventually age and slow down by losing their core fusion process, as our Sun is destined to do in the distant future, they begin to die. The star may go supernova exploding in a massive release of stellar energy which consume most of the planets within its system or star may become a dwarf star of heavy metals. Either way the orbits of some planets change because the gravitational energy release from the star cancels out the orbit of some planets sending out of their solar system, drifting through endless space. This is probably why Venus is the only one of the solar planets that doesn't have a magnetosphere or plasmasphere and rotates backward compared to all the other planets. This makes Venus a quirky planet in our Solar System which could have come in from an outer orbit or from out of the Solar System as an interloper in a temporary orbit that has now appeared to settle in its permanent orbit. Such is the nature and possible origin of Venus.

To date, astrobiologists are still looking at life on other planets with an anthropocentric point of view, that is, for life to exist on a planet to exist, it must have water, be in the **Zone of Habitability**, and be conducive to human habitability! But arrogantly and erroneously they state that there is no definitive proof that has been found of past or present life on Venus. Extreme surface temperatures and an atmospheric pressure 90 times that of Earth make the conditions on Venus for water-based life as we know it unlikely on the surface of the planet.
https://en.wikipedia.org/wiki/Life_on_Venus

However, that very reason may be exactly what a very technically advanced ET civilization may actually consider doing in order to keep less advanced civilizations like humans from exploring too intensely such planets with hostile environments.

Venusians Visit the White House

Early contactees believed that there were Venusian emissaries who came to Earth in 1957 to meet with **US President Dwight Eisenhower** and some of his administration including **Vice President Richard Nixon**. The Venusians in question are **Commander Valiant "Val" Thor** and his brother **Donn Thor, Teel (or Tanyia) and Jill** are said to resemble very attractive humans, but with psychic abilities like telepathy, to walk through walls, to dematerialize and rematerialize at will, and to leave no fingerprints. Depending on who tells the story and its source on the internet, the accounts are similar with some variations. One cannot help but also, recall the similar account of President Eisenhower's meeting with Extraterrestrials in which his abrupt departure from a golf game caused national concern for his whereabouts and later became headline news, nationally.

Valiant Thor and his entourage landed in a flying saucer on March 16, 1957, in Alexandria, Virginia and were immediately greeted by police with guns drawn followed quickly by the CIA who took charge of the situation. Under heavy security and numerous patrol cars from various departments and government agencies, the Venusians were taken to the **White House** as requested by Val to meet with the US President. **Eisenhower** was cordial but cautious and required proof of his Thor's planetary origins. Thor said that he offered the President a ride aboard his ship if he wished but, this offer was politely refused for security reasons.
http://madreterra.myblog.it/2015/06/14/extraterrestri-noi-il-venusiano-valiant-thor/

**This photo taken by preacher Frank E. Stranges shows allegedly three Venusians:
Commander Valiant Thor (right), facing the camera (middle)
Donn Thor and to his right, Jill**

http://madreterra.myblog.it/2015/06/14/extraterrestri-noi-il-venusiano-valiant-thor/

Val gave the US President a message from the **High Council of Venus** which offered to help the human family in its various social and technical developments, however, once again, President Eisenhower refused Val's generosity stating that it would upset the economy of the United States and could plunge her into an abyss of chaos. In brief, he politely told Val that the people of this planet were not ready to cope with such conditions as would come into existence if the recommendations of this unearthly visitor were put into action.

Nevertheless, he was invited to assist a number of scientists who were out working on medical projects directly associated with the space sciences. He and his crewmates were made "guests at an apartment" at the Pentagon (*this sounds more like a "political in-house imprisonment"*). His allotted time to acquaint the leaders of the United States with his suggestions was limited to three years. (Bold italics added by author for emphasis).

During this time, he refused to advise them regarding a certain "bomb in the sky" which we now know as the Star Wars system.

Sightings of "**Venusians**" began in the 50s in Hunterdon County, NJ with **Howard Menger,** who claims they frequently came to his farm. Menger went on talk shows to talk about his experiences and also, hosted "conventions" at his home. At one of these conventions, Commander Valiant Thor appeared, claiming he was the leader of all Venusians on Earth. He wowed his audiences by speaking any language known to man, including dialects. Thus, began Thor's mystery.

Thor said he came from a race of Venusians that lived within the planet itself, not on its surface. They were superior to humans: more intelligent, better-looking, less-violent. They were sent here, he said, to protect humans from themselves and to prevent nuclear war. He asked to speak with the President, but instead, according to Thor, he met with **Secretary of State John F. Dulles** (later, he became CIA Director). He was held by the **Pentagon** for questioning. However, Thor's story gets stranger with the introduction of **Frank E. Stranges**, an ordained minister (preacher).

Stranges said that agents of the Venusians escorted him to the Pentagon in 1959 to meet with Val, and he claimed to have seen Val's brother, Donn Thor, walk through walls. He also said that, upon meeting Valiant for the first time, the alien showed him a suit that would only zipper up or open when Val waved his hand; it was made from "fabric" that could not be destroyed. The alien said to **Stranges** that the government was doing tests on him, and his IQ was determined to be close, or over, 200. *(An IQ of 200 or more in humans is rare, although, it is not uncommon in this current age. Dr. Steven Greer has an IQ of 186 + and I know of one teacher locally, who told me he had an IQ of 193 - 196!).* Stranges said that **Valiant** and his shipmates) are a race of people created by God before **Adam**, and never fell from grace like humankind did. (Bold italics added by author for emphasis).

The Venusian spaceship, **Victory One (or Victor one)** is said to contain **Valiant, Donn, Teel, Jill** and a medical doctor called **"Doc."** Stranges says he still meets with Thor to this day. Sometimes they go back to Venus, but they always return here to Earth. Val said that he has a specific timeline to expose himself to the public.
http://www.bibliotecapleyades.net/bb/stranges.htm and
http://www.nextagemission.com/valthor1.html

Needless to say, this story from a scientific point of view with regard to the absence of any life makes it impossible, given the planetary nature of Venus as already described above, it prohibits any kind of human being surviving there in such harsh conditions. For this reason, alone the stories of Valiant Thor are suspect and viewed as an elaborate hoax.

With **Frank E. Stranges'** contribution to the **Valiant Thor** legend, there is a decidedly new age religious connotation added to Thor's mission with a distinct atmosphere of re-born Christian concepts. This too makes this story suspect and possibly disingenuous.
https://www.youtube.com/watch?v=Zq21hb4v89Y

It would be interesting to know if any contactees having communications with visiting ETs have received confirmation of other Earth-based religious belief systems such as **Islam** or **Buddhism** or even the **Baha'i Faith**. Nowhere in all my 65+ years of UFO and ETI investigation and research have I come across any ET communications within the UFO/ETI literature that seems to validate one particular belief system over another, more than Christianity. In fact, it would seem that the eyewitness or the contactees, whether consciously or subconsciously does put his or her religious perspective or bias on the experience. It is one more aspect of anthropocentric thinking that we humans have placed upon this phenomenon. We need to stop doing this until it can be proven otherwise. It would be the height of Christian arrogance to think that they would baptize an advanced interstellar intelligence into the antiquated faith of Christianity. In all likelihood, visiting ETI could teach human a thing or two about religion and spirituality, rather than we trying to teach them.

Could an Extraterrestrial Intelligence cloak themselves on or beneath the surface of another planet without physical detection by our scientists? Anything is possible if that ETI civilization is highly advanced technologically.

Venus at this present time does not appear to have any evidence of past or current ET presence on its surface; in fact, Venus is often the culprit in a lot of misidentified UFO sighting accounts that are reported by both the lay public and from many experienced and well-trained observers like pilots.

U.S. President Jimmy Carter reported having seen a UFO in 1969, which later analysis *"suggested"* was probably the planet although, this analysis is not conclusive. Countless other people have mistaken Venus for something more exotic. When military officials or government authorities tell you that what you saw landed in your backyard or a nearby field was not a UFO but the planet Venus then, it becomes obvious that Venus is now the convenient go-to explanation for most nighttime UFOs sightings. The reality is usually something far more unearthly requiring still further cover-up of the truth by the **Power Controlling Elite** aka, **Wealthy Corporate Elite!** http://en.wikipedia.org/wiki/Venus

CHAPTER 79

THE PLANET MARS ("THE GOD OF WAR")

Like so many other planets in our star system that are named after the Roman pantheon of gods, **Mars**, the fourth planet from the Sun is named after the god of war. No other planet in our Solar System has invoked more speculation and controversy in the last hundred years than has Mars. When it comes to the possibility of life existing elsewhere, Mars is the number one choice picked by scientist and the general public. Mars more than any other planet in the Solar System shares similar geological features with the Earth and may once have had oceans on its surface billions of years ago. It is often described as the **"Red Planet"** because of the iron (lll) oxide, more commonly known as hematite, or rust which is prevalent on its surface giving the planet its reddish appearance. It is also an indication of the presence or former presence of water on its surface. Mars is a terrestrial planet with a very thin atmosphere, having surface features reminiscent both of the impact craters of the Moon and the volcanoes, valleys, deserts, and polar ice caps of Earth. The rotational period and seasonal cycles of Mars are likewise similar to those of Earth, as is the tilt that produces the seasons. Mars sidereal rotational period or day is (24.622 hr.), almost 40 minutes longer than the Earth. Mars is home to **Olympus Mons**, the largest known volcanic mountain within the Solar System and home to **Valles Marineris**, one of the largest canyons known, larger than the Grand Canyon in southwest USA. The smooth **Borealis Basin** in the northern hemisphere covers 40% of the planet and may be a giant impact feature.

Phobos and Deimos are the two known moons of Mars which are small and irregularly shaped and may have been captured asteroids, similar to **5261 Eureka**, a Martian trojan asteroid.

Until the first successful Mars flyby in 1965 by *Mariner 4,* many speculated about whether there was any presence of liquid water on the planet's surface. This speculation was based on observational error determined from periodic variations in light and dark patches, noticeably in the polar latitudes, thought to be seas and continents; long, dark striations were misinterpreted to be irrigation channels for liquid water. These straight line features were later explained as optical illusions, though geological evidence gathered by unmanned missions suggest that Mars once had large-scale water coverage on its surface. In 2005, radar data revealed the presence of large quantities of water ice at the poles and at mid-latitudes. The Mars rover *Spirit* sampled chemical compounds containing water molecules in March 2007. The *Phoenix* lander directly sampled water ice in shallow Martian soil on July 31, 2008. [Marshttp://en.wikipedia.org/wiki/Mars](http://en.wikipedia.org/wiki/Mars)

Initially, Mars hosted five functioning spacecraft: three in orbit – the *Mars Odyssey, Mars Express,* and *Mars Reconnaissance Orbiter*; and two on the surface – **Mars Exploration Rover Opportunity** and the **Mars Science Laboratory** *Curiosity*. Defunct spacecraft on the surface include **MER-A** *Spirit*, and several other inert landers and rovers, both successful and unsuccessful, such as the *Phoenix* **lander**, which completed its mission in 2008. Observations by the *Mars Reconnaissance Orbiter* have revealed possible flowing water during the warmest months on Mars. [Marshttp://en.wikipedia.org/wiki/Mars](http://en.wikipedia.org/wiki/Mars)

True colour image of Mars with its two asteroid-size moons:
Phobos (upper left) and Deimos (upper right). 1995 Hubble Image
https://www.universetoday.com/129026/hubble-sees-mars-close/ and
https://astrobob.areavoices.com/2013/08/17/curiosity-sees-unearthly-moondance-in-martian-skies/

As of February 2021, there is now a growing number of nations sending space craft to explore Mars whether in orbit, landing or with land rovers and now the USA as a small helicopter flying over the around the Martian terrain. Below is a chart listing space probes sent to Mars.
https://en.wikipedia.org/wiki/List_of_missions_to_Mars

International Space Missions to Mars 1960 to 2021

Mission	Spacecraft	Launch Date	Operator	Mission Type[1]	Outcome[2]	Remarks	Carrier rocket[3]
1M No.1	1M No.1	10 October 1960	OKB-1 ⬛ Soviet Union	Flyby	Launch failure	Failed to orbit	Molniya
1M No.2	1M No.2	14 October 1960	OKB-1 ⬛ Soviet Union	Flyby	Launch failure	Failed to orbit	Molniya
2MV-4 No.1	2MV-4 No.1	24 October 1962	⬛ Soviet Union	Flyby	Launch failure	Booster stage ("Block L") disintegrated in LEO	Molniya
Mars 1	Mars 1 (2MV-4 No.2)	1 November 1962	⬛ Soviet Union	Flyby	Spacecraft failure	Communications lost before flyby	Molniya
2MV-3 No.1	2MV-3 No.1	4 November 1962	⬛ Soviet Union	Lander	Launch failure	Never left LEO	Molniya
Mariner 3	Mariner 3	5 November 1964	NASA 🇺🇸 United States	Flyby	Launch failure	Payload fairing failed to separate	Atlas LV-3 Agena-D
Mariner 4	Mariner 4	28 November 1964	NASA 🇺🇸 United States	Flyby	Successful	**The first flyby of Mars** on 15 July 1965	Atlas LV-3 Agena-D
Zond 2	Zond 2 (3MV-4A No.2)	30 November 1964	⬛ Soviet Union	Flyby	Spacecraft failure	Communications lost before flyby	Molniya
Mariner 6	Mariner 6	25 February 1969	NASA 🇺🇸 United States	Flyby	Successful		Atlas SLV-3C Centaur-D
2M No.521	2M No.521 [4] (1969A)	27 March 1969	⬛ Soviet Union	Orbiter	Launch failure	Failed to orbit	Proton-K/D
Mariner 7	Mariner 7	27 March 1969	NASA 🇺🇸 United States	Flyby	Successful		Atlas SLV-3C Centaur-D
2M No.522	2M No.522 (1969B)[4]	2 April 1969	⬛ Soviet Union	Orbiter	Launch failure	Failed to orbit	Proton-K/D

Mission	Spacecraft	Launch Date	Operator	Mission Type[1]	Outcome[2]	Remarks	Carrier rocket[3]
Mariner 8	Mariner 8	9 May 1971	NASA United States	Orbiter	Launch failure	Failed to orbit	Atlas SLV-3C Centaur-D
Kosmos 419	Kosmos 419 (3MS No.170)	10 May 1971	Soviet Union	Orbiter	Launch failure	Never left LEO; booster stage burn timer set incorrectly	Proton-K/D
Mars 2	Mars 2 (4M No.171)	19 May 1971	Soviet Union	Orbiter	Successful	Entered orbit on 27 November 1971, operated for 362 orbits[5]	Proton-K/D
	Mars 2 lander (SA 4M No.171)			Lander	Spacecraft failure	**First impact on Mars,** deployed from Mars 2, failed to land during attempt on 27 November 1971.[6]	
	Prop-M			Rover	Failure Lost with Mars 2	Lost when the Mars 2 lander crashed into the surface of Mars.	
Mars 3	Mars 3 (4M No.172)	28 May 1971	Soviet Union	Orbiter	Successful	Entered orbit on 2 December 1971, operated for 20 orbits[7][8]	Proton-K/D
	Mars 3 lander (SA 4M No.172)			Lander	Partial success[9][10]	**The first lander on Mars**, soft landed on 2 December 1971. A first partial image (70 lines) was

Mission	Spacecraft	Launch Date	Operator	Mission Type[1]	Outcome[2]	Remarks	Carrier rocket[3]
	Prop-M			Rover	carrier vehicle failed before rover was deployed	transmitted. Contact lost 104.5 seconds[11] after landing.[12] **First rover on another planet**, 4.5 kg (9.9 lb) rover connected to the Mars 3 lander by a tether. Deployment status unknown due to loss of communications with the Mars 3 lander.[11]	
Mariner 9	Mariner 9	30 May 1971	NASA United States	Orbiter	Successful[13]	**The first orbiter of Mars.** Entered orbit on 14 November 1971, deactivated 516 days after entering orbit	Atlas SLV-3C Centaur-D
Mars 4	Mars 4 (3MS No.52S)	21 July 1973	Soviet Union	Orbiter	Spacecraft failure	Failed to perform orbital insertion burn	Proton-K/D
Mars 5	Mars 5 (3MS No.53S)	25 July 1973	Soviet Union	Orbiter	Successful	Contact lost after 9 days in Mars orbit. returned 180 frames	Proton-K/D
Mars 6	Mars 6 (3MP No.50P)	5 August	Soviet Union	Flyby	Successful	Flyby bus collected	Proton-K/D

Mission	Spacecraft	Launch Date	Operator	Mission Type[1]	Outcome[2]	Remarks	Carrier rocket[3]
		1973				data.[14]	
	Mars 6 lander			Lander	Spacecraft failure	Contact lost upon landing, atmospheric data mostly unusable.	
	Mars 7 (3MP No.51P)			Flyby	Successful	Flyby bus collected data.	
Mars 7	Mars 7 lander	9 August 1973	Soviet Union	Lander	Spacecraft failure	Separated from coast stage prematurely, failed to enter Martian atmosphere.	Proton-K/D
	Viking 1 orbiter			Orbiter	Successful	Operated for 1385 orbits. Entered Mars orbit on 19 June 1976.	
Viking 1	Viking 1 lander	20 August 1975	NASA United States	Lander	Successful	The second lander successfully returning data, deployed from Viking 1 orbiter. Operated for 2245 sols. Landed on Mars in 20 July 1976.	Titan IIIE Centaur-D1T
Viking 2	Viking 2 orbiter	9 September 1975	NASA United States	Orbiter	Successful	Operated for 700 orbits. Entered Mars orbit on 7 August 1976.	Titan IIIE Centaur-D1T
	Viking 2 lander			Lander	Successful	Deployed from Viking 2 orbiter, operated for	

Mission	Spacecraft	Launch Date	Operator	Mission Type[1]	Outcome[2]	Remarks	Carrier rocket[3]
						1281 sols (11 April 1980). Landed on Mars on September 1976.	
Phobos 1	Phobos 1 (1F No.101)	7 July 1988	Soviet Union	Orbiter	Spacecraft failure	Communications lost before reaching Mars; failed to enter orbit	Proton-K/D-2
	DAS			Phobos lander	Failure Lost with Phobos 1	To have been deployed by Phobos 1	
Phobos 2	Phobos 2 (1F No.102)	12 July 1988	Soviet Union	Orbiter	Mostly successful	Orbital observations successful, communications lost before lander deployment.	Proton-K/D-2
	Prop-F			Phobos rover	Failure Lost with Phobos 2	To have been deployed by Phobos 2	
	DAS			Phobos lander	Failure Lost with Phobos 2	To have been deployed by Phobos 2	
Mars Observer	*Mars Observer*	25 September 1992	NASA United States	Orbiter	Spacecraft failure	Lost communications before orbital insertion	Commercial Titan III
Mars Global Surveyor	*Mars Global Surveyor*	7 November 1996	NASA United States	Orbiter	Successful	Operated for seven years	Delta II 7925
Mars 96	Mars 96 (M1 No.520)(Mars-8)[4]	16 November 1996	Rosaviakosmos Russia	Orbiter Penetrators	Spacecraft failure	Never left LEO	Proton-K/D-2
	Mars 96 lander			Lander	Failure Lost with Mars 96	Two Mars landers to	

Mission	Spacecraft	Launch Date	Operator	Mission Type[1]	Outcome[2]	Remarks	Carrier rocket[3]
	Mars 96 lander			Lander	Failure Lost with Mars 96	have been deployed by Mars 96.	
	Mars 96 penetrator			Penetrator	Failure Lost with Mars 96	Two Mars Penetrators to have been deployed by Mars 96.	
	Mars 96 penetrator			Penetrator	Failure Lost with Mars 96		
Mars Pathfinder	Mars Pathfinder	4 December 1996	NASA 🇺🇸 United States	Lander	Successful	Landed at 19.13°N 33.22°W on 4 July 1997,[15] Last contact on 27 September 1997	Delta II 7925
	Sojourner			Rover	Successful	**The first rover to operate on another planet**, operated for 84 days[16]	
Nozomi	Nozomi (PLANET-B)	3 July 1998	ISAS 🔴 Japan	Orbiter	Spacecraft failure	Performed a Mars flyby. Later contact lost due to loss of fuel.	M-V
Mars Climate Orbiter	Mars Climate Orbiter	11 December 1998	NASA 🇺🇸 United States	Orbiter	Spacecraft failure	Approached Mars too closely during orbit insertion attempt due to a software interface bug involving different units for impulse and burned up in the atmosphere	Delta II 7425
Mars Polar	Mars Polar	3	NASA	Lander	Spacecraft	Failed to land	Delta II

Mission	Spacecraft	Launch Date	Operator	Mission Type[1]	Outcome[2]	Remarks	Carrier rocket[3]
Lander/Deep Space 2	Lander Deep Space 2	January 1999	🇺🇸 United States		failure		7425
				Penetrator	Spacecraft failure	No data transmitted after deployment from MPL.	
	Deep Space 2			Penetrator	Spacecraft failure		
Mars Odyssey	Mars Odyssey	7 April 2001	NASA 🇺🇸 United States	Orbiter	Operational	Expected to remain operational until 2025.	Delta II 7925
	Mars Express		ESA 🇪🇺 European Union	Orbiter	Operational	Enough fuel to remain operational until 2026.	
Mars Express		2 June 2003	ESA			No communications received after release from Mars Express.	Soyuz-FG/Fregat
	Beagle 2		🇬🇧 United Kingdom	Lander	Lander failure	Orbital images of landing site suggest a successful landing, but two solar panels failed to deploy, obstructing its communications.	
Spirit	Spirit (MER-A)	10 June 2003	NASA 🇺🇸 United States	Rover	Successful	Landed on 4 January 2004. Operated for 2208 sols	Delta II 7925
Opportunity	Opportunity (MER-B)	8 July 2003	NASA 🇺🇸 United States	Rover	Successful	Landed on 25 January 2004. Operated for 5351 sols	Delta II 7925H
Rosetta	Rosetta	2 March	ESA	Flyby	Successful	Flyby in	Ariane

Mission	Spacecraft	Launch Date	Operator	Mission Type[1]	Outcome[2]	Remarks	Carrier rocket[3]
	Philae	2004	European Union	(Gravity assist) Flyby (Gravity assist)	Successful	February 2007 en route to 67P/Churyumov–Gerasimenko[17]	5G+
Mars Reconnaissance Orbiter	Mars Reconnaissance Orbiter	12 August 2005	NASA United States	Orbiter	Operational	Entered orbit on 10 March 2006	Atlas V 401
Phoenix	*Phoenix*	4 August 2007	NASA United States	Lander	Successful	Landed on 25 May 2008. End of mission 2 November 2008	Delta II 7925
Dawn	*Dawn*	27 September 2007	NASA United States	Flyby (Gravity assist)	Successful	Flyby in February 2009 en route to 4 Vesta and Ceres	Delta II 7925H
Fobos-Grunt/Yinghuo-1	Fobos-Grunt	8 November 2011	Roskosmos Russia	Orbiter Phobos sample return	Spacecraft failure	Never left LEO (intended to depart under own power)	Zenit-2M
	Yinghuo-1		CNSA China	Orbiter	Failure Lost with Fobos-Grunt	To have been deployed by Fobos-Grunt	
Mars Science Laboratory	*Curiosity* (Mars Science Laboratory)	26 November 2011	NASA United States	Rover	Operational	Landed on 6 August 2012	Atlas V 541
Mars Orbiter Mission	Mars Orbiter Mission (*Mangalyaan*)	5 November 2013	ISRO India	Orbiter	Operational	Entered orbit on 24 September 2014. Mission extended to 2020.[18]	PSLV-XL
MAVEN	MAVEN	18 November 2013	NASA United States	Orbiter	Operational	Orbit insertion on 22 September 2014[19]	Atlas V 401

Mission	Spacecraft	Launch Date	Operator	Mission Type[1]	Outcome[2]	Remarks	Carrier rocket[3]
	ExoMars Trace Gas Orbiter		ESA/Roscosmos European Union/Russia	Orbiter	Operational	Entered orbit on 19 October 2016	
ExoMars 2016	Schiaparelli EDM lander	14 March 2016	ESA European Union	Lander	Spacecraft failure	Carried by the ExoMars Trace Gas Orbiter. Although the lander crashed,[20][21] engineering data on the first five minutes of entry was successfully retrieved.[22][23]	Proton-M/Briz-M
	InSight			Lander	Operational	Landed on 26 November 2018.	
InSight	MarCO A	5 May 2018[24][25]	NASA United States	Flyby	Successful	Flyby 26 November 2018. Last contact 29 December 2018.	Atlas V 401
	MarCO B			Flyby	Successful	Flyby 26 November 2018. Last contact 4 January 2019.	
Emirates Mars Mission	Hop	19 July 2020[26]	MBRSC United Arab Emirates	Orbiter	Operational	Entered orbit on 9 February 2021.[27][28][29]	H-IIA
Tianwen-1	Tianwen-1 orbiter	23 July 2020[30][31]	CNSA China	Orbiter	Operational	Entered orbit on 10 February 2021	Long March 5
	Tianwen-1			Lander	En route	Proposed	

Mission	Spacecraft	Launch Date	Operator	Mission Type[1]	Outcome[2]	Remarks	Carrier rocket[3]
	lander					landing: NET May 2021	
	Tianwen-1 rover			Rover	En route	Proposed landing: NET May 2021	
Mars 2020	Perseverance	30 July 2020[32]	NASA 🇺🇸 United States	Rover	Operational	Landed on 18 February 2021[33]	Atlas V 541
	Ingenuity			Heli-copter			

The current understanding of **planetary habitability** – the ability of a world to develop and sustain life – favors planets that have liquid water on their surface. This most often requires that the orbit of a planet lie within the **habitable zone,** which in our star system currently extends from just beyond Venus to about the **semi-major axis of Mars**. During perihelion, Mars (closest approach) dips inside this region, but the planet's thin (low-pressure) atmosphere prevents liquid water from existing over large regions for extended periods. The past flow of liquid water demonstrates the planet's potential for habitability. Some recent evidence has suggested that any water on the Martian surface may have been too salty and acidic to support regular terrestrial life.

Author's Rant: This whole concept of the zone of habitability in which life is thought to exist may have to be re-written if life is discovered on the Jovian moon of Europa or even on the methane lake covered Saturnian moon of Titan. Scientists need to think outside the box and stop thinking in anthropocentric terms and concepts or they may overlook important areas of discovery. The habitable zone should only be considered as a starting point that would inevitably evolve beyond this basic theory for finding life, elsewhere in the universe. Life is diverse and extremely adaptable to almost any harsh environment, as we are currently discovering right here on Earth. The extremophile forms of life should be our starting point.

It is important to cover some of the basic planetary science for the current conditions that NASA scientists are *"reporting to find"* on Mars so that we may appreciate the incredible discoveries uncovered by the various Mars probes and rovers sent by NASA and Russian Space Agency, but not reported to the general public.

The lack of a magnetosphere and extremely thin atmosphere of Mars are a challenge: the planet has little heat transfer across its surface, poor insulation against bombardment of the **solar wind** and insufficient atmospheric pressure to retain water in a liquid form (water instead sublimates to a gaseous state). Mars is for all intents and purpose, a geologically dead planet; the end of volcanic activity has apparently stopped the recycling of chemicals and minerals between the surface and interior of the planet. These are the geological and environmental conditions that

NASA wants you to accept as the actual and legitimate discoveries being made with every meter travelled and every rock examined by their Martian rovers. Their chief agenda and mission is the search for water which will mean that there is at least some probability that some microbial life may be discovered on the planet which would culminate in some big public or news media announcement. The fact that NASA's Mars rovers are photographing all kinds of unusual artifacts of artificiality suggesting that the former inhabitants had an intelligence of greater complexity than microbes is totally ignored at least not reported in the public press.

On November 20, 2012, NASA had us all salivating by leaking to the news media, either prematurely or intentionally (depending on your perception of NASA), at the possibility that the Mars rover Curiosity had discovered microbial life on Mars. The media reports suggested that the Mars rover discovery was *"gonna be one for the history books"* according to Curiosity chief scientist **John Grotzinger** of Caltech in Pasadena, *"but we'll have to wait a few weeks to learn what the new Red Planet find may be"*.

Curiosity's **Sample Analysis** at **Mars** instrument **(SAM)** is the rover's onboard chemistry lab, capable of identifying organic compounds - the carbon-containing building blocks of life as we know it. **SAM** apparently spotted something interesting in a soil but, scientists want to check and double-check the results before official announcement was made. Indeed, Grotzinger confirmed to SPACE.com that the news will come out at the fall meeting of the American Geophysical Union on December 3-7 in San Francisco.

In addition to analyzing soil samples, SAM also takes the measures the Martian air. Many scientists want to know if Curiosity would detect any methane, which is produced by many life forms here on Earth. A SAM analysis of Curiosity's first few sniffs found no definitive trace of the gas in the Martian atmosphere, but the rover will keep looking. Was this all hype by NASA to generate more public interest and keep those dollars coming into their budget and programs? http://www.sott.net/article/253871-Mars-Mystery-Has-Curiosity-rover-made-a-big-discovery

The date came and went and NASA sheepishly admitted that it was all premature and nothing significant had been discovered as yet in determines of water, methane or microbial life yet, the anomalies and artifacts kept appearing and were photographed and nothing was mentioned about them.

Evidence suggests that the planet was once significantly more habitable than it is today but, whether living microbes ever existed there remains unknown. The **Viking probes** of the mid-1970s carried experiments designed to detect microorganisms in Martian soil at their respective landing sites and had positive results, including a temporary increase of CO_2 production on exposure to water and nutrients. This sign of life was later disputed by some scientists, resulting in a continuing debate, with NASA scientist **Gilbert Levin** asserting that Viking may have found life. A re-analysis of the Viking data, in light of modern knowledge of **extremophile** forms of life, has suggested that the Viking tests were not sophisticated enough to detect these forms of life. The tests could even have killed a (hypothetical) life form. Tests conducted by the Phoenix Mars lander have shown that the soil has a very alkaline pH and it contains magnesium, sodium, potassium and chloride. The soil nutrients may be able to support life, but life would still have to be shielded from the intense ultraviolet light. Marshttp://en.wikipedia.org/wiki/Mars

NASA was not going to admit to anything that resembled life in any form or any markers to its existence. It would seem that lies and denials were becoming the agency's official mantra when it comes to finding any signs of extraterrestrial life, small or large, simple or complex on Mars.

As previously discussed, at the **Johnson Space Center**, some fascinating shapes have been found in the **meteorite ALH84001**, which is thought to have originated from Mars. Some scientists propose that these geometric shapes could be fossilized microbes extant on Mars before the meteorite was blasted into space by a meteor strike and sent on a 15 million-year voyage to Earth. An exclusively inorganic origin for the shapes has also been proposed.

Small quantities of methane and formaldehyde recently detected by Mars orbiters are both claimed to be possible evidence for life, as these chemical compounds would quickly break down in the Martian atmosphere. Alternatively, these compounds may instead be replenished by volcanic or other geological means, such as serpentinization.

The German space agency discovered Earth **lichens** do survive in simulated Mars conditions. The simulation based temperatures, atmospheric pressure, minerals, and light on data from Mars probes. This is the basis of **Terraforming** a planet for human habitability.
[Marshttp://en.wikipedia.org/wiki/Mars](http://en.wikipedia.org/wiki/Mars)

Will Mars Have a Renascence or a Terraforming Program in Its Future?

A "**Renascence Program"** (or renaissance - rebirth) is a method that could be used to "re-awaken" the dormant plant and microbial life already extant on the Martian surface as it is believed by some planetary scientists and exobiologists to be still present on Mars. This is really no different than a cryogenic or "deep freeze" program currently used to preserve microbial, plant or animal life for a period of time, to be re-activated at a later date. It is even a consideration for long term space flight requiring possible centuries to complete a destination or mission. The difference in a renascence program is the re-vitalization of life particularly dormant life that may be in a stasis of hibernation for many hundreds of millions of years or possibly milliards of time. This is obviously new frontier science but theoretically, it may be possible to re-awaken life after a few years but, is this even possible after eons of time in the "deep freezer of Mother Nature"? Not likely!

Terraforming is a hypothetical planetary engineering process considered more like science fiction than practical science at this current time yet; it has not stopped scientists from exploring and debating the concepts and practicality of such a massive undertaking with all its attending ramifications. Literally, it means "Earth-shaping" of a planet, moon, or other body by deliberately modifying its atmosphere, temperature, surface topography or ecology to be similar to the biosphere of Earth, in order to make it habitable by Earthlings. Whether you use the late Carl Sagan term of *"Planetary Engineering"* or NASA's *"Planetary Ecosynthesis"* or the current *"Planetary Modeling"* of Mars, we are still referring to terraforming for humans unless, someone has decided that we are "reshaping" or "remodeling" or whatever, for some other alien life form. http://en.wikipedia.org/wiki/Terraforming

Lichens, as proposed by some planetary scientists, could be the initial means by which Mars may be *terraformed* into a more hospital planet suitable for human habitation in the not too distant future. A *"Genesis Program"* similar in concept to the Star Trek movie, the "Wrath of Khan" (where future Earth scientists develop a means to **terraform** a dead planetary body to habitable Earth-like conditions) could be implemented with hundreds or thousands of satellites. The satellites would "seed" the upper Martian atmosphere with microbial life along with rudimentary plant life like lichen that would quickly settle out to the surface and adapt to the Martian environment.

With careful monitoring, greater complexities of plant life forms could be introduced as the Martian soil becomes conditioned from the initial seeding program. Hopefully, from the microbial and plant life, an evolving atmosphere will have generated the elements of nitrogen, carbon dioxide, and oxygen, reaching a point of stabilization whereby, simple animal life and eventually higher life forms could be introduced. Within 100 to 500 hundred years, it is theoretically possible to have Mars suitably terraformed to accept a habitation by permanent, sustained and ecological-minded humans. Failing to find any existing or dormant plant or animal life on Mars then, a *Genesis* type program becomes viable option.

As already indicated, Mars is the primary candidate for terraforming and then eventual colonization, Venus has also been considered but the methods for terraforming would be different from Mars. Terraforming Venus requires two major changes; removing most of the planet's dense (**9 MPa carbon dioxide**) atmosphere and reducing the planet's 450 °C (723.15 K) surface temperature. These goals are closely interrelated since Venus' extreme temperature is thought to be due to the greenhouse effect caused by its dense atmosphere. Sequestering the atmospheric carbon would likely solve the temperature problem as well.
http://en.wikipedia.org/wiki/Terraforming and

Jupiter's moon Europa is another candidate with one advantage in its favour and that is the presence of liquid water trapped under a thick surface crust of ice. Jupiter's other moons Ganymede and Callisto also have an abundance of water ice. Saturn's Titan offers several unique advantages, such as an atmospheric pressure similar to Earth and an abundance of nitrogen and frozen water. Other moons and planetary bodies include the Moon, the most obvious place to start this type of megaproject; there is even Mercury, Saturn's moon Enceladus and the dwarf planet Ceres.

Each planet or moon will, of course, have its attendant advantages and disadvantages to terraforming. The long timescales and practicality of terraforming are subject to debate. The big question to terraforming and eventual colonization of any planet is whether we should even do it, let alone all the other inherent unanswered questions related to the ethics, logistics, economics, politics, and methodology of altering the environment of an extraterrestrial world.

At the present time, economic resources required for terraforming are far beyond that which any government or society is willing to allocate for this purpose. Although a Manhattan-style project similar to developing and building the first atomic bombs could be employed particularly if there were a timeline to the eventual extinct of life on this planet. Even then, we've already witnessed the stealing of trillions of dollars world –wide particularly from the U.S. by the wealthy corporate elite for their own personal agendas as previously discussed in this book. ***Maybe this program is already underway somewhere else in our Solar System!*** (Bold italics added by author for emphasis). http://en.wikipedia.org/wiki/Terraforming

Paraterraforming also known as the *"worldhouse"* concept (think of this as a huge greenhouse) that uses domes in the construction of a habitable enclosure on a planet which eventually grows to encompass most of the planet's usable area. The enclosure would consist of transparent roof held one or more kilometers above the surface, pressurized with a breathable atmosphere, and anchored with tension towers and cables at regular intervals.

An artist conception of a possible terraforming program or reclamation program to restore Mars back to its original life sustaining and habitable conditions
https://www.universetoday.com/14883/mars-colonizing/ **(Credit: Daein Ballard)**

Where have we heard of this before?

Proponents would point out that this technology has been known since the 1960s and later implemented with the **Biosphere 2 dome project** which contained a habitable environment but, unfortunately, failed due to a number of reasons. But, we've also heard this before from Apollo astronauts who reported and photographed such dome structures in craters on the Moon that were supported by cables and tall towers! It would appear that someone else has already constructed these types of structures on the Moon! http://en.wikipedia.org/wiki/Terraforming and https://www.youtube.com/watch?v=O5k0MtlWPOs

Signs of a Life Sustaining Atmosphere - "Why is the Sky Blue, Daddy"?

If we are to understand the true nature of Mars, we need to heed the advice of the late Apollo11 astronaut **Neil Armstrong** who stated in one of his last interviews by *"removing one of truth's protective layers"*.

When considering whether there is life on Mars, one must look at this question within the context of what kind of environmental conditions are there that existed in the past and may still be present, that were conducive for life to establish and propagate itself? Is there microbial life lying dormant or fossilized within the Martian rocks and soil? Is there any evidence for plant life frozen and dormant in the Polar regions or still flourishing in the equatorial regions of Mars? Did Mars have complex animal life forms upon it surface in the distant past? Was there any intelligent life forms on Mars at any time and did they leave artifacts and remnants of their civilization behind due to some ancient catastrophe that befell them? Is there an existing Martian intelligence still present on Mars or within the Solar System? What is our relationship, if any, to these Martians and/or Extraterrestrial intelligence?

Author's Rant: As a former university student in astronomy, it struck me that Mars has always been portrayed by the early Viking landers and orbiters as having a red atmosphere, why is that? This got me to thinking about a clever TV commercial back in the "70s that was diabolically funny but also, informative which illustrates a similar predicament of getting the truth out of NASA regarding its discoveries found on Mars.

In the commercial, a young boy holds a **Cadbury's chocolate bar** called a **Bar Six** and he asks hisfather that age-old question: *"Why is the sky blue, Daddy?"*

The father who is busy reading a newspaper answers his son without looking up from his paper with the line: *"**The sky is blue due to the scattering of light from dust particles in the atmosphere which reflects the blue end of the spectrum!**"*

The boy politely says: *"**Thank you, Daddy.**"*

The boy being an inquisitive young lad still holding his Bar Six chocolate bar, again asks his father another question: *"**Why do they call this chocolate bar, a Bar Six?**"*

This time, the father looks up and sees his son holding the chocolate bar: *"Let me see?* Opening the chocolate bar wrapper, the father says *Bar Six…six bars"*.

Then, quite unexpectedly, the father says: *"Now, if I eat one, what is it then"?*

The son replies innocently: *"A Bar Five".*

The father: **"right!"**

Finding the taste of the chocolate bar agreeable, the father says cleverly: *"If I eat one more bar, what is it, now"?*

The son with a look of astonishment replies: *"Bar Four!"*

The father mischievously breaks off another piece of chocolate and says: *"And now?"*

The son now beginning to feel uneasy at the impending circumstances says: *"Bar Three".*

The father greedily breaks off two more pieces of chocolate and shoves it into his already full mouth, and with barely any comprehensibility, now says: **"And again?"**

Son dejectedly responds to the inevitable outcome: **"Bar One!"**

The father pops the last remaining piece of chocolate into his mouth with a smile and picks up his newspaper: *"And now"?*

The son in a sad and dejected voice: *"Bar One. Thank you, Daddy."* He turns away in disappointment.

The lesson that can be taken away from this amusing commercial is that the young polite boy full of innocence and inquisitiveness represents the *general public*, the boy's father, a man of high education, wisdom and in a position of trust is called upon to impart information to an inquiring young mind, represents *NASA* or any other government agency or position of officialdom. But for this purpose of story-telling the father represents NASA. The Bar Six chocolate bar represents the *potential new knowledge* to be gained which the innocent and trusting boy holds within his hands or in this case signifying the *planet Mars*.

The first question asked by the boy is a question to establish trust and confidence. The answer to the question by the father who is really too busy in his own matters of reading *(space exploration)* is given honestly and accurately to his son but with a sense of *don't bother me, can't you see I'm busy with other things*? Honesty and confidence is nevertheless, established. It is not until the son has displayed something of personal interest to himself, the Bar Six *(planet Mars)* chocolate bar (this could even be a *Mars Bar*) does the father *(NASA)* show any real interest in his son and it is not really his son that he is interested in but, what the son has in his hand which could also represent financial resources.

The slow but methodical devouring of the chocolate bar while eliciting a trusting response by the son to an ever changing situation of facts is indicative of how NASA discloses on occasion, some genuine knowledge in planetary science to support their established position as scientists in order to gain the public's trust and financial resources. When the public raise matters of a particular nature such as whether there is life on Mars, now or in the past, does NASA respond with a small insignificant truth but then, alters that truth as they take advantage of what is being discovered there. Each discovery on Mars is devoured or covered up by NASA. Their discoveries are altered to half-truths, denials, obfuscation and even lies, while at the same time getting the public to acknowledge and accept this latest revelation of lies as full truths.

The son *(public),* however, quickly realizes what is going on as he sees his father *(NASA)* devouring *(covering up)* all truth leaving, in fact, little or no new knowledge, evidence or truth behind to share with anyone else, particularly with the general public. The sad state of affairs is that like the son, the public turns away dejected, disappointed and powerless to do anything to correct the situation at this time. Only time is on the side of the public and eventually as they seize or regain power on equal footing with NASA, will they get the suppressed knowledge and hidden truth kept from them for so long.

One researcher, **Ted Twietmeyer** has been steadily gathering data and information that strongly incriminates NASA in a massive cover-up of evidence that supports and confirms at least the past existence of life on Mars and not just microbial life. In his e-Book "What NASA Isn't Telling You…About Mars" published in 2005 is a whole different understanding of what Mars is really like which stands juxtaposed to the current understanding held by public perception.

Twietmeyer addresses the question of why the Martian atmosphere always appears red and surprisingly it has nothing to do with the particles in the sky. It has to do with the colour calibration target sometimes known as the *"Sundial"* that are on each spacecraft sent to Mars over the past 35 years. Images of these *"Sundials"* show blue in one corner when on Earth then oddly, the same corner of the target displays as red when that portion of the Mars rover is shown in the photograph. (See below). e-Book: "What NASA Isn't Telling You About Mars" © 2005 by Ted Twietmeyer

Original colour photo of Mars from the Spirit rover which NASA released to the public. Note the RED corner and electrical connector of the calibration target circled in green compare it to the same corner and electrical connector at right in the Pancam target calibrated on Earth which is BLUE, its true colour! So is Mars really red in colour?
http://www.goroadachi.com/etemenanki/mars-hiddencolors.htm **(Credit: Goro Adachi)**

How can this be unless, there is deliberate tampering going on at the NASA photo labs, back on Earth? Scientists state that it is because they haven't quite got the colour right when it is in the Martian environment so it has to be constantly re-calibrated. This means that the majority of Mars images are either automatically altered by the computers on board the rovers or they are altered after they are transmitted back to Earth. Like the Apollo moon mission photos, any anomalous objects are airbrushed or sanitized out before being released into the public domain. The following photographs are just a small sampling of the tens of thousands of photo images transmitted back to NASA and in most cases what the public sees is a pink or red sky above the reddish brown Martian surface. Beside each original NASA photo is the colour corrected image as you would see it if you were an astronaut on the surface of Mars, based on the Pancam colour target calibrated on Earth. Typically, the landers and rovers photograph the Martian landscape in correct, *normal* colour calibration. The images only become altered or filtered by NASA when a situation requires clarity of detail or scientific measurement for light density, contrast, and other factors. When NASA releases new photo images out to the public, almost immediately investigators check to see what is contained in these new Mars images. Frequently, people do find some strange objects that are unusual which somehow seem to get overlooked by the censors of NASA. At those times, we are lucky enough to see an unaltered Mars images for what they truly are. We can only hope that these kinds of oversights continue.

Twietmeyer explains that this is a psychological process designed to instill and re-enforce the perception that Mars is not only a red planet when viewed through a telescope but, its

364

atmosphere is red as well. It would be like saying that because the Earth appears as blue from the Moon, therefore, the soil and dirt upon the Earth must also be blue.
http://conspiracy2012.files.wordpress.com/2012/05/what_nasa_isnt_telling_you_about_mars.pdf and e-Book: "What NASA Isn't Telling You About Mars" © 2005 by Ted Twietmeyer

Photo on left is the original but altered NASA image and photo on right is the colour corrected image as the human eye would naturally see it, note the sky is blue!
http://www.goroadachi.com/etemenanki/mars-hiddencolors.htm

Twietmeyer, of course, realizes that this is all utter bilge and bad erroneous science that is being perpetrated upon the general public! No matter, what other scientists or the educated public are saying about this aspect, NASA has dug its heels into the ground and refuses to budge. It may be that with their heels dug so far down into the ground, they may have unwittingly dug their own grave! Only time will tell.

Richard Hoagland, who can be considered as the leading pioneer researcher into all things related to NASA cover-ups in lunar and planetary discoveries and certainly, should be viewed as the "father of Mars anomaly research" has stated on his *Enterprise Mission* website that "NASA has a long and curious "history" regarding the "real" Mars … and certainly its color".
http://www.enterprisemission.com/colors.htm

Hoagland recounts an infamous controversy that *still* swirls around the release by JPL of the first "true color Viking Lander image," just one day after Viking touched down in the pre-dawn darkness (Pacific Time) of July 20, 1976. Within a few hours of that historic publication -- the release of the first color photograph *from the surface of Mars* -- another, hurriedly *revised* version of this first color surface image was suddenly produced – *"correcting the initial color engineering problems"* in the *first* image. (Bold italic emphasis added by author).

Decades later, after this first Mars landing, **Richard Hoagland** tells how **Ron Levin**, a physicist at MIT and the son of Viking's chief scientist for biology investigations at JPL, **Dr. Gilbert Levin** relates a *very* different story of this incident:

Back in the summer of 1976, Ron was a high school graduate assisting his father at JPL during that incomparable "Viking Summer" (where Hoagland was also present, covering the extraordinary Viking story for millions of readers of a major magazine, and a couple of broadcast television networks …).

The following is from Levin's first-hand recollections of the whole affair, recounted in a recent book by science writer **Barry DiGregorio** -- the remarkable "overreaction" by JPL that occurred in response to Ron Levin's naive efforts to "correct" what seemed to him that July afternoon to be "a *deliberate* – if perplexing – methodical distortion of the incoming Viking Lander data" *(Mars: The Living Planet, B. DiGregorio, G. Levin and P. Straat, Frog Ltd, Berkeley, CA 1997).*

According to DiGregorio's narrative:

"At about 2:00 P.M. PDT, the first color image from the surface of another planet, Mars, began to emerge on the JPL color video monitors located in many of the surrounding buildings, specifically set up for JPL employees and media personnel to view the Viking images. Gil and Ron Levin sat in the main control room where dozens of video monitors and anxious technicians waited to see this historic first color picture. As the image developed on the monitors, the crowd of scientists, technicians, and media reacted enthusiastically to a scene that would be absolutely unforgettable – Mars in color. The image showed an Arizona-like landscape: blue sky, brownish-red desert soil, and gray rocks with green splotches ...

"Gil Levin commented to **Patricia Straat** [his co-Investigator] and his son Ron, **'Look at that image! It looks like *Arizona'*** [below]. Bold italics added by author for emphasis). http://www.enterprisemission.com/colors.htm

"Two hours after the first color image appeared on the monitors, a technician abruptly changed the image from the light-blue sky and Arizona-like landscape to a uniform orange-red sky and landscape [below]. Ron Levin looked in disbelief as the technician went from monitor to monitor making the change. Minutes later, Ron followed him, resetting the colors to their original appearance. Levin and Straat were interrupted when they heard someone being chastised. It was Ron Levin being chewed out by the Viking Project Director himself, **James S. Martin, Jr. Gil Levin** went immediately and asked, "What is going on?" Martin had caught Ron changing all the color monitors back to their original settings. He warned Ron that if he tried something like that again, he'd be thrown out of JPL for good. The Director then asked a TRW engineer assisting the Biology team, **Ron Gilje**, to follow Ron Levin around to every color monitor and change it back to the red landscape. http://www.enterprisemission.com/colors.htm

On left, the first official NASA *Viking Lander* color picture (with blue sky) from Mars. The same picture on the right is a re-adjusted supposedly "normal red sky' version.

"What Gil Levin, Ron and Patricia Straat did not know (even to this writing) is that the order to change the colors came directly from the NASA Administrator himself, **Dr. James Fletcher**. Months later, Gil Levin sought out the JPL **Viking Imaging Team** technician who actually made the changes and asked why it was done. The technician ***responded that he had instructions*** from the Viking Imaging Team that the ***Mars sky and landscape should be red*** and went around to all the monitors 'tweaking' them to make it so. Gil Levin said, 'The new settings showed the American flag (painted on the Landers – (above) as having *purple* stripes. The technician said that the ***Mars atmosphere made the flag appear that way*** [emphasis added].' (Bold italics added by author for emphasis).

As someone who was also at JPL that afternoon, and vividly remembers a similar shock -- when the "Arizona Mars" initially flashed on the JPL monitors was suddenly transformed into a Martian "Red Light District" – I now kick myself for not asking *lots* more questions.

But, it was 1976 -- and we all *trusted* our Space Agency back then ….
http://www.enterprisemission.com/colors.htm

One of the basic questions that I should have asked involves the *physics* behind JPL's abrupt color alterations. Or, as Gil Levin put it:

"If atmospheric dust were scattering red light and not blue, the sky would appear red, but since the red would be at least partially removed by the time the light hit the surface, its [the direct sunlight's] reflection from the surface would make the surface appear more blue than red. There would be less red light [in the direct sunlight illumination] left to reflect. And what about the sharp shadows of the rocks in the black and white images yesterday? If significant scattering of the light on Mars occurred [from lots of red dust in the atmosphere], the sharp shadows in those

367

images would not be present, or at best, would appear fuzzy because of diffusion by the [atmospheric] scattering [emphasis added]!"

Levin was describing the well-known phenomenon of **"Raleigh scattering"** -- whereby the similar-sized molecules of *all* planetary atmospheres (be it the **primary nitrogen of Earth**; the **carbon dioxide atmosphere of Mars**; or even the predominantly *hydrogen* **atmospheres of Jupiter and Saturn!)** all produce *blue skies* when sunlight passes through them.

On the left is the 1979 NASA Viking raw data showing frost on the surface of Mars with the pinky orange sky. On the right is the true sky colour of Mars: blue! JPL scientists are "mystified" as to why NASA would alter the colour on their Mars photo images
http://xfacts.com/spirit2004/

If you examine the long Martian photographic record – which encompasses *hundreds of thousands* of images, acquired by dozens of observatories even before the Space age dawned – you can see blatant evidence that Levin's right and JPL is *wrong* … regarding the scientifically expected "color" of the Martian atmosphere. http://www.enterprisemission.com/colors.htm

368

Another Mars rover photo on the right and on the left the same image but, colour corrected to bring out more Martian surface detail

http://mars-news.de/color/blue.html

This colour tampering of images is an agenda designed to confuse the public's perception of Martian reality. The images transmitted from Mars may be amazing or pretty to look at but, they are so far removed from reality, that it becomes almost impossible to tell what you are looking at even when it is explained to you. Could certain areas of the planet Mars look like the painted deserts or the red countryside hills and cliff escarpments of Arizona, New Mexico, and Colorado? Of course but, are we to believe the whole planet from pole to pole is like is this?

Land often appears pink or salmon coloured or dark reddish brown. Sometimes, sand dunes are pink with dark streaks on them, are these trees and vegetation or eruptions of dark sand matter due to carbon dioxide ice melting just below the surface? In some photo images sand dunes and vast tracts of the Martian surface, as well as crater basins, appear blue, sometimes with ripples on the surface, are there really blue sand dunes? One wonders, whether he is looking at sand dune ripples or the ripples upon a small lake within a crater basin! This is deliberate photo tampering to obfuscate and hide what is really being photographed on the Martian surface; NASA knows more than what they are telling us. An example of this type of imaging can be seen below which has drawn a lot of internet controversy, as it may show signs of possible Martian vegetation or trees growing in sand dunes.

369

Are these trees or vegetation growing out of pink sand dunes in the south polar region of Mars? NASA says it is dark soil ejecta released by gases from the melting carbon dioxide ice below the sand dune surface. Now, which way is up and which way is down?

The reality is that Mars does not have a red or pink sky but a blue sky, much like what you expect to see on Earth. No rose coloured glasses or filters are required! A big tip off would be the bluish haze seen above the Martian surface by orbiting spacecraft indicating not only an atmosphere of some considerable density and pressure but, it also indicates that *"The sky is blue due to the scattering of light from dust particles in the atmosphere which reflects the blue end of the spectrum!"* (See photos below).

Temperature and air pressure go hand in hand in determining whether Earth, Mars or any planet actually has an atmosphere or not. We are told by NASA that Mars atmosphere is only 1.0% of Earth's but, that would indicate near-vacuum conditions on Mars. This is nonsense, Mars has ice and weather conditions similar to Earth and an atmosphere of 1.0% would cause ice to simply boil away or evaporate. We see seasonal changes in the Martian polar regions where polar caps shrink and grow much like on Earth. In other regions of the planet, vast areas of ground darken, expand then, become lighter in colour and contract at different times of the years, as if there were growing seasons of vegetation. There are dust devils that whip across the Martian deserts and

370

valley floor areas indicating meteorological frontal activity due to variability of temperature and barometric changes timed to spring and fall seasons. As far as NASA scientists know, it doesn't rain on Mars yet, there seems to be weather systems that can blow up tremendous dust storm that can cover three-quarters of the planet but, such weather systems shouldn't exist with a 1.0% atmosphere that is essentially a vacuum.

These photographs clearly show an atmospheric limb around Mars that is blue in colour and not red as is portrayed so often by NASA in many of its rover images
(Google Images)

Twietmeyer says in his e-Book that NASA claims that frozen carbon dioxide (dry ice) is evident in craters and in the open areas of the polar regions. The question is how can a CO_2 gas never evaporate because the polar caps to contract and expand seasonally. On Earth water has a lower boiling point or temperature with a decrease in air pressure and example of this is water boiling more quickly on a mountaintop like Mt. Everest because air pressure is much lower the higher up you go, so water boils quickly compare to areas at sea level. Dry ice has an ambient surface temperature of -79°C, if it rises above this temperature it starts to evaporate but, it can also evaporate even faster below this temperature if the air pressure is lower.

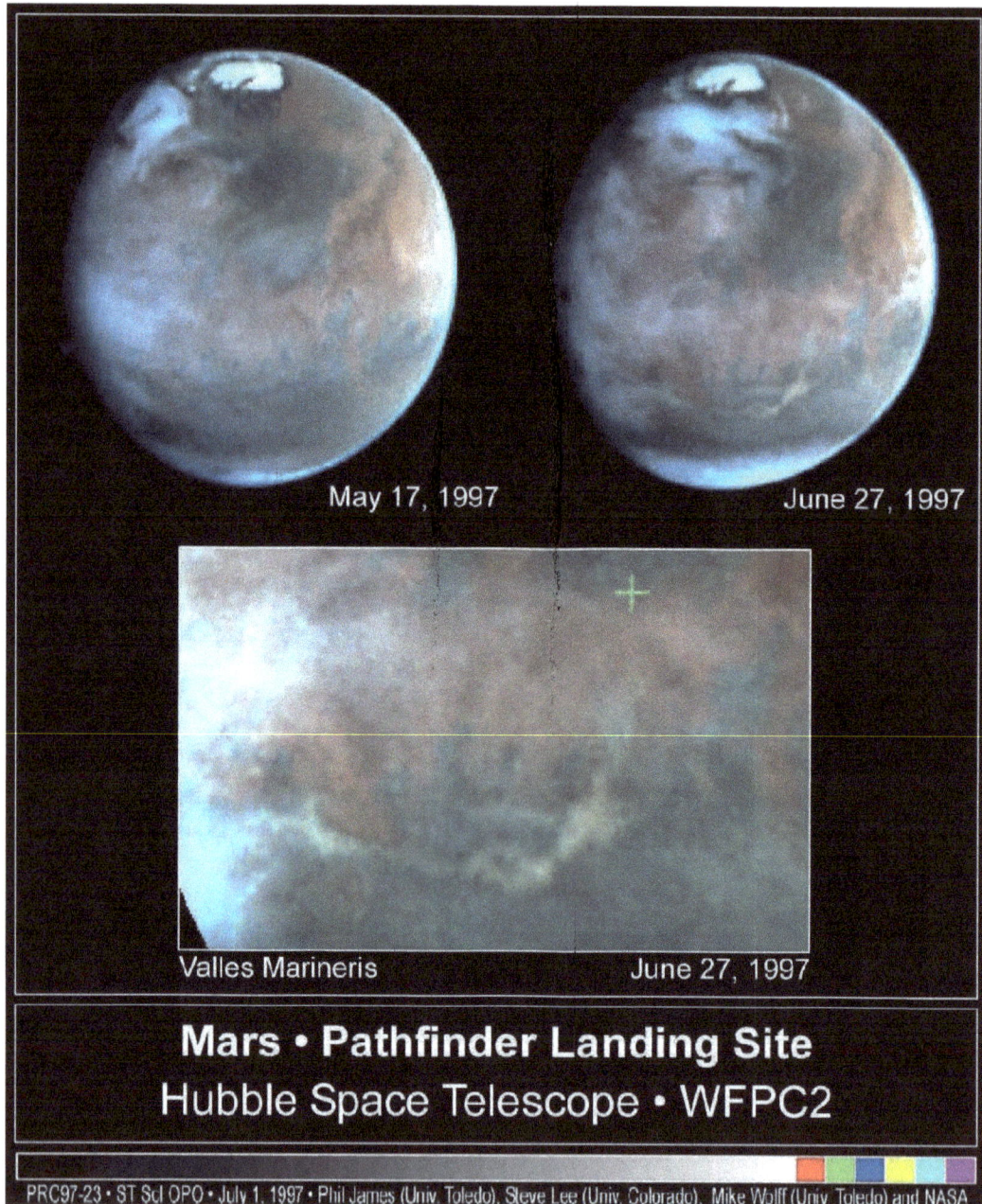

May 17, 1997   June 27, 1997

Valles Marineris   June 27, 1997

Mars • Pathfinder Landing Site
Hubble Space Telescope • WFPC2

PRC97-23 • ST ScI OPO • July 1, 1997 • Phil James (Univ. Toledo), Steve Lee (Univ. Colorado), Mike Wolff (Univ. Toledo) and NASA

These photos taken by the Hubble Telescope clearly show Mars with a blue atmosphere, cloud formations and demonstrates the blue "limb-brightening" consistent with Rayleigh-scattered light! There is no evidence of a reddish suspended "dust layer even at the limb."

https://www.bibliotecapleyades.net/marte/esp_marte_56.htm

What Twietmeyer concludes is that Mars is far warmer than would allow for CO_2 ice to remain in a frozen state for very long, let alone for millennia or eons of time, therefore, there has to be as yet, some undiscovered physics of weather at play on the surface of Mars. An environment that

372

allows for a continuous cycle of sublimation of carbon dioxide ice would require an air pressure greater than 1.0% and would be near impossible for any of the Mars rovers to survive those extremes of temperature based on NASA's operational specifications for solar powered rovers; nuclear-powered rovers are a different story.
http://conspiracy2012.files.wordpress.com/2012/05/what_nasa_isnt_telling_you_about_mars.pdf and e-Book: "What NASA Isn't Telling You About Mars" © 2005 by Ted Twietmeyer

Therefore, there is some other type of weather system in operation in the atmosphere of Mars as carbon dioxide ice would have evaporated off the surface ages ago because Mars is not always consistently cold enough to prevent all the ice from evaporating. Could ice remain in a deep crater for a period of time and is that ice, water based and/or CO2 ice based? As yet, NASA is not certain. The evidence indicates that Mars temperature and atmospheric pressure is much higher than previously thought, that means that Mars has a much denser atmosphere which permits the current weather conditions to exist in the manner in which they have been observed, much like conditions on Earth and in contradiction to what NASA has told us!

"The critical feature of all the above Mars images is the now blatantly obvious, "Raleigh scattered" *bluish limb* of Mars!

This astonishing fact – an unmistakably "Earthlike" sky on Mars – is the exact *opposite* of what NASA has tried to get us to believe all these years, and with a variety of now *clearly altered Martian images.*

This remarkable scientific and political conclusion is confirmed by a close-up view (below) of (obviously, now!) the two most "Earthlike" planets in the solar system … Earth and Mars. Note the distinct *blue* atmospheric layer floating over the russet Martian deserts in this close-up view, from the wide-angle MGS camera … and not a *trace* of reddish atmospheric dust in sight!"
http://www.enterprisemission.com/colors.htm

Composition of the Atmosphere on Mars

Gas	Formula	Distribution (Percent)
Carbon Dioxide	CO_2	95.4
Nitrogen	N_2	2.7
Argon (40)	Ar-40	1.6
Oxygen	O_2	0.13
Carbon Monoxide	CO	0.07
Water	H_2O	0.03*
Argon (36)	Ar-36	0.0005
Neon	Ne	0.0003
Krypton	Kr	0.00003
Ozone	O_3	0.00001*
Xenon	Xe	0.000008

*Abundance varies with season and location

Mars has a ***carbon dioxide based*** atmosphere unlike Earth's ***nitrogen based*** atmosphere due in large part to CO2 molecules and in a lesser degree to water molecules but, both molecules are present and *"water"* whether carbon dioxide based or hydrogen/oxygen based is still water and as the old adage states: *"where there is water, there is life!"* Could this also be true for Mars, only further investigation into other factors will support this premise? The first condition in our argument has been tentatively met for the possible presence of simple organic life on Mars, situated in the zone of habitability!

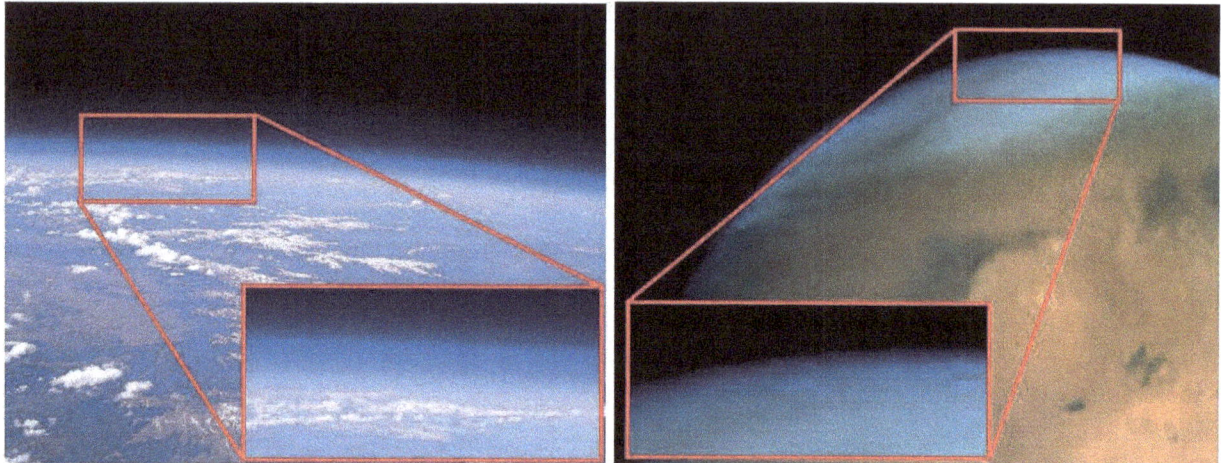

The Blue Rayleigh scattering of light on Earth and on Mars
(c) Terry Tibando

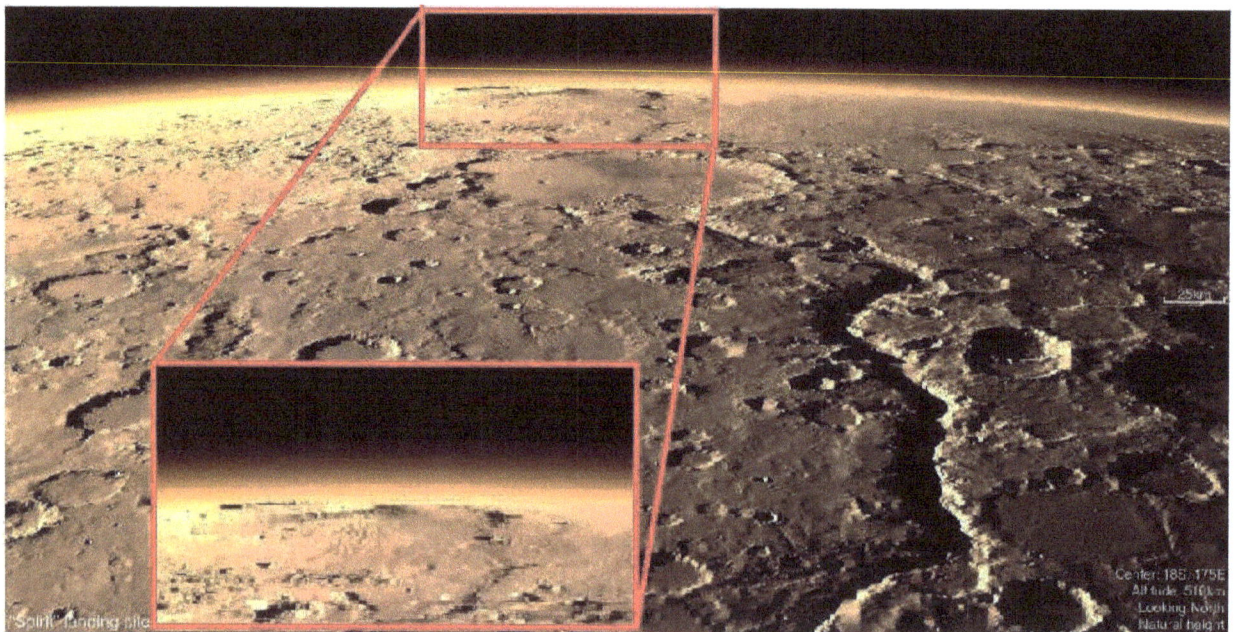

This photo by NASA'S Spirit orbiter shows the false red-orange colour atmosphere that NASA would like the public to believe is actually existing around the planet Mars
(c) Terry Tibando

Richard Hoagland raises some interesting questions as to why the colour of the images from Mars have been tampered with for the past 40 plus years by NASA's JPL **Image Processing** laboratory (IPL). Did NASA know something about Mars that they felt the public wasn't ready for based on the true colour of Mars atmosphere and its surface?

Hoagland recalls a statement written earlier by DiGregario, rather cryptic in nature regarding "the head of NASA's role" at the time of Viking's initial color image of the Martian surface which was based on a remarkable confirmation of this 1976 incident … and from an *official* source -- former JPL Public Affairs Officer, Jurrie J. Van der Woude. In a letter to DiGregario (also reproduced in *"Mars: The Living Planet"*), Van der Woude wrote:

"Both Ron Wichelman [of JPL's Image Processing laboratory (IPL)] and I were responsible for the color quality control of the Viking Lander photographs, and Dr. Thomas Mutch, the Viking Imaging Team leader, told us that he got a call from the NASA Administrator asking that we destroy the Mars blue sky negative created from the original digital data [emphasis added] …. http://www.enterprisemission.com/colors.htm

*Incredibly and almost beyond belief, the NASA Administrator was ordering **official NASA data to be destroyed**!*

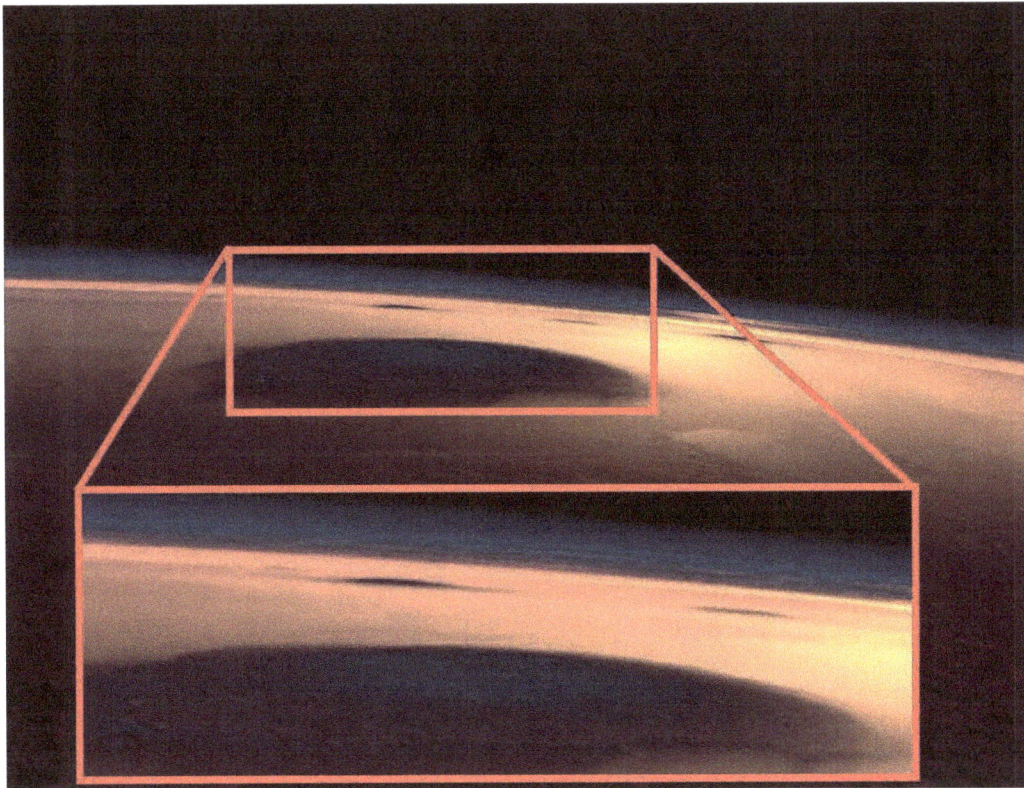

This is the real unaltered colour of the skies of Mars due to the Blue Rayleigh scattering of light! Like it says in the song: *"Nothing but blue skies all day long!"*
(Google Images)

"This bizarre sequence of events raises too many disturbing questions … like … why was the *Administrator* of NASA so determined to conceal the "true" colors of Mars from the American people and the world, in 1976? Why would he order the head of the Viking Imaging Team to literally *eliminate* an important piece of historical evidence from the official Mission archive – the original "blue-sky negative" – if the initial release was only "an honest technical mistake?!" Wouldn't that record be an important part of the ultimate, triumphant story of "NASA scientists eliminating initial scientific errors, in their continued exploration of the frontier and alien environment of another world ...?"

In truth, none of Ron Levin's story (or Van Der Woude's significant confirmation), makes any scientific sense ... unless ... certain individuals in NASA in 1976 felt *compelled* to hide – and at *all costs* – the visible appearance of the *actual* Martian surface ...."
http://www.enterprisemission.com/colors.htm

At this initial stage of our argument and toward establishing proof of life having existed on Mars, it would be safe to say that Mars has conditions that indicate higher temperatures and atmospheric pressures than were officially acknowledged by NASA. Mars has weather conditions that demonstrate planet-wide storms, small dust devils, major cloud formations and atmospheric turbulence, precipitation of CO_2 ice and possibly water molecules that are all indicative of weather patterns and conditions that can be found on Earth. Interestingly enough, when it comes to planet –wide dust storms on Mars as every elementary and high school student knows from basic arts and science that when you mix the *yellow* from a global-wide Martian dust storm to the *blue* Rayleigh-scattering light of Mars, you get a *"green"* atmosphere. Therefore, **Mars has a BLUE sky due to a scattering of light from dust particles in the atmosphere which reflects the BLUE end of the spectrum, and NOT the RED end of the spectrum**!
https://www.youtube.com/watch?v=tk3qLe2wWMA and
https://www.youtube.com/watch?v=Y21Bqc2We5o and
https://www.youtube.com/watch?v=AXTKsdeeNJI

CHAPTER 80

"CAPTAIN, WE DISCOVERED MULTIPLE SIGNS
OF LIFE DOWN HERE ON THE PLANET!"

Signs of Microbial Life. Show Us the Martian Microbes, Already!

Another contrary condition that shouldn't exist on Mars but, which is proof toward the second condition for life on the Red Planet is the discovery of Methane gas in the atmosphere by the **European Space Agency's** instruments on board their orbiting spacecraft, **Mars Express.** **Methane (CH_4)** is an organic molecule present in gaseous form in the Earth's atmosphere. More than 90% of methane on our home planet is produced by living organisms. On Earth, **CH4 gas** is the by-product of every living organism; it floats in air and is a major indicator that life exists on this planet. http://sci.esa.int/science-e/www/object/index.cfm?fobjectid=46038

So, why is there Methane in orbit about Mars, if there are no life forms or any active volcanism on Mars? Methane means life in some form exists nearby. Is this a by-product of Martian microbes or from animals or plant life? Based on photochemical models and on the current understanding of the composition of the Martian atmosphere, methane has a chemical lifetime of about 300-600 years, which is very short on geological time scales. This implies that the methane that is observed today cannot have been produced 4.5 billion years ago when the planets formed. http://sci.esa.int/science-e/www/object/index.cfm?fobjectid=46038

One possibility is a biological origin. The discovery of microbial life 2 to 3 kilometres beneath the surface of the Witwatersrand basin in South Africa led scientists to consider that similar organisms could live, or have lived in the past, below the permafrost layer on Mars. By analogy with Earth, the biological origin of Martian methane could be explained by the existence of micro-organisms, called **methanogenes**, existing deep under the surface, and producing methane as a result of their metabolism.

If the methane on Mars is biotic, two scenarios could be considered: either long-extinct microbes, which disappeared millions of years ago, have left the methane frozen in the Martian upper subsurface, and this gas is being released into the atmosphere today as temperatures and pressure near the surface change or some very resistant methane-producing organisms still survive.

An alternative explanation is that the methane is geological in origin. It could be produced, for example, by the oxidation of iron, similar to what occurs in terrestrial hot springs, or in active volcanoes. This gas could have been trapped in solid forms of water, or 'cages' that can preserve methane of ancient origin for a long time. These structures are known as 'clathrate hydrates'.

A geochemical process called **serpentinization** could also produce the abiotic methane. Serpentinization is a geological low-temperature metamorphic process involving heat, water, and changes in pressure. It occurs when olivine, a mineral present on Mars, reacts with water, forming another mineral called serpentine, in the presence of carbon dioxide and some catalysts. When certain catalysts are also present, the hydrogen combines with the carbon to form methane.

On Mars, it is possible to find all these primary elements: olivine, carbon dioxide, and some catalysts, but the chemical reaction needs liquid water to occur. This implies that, if the Martian methane comes from serpentinization, it could be related to subsurface hydrothermal activity.

Concentrations of methane have been observed in 2003 and 2006 in three specific regions of Mars: **Terra Sabae, Nili Fossae** and **Syrtis Major,** and data suggest that water once flowed over these areas. Deep liquid water areas below the ice layer would be able to provide a habitat for microorganisms, or a favourable place for the hydro-geochemical production of methane. Further processing in the Martian atmosphere may play an important role that accounts for the observed seasonal variability. Whether geochemical or biochemical in origin, the variation in concentrations of methane that has been measured indicates that Mars could still be active today.

Observations from the ESA's Mars Express have detected variable amounts of methane in the atmosphere of Mars. Could this be evidence for life on Mars? International space agencies are planning an ambitious, long-term Mars Robotic Exploration Program in *2016-2018* to search for signs of past and present life on Mars, studying the water and geochemical environment as a function of depth in the shallow subsurface, and investigating Martian atmospheric trace gases and their sources. http://sci.esa.int/science-e/www/object/index.cfm?fobjectid=46038

So, what can explain the presence of this gas on the Red Planet? One wonders if anyone else besides the **ESA** taken scientific measurements of the Martian atmosphere from high in orbit or on the surface?

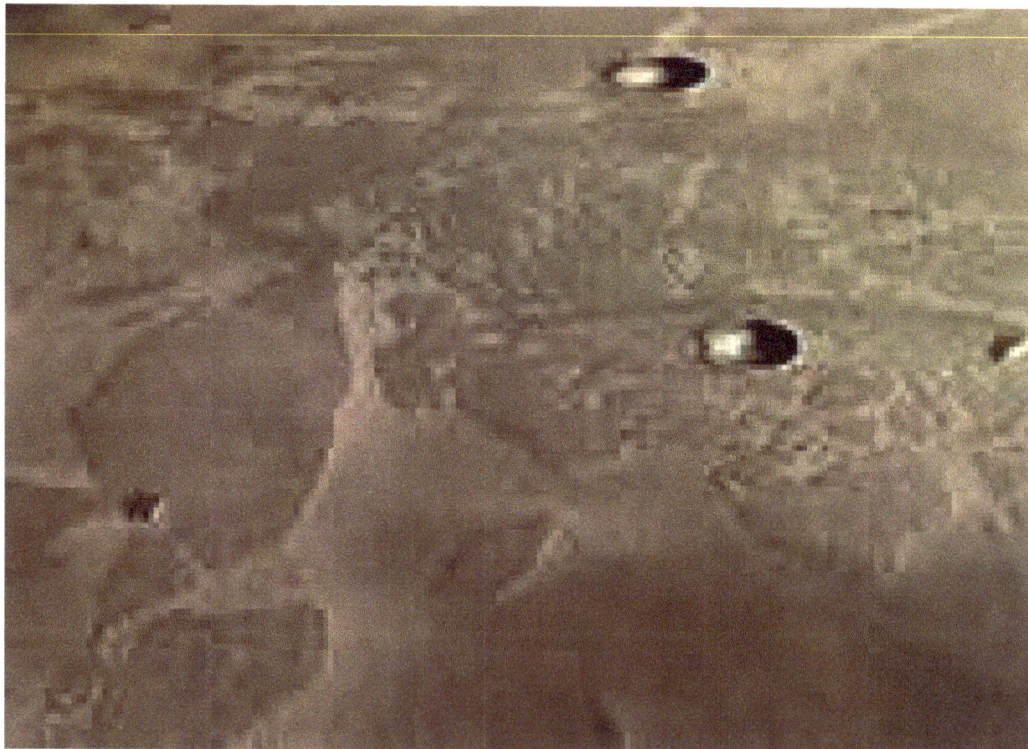

Elysium Planitia - raised levels of methane were detected by a Mars Express instrument
http://www.daviddarling.info/encyclopedia/E/Elysium_Planitia.html

As of this writing, as it turns out, NASA has also confirmed (perhaps, reluctantly) Methane in the upper atmosphere of Mars but, ***only 10.5 parts per billion!*** Once again it must be asked, why did NASA take so long to disclose this little bit of information when the Viking orbiter would have discovered this fact back in 1976? And NASA claims it was only a few parts per billion when ESA has recently detected *"plumes"* of Methane in the Northern hemisphere of Mars which is of great interest because of its potential biological origins. (Bold italics added by author for emphasis). http://sci.esa.int/science-e/www/object/index.cfm?fobjectid=46038

How does NASA account for Methane in the atmosphere and not just in trace amounts but, according to the European Space Agency, there are plumes of Methane, as opposed to the parts per billion as reported by NASA? Who's telling the truth when it comes to Mars and which of these agencies is releasing incorrect data and possibly covering up a major discovery, a *"truth"* that seems to have a *"protective layer"* over it? Any guesses as to who it could be?

We need to backtrack and look at some of the spacecraft that orbited and landed on Mars to find what data was gathered and transmitted back to Earth.

The exploration of Mars in the early '60s by unmanned space probes from the former Soviet Union was met with many failures with their **Mars Zond** and **Cosmos** probe programs. By December 2, 1971, **Mars 3 lander** became the first spacecraft to attain a soft landing, but its transmission was interrupted after 14.5 seconds. Further space probes from the Soviet Union were sent to Mars: the **Mars 4** and **Mars 5 orbiters** and the **Mars 6 and Mars 7 fly-by/lander** combinations. All missions except **Mars 7** sent back data, with Mars 5 being most successful. Mars 5 transmitted 60 images before a loss of pressurization in the transmitter housing ended the mission. **Mars 6** lander transmitted data during descent but failed upon impact. Mars 4 flew by the planet at a range of 2200 km returning one swath of pictures and radio occultation data, which constituted the first detection of the nightside ionosphere on Mars. Mars 7 probe separated prematurely from the carrying vehicle due to a problem in the operation of one of the onboard systems (altitude control or retro-rockets) and missed the planet by 1300 km. http://en.wikipedia.org/wiki/Exploration_of_Mars

By comparison, the US was more successful with their Mariner program with fewer failures than had been experienced by the Russians. In 1964, NASA's **Jet Propulsion Laboratory** made two attempts at reaching Mars. **Mariner 3** and **Mariner 4** were identical spacecraft designed to carry out the first flybys of Mars. Mariner 3 was launched on November 5, 1964, but the shroud encasing the spacecraft atop its rocket failed to open properly, dooming the mission. Three weeks later, on November 28, 1964, Mariner 4 was launched and flew past Mars on July 14, 1965, providing the first close-up photographs of another planet. The pictures sent back to Earth showed impact craters. It provided radically more accurate data about the planet; a surface atmospheric pressure of about 1% of Earth's and daytime temperatures of -100 °C (-148 °F) was estimated. No magnetic field or Martian radiation belts were detected. The new data meant redesigns were required for future Martian landers and showed life would have a more difficult time surviving there than previously anticipated.

NASA continued the Mariner program with another pair of Mars flyby probes, **Mariner 6 and 7**. They were sent at the next launch window and reached the planet in 1969. **Mariner 9**

successfully entered orbit about Mars, the first spacecraft ever to do so, but **Mariner 8** failed after launch. When Mariner 9 reached Mars, it and two Soviet orbiters Mars 2 and Mars 3, found that a planet-wide dust storm was in progress. The mission controllers used the time spent waiting for the storm to clear to have the probe rendezvous with, and photograph, Phobos. When the storm cleared photographs by Mariner 9, of the Martian surface indicated more detailed evidence that liquid water might at one time have flowed on the planetary surface. It also found that **Nix Olympica** was the highest mountain (volcano), on any planet in the entire Solar System, and lead to its reclassification as **Olympus Mons**.
http://en.wikipedia.org/wiki/Exploration_of_Mars

The Viking program launched **Viking 1 and 2 spacecraft** to Mars in 1975. The program consisted of two orbiters and two landers – these were the first two spacecraft to successfully land on Mars. The primary scientific objectives of the lander mission were to search for biosignatures and observe meteorologic, seismic and magnetic properties of Mars. Evidence suggests that the planet was once significantly more habitable than it is today, but whether living organisms ever existed there remains unknown. The Viking probes carried experiments designed to detect microorganisms in Martian soil at their respective landing sites and had positive results, including a temporary increase of CO_2 production on exposure to water and nutrients. This sign of life was later disputed by some scientists, resulting in a continuing debate; with NASA scientist Gilbert Levin asserting that Viking may have found life. ***The results of the biological experiments on board the Viking landers remain inconclusive as far as NASA is concern.***

A re-analysis of the Viking data published in 2012, in light of modern knowledge of extremophile forms of life, has suggested signs of microbial life on Mars, although, the Viking tests were not sophisticated enough to detect these forms of life. The tests could even have killed a (hypothetical) life form. Tests conducted later by the Phoenix Mars lander have shown that the soil has a very alkaline pH and it contains magnesium, sodium, potassium and chloride. The soil nutrients may be able to support life, but life would still have to be shielded from the intense ultraviolet light. http://en.wikipedia.org/wiki/Mars

Were there any signs of life on Mars? NASA may very well have had a conclusive answer to that question but, they weren't going to share that with the public or the rest of the scientific community because they were still following the **Brookings Report** protocols and agenda...

Observations and calculations of Mars orbit, rotational period, size and mass, etc. were well known in most cultures dating back to the ancient Egyptian astronomers, as well as the Babylonian, Chinese, and the Indian astronomers. European astronomers like **Tycho Brahe, Johannes Kepler, Giovanni Domenico Cassini, Galileo Galilei** and **Christiaan Huygens** became interested in Mars around the medieval ages due in part to the discovery and implementation of the telescope.

By the 19th century, telescope resolution had reached a level sufficient for surface features to be identified. On September 5, 1877, Mars had reached a perihelic opposition allowing Italian astronomer **Giovanni Schiaparelli** to use a 22 cm (8.7 in) telescope in Milan to help produce the first detailed map of Mars. These maps notably contained features he called *canali*, which were later shown to be an optical illusion. These *canali* were supposedly long straight lines on the surface of Mars to which he gave names of famous rivers on Earth. His term, which means "channels" or "grooves", was popularly mistranslated in English as "canals".

The oriental's **Percival Lowell** and other astronomers of his time, like **Henri Joseph Perrotin** and **Louis Thollon** in Nice, had also observed these Martian *canali* using one of the largest telescopes of that time. Their publications of these *canali* and the supposed life on Mars had a great influence on the public.

The seasonal changes (consisting of the diminishing of the polar caps and the dark areas formed during Martian summer) in combination with the canals lead to speculation about life on Mars, and it was a long held belief that Mars contained vast seas and vegetation. The telescope never reached the resolution required to give proof to any speculations. As bigger telescopes were used fewer long, straight *canali* were observed even by **Flammarion** in 1909, with his larger telescope, irregular patterns were observed, but no *canali* were seen.

Speculations ran wild in unscientific fashion in the 1960s with articles published on Martian biology which detailed scenarios for the metabolism and chemical cycles for a functional ecosystem, putting aside explanations other than life for the seasonal changes on Mars.

It wasn't until the Viking probes of the '70s that scientists and astronomers became aware of the true nature of Mars, it wasn't just another dead planet with a thin atmosphere and lifeless conditions as first reported. By now, with the successful Apollo Moon missions behind them, NASA had developed the fine art of obfuscation and disinformation which they continue to practice with the first photographic images sent back from Mars. NASA had Mars locked in their cross sights and they weren't about to disclose everything they found on the Martian terra firma.

Ufologists and the public were already claiming that NASA wasn't revealing everything they had found on the Moon and some astronauts were becoming more outspoken in their experiences of having encountered UFOs in space, on the Moon and the discovery of alien artifacts there. NASA was now under public scrutiny viewed with suspicion as less of a public agency and more like an arm of the military; less transparent and not forthcoming in all that it was finding. The public was being told only what NASA felt they needed to know, in hopes that they would buy into the **BIG NASA LIE!**

As incredulous as this may be, many independent researchers are now taking up the mantle of self-investigation for themselves pouring over whatever data and photographs of the Moon and Mars that are available in the public domain in order to get at the truth. When such information was not readily available, the **Freedom of Information Act (FOI)** was used to obtain whatever documents and images, they felt was being denied to them through legitimate channels. When such **FOI** was not forthcoming, it was inevitably always due to issues of **National Security,** a nice catchall phrase to basically say: *"it's none of your business; you don't have a need to know!"*

When it comes to getting the truth in our search for Martian life, we, unfortunately, have to rely upon the very same technology of the agency we hold with some suspicion: NASA! It's a dichotomy of: *"damned if you do and damned if you don't!"* Like everything in this book so far, the truth has to be sifted out from the chaff of lies and nonsense that has entered into the public domain.

Recall the big news media hoopla when scientists revealed that they had found a meteorite in Antarctica that contained fossilized microbes believed to have been blasted off the surface of Mars approximately 15 million years ago by another meteorite or asteroid. NASA scientists and astronomers were convinced that it came from Mars based upon similar chemical composition to elements, minerals and gases compounds discovered on Mars by various space probes and Mars landers. They were even convinced up until recently that the fossilized microbes in the meteorite were indeed microbes but, later, the evidence has been called into question as to whether the microbes were not of Earth origin which had worked their way into the meteorite. Some scientists even felt that the supposed microbes were not microbes but some type of mineral.

What is surprising is that no one apparently has asked why NASA has not yet discovered any Martian microbes fossilized or alive, on or in the Martian soil with all their Mars landers and rovers currently running around on the surface of Mars. NASA keeps us informed that they are still searching for signs of water which would inevitably lead to the discovery of Martian microbial life but, at this current date of March 16, 2013, no evidence for either water or life has been found, at least any that have been reported but, the proverbial tantalizing carrot of promise is within reach and discovery is imminent, says NASA.

This brings us back to that Antarctic meteorite, one of many in fact, over a period of years of searching and recovery that have been found to have contained fossilized bacteria. Why would a few meteorites found on Earth be considered evidential proof, although debatable, to the point that even a presidential confirmation of its discovery by former President Clinton at a press conference, that life indeed, existed on Mars in the distant past, should still be difficult to find on Mars at the present time? There is a whole red planet that literally should be teeming with past microbial life, evident with every scoopful made by the Mars rovers and landing probes. Where are the microbes? NASA should have discovered them by now, given that their rovers have travelled near old riverbeds or in areas where water once existed. So, show us the Martian microbes, already and stop telling us that you are getting close to its discovery! Or is this all, just another **BIG LIE**? https://www.youtube.com/watch?v=Wj4Q1hPRoDs

Condition number two in our argument for life on Mars is fulfilled with the discovery of **Methane (CH4),** in reasonable abundance in the upper atmosphere of Mars. Methane is indicative of life; it is a by-product of every kind of life form we have seen or can imagine here, on Earth. It is a by-product of microbes, plant life, sea life, animal life and even, intelligent life…humans! And yes, it can be just as simple as the geochemical process of the planet *"farting"…*releasing pockets of trapped gas beneath its surface due to a fissure from seismic activity. But that trapped gas still means that *some type of organism had to produce it* in the first place. (Bold italics added by author for emphasis).

Signs of Plant Life - Is that a Martian Weed or a Tree that I See Over There?

If the ESA space probe is seeing and measuring *"plumes of Methane"* in the Martian atmosphere could it all be rising merely from microbes on or below the surface of Mars? Plumes of Methane suggest that other factors are in play here besides microbes. Could there also be plant life on the surface of Mars and where on Mars should we start looking for it? And if there is plant life growing on the surface of Mars what source of nourishment or nutrients feed the plants in order for them to grow? Is it not water, the very sustenance for life, as it is on Earth?

NASA initial success with the Viking Orbiter/Lander spurred other space probes to be launched to Mars with each succeeding one becoming more ambitious than the last one sent. **Thermal Emission Imaging System (THEMIS)** was one such spacecraft launched in April 2001 that was going to transmit back images of the Martian surface. What was received was a high-resolution black and white image which everyone was expecting would be in colour. NASA explained that it was trying to resolve the colour problem with the new image and finally, a year later that coloured imaged appeared which was still basically, what we have come to expect from NASA years of accumulation of hundreds of thousands of Martian surface images, *another rendition of Martian surface colour, salmon pink,* only in high definition.

One year later, NASA releases it "salmon pink" coloured version of the THEMIS image of Mars from the original B&W (insert), but strangely, there is no tell-tale green in the photo image, why?
http://www.enterprisemission.com/colors.htm

To NASA's humiliation yet, without any response from them, a private citizen **Heald**, a carpenter by trade whose past time pursuit was astronomy and as an imaging amateur, who one month after the B&W image released by NASA and nearly a year before NASA's coloured image version became public, released his own coloured version of the same photographic image. He seemed to have resolved the technical colour issue that plagued NASA THEMIS team and may have given reason why NASA was so recalcitrant in their delay of the final colour image to the public.

It seems Heald's amateur imaging abilities revealed something that NASA was trying to cover up in their own image processing labs at JPL, namely that Heald's coloured version of NASA original image showed some ***unusual patches of green*** on the Martian surface! (See photos above). (Bold italics added by author).

Hoagland states that this created quite a stir across the Web, both positive and some negative response from other amateur investigators and only dead silence from NASA and its team. "The existence of widespread areas down there of *tell-tale green* – indicative (one would then easily believe) of current *living organisms ...Martian plant life* ... NOW on Mars!

For these investigators, the Heald (aka. by blog name: Woodlock) three-color image – though, indeed, containing tantalizing (if subtle) hints of "green" -- does not go far enough.

Another "amateur" -- **Holger Isenberg**, himself an investigator of NASA's strange historical habit of arbitrarily altering the colors of *its* own Mars images (the "JPL story") -- within a few hours, produced *another* version of "Woodlock's" color view (below). In it, not only do the reddish tints become more vivid, but the darker areas now take on an *unmistakable* "green hue!"

NASA, it seems, says Hoagland has a ***"dirty little secret"*** which by now should be obvious to the reader. Green on a planet is usually indicative of plant life! http://en.wikipedia.org/wiki/Mars

As one looks at all the images of Mars that have been taken over the years which are in the public domain, one cannot help but notice that Mars surface colour is never consistent. The true colors of Mars are a combination of salmon pink terrain and mountains, dark blue-green plains and valleys with dark chocolate brown mottlings, hints of ochre colour, and white polar caps, all surrounded by an atmospheric blue limb. The true color of Mars has often been likened in comparison to the desert landscape of Arizona. "NASA's apparent inability to present consistent images", has been a source of inaccurate understanding as to what is taking place on the "Red Planet". http://en.wikipedia.org/wiki/Mars

These dark mottlings are the signs of the presence of vegetation, of a world that is living and not dry or barren, nor dead. These blue-green and dark mottlings are in response to seasonal changes like those that we witness on Earth, when vegetation grows in the spring season then, darkens and withers in the fall season.

All that would be needed is oxygen to sustain Martian vegetation but, alas, the atmosphere is nearly devoid of oxygen, however, if the debilitation of weathering is very slow, it is conceivable that life may be in balance with it. As desert plants on Earth have learned to store up water and protect themselves from evaporation, plants on Mars may have developed some means of utilizing the oxygen produced by photosynthesis for their own internal respiration. It may even have found a way to adapt and grow in a carbon dioxide rich atmosphere in a way that we do not as yet understand. There may be more than one way for plants to grow in this vast universe!

Holger Isenberg's improved version of Heald's imaging work that originally came from NASA's B&W THEMIS image in the Terra Meridiani region of Mars
http://www.enterprisemission.com/colors.htm

These changing dark markings thought to be seasonal changes affecting vegetation on the surface of Mars were dismissed as a phenomenon called "simultaneous contrast", a psychological phenomenon causes colored highlights to impart their *complementary hue* to any adjacent low-luminosity features ... Viewed against a bright ochre background (the Martian deserts), a dark neutral gray marking will take on an *illusory* bluish-green cast.
http://www.enterprisemission.com/colors.htm

But, wait a minute! Carbon dioxide is also an essential element in plant growth, it is required by plants to grow, and according to NASA, it is found in abundance on Mars and its chief by-product is Oxygen!

Carbon dioxide (C O$_2$) is a chemical compound that is found as a gas in the Earth's atmosphere. It consists of simple molecules, each of which has one carbon and two oxygen atoms. Thus its chemical formula is **CO$_2$**. It is currently at a concentration of approximately 385 parts per million (ppm) by volume in the Earth's atmosphere and is a major component of the **carbon cycle**. http://www.newworldencyclopedia.org/entry/Carbon_dioxide

On Mars, there is 95.32% Carbon dioxide, 2.7% Nitrogen, 1.6% Argon, 0.13 % Oxygen, and 0.08% Carbon monoxide. Also, minor amounts of water, nitrogen oxide, neon, hydrogen-deuterium-oxygen, krypton, and xenon. http://www.space.com/16903-mars-atmosphere-climate-weather.html

Is this Martian CO_2 atmosphere sufficient for plant and animal life to exist on the surface?

In general, it is exhaled by animals and used for **photosynthesis** by growing plants. Additional carbon dioxide is created by the combustion of fossil fuels or vegetable matter, as well as other chemical processes. It is an important **greenhouse gas** because of its ability to absorb many infrared wavelengths of the Sun's light, and because of the length of time, it stays in the Earth's atmosphere. In its solid state, carbon dioxide is commonly called **dry ice**. Carbon dioxide has no liquid state at pressures under 4 atm.

Plants require carbon dioxide to conduct photosynthesis, and greenhouses may enrich their atmospheres with additional CO_2 to boost plant growth. As a very powerful **greenhouse gas,** it has a large effect upon climate. It is also essential to photosynthesis in plants and other **photoautotrophs.**

Despite the low concentration, CO_2 is a very important component of the Earth's atmosphere because it absorbs infrared radiation and enhances the greenhouse effect to a great degree. http://www.newworldencyclopedia.org/entry/Carbon_dioxide

According to NASA, Mars has CO_2 in vast amounts in its atmosphere which would mean that its mechanism is the same as on Earth and would absorb Infrared radiation at an incredible rate creating what would be expected as a very high greenhouse effect on Mars which would be conducive to plant growth.

Plants remove carbon dioxide from the atmosphere by photosynthesis, also called carbon assimilation. This process uses light energy to produce organic plant materials by combining carbon dioxide and water. Free oxygen is released as gas from the decomposition of water molecules, while the hydrogen is split into its protons and electrons and used to generate chemical energy via **photophosphorylation**. This energy is required for the fixation of carbon dioxide in the **Calvin Cycle** to form sugars. These sugars can then be used for growth within the plant through respiration.

Plants also emit CO_2 during respiration, so it is only during growth stages that plants are net absorbers. For example, a growing forest will absorb many metric tons of CO_2 each year, however, a mature forest will produce as much CO_2 from respiration and decomposition of dead specimens (e.g. fallen branches) as used in biosynthesis in growing plants. Nonetheless, mature forests are valuable carbon sinks, helping maintain balance in the Earth's atmosphere. Furthermore, phytoplankton photosynthesis absorbs dissolved CO_2 in the upper ocean and thereby promotes the absorption of CO_2 from the atmosphere. This process is crucial to life on earth. http://www.newworldencyclopedia.org/entry/Carbon_dioxide

If the process is crucial for life on Earth then, it should be crucial for Mars too, however, Mars has an abundance of CO2, so it should be easier for plants to grow on Mars. In turn, more oxygen would be expected to be produced with each succeeding Martian year providing there are no other chemical forces or outside forces preventing the photosynthesis process. Could this be the case on Mars? Carbon Dioxide is an indicator of plant life like algae or more complex plant life and even life coming from the large bodies of water.

Carbon Dioxide represents a secondary condition for life on Mars, in addition to the abundance of Methane being the primary indicator of life. The presence of these gases in the Martian atmosphere represents potentially strong evidence for life, whether microbial, plant or simple animal life to be discovered on Mars.

Recent images from orbiting spacecraft have shown large areas in the South Polar regions of Mars that appear to be forested areas of trees and shrubs, possibly deciduous trees that are massive in size; the largest of these trees is estimated to be 1,115 yards across! (See below).

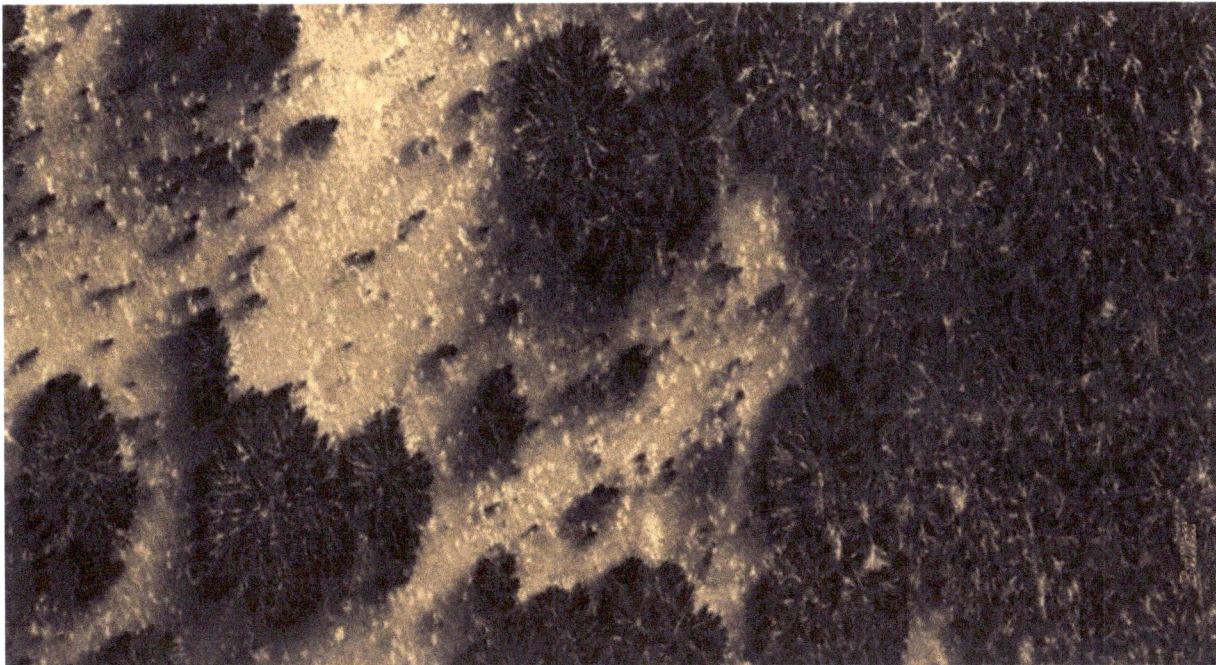

The trees of Mars as seen in this altered colour photo are in fact a forest of trees!
http://www.enterprisemission.com/mpl.htm

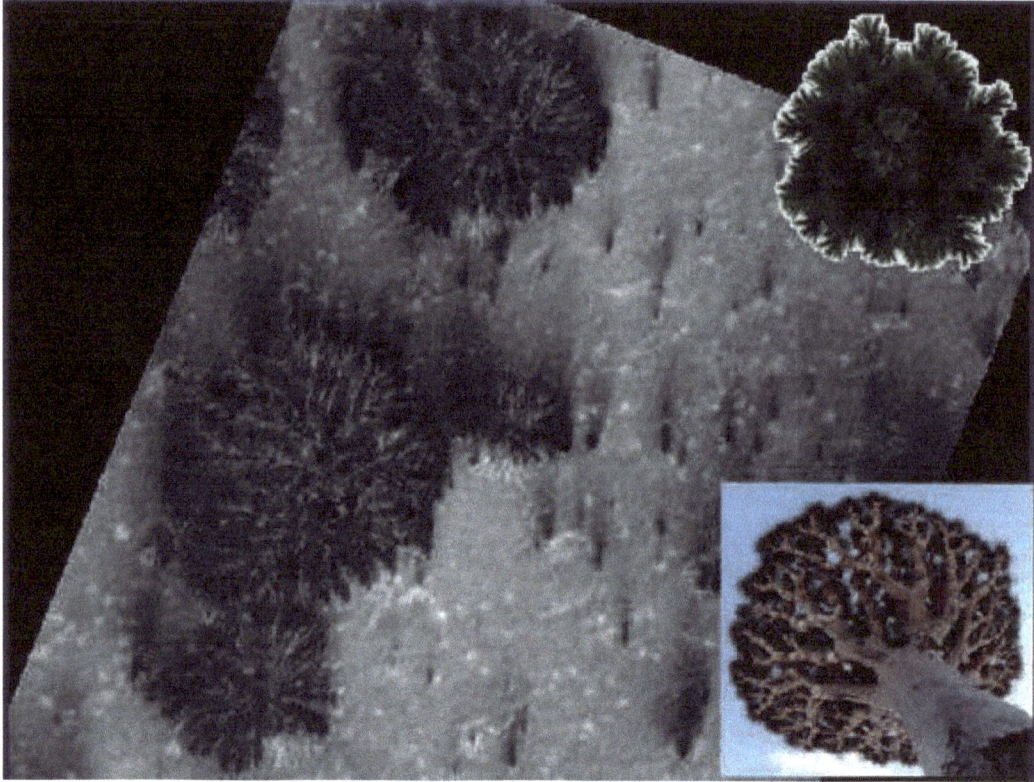

Could this intriguing image be giant Martian trees? Note the fractal pattern in the branches similar to trees on Earth, like the Pine tree and the Dragon Blood Tree inserts.
(c) Terry Tibando)

The alien-like "Dragon Blood Trees" that grow in the archipelago of Socotra.
https://www.pinterest.com/explore/dracaena-cinnabari/?lp=true

And NASA scientists views these supposed trees as nothing more than the trapped gas emissions of carbon dioxide ice that erupt in geyser form or as a slow leakage that flows over the surface in a spidery formation with branch-like tendrils that are reminiscent of trees and shrubbery found on Earth. In the end, NASA may very well be correct in their assessment of this phenomenon, however, what is strange is that these CO2 eruptions seem to take on the flow patterns of many *different types of trees* that can be found on Earth!

Warmer temperatures heat the sand dunes causing the CO2 ice below to melt which violently erupt as geysers spewing gas and darker subsurface material over the landscape
http://www.thelivingmoon.com/20UMLR/03files/Geysers_Discovered_on_Mars.html

This begs the question: are all these flow patterns simply C02 geyser eruptions or are they Martian trees and vegetation or a combination of both geological and organic phenomenon?

A point of consideration: if Mars has warmer temperatures and a higher atmospheric pressure during its seasonal changes like on Earth then, carbon dioxide could sublimate from a gas to a solid and back again to a gas upon the Martian surface. With massive global dust storms that arise from time to time lasting for months, the temperature variations would allow the CO2 gas to precipitate out of the atmosphere and onto the surface.

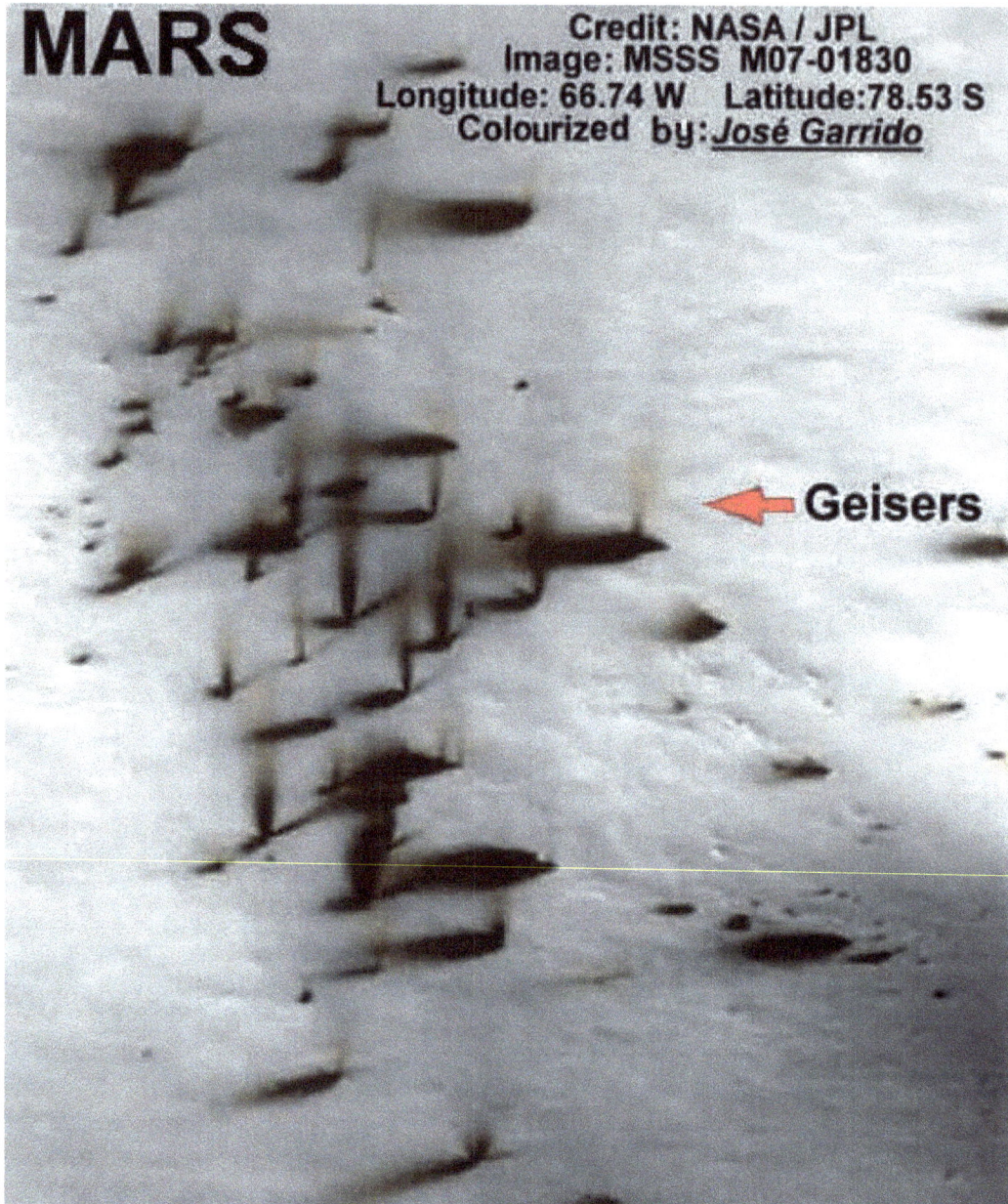

This colourized image of CO2 geysers (Geisers) shows ejecta material as well as plumes.
Note that there are no tree-like structures visible in this photo

http://www.thelivingmoon.com/20UMLR/03files/Geysers_Discovered_on_Mars.html

Storm fronts blowing dust would covered the solid CO2 on the ground in the fall season until spring, when it would melt again from the warmer temperatures heating the surface of sand dunes sufficiently to cause the solid carbon dioxide beneath to vapourize, violently erupting to the surface as geysers mixing with darker ground material spraying it as ejecta over the lighter colour surface. The run-off of material stains the surface soil in small streams and fan patterns creating the illusion of vegetation or trees. But, bear in mind that NASA says that Mars only has a 1.0% (near vacuum) atmosphere so, how is this CO2 sublimation or geyser action even possible?

This explanation by NASA seems reasonable providing that the original raw data has not been tampered with so that an alternative explanation can be introduced that appears to give an air scientific authenticity. The problem, however, of getting everyone to buy this explanation is the oddly coloured photo images and the unusual way they are presented and explained. This makes the viewer wonder what he is looking at, what direction is up or down, where the top and the bottom of the dune or valley floor is located and what direction is the sun shining. Without knowing these parameters, one can only guess at what one is looking at or how to interpret the photo images.

It could be a deliberate obfuscation to confuse a lay public and blindly force them into accepting an explanation that may not be correct at all or it is an attempt at falsification of the data. In most situations gravity determines the direction of ejecta flow so, in the photos, why does the stain flows travel upwards? Are these photos altered in any way similar in manner to other tampered photo images that NASA has released?

The reader should look at the images of the pink desert with sand dunes and ask himself how best to interpret what he is looking at then, he should view the other close up images of what appears to be trees or vegetation, and compare these with the tree images taken on Earth. Then and only then, decide upon what he sees in the Martian images. Does he see dark soil stains on Martian dunes from CO_2 emissions or does he see trees and vegetation? Perhaps, both of these phenomena are occurring in the same place!

Careful examination of NASA photos in the public domain reveals that there are many examples of trees and forested areas even, of different species which argues in favour for their existence as oppose to NASA official position, that what is being observed is nothing more than as CO_2 geyser eruptions. In fact, if one were to view the polar caps of Mars and search areas that appear to be areas of vegetation, NASA will state that all this is due to CO_2 geyser eruptions and soil stains on the surface landscape. Officially vegetation does not exist on Mars as far as NASA is concern.

As the reader looks at the photo of the pink sand dunes (below), notice the dark streaks that NASA says are CO_2 emissions coming up from beneath the sand dunes crests, there is even some dark wispy CO_2 plume material seen at the top centre and slightly left of the picture yet, just in front of these streaks are other streaks with bulbous stems or trunks. Some of these bulbous trunks have branches with feathered-like leaves and some have no leaves just branches. Now compare this photo and the close up of the one below with the bulbous trees found on Earth on the Arab Emirates Island of Socotra.

Close up of the Martian sand dunes. Are these plant-like structures trees or flow patterns from CO2 gas eruptions? Comparison to the images below suggests that they are trees!

Are these images the first real evidence of plant life on Mars? Given that (CO2) Methane is abundant in the atmosphere of Mars and Oxygen is also the by-product of photosynthesis of plants, and water is also a constituent of the atmosphere necessary for plant growth, it would by de facto, be evidence of plant life!

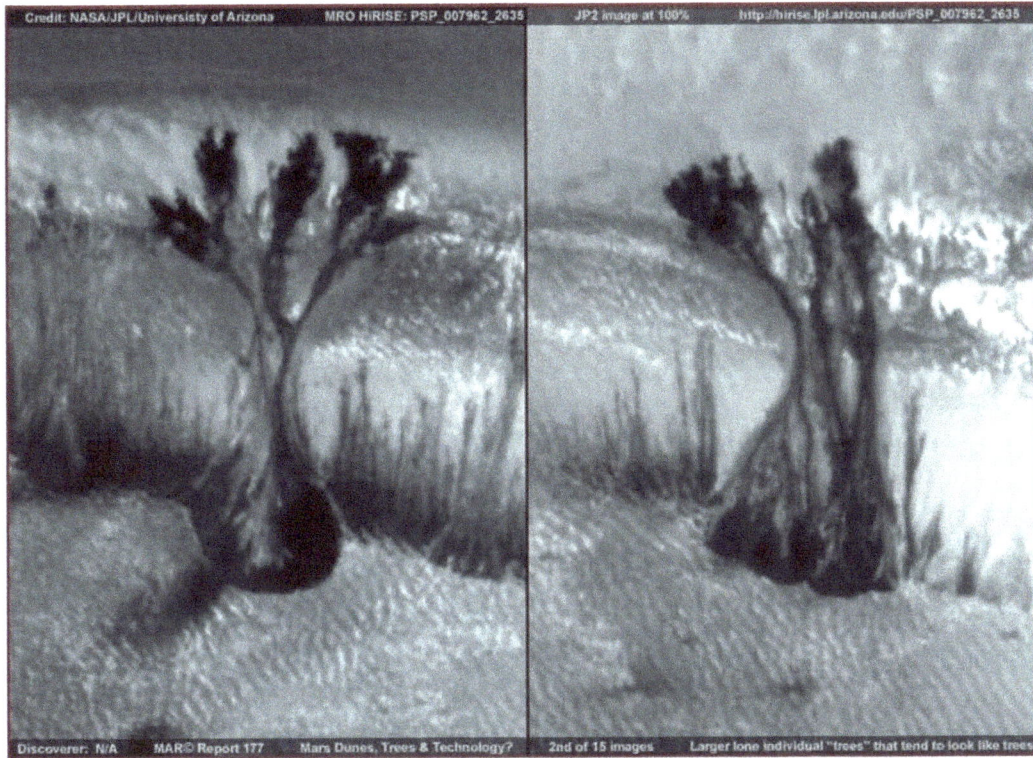

Compare these CO2 bulbous gas patterns to the bulbous trees of Socotra below. Are they really Martian trees or do both phenomena exist side by side?

http://www.marsanomalyresearch.com/evidence-reports/2010/177/dunes-trees-tech.htm

The bulbous alien-like flowering trees of Socotra

https://www.whowho.io/2016/12/11/alien-like-looking-socotra-island-yemen/

**More flowering and leaf bearing bulbous trees of Socotra which
can cling precariously to almost any ledge or surface**
(Google Images and Pinterest)

It seems hard to imagine that the phenomenological aspects of Carbon Dioxide ice which in a gaseous state seems to have the ability to mimic, almost perfectly, the biological aspects of tree life on Earth! This fact on its own would not really be enough to reject outright the scientific explanation that NASA has given for this phenomenon. But, when many other areas of Mars also exhibit this same type of CO2 phenomenon with plumes and ejecta that mimic different species of trees found on Earth then, we have to question the scientific explanation given by NASA. Their explanation is untenable for all such similar circumstances, as it does not explain this mimicry to various **dendriforms** found on Earth. How is this possible on another planet?

The variety of tree photos above provide good comparisons to these strange Martian "trees" found growing in the cold desert areas of Mars but, there are more examples. There are photographs that show large deciduous forest areas as well as coniferous trees (pine trees) found in the polar hemispheres of Mars. (See below).

NASA-JPL-MSSS MGS MOR R06-00983 Mars South Polar area depicting spouts, plants, and forests in a densely packed forest area
http://mars.galactic.to/

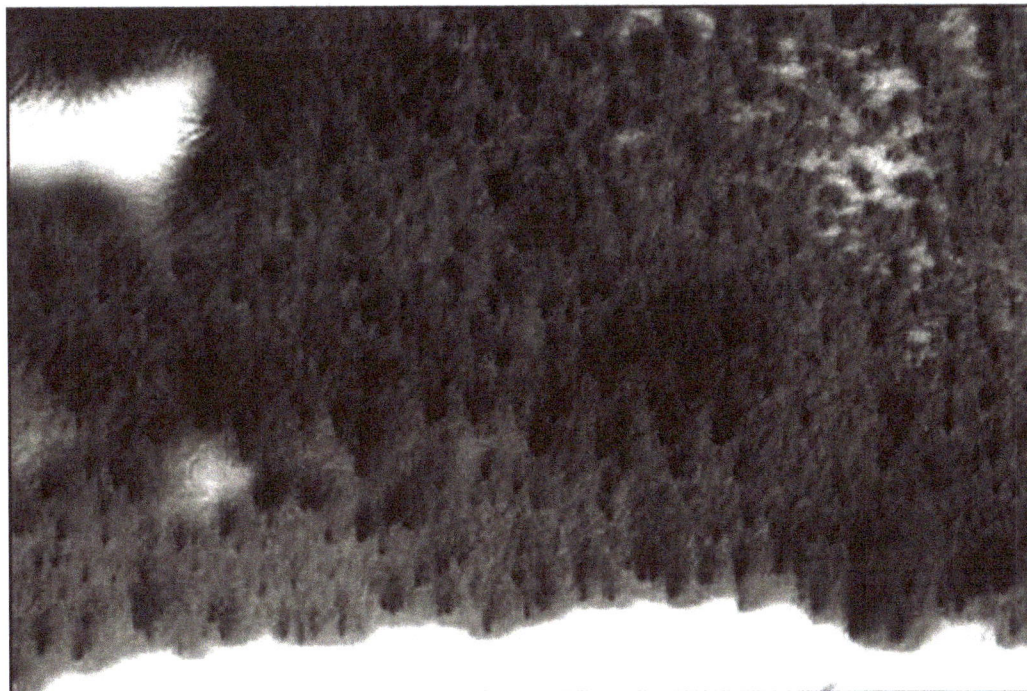

NASA-JPL-MSSS MGS MOC 506-00607
More bio-diversity of Martian forests showing a shorter forest species
https://www.sott.net/article/127698-Trees-on-Mars

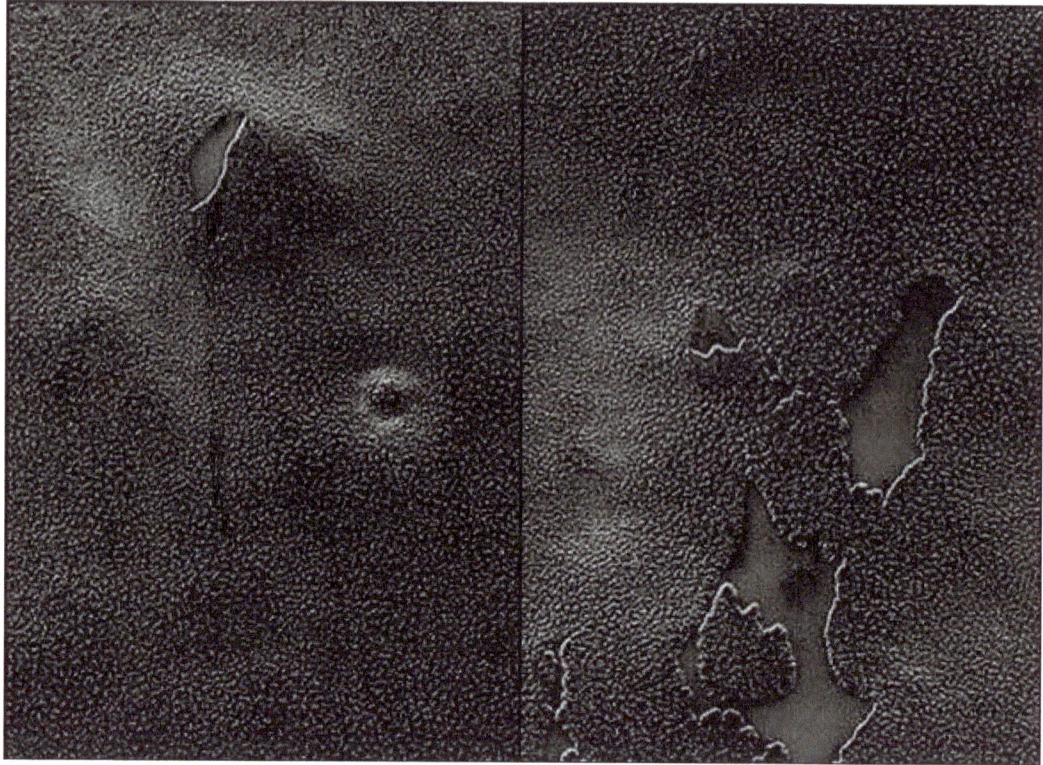

NASA-JPL-MSSS MGS MOC M09-02042
A Highly dense and extensive forested area surrounding unfrozen lakes of water!
http://www.marsanomalyresearch.com/evidence-reports/2006/100/mars-tree-recognition4.htm

The forested area around Devil's Tower, Wyoming, USA. Compare this photo to the images above and below of tightly forested areas on Mars
https://earthobservatory.nasa.gov/IOTD//view.php?id=38904

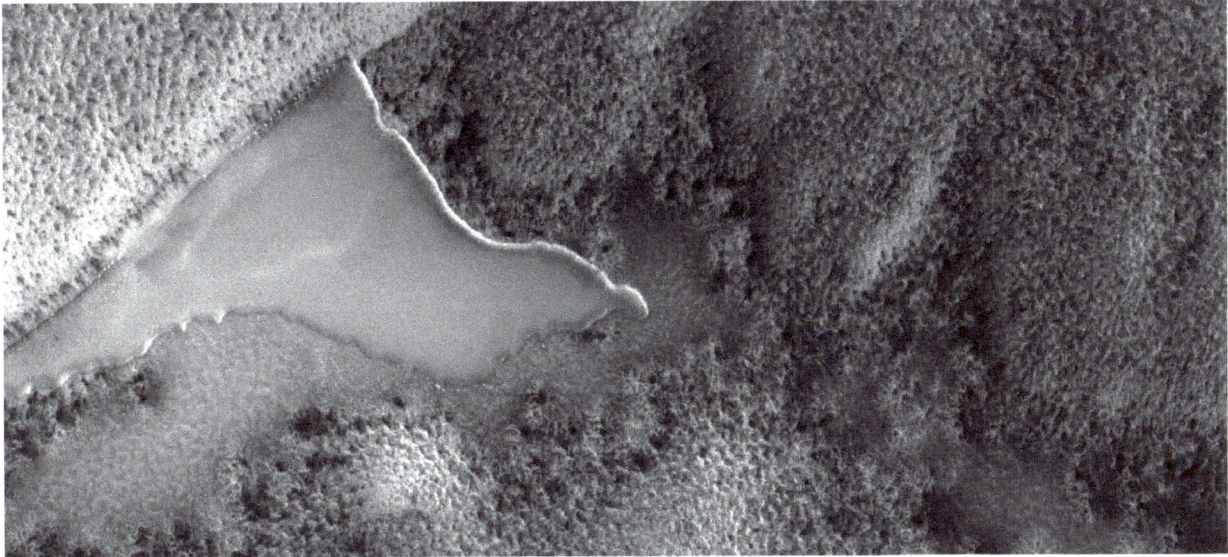

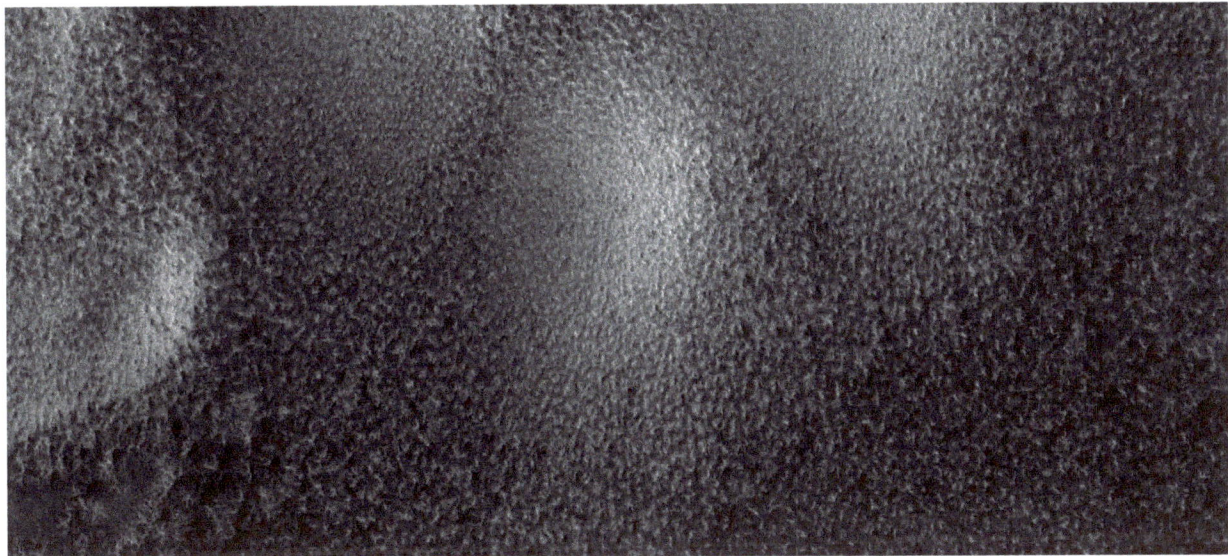

**A reservoir lake, possibly of artificial construction with a forest along its banks
that extends beyond to the right and into the bottom photo image.
Notice the similarity to the forest area of Devil's Tower above.**
http://www.marsanomalyresearch.com/evidence-reports/2006/100/mars-tree-recognition4.htm

Once again, in all the above photographs , NASA has outrightly denied that these areas mostly found in the south polar area are not trees but the results of massive geological processes caused by CO_2 gas erupting through the frozen ground as geysers discolouring the surface with darker soil material. This, NASA says is where the Martian atmosphere goes in the winter season and then erupts as a gas in spring to become an atmosphere again. The odd thing is that NASA has never explained how all this geological thermal/sublimation of CO_2 and possibly ice water is able to reproduce the same structure of various tree species that can be found on Earth.

Is the common lay people simply not trained in the science of geology and photo imaging interpretation and thus simply don't know a geyser from a tree? Officially, NASA considers the Mars a dead planet with a past history of oceans and possibly life, that is its position.

Well, it seems other space agencies don't agree with that stated position because their data supports that life may indeed exist on the surface of Mars as the green hills and valleys in some of their photo images attest. The European Space Agency Mars Express orbiter has taken photos that show these deep valleys and rolling hills with patches of green that as stated before may be indicative of vegetation. (See below).

Is there green algae in the valley floor of Juventae Chasma of Mars
as imaged by ESA Mars Express orbiter?
(Google Image)

Many space agencies have come into being in the last fifty years due mainly in part that they don't feel that NASA has disclosed all that it knows or all of its discoveries. For this reason, many nations have sent their own space probes and orbiters to the Moon and to other planets such as Mars, to get to that undisclosed truth. What these nations are discovering is that their data does not confirm the findings that NASA has reported to exist there. What they had been told by NASA is not entirely correct, or is a distortion of fact or a deliberate cover-up of knowledge for some unknown reason, Could it be that NASA was withholding evidence from the discovery of the existence of Extraterrestrial life on another planet, namely Mars.

It is no secret that NASA leads all other nations in the exploration of planets, however, what they discover is not always shared with other nations even, though the US claims to do so. This may be true from the government's perspective of NASA but, from the perspective of the public and other nations, NASA is not forthcoming with all that it knows. There seems to be a double standard in place, a truth embargo when it comes to what NASA really knows; perhaps, NASA has been deservedly labeled with the moniker: *Never A Straight Answer!*

Another ESA Mars Express photo January 15, 2004, showing a deep blue channel (Reull Vallis) once formed by flowing water east of the Hellas basin by tributaries at the north and south ends. Note the three-sided pyramid at the south end of the lake
http://www.spacetoday.org/SolSys/Mars/MarsThePlanet/MarsWater.html

As stated earlier in this section, *"where there is water, there is life";* in fact, Europe's Mars orbiter has detected water molecules vapourizing from the Red Planet's south pole, scientists announced it the most direct evidence yet of water in the form of ice on the Martian surface.

"The quest for water on Mars - which could indicate life - has fascinated scientists for centuries. Mars watchers have long believed that the planet's poles contain frozen water, but previous scientific findings - including NASA's Mars Odyssey orbiter's evidence of large amounts of ice - were based more on inferences, European Space Agency scientists said."
http://www.chinadaily.com.cn/en/home/2004-01/24/content_300860.htm

While Mars Odyssey has been able to indirectly show the presence of water at the pole using temperature monitors, the European camera has for the first time been able to "literally map the polar cap" using infrared technology that shows where water molecules are present, said scientist **Jean-Pierre Bibring.**

"You look at the picture, look at the fingerprint, and say this is water ice," said agency scientist **Allen Moorehouse**. "This is the first time it's been detected on the ground. This is the first direct confirmation."

NASA's Mars Odyssey, which has orbited the planet for two years, previously turned up evidence of lots of ice mixed with the soil, as little as 40 centimetres from the surface.

James Garvin, the lead scientist for NASA's Mars exploration program, told The Associated Press today that Mars Express had offered further confirmation of what scientists have long known: ***"Mars is a water planet."*** (Bold italics added by author).

As far back as 1940, scientists using telescopes saw vapours they believed indicated the presence of water. But in the 1960s the first Mars mission revealed the planet to be frozen, dry and covered with craters and deep ravines. Conflicting and inconclusive information has been coming in ever since. Now, why do you think that is?
http://www.chinadaily.com.cn/en/home/2004-01/24/content_300860.htm

**Frozen water in a Mars polar crater taken by NASA Mars global Surveyor orbiter.
A rare admission by NASA that water exists in some measurable volume**
http://www.iceandclimate.nbi.ku.dk/research/ice_other_planets/ice_on_mars/

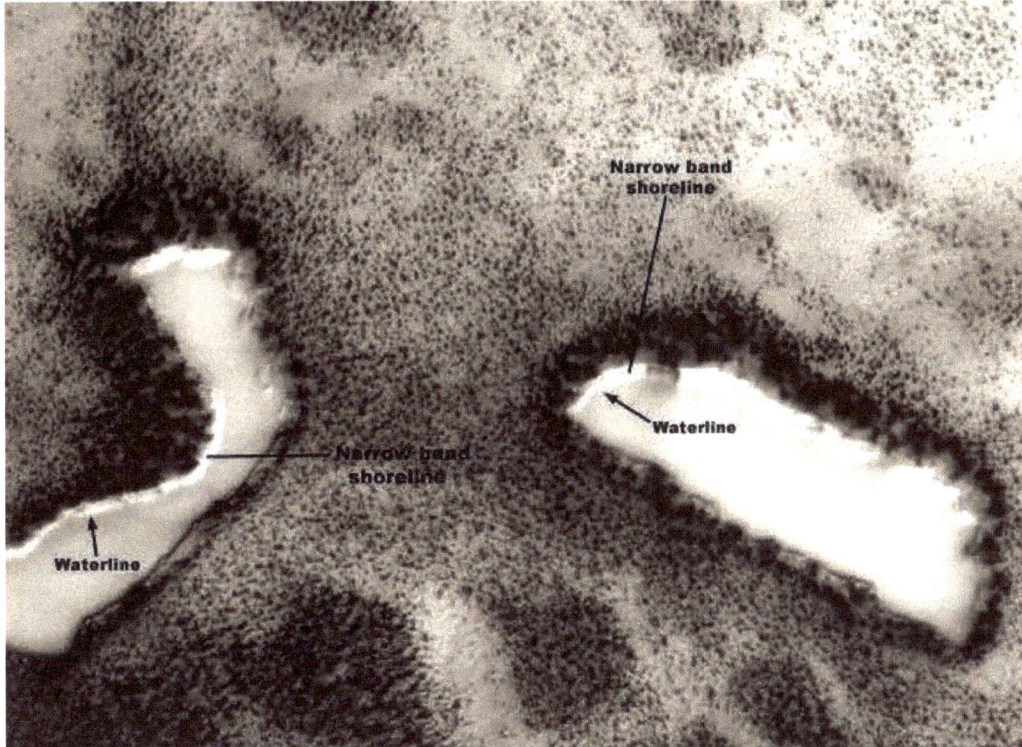

NASA-JPL-MSSS MGS MOC R07-01100
More Martian lakes with surrounded by a ring of forests

Knowing and understanding the basic processes of weather as is taught in high school and in its more advanced mechanics in university as it applies to our own planet becomes the basis of understanding for the weather on Mars. There are obvious differences as has already been pointed out in this section, such as the rich Nitrogen based atmosphere of Earth and the rich Carbon Dioxide based atmosphere of Mars.

Precipitation on Earth is in the form of large droplets of water molecules due to evapourization and condensation whereas, Mars precipitation is a process of sublimation of CO_2 with some evapourization and condensation of water that precipitates into small droplets. At specific times of the year, in similarity to the weather patterns of Earth, Mars has seasonal changes where this sublimation and precipitation process takes place resulting in a change of landscape features. The increase or decrease in polar capsize occurs with the blanketing of the Martian surface with snow and ice filling many craters, valleys, and mountain tops. Large dust storms assault the Martian surface covering areas where snow and ice in the way of CO_2 ice and water ice have accumulated covering this settlement of earlier precipitation.

Come the spring equinox, much of the snow and ice is released back into the atmosphere through the melting and sublimation processes of higher thermal activity as a result of increases in the atmospheric pressure and temperatures. On Earth life is renewed and reinvigorated, colour changes appear all over the surface of the planet as plant life re-awakens from hibernation, revived from the stored energy of the previous season; on Mars the processes are no different.

401

With these changes in seasonal weather patterns, life has found a way to establish itself on Mars in the way of microbes and plant life such as trees, simple grasses, algae and lichen sustained by sources of water. It is a precarious balance of nature that has re-asserted itself from a former catastrophe in its dim past. Life, it would seem has returned to Mars, once again!

The third condition in the search for life on Mars has been fulfilled and it would seem that we are well on our way to proving that life not only was a reality of the distant past but, life is also a reality occurring at this present time on Mars! **It would seem to be the biblical equivalent of a Genesis II (revival) on Mars, at least from this author's perspective.**

Signs of Ancient Martian Sea Life

If plant life exists on Mars, could there also be a sea life or freshwater life and are there any other complex animal life forms present? Apparently, there is! The Martian rovers Spirit and Opportunity, once more appear to have uncovered evidence of such ancient sea life in its exploration of the Martian terrain.

Sir Charles W. Shults III at **Xenotech Research** reports some of the best images from the Opportunity Rover were released very early in the mission. Sol 028 microscopic images below right contained some of the best and clearest features and are still outstanding in content and detail. A terrestrial urchin that matches the most prominent features of this particular fossil organism is used for comparison for the uninitiated to see for themselves.

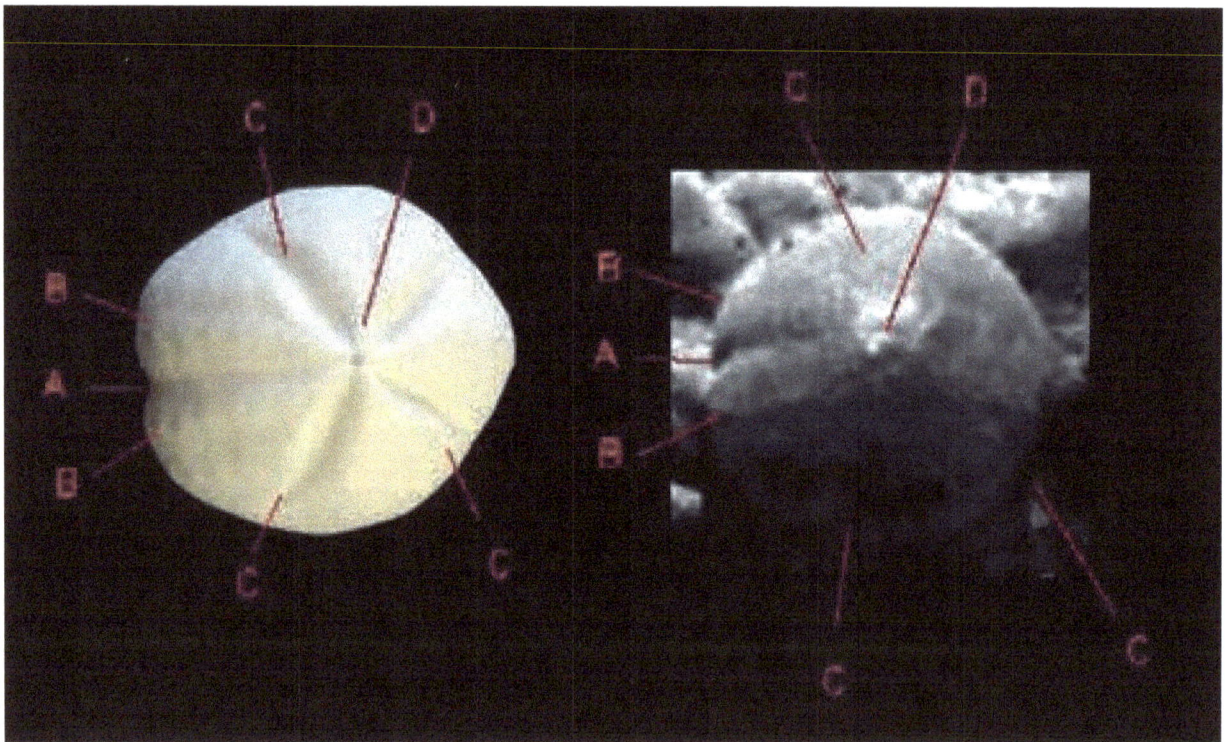

Terrestrial sea urchin at left and a similar Martian sea urchin at right
http://www.nationalufocenter.com/artman/publish/article_86.php

A - the cleft that appears on so many of the spherules, similar to the cleft on a peach
B - the margins around the cleft that are raised and divided by sutures, almost like lips
C - shallower clefts that surround the major cleft in a roughly pentagonal manner
D - a prominent single bump at the apex of the shell on both the terrestrial Sea Gopher (left) and the Martian urchin (right)

There are other features that are fainter but that match perfectly. Note that many trillions of the "blueberries" have this cleft so prominently that it can be seen in the panoramic images as well as the microscopic images. Sea urchins are close relatives of sea biscuits and sand dollars, and all share many features of their anatomy. Many of the Martian specimens are clearly sand dollars, others are like sea biscuits, and some are clearly urchins. All three classes existed on Mars, just as they do in our seas here on Earth. Thanks to Sir Charles W. Shults III, K. B. B.
http://www.xenotechresearch.com/seagoph1.htm and
http://www.nationalufocenter.com/artman/publish/article_85.php

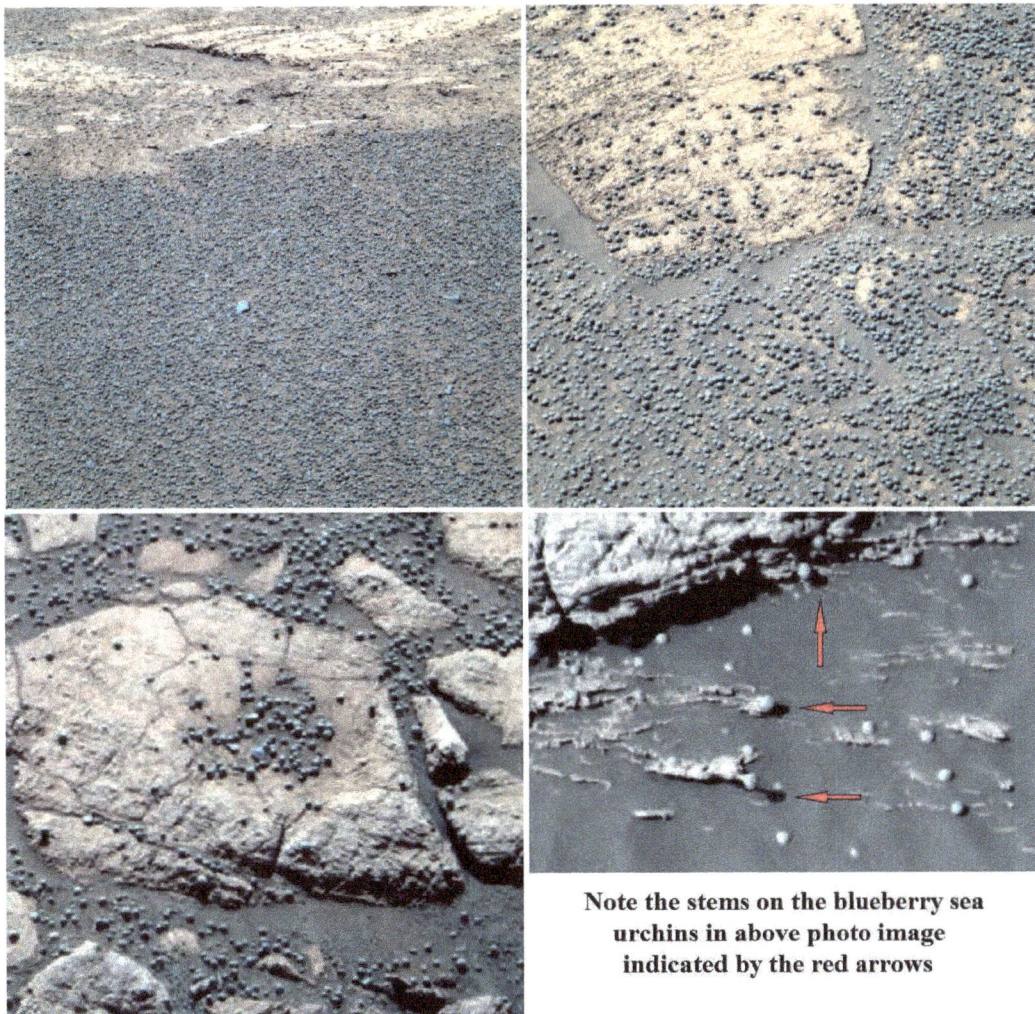

**Note the stems on the blueberry sea
urchins in above photo image
indicated by the red arrows**

**Ancient sea life abounds in the above photos indicating a proliferation of "Blueberry"
sea urchins that once existed at the bottom of a now dried up Martian seabed**
http://www.abovetopsecret.com/forum/thread611659/pg1&mem=

Seemingly, with all the brilliance behind the remote operation of the Mars rovers to uncover life in every meter of its travel across the Martian terrain, there also, is the bizarre nature of NASA to either ignore obvious artifacts or to destroy what it finds and then to strangely remind silent about it or to come up with some alternative explanation for its actions *(Never A Straight Answer)*.

Two astrobiologists have also made similar charges that when NASA's Viking Lander when it first touched down in 1976 upon Mars and scooped into the Martian subsurface, it not only found microbial life still extant but, in the process of chemical analysis killed that rudimentary Martian life and has been doing it ever since with other Mars probes.

Scientist **Gilbert Levin,** chief scientist of the biology portion of the Viking Landers insists his biology experiment proved life is in the Martian soil.

Dirk Schulze-Makuch, an astrobiologist at Washington State University and his German colleague, **Joop Houtkooper** of Justus Liebig University charges NASA as having killed it.

Their explosive allegation—raised at the **American Astronomical Society** meeting in Seattle, Washington—*raises one more doubt about the US space agency*. (Bold italics added for emphasis by author).

When the Viking mission tested for life it detected hydrogen peroxide from oxidation. The two scientists claim the oxidation process didn't come from the soil, but in the killing of Martian life scooped up and chemically destroyed by the robot space explorer.

The two have stirred up controversy—nothing new to NASA—with their report alleging the death of Martian life.

"It is interesting to note that the Viking experiments were conducted under too warm and wet conditions from the perspective of the hydrogen peroxide hypothesis for Martian life," they concluded in their in-depth report. *"If the hypothesis is true, it would mean we killed the Martian microbes during our first extraterrestrial contact."*

No doubt Gilbert Levin would agree with them. *"We obtained positive data corresponding with all the pre-mission criteria, which proved the existence of microbial life in the soil of Mars,"* Levin told **National Geographic**

In essence, the two scientists argue, NASA unwittingly killed exactly what they were seeking: life. Some scientists think that a reddish bacterium like this tints the Martian soil. No doubt, a strong argument could be made for this case.

NASA didn't do it on purpose, Schulze-Makuch and Houtkooper rush to explain. ***The agency was just ignorant... (or possibly just in a haze of scientific ineptitude!)*** (Bold italics and parenthesis added by author for emphasis).
http://beforeitsnews.com/space/2011/09/scientists-charge-nasa-killed-life-on-mars-1153548.html

Sir Charles Shults has cataloged stacks of images revealing the evidence of Martian life that appears to be scattered across Mars. The scientist has gathered proof—proof that NASA rigorously seems to ignore in uncharacteristic fashion, as if to silently say: ***"Seen that, done that, we know what it is, so what! We're on a bigger, more important mission!!"*** Pray tell, what could that bigger mission be? (Bold italics added by author for emphasis).
http://www.shultslaboratories.com/index.htm

(C) American Museum of Natural History

**Crinoids, sea creatures (sometimes referred to as sea flowers)
commonly found in the depths of Earth's oceans**
http://www.enterprisemission.com/_articles/03-08-2004/crinoid_cover-up.htm **and**
https://www.slideshare.net/Charlotte122899/paleozoic-era-11541743

Shults photographic collection, all courtesy of NASA include various Martian fossils, seashells, a Martian sand dollar with telltale star marking, a fossil of an ancient segmented sea creature, fossils of trilobite, and fossilized bacteria.

During the Mars rover missions, scores of armchair astronauts screamed in disbelief as NASA systematically steered their little juggernauts towards fossil evidence and then blithely pressed the little rovers ahead, mindlessly crushing the precious fossils under their grinding treads.
http://beforeitsnews.com/space/2011/09/scientists-charge-nasa-killed-life-on-mars-1153548.html

One has to wonder what is going on in the minds of NASA as their Mars rover Opportunity comes across yet, another possible sign of ancient sea life that looks for all the world like a crinoid, a sea creature (sometimes referred to as "sea flowers") which are commonly found in the depths of Earth's oceans. It would seem as if the Mars rovers are on a search and destroy mission, instead of a mission of exploration and documentation as any reasonable person would expect from a "publicly owned" space agency.

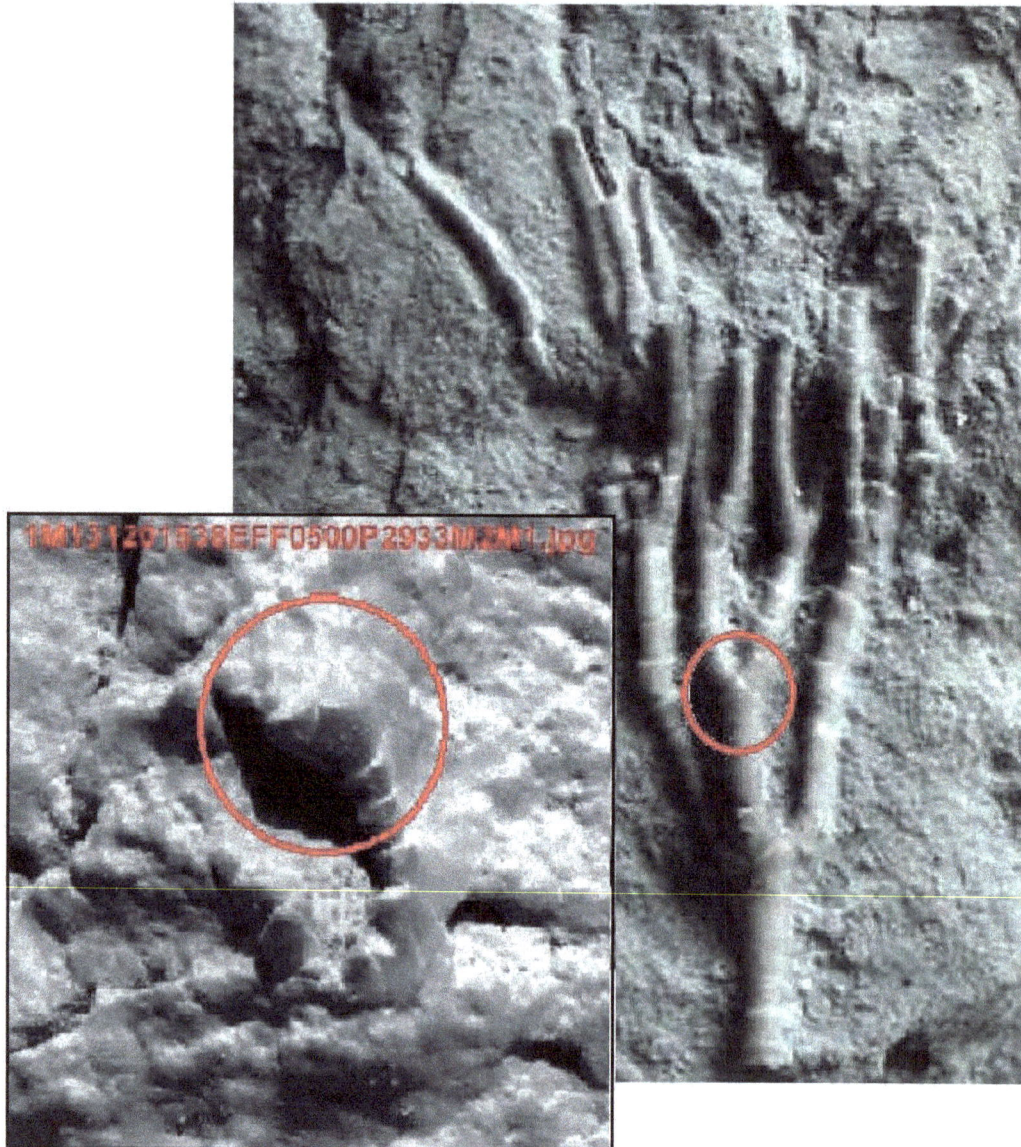

In these two images, there is evidence of similar organic structure. The image on the right is a fossilized crinoid from Earth and the image on the left is a crinoid from Mars

http://www.enterprisemission.com/_articles/03-08-2004/crinoid_cover-up.htm

Richard Hoagland, along with many other biologists and scientists felt with immense satisfaction that here, at long last was the proof of life that they have been searching for on Mars. NASA's rover photographed what clearly showed the fossilized structure of a sea creature, a Martian crinoid. Then, in an act of absolute incredulity and disbelief, NASA callously signaled the rover to grind the Martian crinoid out of existence!

NASA then, re-photographs their rover's handiwork and then, the rover moves on to its next intended *"target area!" "Nothing here to see folks, just another bit of unusual rock. Move along now!"* (Bold italic added by author for emphasis).

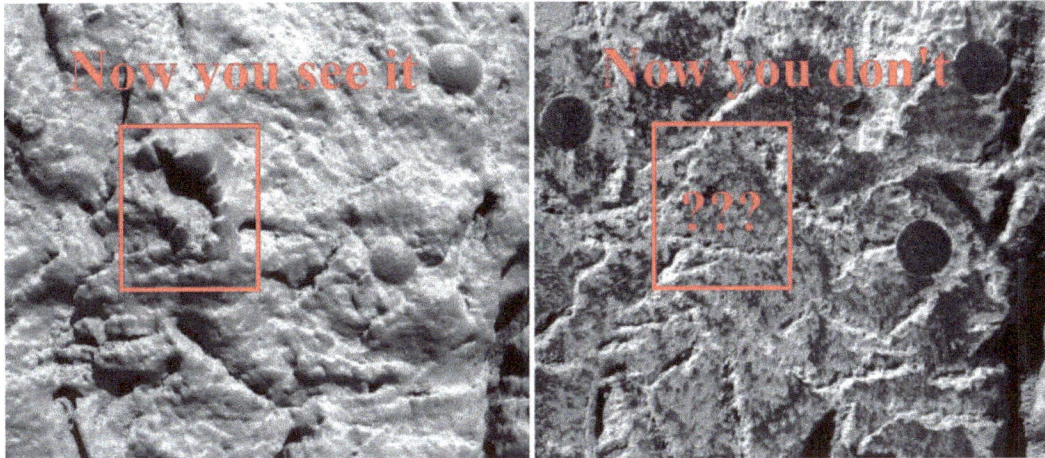

The same little fossilized Martian Crinoid above meeting an unusual end, to be ground out of existence by the robotic arm of the Mars rover Opportunity from Earth!

http://www.enterprisemission.com/_articles/03-08-2004/crinoid_cover-up.htm

This act seems to indicate either, the mind of stupidity behind the operational controls of the Mars rover or a deliberate agenda intended to hide or remove any evidence of ancient life or anomalous artifacts. Such actions do not bode well for the future of NASA, should they continue to operate in this manner!

Given the growing concern that the public is not being informed of everything that NASA finds on Mars, still there are some tantalizing images being transmitted back to the photo labs of JPL and somehow these very revealing images make their way into the public domain for further scrutiny and investigation.

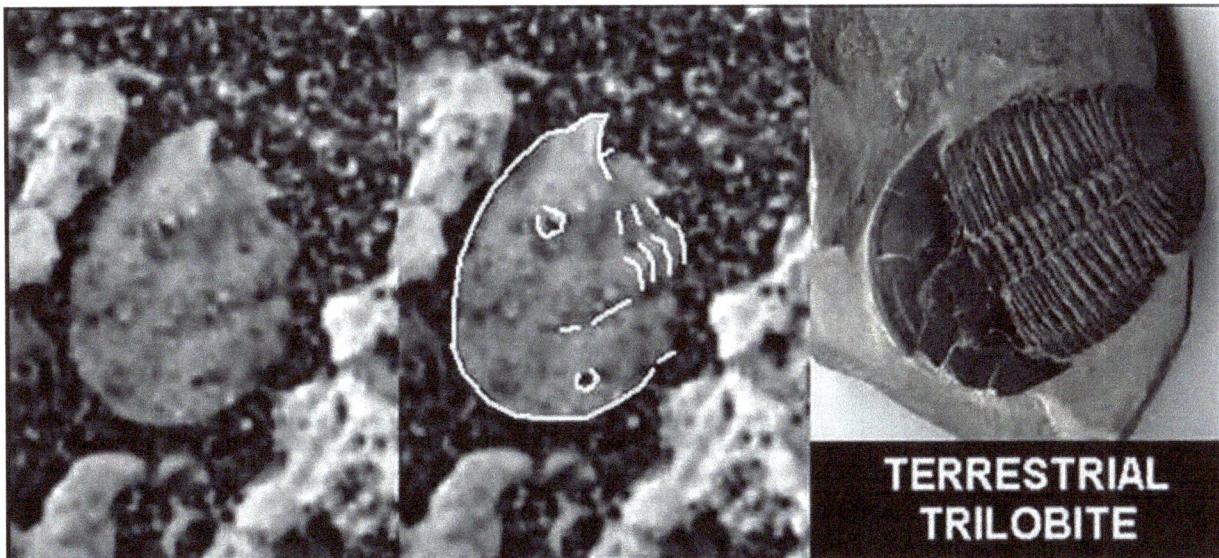

A fossilized Martian Trilobite (on left and highlighted in middle) discovered on Sol 158 in Endurance Crater compared to a fossilized Trilobite found on Earth

http://www.shultslaboratories.com/CoastPage16F.htm

Blastoids found on numerous occasions by Mars rovers Spirit and Opportunity
http://www.shultslaboratories.com/CoastPage16E.htm and (Google Images)

Fossilized blastoids found on Earth. Compare to the many Martian blastoids images above
https://www.britannica.com/animal/blastoid/images-videos

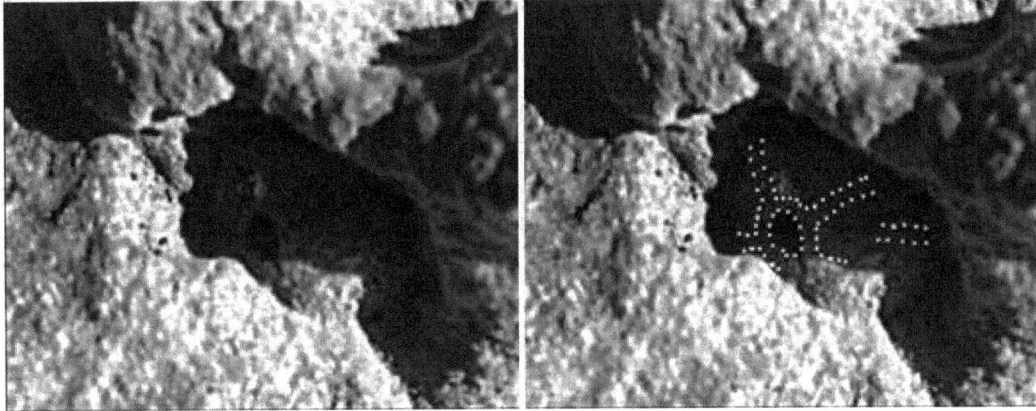

Another type of Martian fossilized sea urchin, evidence for an ancient Mars sea life!
(Google Images)

Sea life is evident in the many fossilized structures photographed by the Mars rovers Spirit, Opportunity and more recently by Curiosity found on the ancient Martian sea beds and rivers. It is proof positive that Mars did at one time in its distant past have water in its lakes, rivers, and oceans! Water, in this case, is proof of life!

More and more scientists from biologists to geologists are swelling the ranks of those, who not only believe that an ancient life once existed on Mars but, that many of those elements necessary for the existence life may, in fact, still be present in sizeable quantities and volumes, that life on Mars may still exist, today!

"Where there is water, there is life", this age old adage has become our modern day mantra which up to this point has served us well and so far Mars seems to have "lived up" to it in abundance! As we dig further through the Martian terrain via the Mars rovers, we wonder, could there be signs of ancient complex animal life that also, once existed on land? Did NASA's rovers find and photograph anything resembling land dwelling creatures that slithered, crawled, walked or even flew?

Bear in mind that NASA is very slow or even, very recalcitrantly tight-lipped to admit that they have already determined that Mars is a live planet, contrary to their repeated assertions that Mars is a dead, dry, barren planet that *"may have at one time supported life"!* **Yeah, right!!**

However, the more we investigate some of those intriguing photos that NASA has somehow, let slip through their hands into the public domain, the more contradictions in the photographic evidence that supports the hypothesis that life once existed on Mars in the distant past. What we see is astounding evidence that Mars was/is an Earth –like planet!

Signs of Fossilized Animal Life

Complex life forms require instinctual adaptability to survive on land and given that NASA scientists state that the atmosphere of Mars in the distant past was very similar to the kind of atmosphere that Earth had in its Cretaceous Period, they, therefore, conclude life was also abundant in the ancient Martian seas and also upon the land. NASA as of April 2013 feels very confident that fossilized evidence of sea life and probably land dwelling animal life will be discovered by their nuclear-powered Mars rover Curiosity in the Gale Crater, in which it is currently exploring.

To most people with a passing interest in NASA's exploration of Mars this would be good news, particularly for the advancement of science in general and to the many public and private financial supporters of NASA but, to the many armchair amateur astronomers and investigators of Mars, they strongly believe that NASA has already found the fossilized evidence and is deliberately sitting on the information. NASA it seems is following a carefully orchestrated agenda inspired by the Brookings Institute Briefing of slowly releasing its discoveries to the public in a well-controlled disclosure of information. Such information that would be enough to titillate the public interest at the right time but, giving NASA plenty of time for itself to examine thoroughly their new discoveries for any incriminating evidence of life or Martian artifacts, before releasing a sanitized version of their discoveries into the public domain through live press conferences or the news media.

We have seen already from the above photo images that an atmosphere and water are present in sufficient quantities and volume that former sea life in ancient times existed on Mars. The presence of liquid water and not just ice water, gave us reasonable probability to suspect that there is plant life growing seasonally upon the Martian terrain in different climates. Once again, anomalist researchers of Mars have uncovered through NASA raw photographs, many examples of plant life in and around surface bodies of water in the form of trees and some shrubbery. The next logical question of the evolutionary process to be asked, could there be some vital piece of evidence that would indicate animal life once existed on Mars, perhaps some higher forms of life such as animal or reptilian?

Viola! There is!! Mars so far doesn't seem to disappoint us in our research for life!!!

Could this oddly shaped rock be a fossilized creature? Could it be exactly what it appears to be, namely a lizard or is this another case of pareidolia running wild?

410

Now, obviously from the photograph above of the lizard shape rock, one unusual rock doesn't constitute proof that such animal life existed on Mars but, what if there were two or three such images of animal life whether fossilized or actual bone fragments or partial skulls and body parts that were easily recognizable? What if there were numerous photographic examples of former fossilized animal life scattered all over the Martian terrain would this be enough to satisfy the scientists, the paleontologists, the astro or exo-biologists, the curiosity seekers, and researchers?

Buckle up Dorothy! This is about to be a bumpy and exciting ride over the Martian terrain!!

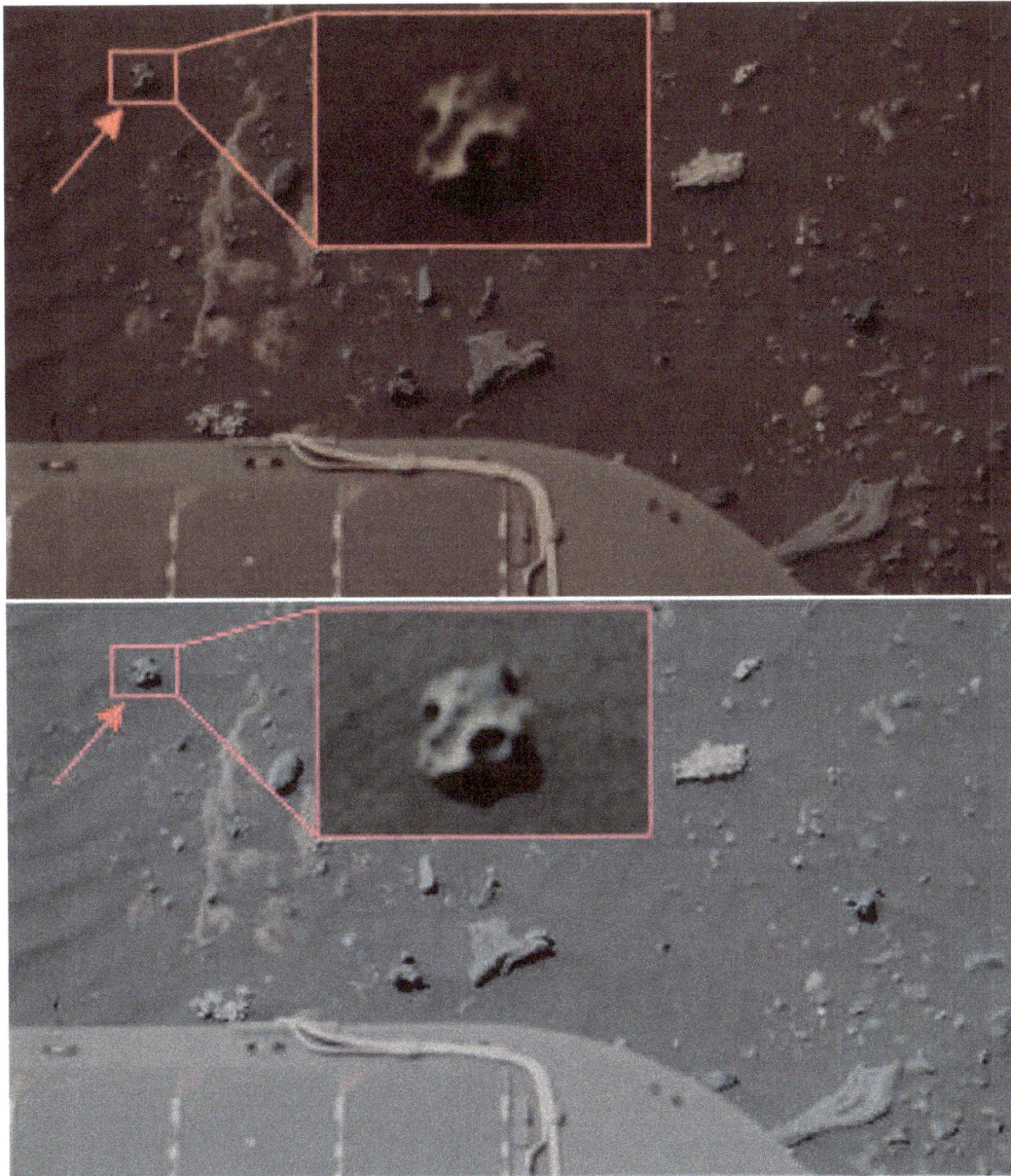

Photo of an animal skull taken by the Pan Cam on the Mars rover Spirit on Sol 820
Rocks simply do not have this type of organic symmetry unless they are fossils.
Bottom image has been auto-colour corrected
http://mars.map-base.info/mars_life/index_en.shtml

These resourceful amateur astronomers and investigators have spent much of their time pouring over the thousands of publicly released photographs sent back to Earth by the rovers. Sometimes, during those long, tedious eye-straining occasions, something unusual yet, familiar pops out from among those many Martian images of rock, soil, rolling hills and impact craters that shouldn't be there. Across the Martian landscape, perhaps partially hidden behind another rock or just out in the open are a few inconspicuous anomalies, telltale signs of an ancient sea creature or a land animal life or even, artifacts of higher intelligence. Below is one such ancient animal fossil!

Most people who examine this photograph see similarities to various Earth rodents primarily based upon the resemblance and the positioning of what appears to be the incisor teeth in the front of the Martian fossilized animal head. This assumption is understandable and it may be that this ancient fossil is an extinct rodent but, keep in mind that appearances can be deceiving and not everything on Mars is **EXCATELY** like what is found on Earth, only that it bears a close resemblance to Earth life. The cranial structures of the Martian skull strongly resembles that of a *small feline (cat) species* than it does to the rodent family at least that is the opinion of this author, having spent thirty years in the pest control industry removing dead animals like mice, rats, squirrels and cats!

This Earth-Mars similarity between skulls should not be viewed as a matter for dispute with regard to what type of animal skull the Martian skull resembles. The only final conclusion we should agree upon at this time should be that *this is an animal skull discovered on Mars* by the Mars rover Spirit! Proof positive of a higher animal life form! That fact is indisputable and cannot be denied even by NASA. (See comparison photos below).

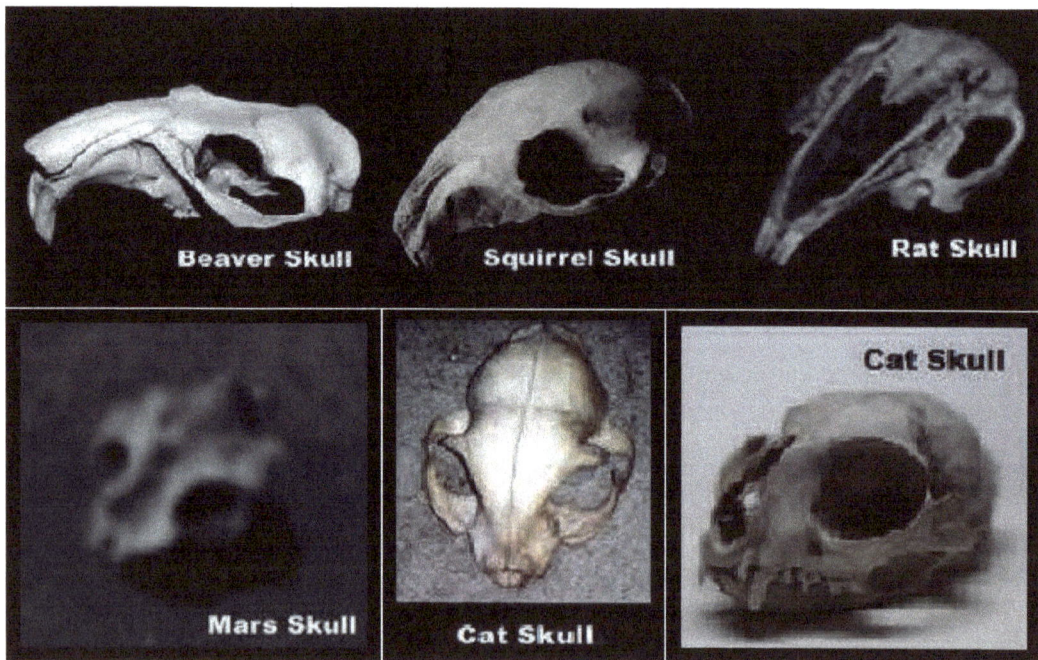

A comparison of the Mars Skull to various Earth-based feline and rodents skulls
http://thecydoniainstitute.proboards.com/thread/50/traces-past-life-on-mars and (**Google Images**)

412

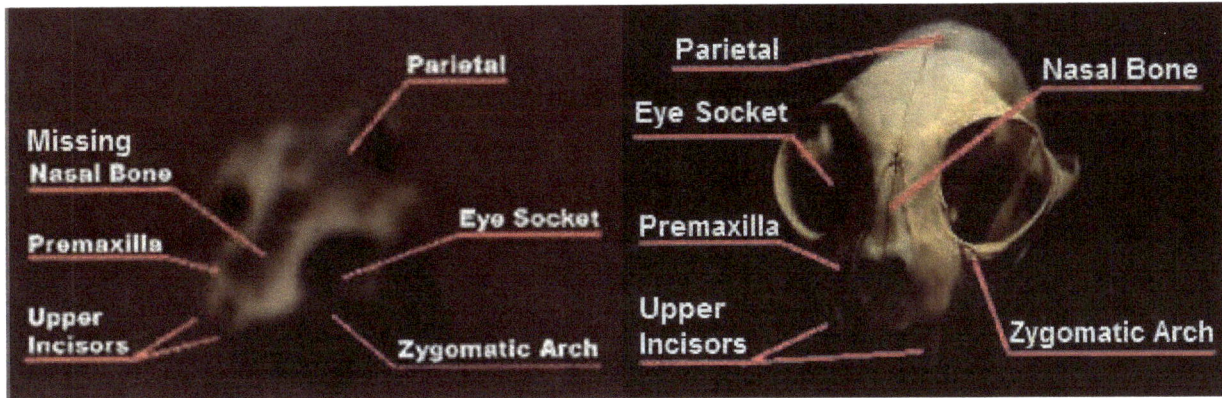

A photo comparison study of the Martian skull on the left and an Earth feline skull on the right. The point here is not what type of skull this is but, that it is of Martian origin!
(c) Terry Tibando

The little Martian animal skull is a proof of former life but can be easily interpreted as an unusual rock with some holes in it and therefore is dismissible as an oddity. If the rover could only find more fossilized animal parts then, this would be a red-letter day in the history books for NASA. It would appear that NASA had hit the jackpot in its search of the Martian terrain for unusual rocks and whether admitted to or not, ancient fossilized life!

The amazing Mars rover Spirit seems to have come across an unusual piece of Martian real estate that was a treasure trove of ancient life forms located in a single site. **Gusev crater** may very well be the discovery site that NASA scientists had been secretly looking for to prove with absolute certainty that Mars did indeed support life in its distant past. Not only does that life shows up in abundance in the Spirit photo image that can be seen at the NASA JPL raw image website but, there is a diversity of fossilized life that litters the surface and many of the fossils appear very similar to the life forms that can be currently found on Earth! (http://marsrover.jpl.nasa.gov/gallery/all/2/p/016/2p127793693eff0327p2371r1m1.html)

The photo image depicts what could be best described as an animal graveyard or maybe, the area in question was once a Martian zoo. The Gustev Crater site is truly remarkable that needs to be vigorously searched; it requires an in-depth scientific study to determine the cause of so much death in one area, to so many animals. What appears to be just unusual rocks with curves and holes in them upon, closer examination reveals organic features. The fossils are clearly visible, they pop out from the Martian debris field on their own, particularly items **"c"** and **"d"** as seen in the first photo image below. Skull "d" is thought to be a **"diapsida skull" (lizard),** however, this author believes its appearance is more avian than lizard and a comparison below is made to show the similarity. It is also remarkable that from the type of pulverization of the terrain that these fossils still bear recognizable familiarity. If a layman such as this author can recognize familiar shapes and objects as can be found in this debris field then, it stands to reason that NASA scientists trained in photographic image analysis would also recognize the same objects. To date, there does not seem to be any other detailed photos of this debris site that NASA has released into the public domain nor any official statement of its existence.

413

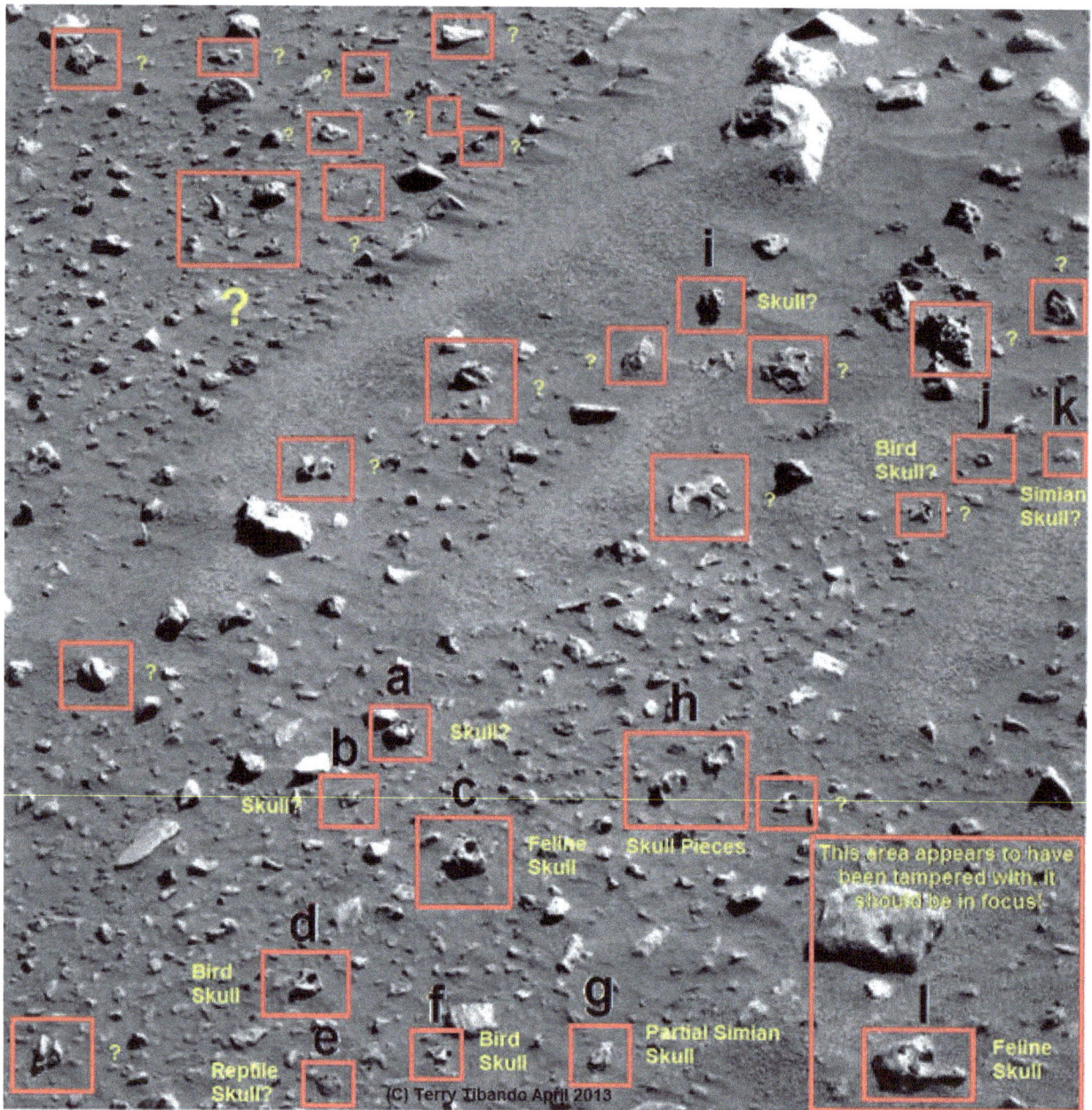

The original black and white Panoramic Camera image taken by Spirit's mission to Gusev Crater on Sol 016. Animal Bone yard?

Below is a colour enhanced photo of the black and white image above and further below it are some of the more outstanding fossilized features from the coloured photo image which have been enlarged. Beside them is their Earth counterparts illustrated for comparison purposes. Needless to say, such comparisons have limitations, Mars is an entirely different planet with its own unique environmental conditions than Earth so, it is natural to see some similarities as well as differences in animal life. Room still exists for further interpretative analysis given that this is an exploration of an untested understanding in the new science of exobiology. Therefore, this current photo interpretation should be considered as a starting point for further research.

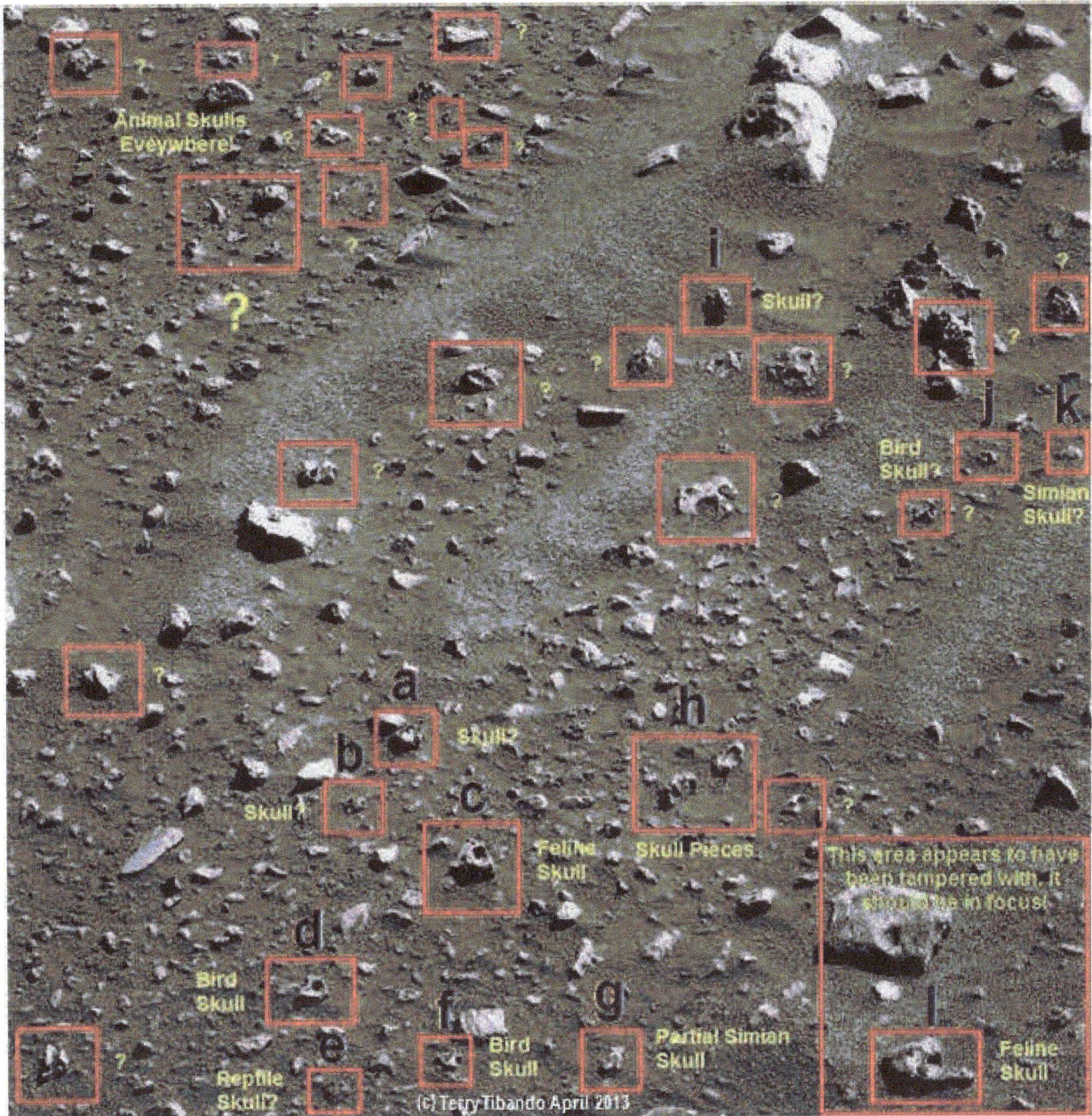

Did the Mars rover Spirit discover an animal graveyard or zoo in this colour enhanced photo of Gusev Crater? The terrain is littered with fossilized or mummified skulls and bones of animals, birds, and reptiles.

https://mars.nasa.gov/mer/gallery/all/2/p/016/2P127793693EFF0327P2371L7M1.HTML **(Credit: NASA/JPL/Cornell**

Note the photo images below illustrate Martian fossilized skulls with red borders and Earth animal skulls on the right. The author considered the bone yard found in **Gusev Crater** with an eye to what appeared obvious to most people without over analyzing or interpreting the images as is often done by researchers who post their finds on YouTube and strive to convince you, the viewer that a pile rock or some stones represent an alien body or a limb or a building. Readers and investigators are as always encouraged to do further research on all the images in this book.

"C" - A Martian feline skull compared with a mummified Earth cat

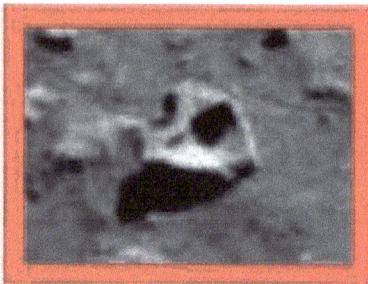

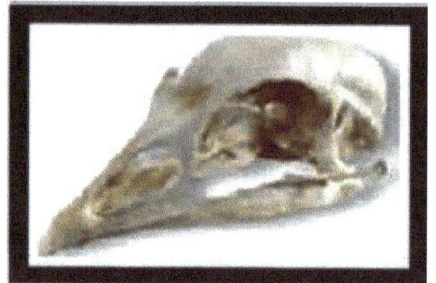

"D" - A Martian bird skull compared with a bird skull from Earth skull

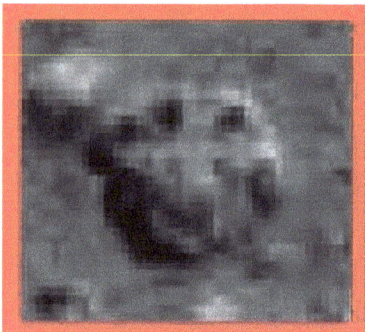

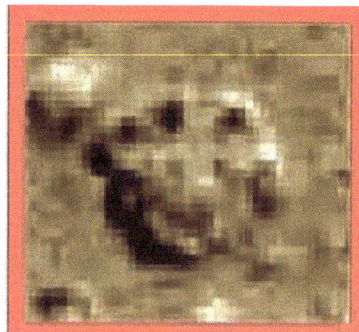

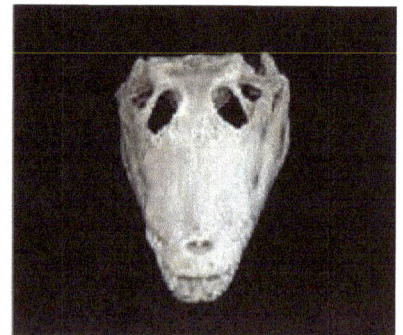

"E" - A Martian reptile (lizard or alligator) compared to an Earth type alligator skull

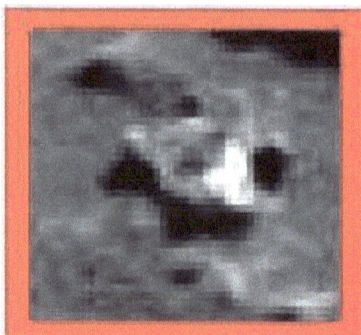

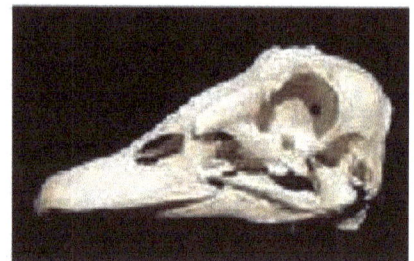

"F" - Another possible Martian bird or duck skull compared to an Earth duck skull

("C - F") https://mars.nasa.gov/mer/gallery/all/2/p/016/2P127793693EFF0327P2371L7M1.HTML (c) Terry Tibando

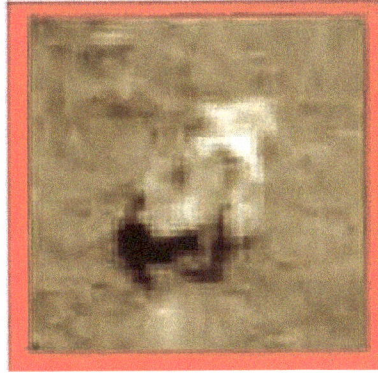

"G" – A Martian Simian or Ape skull compared with an Earth gorilla skull

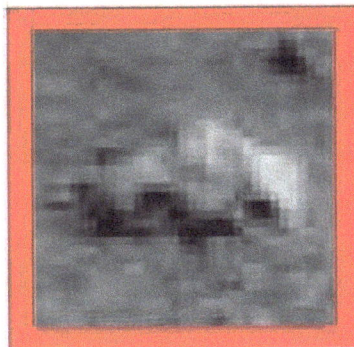

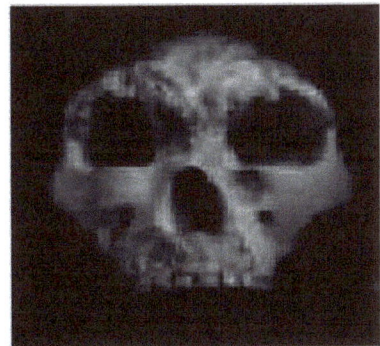

"K" – Another partially buried Martian hominid skull compared to an Earth simian skull

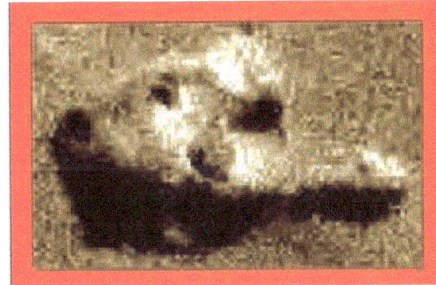

"L" – Another larger Martian feline skull compared to an Earth mummified house cat

("G, K, L") https://mars.nasa.gov/mer/gallery/all/2/p/016/2P127793693EFF0327P2371L7M1.HTML (c) Terry Tibando

At this current time, the disclosure of many discoveries found by the Mars rovers with regard to the possible existence of ancient life on Mars has come from the investigation and research work from non-NASA scientists and the growing community of amateur astronomers and planetary anomalists. NASA continues to be silent on what it finds and in many press conferences is either, re-directive and/or selective of the information and discoveries that they disclose to the public. The official handling of information for the possible discovery of life on other planets by NASA is no different than the lack of acknowledgement and dissemination of information by the US government and military on the existence of UFOs and ETI.

The **Mars rover Curiosity** continues to send back remarkable images as it travels over the rockyterrain within **Gale Crater** towards **Mount Sharp** photographing everything in sight and accessibility. On Sol 109, (2012) Curiosity took some photos of what appears to be a fossilized spine sticking out of the ground which seems to be a part of an even larger fossilized skeleton that could have been an ancient sea creature. Given that Curiosity is exploring what use to be an ancient Martian lakebed; this would seem to be an appropriate a location given that on Earth, many such similar fossil finds are constantly found in ancient dry lakes and sea beds all the time.

On Sol 109, Curiosity photographed (Top) a fossilized spine-like structure of a large creature in Gale Crater. A dinosaur skeleton (Middle) and a whale skeleton (Bottom) are shown for comparison purposes
http://www.gigapan.com/gigapans/126202 and http://www.paulbeckers.com/eg_02/westerndesert-home_hg.htm
(c) Paul Beckers and http://www.charismaticplanet.com/wadi-al-hitan-whale-valley-egypt/

418

**An enlargement of the fossilized spine of the "Martian dinosaur"
from the photo, image above**

http://www.gigapan.com/gigapans/126202

Can it be stated with absolute certainty that this unusual rock formation is actually the vertebrae of some type of prehistoric Martian beast? The claim by many researchers is that these formations on Mars are the vertebrae of large, extinct creatures that once inhabited the Red Planet, washed into the crater when it once contained water.

This find conjures theories of how the bones of this creature ended up in the crater. Perhaps, from an ancient cataclysm that occurred on Mars causing the animal to die a slow, gradual death as the planet became dusty and dry unable to sustain life of any kind. For tens of thousands of years or much longer, it fossilized lying in the open on the Martian surface and then, being gradually covered by the incessant seasonal Martian winds that can blow over the surface for months on end. Such global sandstorms would act like a blanket slowing the decaying process of organic life or simply cause it to dry and become mummified over the centuries or millennia. With repeated cyclical sandstorms blowing over the planet covering and uncovering, revealing the former life that once existed on the planet beneath the surface and then, once more burying it, eventually causing fossilization of organic life to occur or weathering all down into mere dust to be added to Mars annual and cyclical sand storms.

419

We seen from the above photos that large mammals such as whales may have swam in the ancient seas or oceans of Mars, could other large creatures have also roamed on land at one time perhaps, millions of years ago or in more recent times? From the images seen below that were taken by the Mars rover Opportunity on Sol 121, it shows what looks like the skeletal remains of a dinosaur, a Tarbosaurus to be exact, at least this is its general appearance when compared to a terrestrial version of the same creature.

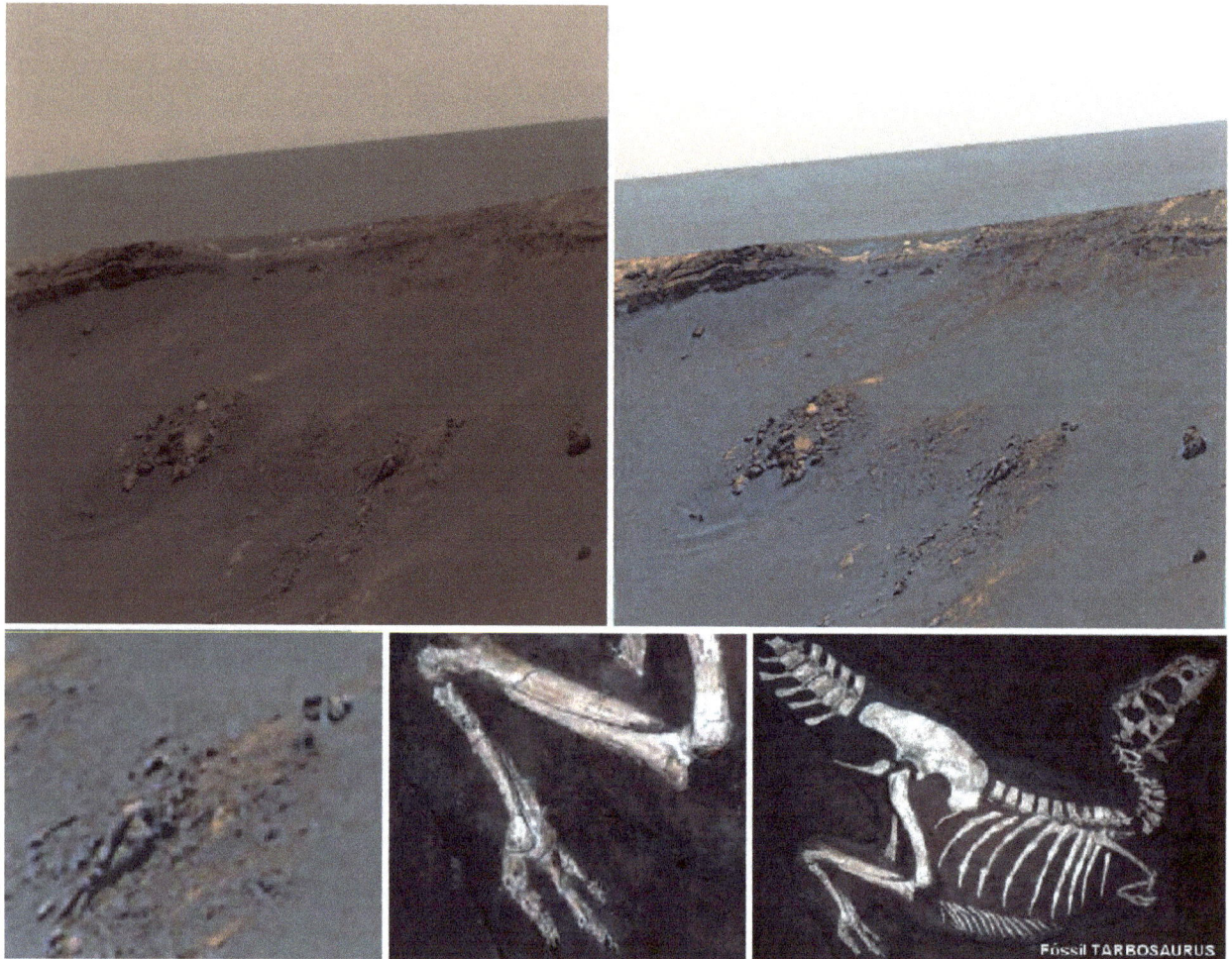

Mars rover Opportunity photographs with its Pan Cam what appears to be rock and rubble debris in a crater on Sol 121, but closer inspection reveals that it may have captured the fossilized remains of a Tarbosaurus dinosaur. Bottom right two images show a terrestrial Tarbosaurus leg
https://mars.nasa.gov/mer/gallery/all/opportunity_p121.html and https://en.wikipedia.org/wiki/Tarbosaurus

With so much ancient life scattered over the Martian terrain and with so much similarity to life on Earth, We are forced to ask, what is going on that two planets can have so many of the same, nearly identical creatures and possibly the same evolution. The hypothesis of Panspermia begs our attention as a real possibility for life arising on Mars. Did life arise on Earth before Mars and through the process of panspermia did Mars receive the spark of life from Earth or was it the other way around? Could life have originated elsewhere in the universe and somehow was carried here to our solar system by a comet or meteor seeding both Mars and Earth

420

simultaneously? Could the similarity of life between Earth and Mars be that life originated on Mars? Could an ancient but advanced Martian civilization faced with the inevitable extinction of its whole civilization, possibly by an impending global cataclysm, either natural or self-inflicted, have been able to build through the collective will and efforts of its people, massive space arks to save itself and some of its indigenous Martian life forms by relocating to Earth? It would give new meaning to the biblical story of the flood and the ark of salvation. Or is there some other possible origin myth that has yet to be uncovered that can explain the similarities of life on two neighbouring planets?

If there is one treasure trove site of ancient animal artifacts, no doubt there have to be other sites waiting to be discovered by the Mars rovers. The old adage that *"life is where you find it"* is certainly true for life on Earth, particularly in extreme and harsh environments and on Mars this adage seems to hold just as true for this planet.

The few photographic examples presented thus far of ancient fossilized creatures discovered on Mars only scratches the surface in photographic evidence. There is much more on NASA/JPL websites and on other internet sites for the reader to explore for himself and the author encourages the reader to do so.

Signs of Extant Martian Animal Life – The Discovery of New Martian Life Forms

We have proven that ancient life once existed on Mars but could there still be some complex life forms currently existing on Mars? Ancient fossilized life is one type of proof in our search for life on Mars but now, we will look at a few amazing examples imaged by NASA's Mars rovers, evidence of real Martian life that not only survives but, thrives in the Martian environment.

Most people will remember the quizzical photograph taken by Mars rover Opportunity that had just landed and started taking panoramic images of its landing site. The photo image shows what appears to be a small white object in the sandy crater, on the Meridiani Planum landscape which has been dubbed by NASA as the **"white bunny or rabbit"** creature. (See photo below).

Jeff Johnson, a scientist with the U.S. Geological Survey and a member of the panoramic camera team, heard from others about a small, fuzzy-looking object in the mission success panorama. Viewing the image on his computer screen, Johnson wondered aloud, "What in the world is this?" Colleagues gathered around his computer table, trying to make sense of the oddity.

Most team members agreed that the "bunny ears" had been, at some point, part of the rover or its lander. The yellowish color led many to conclude that the object was a piece of airbag material.

The Mars Pathfinder mission set a precedent in 1997 for the puzzling pieces around the landing site. An object dubbed "Pinky" caught the attention of the Pathfinder science team and the public. ***Although never positively identified***, it was thought to be a piece of Kapton tape – an adhesive used often in aerospace applications. (Bold italics added for emphasis by author).

In trying to get a higher resolution image of the small strange object, Johnson found that it appeared to be now up the slope of the crater further away! It was after all small about 5cm

(2 in.) in size and light that the slight wind could have blown it about away from the lander. At another point, the "bunny" appeared to have been blown under the egress ramp of the lander as imaged by the lander's camera. This time, it had moved about 15 to 16 feet from its last position on the crater slope!

"There's no evidence of a mark that it left in the soil as it moved," Johnson noted. ***"It was light enough and small enough to not leave any 'footprints'.*** True enough but, what was it? (Bold italics added for emphasis by author). This is an important statement to remember in this story.

Without seeing the "bunny ears" object up close with their own eyes, it's difficult for NASA scientists to provide a positive identification. However, scientists and engineers are quick to deflate the myth that it is anything inexplicable.

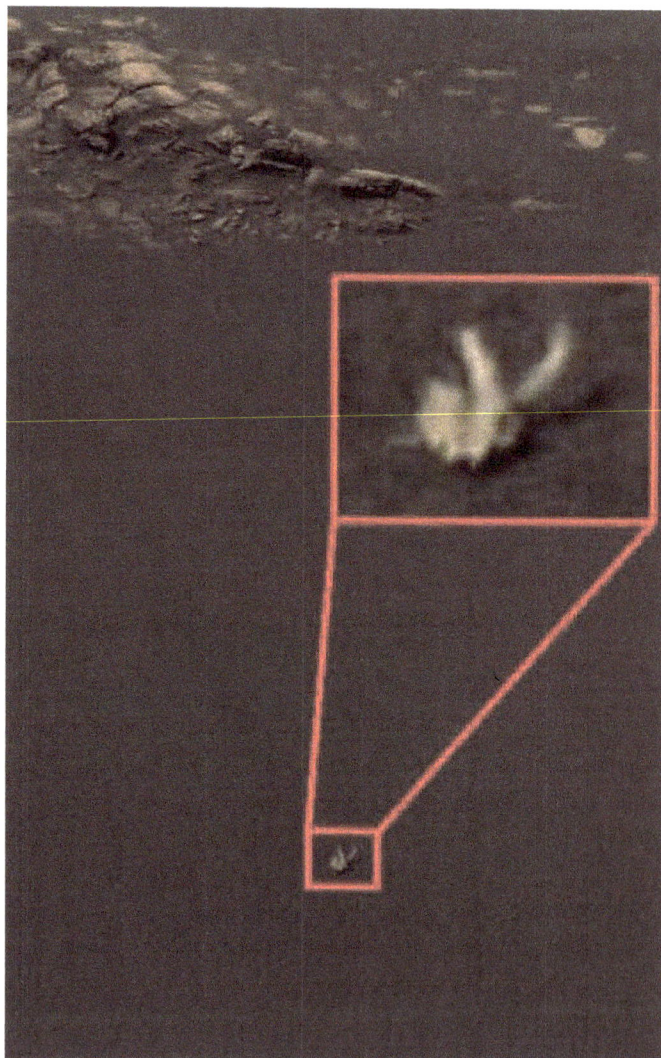

The Martian "white bunny" greets the rover Opportunity after its landing on Mars! This Martian creature has now been officially named by the this author
https://mars.nasa.gov/mer/spotlight/opportunity/b19_20040304.html

This seems to be a typical NASA ploy, to not consider the possible reality for something unusual but, to come up with an alternative explanation that is more scientifically "palatable" even, if your gut instinct tells you otherwise that you may be looking at something Extraterrestrial! *"Our team believes that this odd-looking feature is a piece of soft material that definitely came from our vehicle,"* said **Rob Manning**, lead engineer for entry, descent, and landing. *"We cannot say exactly where it came from but we can say that there are several possibilities: **cotton insulation, Vectran covers, and wraps from the airbag, Zylon bridle tensioning ties, or felt insulation from the gas generators...** The list goes on. We do not think this is parachute material, however, due to its color (it does not look blue enough to be the undyed nylon or red enough to be the dyed nylon)."*

Knowing the possibility that we could have left a bit of a mess nearby, once we saw this feature we only marveled at how clean everything looked and we have not given it another thought. We try to make sure that bits do not fall off, but they do, and we were not at all surprised."
https://mars.nasa.gov/mer/spotlight/opportunity/b19_20040304.html

NASA admission to this litter and debris falling off their probes, landers and rovers is actually more of a problem than the public realizes as there is ample proof from photographic images showing mechanical bits, pieces, and tanks coming off and laying on the ground beside the rovers that appear in photographs which are manmade and not extraterrestrial in origin. Once again, we call into question the quality of workmanship from NASA's engineers, specialists, and mechanics!

Johnson believes, after spectrum analysis with comparable airbag material that the "bunny ears" are, indeed, a wayward piece of airbag material or something similar.
https://mars.nasa.gov/mer/spotlight/opportunity/b19_20040304.html

This rational explanation satisfied NASA officials and by itself, this strange little "bunny" creature could be explained away as something quite normal and very terrestrial in appearance... a part of the lander's airbag. However, what makes this explanation doubtful is that there are more than one of these "white bunny creatures" in the original panoramic photograph taken by the pan camera on Opportunity!

There are possibly two or more unusual white pieces of something within the crater and one of them looks very much like the first white bunny that the public has become so familiar with from news releases. In the panoramic photo image (see below), the middle **(b)** white object is round without much detail which cannot be discerned clearly, even when enlarged but, the third **(c)** white object further to the right on the crater slope appears most like the "white Martian bunny" only facing away from the camera! While object **(d)** shows other smaller objects that may also be a Martian "bunny". https://www.youtube.com/watch?v=9E4fq9Cgz6o

Somehow, NASA doesn't explain or acknowledge these other white bunnies in the photo image. Could these "white pieces of stuff" be nothing more than pieces of airbag material or are they a form of small Martian life indigenous to that region of Mars?

The anatomical structure of this creature suggests it has a long conical head, like an echidna found on Earth, it has two large beady eyes similar to a rat or a house mouse in which its vision may be keener for nighttime activity than during the day, and it has two long ear-like appendages above its head possibly as auditory organs for hearing like a rabbit. Its mouth is probably small like a rodent or maybe an anteater, therefore, it may have a long retractable tongue to collect food with or it may also use it tongue as an olfactory organ like a snake or lizard. It has two short but, powerful front legs and like a mole, it front paws may have long, strong claws used for digging and burrowing. This little Martian creature has probably four legs in all but, it is conceivable that it may have as many as six little legs, altogether (it is hard to tell how many legs or feet it has from the photographs, even when the image is enlarged). This is all speculation of course.

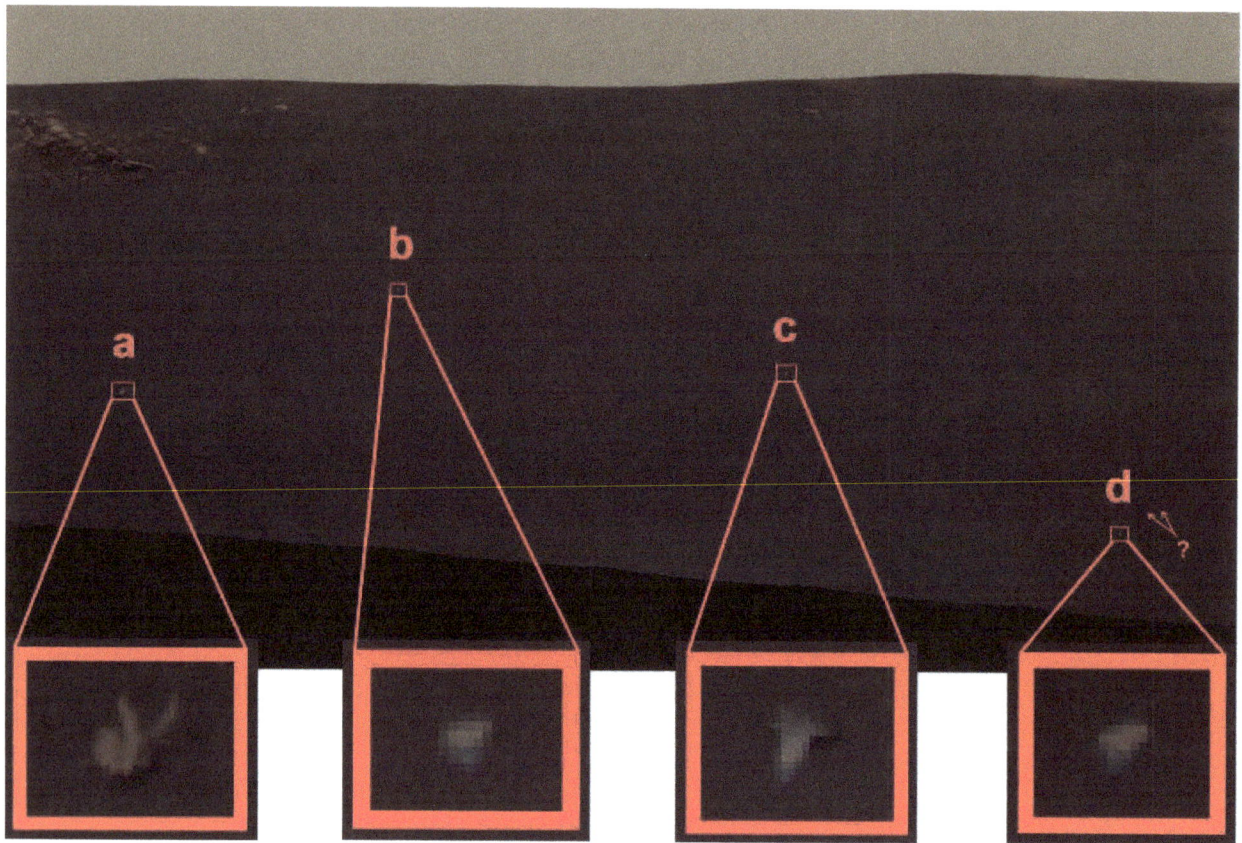

Part of the panoramic photograph taken by Opportunity showing Martian White "Bunnies" on the crater slope. Note the similarity of image "a" with image "c" while "b" and "d" may be emerging from the sand or digging down under
https://www.youtube.com/watch?v=9E4fq9Cgz6o

However, after some careful sketching, some terrestrial research and some intuitive insight into similar life forms native to the Earth, this author soon realized that this piece of "white fluff" thought to be pieces of lander airbag *blowing about in the Martian wind* was, in reality, a living creature which apparently has a terrestrial cousin! The **Australian Lesser Bilby**!!!!!

Now, it should be understood that if an amateur anomalist researcher of Mars can find Martian oddities, artifacts, and possible fossils or life forms, then you have to know that NASA has also found these artifacts. But, because NASA cannot possibly admit that it has found life forms or fossils or artifacts on Mars as these would represent tangible proof that we are not alone in the universe. It would prove that other life exists within our own Solar System. Such acknowledgement of other life forms and those that are still extant on Mars exposes their hiddenagenda which does not include open transparency or disclosure of their discoveries as per **Brookings Institute Report,** therefore; they have enforced a self-imposed silence upon themselves at this time. This leaves the amateur planetary anomalist researcher to make discoveries and publicize their finds via the internet and other news media avenues.

(c) Terry Tibando April 15, 2013

(C) Terry Tibando April 15, 2013

Because NASA refuses to acknowledge this first evidence of a living Martian creature, I have, therefore, taken the liberty to name this Martian creature after myself: *"Martis, Animalia, Chordate, Mammalian, Marsupialia, Peramelemorphia, Thylacomyidae, Macrotis, M. Leucrura, Tibando"* **or the "Tibando Long-Eared Martian Bilby"**(c) Terry Tibando - April 15, 2013
- 9:00 PM

In keep with this current policy of public disclosure in which NASA is unwilling to do as a routine part of its mandate which was originally entrusted to it through the funding of a taxpaying public, this author has therefore taken the liberty to name his discovery of a newly found Martian creature after myself as its first discover: ***"Martis, Animalia, Chordate, Mammalian, Marsupialia, Peramelemorphia, Thylacomyidae, Macrotis, M. Leucrura, Tibando"*** or the **"Tibando Long-Eared Martian Bilby"** (April 15, 2013)

When NASA hides things in plain sight of objects in their raw photographic data whenever the Mars rovers and orbiting satellites images the Martian terrain, inevitably some amateur researcher will find it, analyze it and draw some amazing conclusions which will lead them one step closer to understanding what NASA knows and has discovered on Mars.

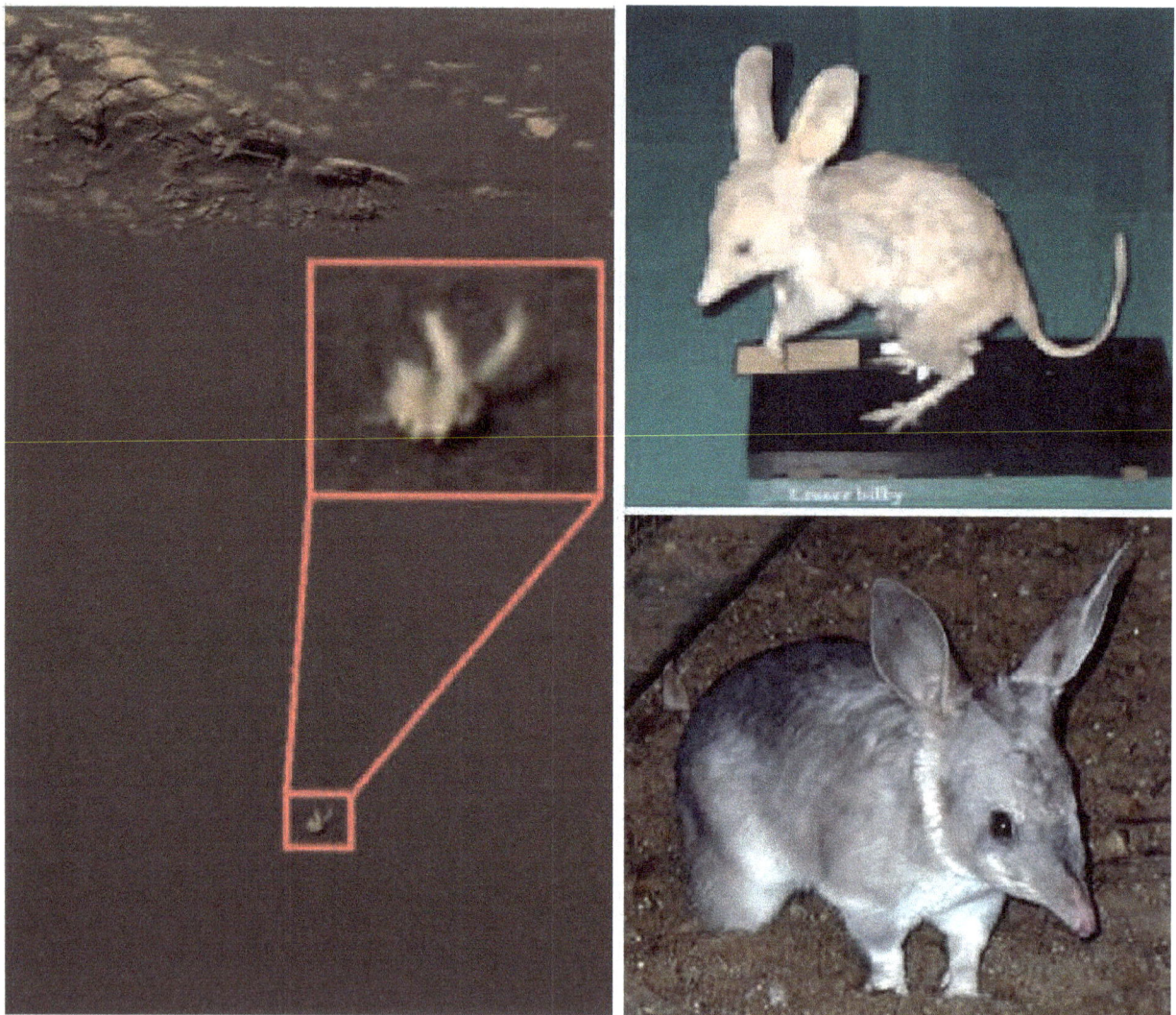

Here is what NASA has denied exists on Mars, a "Martian Bilby" similar to the Australian Lesser Bilby (top right) and the Greater Bilby (bottom right) which are found in the "DRY HOT DESERT" areas of the Central and Northwestern Australian outback

https://mars.nasa.gov/mer/spotlight/opportunity/b19_20040304.html and https://en.wikipedia.org/wiki/Lesser_bilby and http://members.optusnet.com.au/bilbies/Charleville_Photos.htm

If this is an extant life form then, this would be the first positive proof of another life form on a planet other than Earth. Proof that Mars is no longer a barren, dry and dead planet as officially posited by NASA, but is very much a living planet! This fact alone should have all scientists clambering to implement a program of protection to all indigenous life forms on Mars which would then obviate any need for terraforming missions and agendas until the life cycle behaviours of such Martian life forms is completely understood and fully documented.

Is this the end of the *"white rabbit"* story? The saga apparently continues with a tragic ending! When we think of NASA, particularly in its early days of space exploration, we perceived its missions with innocent eyes and with an imagination of wonderment. Unfortunately, long gone are those days of innocence. NASA and JPL's overarching mission is not, in the opening words of the sci-fi TV show **Star Trek**: "*...to seek out new life and new civilizations"*, but, increasingly it appears that the search for life is a secondary mission, no matter how much they would try to convince us to believe otherwise, this is not their sole primary mission.

Even, their NASA employees and supporters, particularly those who have not been around for a few well seasoned years, still cling to those noble aspirations of a peaceful exploration of space for the benefit of all mankind. NASA's institutions over the years have demonstrated a persistent and an unscientific reluctance to actually examine certain things on the Martian surface that they come across. Over the years of exploring the Martian terrain, a disturbing agenda is emerging as recent events in the last decade on the Red Planet have revealed. NASA and JPL's agenda seems more hell bent on killing extraterrestrial life than examining it.

Now, one could argue that what was thought to be life was just an optical illusion, a pareidolia of the imagination from the untrained eyes and minds of amateur investigators and that the Mars rovers had not kill anything that was animated, but merely traversed over common rocks and stones on the surface.

However, investigators and researchers and some scientists believe that life does exist on Mars and strongly feel that NASA has betrayed our trust and their mission quest — *and their own raw photographic data arrogantly or unwittingly proves it*! It seems the **Mars rover Opportunity** has been ruthlessly and mindlessly used in true robotic fashion to grind any possible evidence of life into the Martian dust or under its wheels in the **Meridiani Planum** area.

In one photo image (see below) taken by Opportunity, the rover moves off the egress ramp of its lander to move toward some rock outcroppings which it had photographed earlier, the order is sent by JPL to Opportunity do a little detour maneuver toward a "white rabbit" creature and apparently runs over it, crushing it under the rover's tire treads into the Martian sand.

NASA's explanation for this little maneuver was that in climbing out of the crater, the rover lost traction and slipped on the crater slope but the photo below does not lie. The direction of travel was deliberate, with malicious of intent; to eliminate all traces of life that the rover comes across!

Now, before people start to shout that this is nothing more than the author trying to look for conspiracies against NASA, further proof came from other photographs taken by Opportunity while exploring the crater in which it had landed. Opportunity travelled back and forth within the this crater photographing the terrain, its landing ramp and it's tire tracks when it snapped another Martian "white bunny creature" object moving rapidly across the sandy crater floor leaving dust and a track of its own in its wake! Was NASA's Mars rover merely photographing its own wheel tracks or was it trying to kill another Martian life form by running over it!?

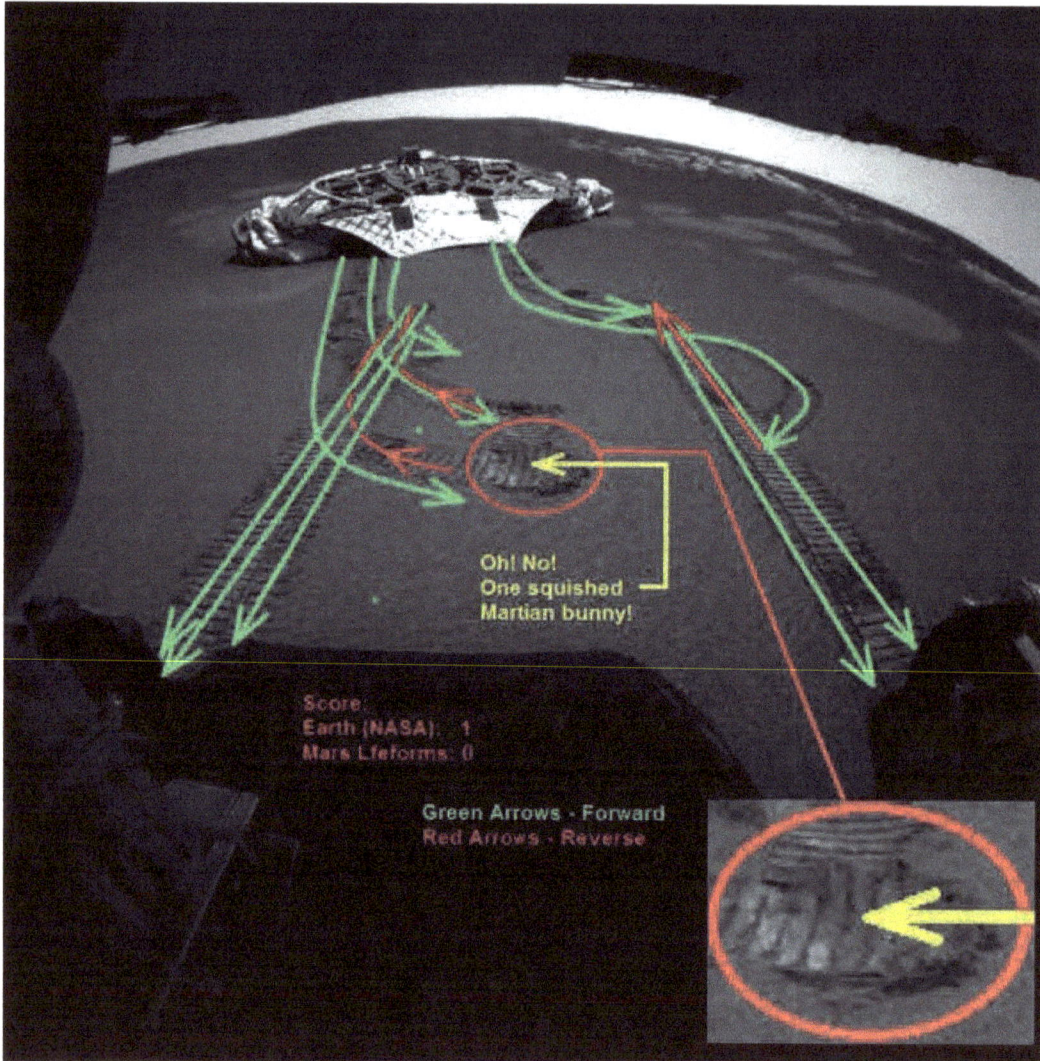

On Sol 012 Opportunity's Rear Hazcam images its tracks, showing how the rover detours to ruthlessly crush what may be a Martian life form then, continues straight ahead. The white residue in the track marks is all that remains of the "Martian Bunny"
https://mars.nasa.gov/mer/gallery/all/1/r/012/1R129255699EFF0242P1211R0M1.HTML and (c) Terry Tibando

Jeff Johnson said that the object was not alive, that *"it was light and small enough not to leave a track or any footprints"* yet, in the photograph below, it indicates just the opposite of that statement. That *"little piece of airbag material"* that everyone at NASA agreed was not alive, seems to be very animated as it appears to be kicking up dust and leaving tracks as it desperately ran for its dear life, to get the hell out of the way of the oncoming wheels of the **Mars rover Opportunity!** (See photos below).

What are the chances that this is another Martian creature, a "cosmic cousin" to its earth-bound terrestrial relative, the **White Marsupial Mole**? This Martian creature leaves tracks similar to the White Marsupial Mole of Australia which lives in the Hot, Dry Desert areas of the Australian Northwestern outback. The odds would have to be astronomical that it is merely a freakish coincidence of nature unless there is an intelligent purposeful design behind such synchronicity on two different worlds!

(C) Terry Tibando April 15, 2013

A Little Critter Scurries Across the Track Laden Crater Floor

(C) Terry Tibando April 15, 2013

Is this another piece of lander airbag debris blowing across the Martian sand or is it something live scurrying across the track laden crater floor?

http://www.blastr.com/2017-4-26/caught-camera-martian-opportunity-hole-one

Dust — Track

White "Bunny"

Direction of travel

Rover tire tracks

(C)Terry Tibando April 15, 2013

The "White Bunny" is really a White Martian Marsupial Mole

Earth White Marsupial Mole

(C) Terry Tibando April 15, 2013

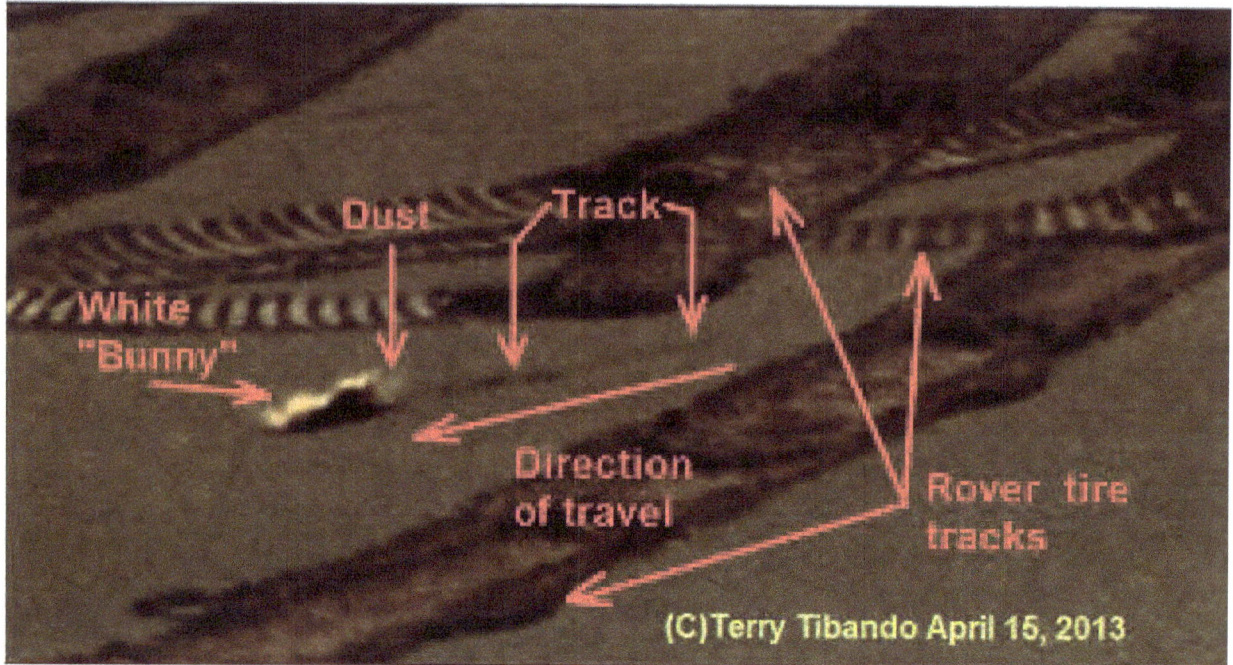

Is this another Martian "white bunny" creature kicking up dust in order to get out of the way of the rover Opportunity's wheels that are running back and forth within the crater? Note the comparison of the terrestrial White Marsupial Mole insert at bottom right
https://photojournal.jpl.nasa.gov/catalog/PIA05755 and **(c) Terry Tibando**

Once again, NASA must know something about this and is probably pondering these Martian oddities and possible indigenous life forms. Such images must be creating some sleepless nights for some of NASA's scientists. How do they explain this similarity with terrestrial creatures and do they continue to suppress and cover up such discoveries?

In the above images, the little Martian creature seems to be digging itself down and under the surface in order to save itself from a certain death by a mechanical "alien invader". This instinctual life-saving behavior is universally inherent in all animal species whether on Earth or on Mars and probably on any planet in the universe where life has arisen. Naturally, an indigenous life form never having seen another life form, whether organic or mechanical in origin which suddenly invades its territory, would not know what to make of it and perhaps, would express a moment or two of curiosity before high-tailing it to safety. That moment or two could be the difference between life and death as indicated by the photos above!

Again, we see another Martian creature that appears to be somewhat different from the **"Tibando Long-eared Martian Bilby"** that was imaged earlier by Opportunity. It now appears that NASA has unexpectedly discovered another "white scurrying Martian creature" on Mars, the second animal in such close proximity to the first discovered creature.

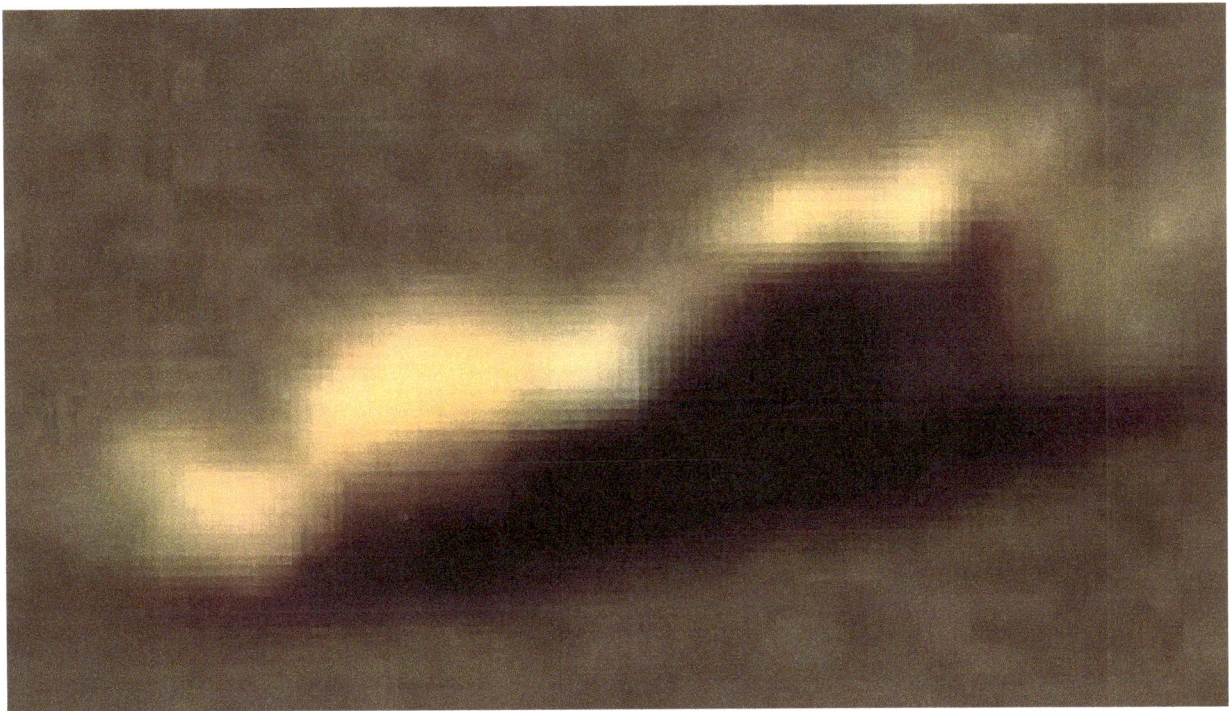

This photo shows a close-up of a white Martian "Mole-like" creature. Note the rippling motion of the creature's body (head is at left) as it starts to dig into the Martian soil and the dust being kicked up by the hind legs
https://photojournal.jpl.nasa.gov/catalog/PIA05755and (c) Terry Tibando

Like many mole species on Earth which can inhabit a variety of environments, on Mars, it seems to prefer arid or desert areas. It spends much of its time burrowing under loose sand and dirt where their food (which may be a type of small Martian insects, slugs, and grubs) is also found. The white furry creature has a fine fur or hair covering its body and it is conceivable that this fur may change colour during the equinox seasons like Earth rabbits, weasels, and ermine, etc. Like its Earth cousins, it probably goes deeper into the ground during late fall and winter seasons on Mars as the surface temperatures drop and the land cools or freezes. In order to survive winter conditions, it may survive underground, perhaps in hibernation. It may have a significant

431

tolerance to high levels of carbon dioxide and low levels of oxygen assuming that Mars atmosphere is as NASA indicates. Its body is larger and muscular near the head and shoulder area gradually tapering along the dorsal of its back ending with a short pointed tufted tail.

Examination of the marsupial moles found on Earth indicates some close comparisons to body morphology. (See photos below for comparison). It's back and spine give the little creature an undulating up and down motion (see photo above), particularly when it digging through the ground using it's two very strong front mole-like paws in which case, its hind legs may only be rudimentary limbs for limited movement. http://www.molecatchers.com/meet_the_mole.html

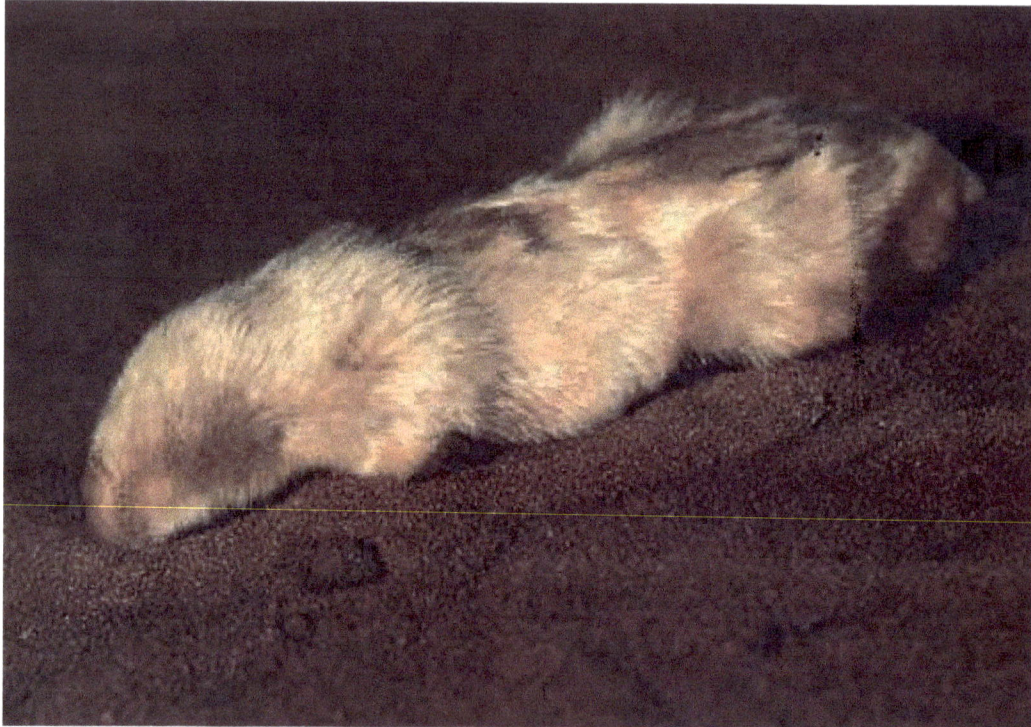

An Earth marsupial mole digging into the soil, its head is left. Compare this image to the one above of the Martian "Marsupial Mole" moving rapidly, kicking up dust with a rippling body motion in both photos.
http://www.factzoo.com/mammals/marsupial-mole-creamy-underground-fur-pellet-claws.html

Being mammalian in appearance, the Martian creature no doubt has both male and female of the species and thus, bear live young, or it may possibly be asexual thus, parthenogenetic, laying eggs in clutches like the echidna perhaps, once a year depending in accordance to the warm Martian spring and summer seasons. It is also conceivable that it too is a marsupial creature like Earth's marsupial mole. Part of its behavior is to come to the surface to "sun itself" during warm days in the spring and summer which explains how it was photographed on several occasions by the Mars rover Opportunity, right after it landed on Martian terrain.
http://en.wikipedia.org/wiki/Mole_%28animal%29

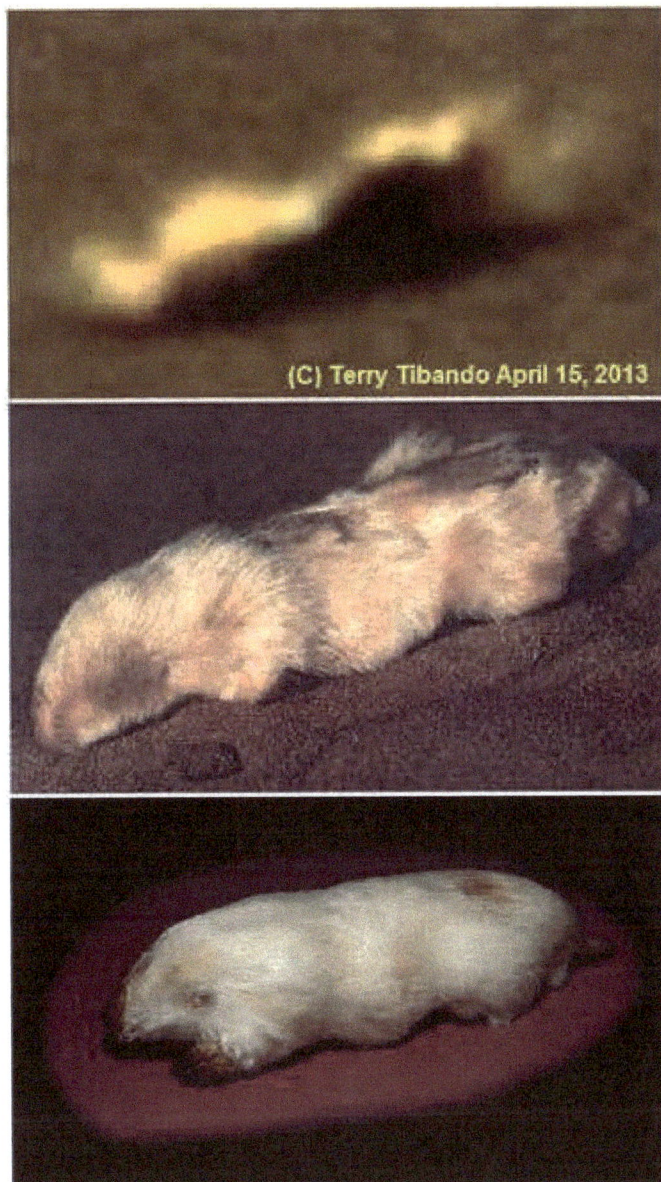

A comparison of the Martian Mole and two White Earth Marsupial Moles from the desert area of Western Australia. This is yet another Martian creature that NASA has refused to acknowledge its existence so, I have also named this creature after me: *"Martis, Animalia, chordate, Mammalia, Marsupialia, Australidelphia, Notoryctemorphia, Notoryctidae, N. Typhlops (N. Caurinus), Tibando"* **or the "Tibando White Marsupial Martian Mole"**
(April 15, 2013 – (9:00 PM)

https://photojournal.jpl.nasa.gov/catalog/PIA05755 and http://www.flickriver.com/photos/centralaustralia/4723352861/ and http://www.factzoo.com/mammals/marsupial-mole-creamy-underground-fur-pellet-claws.html

This remarkable little Martian critter is another proof that life exists elsewhere besides on Earth. The taxonomy of the "White Marsupial Martian Mole" is a combination of several Earth

mammalian type creatures: mole, rodent, and marsupial in appearance. Admittedly, this is subjective information that is definitely open to interpretation and correction by biologists or exo-biologists, as no one has yet stepped foot on Mars from Earth to scientifically confirm this one way or the other, as far as we know. In the Black world of Science with its covert black space program, this would be a whole other matter.

Therefore, in keeping with a spirit of fair play in designating new names upon extraterrestrial creatures that NASA will not admit to these creatures existence, this author has taken it upon himself to name this creature after himself as its first discoverer: *"Martis, Animalia, chordate, Mammalia, Marsupialia, Australidelphia, Notoryctemorphia, Notoryctidae, N. Typhlops (N. Caurinus), Tibando"* **or the "Tibando White Marsupial Martian Mole"**

The mystery of the newly discovered Martian creatures may never be resolved to everyone's satisfaction but fortunately, however, other photo images taken by the Mars rovers Spirit, Opportunity and now, Curiosity clearly shows evidence of not only large fossilized animal life forms but also, living animal and reptilian life forms. Under minor scrutiny, these images are so obvious to anyone looking at them that they practically leap off the photographs to such a degree that NASA has decidedly remained silent on the whole matter, preferring to talk about the possibly of finding the first evidence of water or the future discovery by one Mars rover of microbes in the Martian soil.

With more of the public becoming interested in knowing if other life in the universe exists besides on Earth, Mars becomes the neighbourhood planet of choice to start investigating for those signs of past or current extraterrestrial life. The armchair astronomer through the power of the computer is now able to access tens to hundreds of thousands of NASA photographs of Mars and where NASA comes up short or is arrogantly silent on the obvious anomalies discovered on Mars, these amateur planetary explorers are revealing to the public what NASA is failing to do through honest and truthful disclosure.

 "While Mars was likely a more hospitable place in its wetter, warmer past, the Red Planet may still be capable of supporting microbial life today", some scientists say.

Ongoing research in Mars-like places such as **Antarctica** and Chile's **Atacama Desert** shows that microbes can eke out a living in extremely cold and dry environments, several researchers stressed at **"The Present-Day Habitability of Mars" conference** held here at the University of California Los Angeles this month.

And not all parts of the Red Planet's surface may be arid currently — at least not all the time. Evidence is building that liquid water might flow seasonally at some Martian sites, potentially providing a haven for life as we know it.

"We certainly can't rule out the possibility that it's habitable today," said **Alfred McEwen** of the University of Arizona, principal investigator for the **HiRise camera** aboard NASA's **Mars Reconnaissance Orbiter** spacecraft." http://www.space.com/19928-mars-habitable-life-possible.html?cid=dlvr.it

"McEwen discussed some intriguing observations by HiRise, which suggest that briny water may flow down steep Martian slopes during the local spring and summer.

Sixteen such sites have been identified to date, mostly on the slopes of the huge **Valles Marineris** canyon complex, McEwen said. The tracks seem to repeat seasonally as the syrupy fluids descend along weather-worn pathways.

Water seeps and flows down the slopes of the huge Valles Marineris canyon complex
http://www.space.com/19928-mars-habitable-life-possible.html?cid=dlvr.it (Credit: NASA/JPL-Caltech/Univ. of Arizona)

"Briny water on Mars may or may not be habitable to microbes, either from Earth or from Mars," McEwen said.

Martian life may be able to survive even in places where water doesn't seep and flow, some scientists stressed. For example, microbes here on Earth make a living in the Atacama and the dry valleys of Antarctica, both of which are extremely cold and arid, said **Chris McKay** of NASA's Ames Research Center in Moffett Field, Calif." http://www.space.com/19928-mars-habitable-life-possible.html?cid=dlvr.it

As it was stated earlier, NASA prefers to play it safe by looking for water and microbes and at this stage of space exploration on another planet, playing it safe simply won't cut it at this time

435

when larger things are lurking around every bend, in every crater, on every hill and under every rock on Mars.

Another example of a living creature is a photograph (see below) captured by the Mars rover Curiosity on Feb 27, 2013, or on the Martian day: SOL 0193 by the Right Mastcam: 2013-02-20 14:01:39 UTC. It shows what appears to be another live, but larger animal, possibly a ferret or lizard-like creature seen mid-range from the camera across a rocky terrain in the Gale Crater. Raw photo image can be found at: **msl-raw-images/msss/00193/mcam/0193MR1024018000E1**

The open, but rocky plain of the Gale Crater shows what appears to be a rodent or lizard-like creature among the rocks circled in red
http://www.digitaljournal.com/article/351083

What is interesting so far in our search for Martian life is that most of the NASA photo images where there is Martian life; it appears to be able to blend or camouflage itself into the Martian landscape. This could, however, also be due to NASA altering or filtering the colour of each rover image sent back to the JPL photo labs. A preliminary examination could easily overlook

436

the oddity of some of the rocks as just rocks but a closer inspection quite often reveals that there is more to the landscape than first meets the eye.

The above photograph is a classic example of this cursory camouflaging whether natural to the Martian indigenous life or by deliberate cover-up by NASA. On the other hand, NASA has lately let slip a lot of photo images with incredible evidence of animal life, fossils and artifacts of artificiality out to the public domain, almost as if they want the public to draw their own conclusions and disclose whatever information they find whether correct or false. This would allow NASA to step up at anytime to give the "corrected" official version of the facts.

The above circled image enlarged to show detail of lizard-like creature with long tail
http://www.digitaljournal.com/article/351083

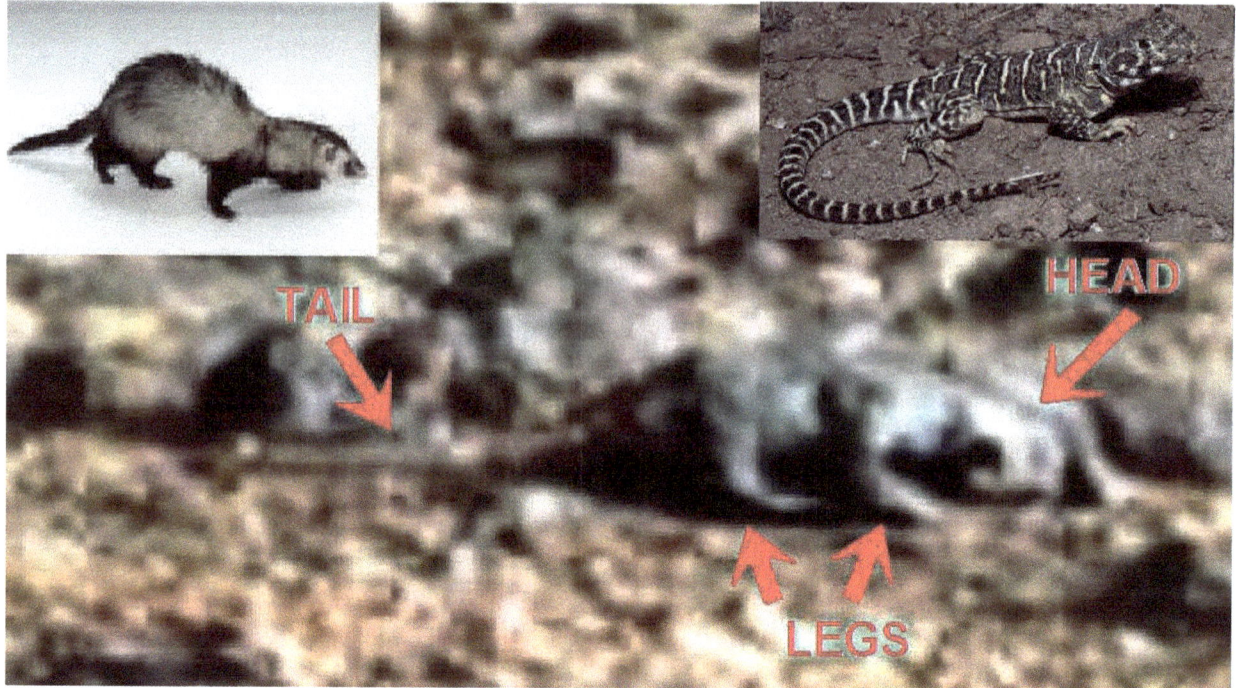

The creature has some aspects of an Earth Ferret (top left), but also shares a closer resemblance to a lizard (top right) as indicated by the highlighted body parts

http://www.digitaljournal.com/article/351083 (c) Terry Tibando

In the above photographs, we see what looks like a small animal or lizard that has a head with two dark eye spots on either side of the head, a wide mouth area, two sets of legs of which three can be seen clearly. There is a sizeable body area and attached is a long tail reminiscent of a rodent or lizard tail in length which appears hairless.

The creature appears to be looking directly at the cameras on the Mars rover and one could argue that this is merely a picture of small rocks and stones. However, it would also be unusual if the legs of lizard closest to the camera merely appeared to be two small rocks which just happened to be identically shaped like two miniature cowboy boots beside each other.

In other words, however, we perceived these rocks to be, they are unusual and indicative of something either organic or artificial in construction worthy of further investigation by the rover's cameras.

In the picture below, there is another example where rocks appear to look like something organic, alive and familiar to our eyes. When some colour is added to the rocks in question, we see that they appear to form yet another small reptile-like creature. Is this all just fanciful delusions to justify the need to see something familiar on the surface of an unfamiliar planet that is otherwise considered by NASA scientists as a cold, dry barren and lifeless world?

If these pictures merely show a trickery of light and shadows upon a pile of rocks, then why is it that the rocks seem to portray objects of familiarity in a mathematical symmetry of curves, sharp right angles, orthogonal shapes, and organic forms, etc.? Pareidolia? This author thinks not!

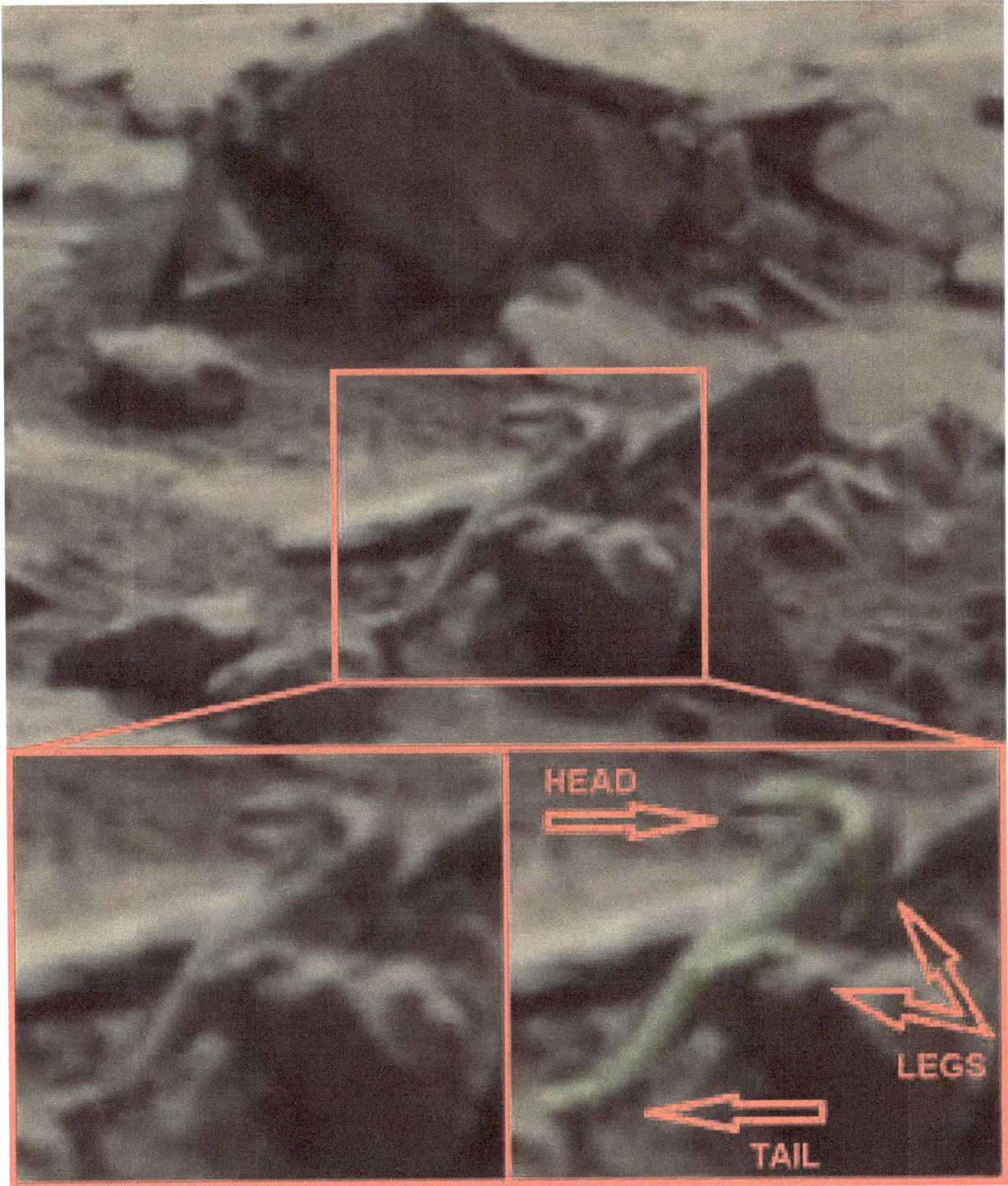

The image was taken by Curiosity on Sol 198, possibly another Martian Lizard crawling across the rocks? Bottom images were enlarged and color enhanced to show detail

Could there still be life running around on and under the surface of Mars at this time? If we accept the original unaltered raw images from NASA and look objectively at what is obvious while avoiding the subjective pareidolia mindset in which most people perceive vague or unfamiliar objects as something recognizable then, the only reasonable conclusion remaining is the existence of Martian animal life! Life that is similar to terrestrial life on Earth yet, uniquely different in its own exobiological away. An example of this is the Martian "Squirrel" seen below.

**Is this a Martian rodent, perhaps a squirrel, lemma or merely
a trick of light and shadow upon a rock?**
http://www.zdnet.com/pictures/was-a-squirrel-discovered-on-mars/2/

Amazingly, this Martian life has found a way to survive in an atmosphere that NASA says is very thin and mostly comprise of carbon dioxide, where the landscape is dry, barren and probably dead, an environment that is basically hostile and poisonous to human life without a space suit, yet, here is proof that life on Mars does exist! Life has found a way to exert itself and thrive!!

One must seriously consider with high suspicion that the atmosphere of Mars as officially posited by NASA is **NOT** what we have been told for the last 60 years. Mars atmosphere, although thin and less oxygenated than on Earth, **IS** actually a life-sustaining atmosphere comparable to the rarified air experienced by mountain climbers, where some breathing apparatus would be needed for long-term exposure. In other words, the Martian atmosphere is

440

breathable by humans on short non-exertive walks or non-labouring activities for a few minutes before feeling shortness in breath.

Keep in mind that you breathe in carbon dioxide with every breath. A very small % of the air we breathe is carbon dioxide. The important thing is to not let the air we breathe contain less than about 18% oxygen or you will pass out. The carbon dioxide is taking the place of the oxygen that you need. The air we breathe in contains 20.95% oxygen and that does not have to fall a great deal before we pass out from lack of oxygen. Even when we breathe in the 20.95% oxygen our body doesn't use all of it because when we exhale 18% of the oxygen comes out. We need that much oxygen to make sure that our blood is at least 92% saturated.

The Russians are training their astronauts to learn to breathe in a rarified oxygen atmosphere in preparation for near future trips to Mars, what do the Russians know that Americans supposedly don't know? It would seem that the Russians believe that it is possible to breathe the Martian atmosphere, even for short periods of time, maybe even, longer. The Russian news service, "Russia Today" reports that volunteers locked inside a sealed capsule are spending ten days breathing a combination of argon, nitrogen, and oxygen, as part of a research program designed to simulate a manned trip to Mars.

"Our experiments show that argon, combined with the right portion of oxygen, is quite safe for humans," says **Aleksandr Dyachenko**, a scientist at the **Institute of Biomedical Problems** at the **Russian Academy of Science**. "I tested it on myself and I'm OK. Our volunteers are also doing fine".

Of course, the reason the volunteers are fine is that the air in their modified deep-sea diving chamber bears only a mild resemblance to the Martian atmosphere. Russian TV station, Akado, reports that the chamber contained 60 percent argon, 15 percent oxygen, and 25 percent nitrogen. The real Martian atmosphere is 95 percent carbon dioxide, 2.7 percent nitrogen, 1.6 percent argon, and less than a percent of oxygen. Earth's atmosphere contains primarily nitrogen (78 percent) and oxygen (21 percent).

The purpose of the experiment appears to be to test if high levels of argon can protect against nerve damage from hypoxia or oxygen starvation. Some studies have suggested that the noble gases, including argon, have that property.

The experiment is one of the first in a series of investigations that will culminate in the Mars 500 program in which six volunteers will be locked in a chamber designed to simulate a Martian-bound spacecraft. http://www.wired.com/wiredscience/2008/04/russian-volunte/ and http://www.youtube.com/watch?feature=player_embedded&v=wvYHP2WzJwk or https://www.youtube.com/watch?v=wvYHP2WzJwk

This training of Russian cosmonauts to breathe Martian atmosphere runs counter to what NASA has been telling the US public and the rest of the world. The question is what is going on with NASA and its disclosure of information which is in direct contradiction to what other international space agencies are reporting about Mars?

There have been far too many facile explanations from NASA *"experts"* that an air of smugness and arrogance among these rocket scientists, who seem self-assured of an exclusivity of scientific truth that too frequently, parallels the piety of a priesthood confident in its possession of divinity at a fundamentalist religious rival. In the church of big news media, their sermons of recent Martian discoveries according to the "holy" guiding doctrine of the **Brookings Institute Report** are liberally offered up as blessings to the faithful supporters of NASA.

However, NASA's failure to dispense the honest truth of its discoveries to its taxpayers and to the whole human race would be enough to make any of the faithful want to change their religion and find another church of true science in order to hear the unabridged, unadulterated truth in space exploration discoveries. From its early beginnings, there have been signs that the NASA space program was just a diversionary distraction for a much larger covert black space program.

We covered every major aspect to the question of whether life could exist on Mars and so far all conditions indicate that it could. It is believed by many that all the evidence points to the inescapable fact that life not only did existed in prehistoric times but, it still exists at this current time on Mars! And that NASA hasn't been honest with the general public about its discoveries on Mars!

In fact, we must seriously consider if NASA's true intention in space exploration is not the search for life, which most likely is its secondary objective, but perhaps its primary focus is in keeping with its real mission objective, as an agency of the **DoD (Department of Defence),** the search, cataloging and eventual acquisition of alien technology!!!

BIBLIOGRAPHY, WEBLIOGRAPHY AND VIDEOGRAPHY

The following list includes all books and major journals, newspapers and web based material, including other reference ebooks and materials such as web links and video links found on the internet in researching this book. Not all chapters use reference material and therefore, these chapter are not listed.

Bibliography listed refers to books marked in RED,
Webliography refers to websites marked in BLUE,
Videography refer to video websites marked in GREEN,
News Service websites are marked in LIGHT BLUE, and
Newspaper websites, magazines and professional papers are marked in PURPLE.

It does not include specific government documents, archival repositories, or various journals or other web based material.

CHAPTER 59
"WE ARE NOT ALONE, NEVER HAVE BEEN"!!!

https://www.youtube.com/watch?v=lkswXVmG4xM

CHAPTER 60
THE HOLOGRAPHIC UNIVERSE AND
THE HOLOGRAPHIC PARADIGM

http://www.crystalinks.com/holographic.html

Selections from the Writings of Abdu'l-Baha; Compiled by the Research Department of the Universal House of Justice; © 1978The Universal House of Justice; ISBN 85398 081 0; Printed in Great Britain by W & J MacKay Ltd, Chatham

https://www.bahai.org/library/authoritative-texts/search#q=life%20in%20the%20universe

https://siriusdisclosure.com/wp-content/uploads/2018/07/Petaluma-Slides.pdf

(The Seven Valleys and the Four Valleys) by Baha'u'llah

http://en.wikipedia.org/wiki/Transpersonal_psychology

CHAPTER 61
THE PRESENCE OF EXTRATERRESTRIAL INTELLIGENCE IN OUR SOLAR SYSTEM: A RESOLUTION TO THE FERMI PARADOX

The Hidden Words of Baha'u'llah; Baha'u'llah; copyright 1954; by National spiritual assembly of the Baha'is of the United States; printed in the USA; Library of Congress Card No. # 54-7328

The Seven Valleys and the four Valleys; Baha'u'llah; copyright 1945 and 1952; by National spiritual assembly of the Baha'is of the United States; printed in the USA; Library of Congress Card No. # 53-12275

http://waitbutwhy.com/2014/05/fermi-paradox.html

http://abyss.uoregon.edu/~js/cosmo/lectures/lec28.html

http://www.daviddarling.info/encyclopedia/V/vonNeumannprobe.html

http://www.rfreitas.com/Astro/ResolvingFermi1983.htm

The Hidden Words of Baha'u'llah by Baha'u'llah; copyright 1954; The National spiritual Assembly of The United States; Printed in the USA; Library of Congress Catalog Card No.54-7328

https://www.youtube.com/watch?v=aYk_nPoIu8E

CHAPTER 62
I SPY AN ASTRONOMICAL COVER-UP

http://en.wikipedia.org/wiki/Carl_Friedrich_Gauss

http://en.wikipedia.org/wiki/Nikola_Tesla

http://en.wikipedia.org/wiki/Communication_with_Extraterrestrial_Intelligence

http://en.wikipedia.org/wiki/Mars_Exploration_Rover

Gleanings from the Writings of Baha'u'llah; No. LXXXII

Gleanings from the Writings of Baha'u'llah; 1939; Baha'i Publishing Trust; Wilmette, Illinois, USA; Library of Congress Card No. 52-14896

http://www.bibliotecapleyades.net/brooking/brookings_report.htm

http://www.daviddarling.info/encyclopedia/B/BrookingsStudy.htmlnd

CHAPTER 63
DO ASTRONOMERS SEE UFOS?

J. Allen Hynek; The UFO Experience; published 1972.

"November 19, 2012; Huff Post – weird News; Canada
http://www.huffingtonpost.com/2012/09/19/lord-martin-rees-aliens
ufos_n_1892005.html?utm_hp_ref=mostpopular

http://www.huffingtonpost.com/dan-mack/astronomers-ufo_b_1901480.html

http://en.wikipedia.org/wiki/Ufology#Surveys_of_scientists_and_amateur_astronomers_co
ncer

http://www.scribd.com/doc/16805639/A-List-of-UFO-Sightings-by-Astronomers.

CHAPTER 64
RADIO TELESCOPES (SETI), SATELLITES, SPACECRAFT,
RADIO AND TELEVISION, ETC.

http://zidbits.com/2011/07/how-far-have-radio-signals-traveled-from-earth/

http://www.ufoevidence.org/documents/doc270.htm

http://en.wikipedia.org/wiki/Kardashev_scale

http://www.youtube.com/watch?feature=player_detailpage&v=4Imd_0iCucg

https://www.youtube.com/watch?v=6GooNhOIMY0
https://www.youtube.com/watch?v=1w6u3ZVhSvY

CHAPTER 65
SETI (SEARCH FOR EXTRATERRESTRIAL INTELLIGENCE)
OR IS IT THE SILLY EFFORT TO INVESTIGATE?

http://en.wikipedia.org/wiki/Search_for_extraterrestrial_intelligence

http://en.wikipedia.org/wiki/SETI_Institute

http://en.wikipedia.org/wiki/Seth_Shostak

http://en.wikipedia.org/wiki/Rio_Scale#Post_detection_disclosure_protocol

http://en.wikipedia.org/wiki/Search_for_extraterrestrial_intelligence

https://www.youtube.com/watch?v=SFFi3sWq5jE

http://en.wikipedia.org/wiki/Arecibo_message

http://www.examiner.com/article/seti-pioneer-fails-to-debunk-crop-circles-as-et-messages

http://www.bibliotecapleyades.net/circulos_cultivos/esp_circuloscultivos12.htm

http://en.wikipedia.org/wiki/Communication_with_Extraterrestrial_Intelligence

http://en.wikipedia.org/wiki/Prime_number

http://en.wikipedia.org/wiki/Stephen_Hawking

CHAPTER 66
THE SPACE RACE AND THE QUEST FOR ALL THINGS EXTRATERRESTRIAL

http://en.wikipedia.org/wiki/Spaceport

http://en.wikipedia.org/wiki/Category:Spaceports_in_the_United_States

http://www.spacefoundation.org/media/space-watch/spaceports-can-be-found-throughout-world

http://en.wikipedia.org/wiki/Space_center

http://en.wikipedia.org/wiki/NASA

http://www.eisenhowermemorial.org/onepage/IKE%20&%20Science.Oct08.EN.FINAL%20%28v2%29.pdf

http://en.wikipedia.org/wiki/Space_Race

http://en.wikipedia.org/wiki/Soviet_space_program

http://en.wikipedia.org/wiki/Lost_Cosmonauts

http://www.cracked.com/article_19142_5-soviet-space-programs-that-prove-russia-was-insane.html

CHAPTER 67
NASA, ASTRONAUTS, COSMONAUTS AND UFONAUTS

http://www.syti.net/UFOSightings.html

http://ronrecord.com/astronauts/lovell-borman.html

http://www.syti.net/UFOSightings.html

http://www.openminds.tv/russian-cosmonauts-ufo-sightings-and-statements/

http://www.youtube.com/watch?v=IiDvkB_rG-Q

http://www.thetruthbehindthescenes.org/2011/08/28/ufo-filmed-by-astronauts-flying-over-hurricane-irene-aug-26-2011/

http://www.youtube.com/watch?feature=player_detailpage&v=jXzH6xfqjU4

http://www.youtube.com/watch?feature=player_detailpage&v=X-RPWhigpQg

https://www.youtube.com/watch?v=FF89H4S19UE

CHAPTER 68
ARE ASTRONAUTS DEBRIEFED AFTER EACH MISSION USING MIND CONTROL METHODS

http://www.hq.nasa.gov/office/oig/hq/old/inspections_assessments/10-23-97.html

http://news.xinhuanet.com/english/2007-02/07/content_5711503.htm

http://aangirfan.blogspot.ca/2012/06/nasa-brainwashing.html

"We Discovered Alien Bases on the Moon II" by Fred Steckling; 1981; by G.A.F. International; Vista, CA. USA; ISBN 0-942176-00-6

http://rense.com/general70/rep.htm

http://www.shockmansion.com/2012/08/27/video-buzz-aldrin-punches-reporter-for-calling-the-moon-landing-fake/

http://www.break.com/usercontent/2011/4/20/alien-encounters-2047497
http://www.youtube.com/watch?feature=player_embedded&v=tRBesDx1WQc

https://cassiopaea.org/forum/index.php?topic=13462.25;wap2

CHAPTER 69
THAT'S ONE ITTY BITTY, SMALL STEP FOR A MAN, ONE HUGE, GIGANTIC LEAP FOR WHOM… NASA?

http://www.nasa.gov/50th/50th_magazine/coldWarCoOp.html

http://en.wikipedia.org/wiki/Apollo%E2%80%93Soyuz_Test_Project

CHAPTER 70
ONE OF THESE DAYS ALICE, JUST ONE OF THESE DAYS
--- POW! RIGHT TO THE MOON!!

http://www.openminds.tv/did-buzz-aldrin-see-a-ufo-on-his-way-to-the-moon/

Above Top Secret by Timothy Good; 1988; William Morrow and Company, Inc.; New York, N.Y. USA; ISBN 0-688-09202-0

"Celestial Raise" by Richard Watson and ASSK [P.O. Box 35 Mt. Shasta CA. 96067 (916)-926-2316); 1987; page 147-148]

http://www.ufos-aliens.co.uk/cosmicphotos.html

Our Mysterious Spaceship Moon" by Don Wilson (Dell, 1975)
home.mytelus.com/telusen/portal/index.aspx

http://www.greatdreams.com/moon/darkmoon.htm

http://www.jsc.nasa.gov/history/mission_trans/AS12_LM.PDF

http://spacetime.forumotion.com/t741-apollo-12-and-the-indians-on-the-moon-bizarre-astronaut-quote-reveals-censored-transcripts-audio

http://www.hq.nasa.gov/alsj/a15/a15.html

https://www.youtube.com/watch?v=Kjo7d5W3ic8

CHAPTER 71
WAS THE APOLLO MOON LANDINGS HOAXED?

http://en.wikipedia.org/wiki/Apollo_hoax#Ionizing_radiation_and_heat

Dark Mission: The Secret History of NASA by Richard C. Hoagland and Mike Bara; 2007; a Feral House Book; Los Angeles, CA; ISBN: 978-1-932595-26-0

http://en.wikipedia.org/wiki/Bill_Kaysing

http://en.wikipedia.org/wiki/Apollo_hoax#Ionizing_radiation_and_heat

http://web.archive.org/web/20070109202315/http://www.thelastoutpost.com/site/1362/default.aspx

"The Facts don't Lie, But The Camera May" by Alex Strachan, Vancouver Sun, Nov 15th 2003

CHAPER 72
"REMOVING ONE OF TRUTH PROTECTIVE LAYERS"
- AN ALIEN PRESENCE ON THE MOON!

https://www.youtube.com/watch?v=u10UPBdSiV8

Dark Mission: The Secret History of NASA by Richard C. Hoagland and Mike Bara; 2007; a Feral House Book; Los Angeles, CA; ISBN: 978-1-932595-26-0

http://en.wikipedia.org/wiki/Lunokhod_programme

http://www.disclose.tv/forum/moon-anomalies-ruins-and-structures-on-the-moon-t4823.html#ixzz2IUbanD7x

http://blog.hallofthegods.org/2009/06/moon-anomalies-castle.html

CHAPTER 73
SEARCHING FOR EARTH-LIKE WORLDS
IN THE GOLDILOCKS ZONE

http://www.space.com/19522-alien-planet-habitable-zone-definition.html

http://www.bibliotecapleyades.net/luna/esp_luna_35.htm

http://www.anomalics.net/archive/cni-news/CNI.0818.html

http://www.marsanomalyresearch.com/evidence-reports/2008/153/moon-tower-shadows.htm

http://en.wikipedia.org/wiki/Martian_meteorite

http://www.washingtonpost.com/wpdyn/content/article/2010/04/30/AR2010043002000.html

http://www.theforbiddenknowledge.com/hardtruth/evidence_life_moon.htm

Dark Mission: The Secret History of NASA by Richard C. Hoagland and Mike Bara; 2007; a Feral House Book; Los Angeles, CA; ISBN: 978-1-932595-26-0

http://www.enterprisemission.com/datashead.htm

http://www.bibliotecapleyades.net/luna/esp_luna_32.htm

http://www.youtube.com/watch?feature=player_embedded&v=MWkGTJEK0Mc

http://www.youtube.com/watch?v=NH7XPMWc8ig

http://www.bibliotecapleyades.net/luna/esp_luna_32.htm

http://en.wikipedia.org/wiki/Apollo_17

http://www.spacefoundation.org/media/space-watch/spaceports-can-be-found-throughout-world

http://www.angelismarriti.it/ANGELISMARRITI-ENG/REPORTS_ARTICLES/Apollo20-InterviewWithWilliamRutledge.htm

http://www.section508.nasa.gov/

http://www.bibliotecapleyades.net/luna/esp_luna_36a.htm#Did%20the%20USA/USSR%20Fly%20a%20Secret%20Joint%20Mission%20to%20the%20Moon

http://www.angelismarriti.it/ANGELISMARRITI-ENG/REPORTS_ARTICLES/Apollo20-InterviewWithWilliamRutledge.htm

http://www.viewzone.com/monalisa.html

http://transition888.heavenforum.org/t568-mona-lisa-ebe

http://en.wikipedia.org/wiki/Moon

http://wattsupwiththat.com/2012/01/08/the-moon-is-a-cold-mistress/

http://www.youtube.com/user/moonwalker1966delta

CHAPTER 74
SOME LUNAR CURIOSITIES, ODDITIES AND ANOMALIES – IS THE MOON A GIANT SPACESHIP?

http://www.bibliotecapleyades.net/luna/esp_luna_16.htm

"The Alien Chaser" by Ronald Regehr from an article by Don Ecker "Long Saga of Lunar Anomalies"; UFO magazine, Vol. 10, No. 1 2 (March/April 1995),

https://www.youtube.com/watch?v=WqwJvw6as9U

http://www.bibliotecapleyades.net/luna/esp_luna_16.htm

Informant News Website

https://www.youtube.com/watch?v=-gBQTjPxaK4

http://en.wikipedia.org/wiki/Dyson_sphere
http://www.youtube.com/watch?v=_ngvIP0Za9M.

CHAPTER 76
THE SUN – ("SOL / HELIOS" – THE GIVER OF LIFE)

http://sohowww.nascom.nasa.gov/about/about.html

http://www.nasa.gov/mission_pages/stereo/main/index.html

http://www.scholarpedia.org/article/Solar_Satellites

http://www.thetruthbehindthescenes.org/2012/10/18/giant-ufo-near-sun-when-prominence-breaks-away-from-the-sun-october-2012/

http://erigia.blogspot.be/2012/07/incroyable-ovni-geant-de-lumiere-sur-le.html

http://sohowww.nascom.nasa.gov/explore/faq.html

http://www.enterprisemission.com/soho.htm

https://www.youtube.com/watch?v=Vw6Hcbw3XOY

https://www.youtube.com/watch?v=aZUGuex5XYY

"Divine Philosophy," compiled by I.F. Chamberlain (Boston: The Tudor Press, 1918), pp. 114-15

https://www.nature.com/articles/d41586-019-03087-1

https://pubs.acs.org/doi/10.1021/acs.est.9b03982ι

https://www.ncbi.nlm.nih.gov/pmc/articles/PMC6476344/

http://articles.adsabs.harvard.edu/cgi-bin/nph-iarticle_query?bibcode=1913PA....21..321M&db_key=AST&page_ind=0&data_type=GIF&type=SCREEN_VIEW&classic=YES

CHAPTER 77
PLANET MERCURY (THE "MESSENGER")

http://www.nasa.gov/mission_pages/messenger/main/index.html

http://www.lifeslittlemysteries.com/1966-cloaked-mothership-mercury-ufo.html

CHAPTER 78
PLANET VENUS (THE "GODDESS OF LOVE")

http://en.wikipedia.org/wiki/Venus

https://www.nytimes.com/2020/09/14/science/venus-life-clouds.html

https://solarsystem.nasa.gov/planets/venus/in-depth/

https://en.wikipedia.org/wiki/Life_on_Venus

http://madreterra.myblog.it/2015/06/14/extraterrestri-noi-il-venusiano-valiant-thor/

http://www.bibliotecapleyades.net/bb/stranges.htm

http://www.nextagemission.com/valthor1.html

https://www.youtube.com/watch?v=Zq21hb4v89Y

CHAPTER 79
THE PLANET MARS ("THE GOD OF WAR")

Marshttp://en.wikipedia.org/wiki/Mars

https://en.wikipedia.org/wiki/List_of_missions_to_Mars

http://www.sott.net/article/253871-Mars-Mystery-Has-Curiosity-rover-made-a-big-discovery

http://en.wikipedia.org/wiki/Terraforming

https://www.youtube.com/watch?v=O5k0MtlWPOs

e-Book: "What NASA Isn't Telling You About Mars" © 2005 by Ted Twietmeyer

http://conspiracy2012.files.wordpress.com/2012/05/what_nasa_isnt_telling_you_about_mars.pdf

http://www.enterprisemission.com/colors.htm

(*Mars: The Living Planet,* B. DiGregorio, G. Levin and P. Straat, Frog Ltd, Berkeley, CA 1997).

https://www.youtube.com/watch?v=tk3qLe2wWMA

https://www.youtube.com/watch?v=Y21Boc2We5o

https://www.youtube.com/watch?v=AXTKsdceNJI

CHAPTER 80
"CAPTAIN, WE DISCOVERED MULTIPLE SIGNS
OF LIFE DOWN HERE ON THE PLANET!"

http://sci.esa.int/science-e/www/object/index.cfm?fobjectid=46038

http://en.wikipedia.org/wiki/Exploration_of_Mars

http://en.wikipedia.org/wiki/Mars

? https://www.youtube.com/watch?v=Wj4Q1hPRoDs

http://www.newworldencyclopedia.org/entry/Carbon_dioxide

http://www.space.com/16903-mars-atmosphere-climate-weather.html

http://www.chinadaily.com.cn/en/home/2004-01/24/content_300860.htm

http://www.xenotechresearch.com/seagoph1.htm

http://www.nationalufocenter.com/artman/publish/article_85.php

http://beforeitsnews.com/space/2011/09/scientists-charge-nasa-killed-life-on-mars-1153548.html

http://www.shultslaboratories.com/index.htm

(http://marsrover.jpl.nasa.gov/gallery/all/2/p/016/2p127793693eff0327p2371r1m1.html)

http://marsrovers.nasa.gov/spotlight/opportunity/b19_20040304.html

https://www.youtube.com/watch?v=9E4fq9Cgz6o

http://www.molecatchers.com/meet_the_mole.html

http://en.wikipedia.org/wiki/Mole_%28animal%29

http://www.space.com/19928-mars-habitable-life-possible.html?cid=dlvr.it

msl-raw-images/msss/00193/mcam/0193MR1024018000E1

http://www.wired.com/wiredscience/2008/04/russian-volunte/

http://www.youtube.com/watch?feature=player_embedded&v=wvYHP2WzJwk

https://www.youtube.com/watch?v=wvYHP2WzJwk

INDEX (VOLUME FOUR)

A

H_____

I_____

USAF's fleet of antigravity vehicles, 292

V

W

X

Y

Z

About the Author

Terry Tibando's background experience and understanding of this phenomenon spans 65 years of personal UFO sightings and ET contact that began at the age of five years. This childhood experience initiated a lifetime of many other-worldly sightings and encounters into a mysterious universe of Unidentified Flying Objects, Extraterrestrial Intelligence and the paranormal. As an experiencer, researcher, and investigator in Ufology he brings a unique and refreshing perspective on this subject based on a world view.

While attending Victoria High School in the mid sixties, Terry began attending UFO lectures meeting such people as Dr. Edward Edwards, a linguist from the University of Victoria and a fellow member of APRO (Aerial Phenomenon Research Organization and also Daniel Fry from New Mexico, USA, well known contactee and UFO author.

Terry attended the University of Victoria majoring in astronomy, physics, math and other sciences. During those university years other alien craft were sighted near his family's home in Victoria leading Terry to theorized that a possible undersea ET base existed off the coast of Vancouver Island which may account for the numerous UFO sightings seen over the Island.

He was a former member of APRO and its Canadian sister organization CAPRO during the sixties. His investigative research culminated back in the summer of 1996 when he met with Dr. Steven M. Greer during a one week "Ambassadors to the Universe" training seminar. They soon discovered that they shared similar UFO/ETI experiences during their early life.

470

Terry was a speaker at the Bellingham UFO Group (BUFOG) UFO seminar in 1996, and as a panel speaker along with Peter Davenport from NUFORC and Sharon Filip, alien abduction researcher.

He has talked on the Grimerica blog talk radio and been interviewed on the Discovery Channel during their "Alien Week" series in 1997 which had two ET spacecraft show up during the TV interview; he has been interviewed on BCTV News and appeared briefly in Dr. Greer's successful documentary movie "Sirius" and was a major financial contributor to the current documentary "Unacknowledged"!

He was instrumental in coordinating, hosting and emceeing the first Disclosure Project event on UFOs and ETS in Canada as a part of Dr. Greer's Disclosure Witness Tour held at Simon Fraser University in Vancouver on September 9, 2001, which included guest speakers Dr. Steven Greer, Dr. Carol Rosin and Dr, Alfred Webre.

For the last 28 years, Terry has been the field coordinator of CSETI Vancouver leading teams of people on field expeditions to successfully establish contact and communications with extraterrestrial intelligences visiting the Earth.